A Literary History of Medicine

Volume 1

A Literary History of Medicine

The ʿUyūn al-anbāʾ fī ṭabaqāt al-aṭibbāʾ of Ibn Abī Uṣaybiʿah

VOLUME 1

Essays

Edited and translated by

Emilie Savage-Smith
Simon Swain
Geert Jan van Gelder

With

Ignacio Sánchez, N. Peter Joosse, Alasdair Watson, Bruce Inksetter, and Franak Hilloowala

BRILL

LEIDEN | BOSTON

Originally published in hardback in 2019 as Volume 134/1 in the series Handbook of Oriental Studies – Handbuch der Orientalistik, Section 1, The Near and Middle East.

Cover illustration: Physician with two patients, illustrating the Arabic translation of Dioscorides' treatise on medicinal substances. Copied in Baghdad, 1224 (Arthur M. Sackler Gallery, Smithsonian Institution: Unrestricted Trust Funds, Smithsonian Collections Acquisition Program, and Dr. Arthur M. Sackler, S1986.97).

The Library of Congress has cataloged the hardcover edition as follows:

Names: Savage-Smith, Emilie, editor. | Swain, Simon, editor. | Gelder, G. J. H. van, editor. | Sánchez Rojo, Ignacio Javier, editor.
Title: A literary history of medicine : the 'Uyūn al-anbā' fī ṭabaqāt al-aṭibbā', of Ibn Abīuṣaybi'ah / edited and translated by Emilie Savage-Smith, Simon Swain, Geert Jan Van Gelder ; with Ignacio Sánchez [and four others].
Other titles: 'Uyūn al-anbā' fī ṭabaqāt al-aṭibbā'
Description: Leiden ; Boston : Brill, [2020]- | Series: Handbook of Oriental studies. Section One, Near and Middle East, 0169-9423 ; volume 134/1 | Includes index. | Contents: volume 1. Essays | Mainly in English; includes Latin, Greek, Cyrillic, and Arabic text.
Identifiers: LCCN 2019040086 | ISBN 9789004410312 (hardback) | ISBN 9789004410558 (v.1 ; hardback) | ISBN 9789004410541 (v.2-1 ; hardback) | ISBN 9789004417830 (v. 2-2 ; hardback) | ISBN 9789004410534 (v. 3-1 ; hardback) | ISBN 9789004410527 (v. 3-2 ; hardback)
Subjects: LCSH: Physicians–Arab countries–Biography. | Ibn Abī Uṣaybi'ah, Aḥmad ibn al-Qāsim, -1269 or 1270. | Physicians–Biography. | Medicine, Arab. | Medicine, Medieval. | Medicine, Greek and Roman. | Medicine–History.
Classification: LCC R144.A1 L58 2020 | DDC 610.9–dc23
LC record available at https://lccn.loc.gov/2019040086

Typeface for the Latin, Greek, and Cyrillic scripts: "Brill". See and download: brill.com/brill-typeface.

ISBN 978-90-04-69655-6 (paperback, 2024)
ISBN 978-90-04-54556-4 (e-book, 2024)
ISBN 978-90-04-41055-8 (hardback)

This book is printed on acid-free paper and produced in a sustainable manner.

Contents

VOLUME 2-1
Arabic Edition

VOLUME 2-2
Arabic Edition

VOLUME 3-2
Annotated English Translation

Acknowledgements

A Literary History of Medicine has been an enormous project encompassing a wide range of topics, and so it is inevitable that we are indebted to many people for their kind and generous assistance.

Foremost, of course, is the financial support provided by the Wellcome Trust and the institutional support given by the University of Oxford and the University of Warwick. The Oriental Institute provided meeting rooms when we held group meetings of the entire team, as did St Cross College, which also provided some very nice lunches for us. St John's College as well as St Cross College on occasion provided accommodation for members of the team coming from overseas. In Oxford, the management of the Wellcome grant was very efficiently conducted by the financial officers at the Oriental Institute, Stephanie Yoxall and Chris Williams, and the everyday but essential needs of the project were aided by the ever-cheerful assistance of Trudi Pinkerton. For the Warwick portion of the Wellcome Trust grant, Katie Klaassen, Colette Kelly, and Fiona Slater have all helped with various organizational requirements.

The IT assistance of Daniel Burt, Computing Officer at The Khalili Research Centre, was essential to the coordination of a project in which members of the team were based on different continents. We could not have managed without him. He also was responsible for drawing up the map that accompanies this publication.

The team is indebted to Dr Peter Starr, Istanbul, for obtaining copies of Istanbul manuscripts, particularly the precious Süleymaniye, Şehid Ali Paşa MS 1923, but others as well. Professor Amr Abd al-Aziz Moneer, Egypt, assisted us greatly by obtaining copies of the edition by ʿĀmir al-Najjār printed in Cairo.

Professor Dimitri Gutas, of Yale University, provided invaluable advice about Avicenna, while Dr Yossef Rapoport, Queen Mary University of London, helped us with some obscure place names and terms for the Copts of Lower Egypt. Professor Richard Sorabji, King's College London, assisted us in interpreting a number of complicated philosophical arguments, while Professor Gerrit Bos, Universität zu Köln, provided assistance with some terminology and the biography of Maimonides. Professor Charles Burnett, of the Warburg Institute in London, gave us references to various important publications, and the identification of certain foodstuffs was aided by suggestions from Professor David Waines, of Lancaster University. Dr Uwe Vagelpohl, University of Warwick, has provided excellent advice on all aspects of Greco-Arabic medical history.

Closer to home, Professor Sebastian Brock, of the Oriental Institute in Oxford, was very helpful in sorting out sources for Melkite physicians, and Nigel Wilson, of Lincoln College, University of Oxford, helped us identify some obscure Greek names. There are, of course, others who have also helped with various points, and those kind people will be mentioned in the course of the annotations to the translation.

The Bodleian Library, as always, provided access to many manuscript and print resources that were required, and Dr. Gillian Evison, Keeper of the Oriental Collections, aided the project greatly by allowing a sabbatical leave for Alasdair Watson, thus enabling him to participate in this project.

For the 'public engagement' part of the project – titled 'Medieval Medicine in Board and Card Games', also supported by the Wellcome Trust – the History of Science Museum here in Oxford was an essential partner. We are indebted to the support of its Director, Dr Silke Ackermann, as well as to the Administrator, Christina Lee. The project was designed and coordinated by Daniel Burt, himself an experienced board-game designer, in collaboration with the Lead Education Officer at the History of Science Museum, Chris Parkin. They were then joined for the duration of the project by Susannah (Sukie) Trowles, part-time education officer. The project produced four working prototypes of board games or card games for a range of age groups from Primary School age to Adult. Through these board and card games, it is intended that the players learn of some of the physicians and their medical care that Ibn Abī Uṣaybiʿah describes in his large history of medicine.

We also wish to thank Luciana O'Flaherty, editor of the Oxford World's Classics, for undertaking to include in this prestigious series a volume of selections from Ibn Abī Uṣaybiʿah's *Best Accounts*. Henrietta Sharp Cockrell made the selections and annotated them for a general reader; we are indebted to her for the extraordinary amount of work she put into this paperback volume that we hope will bring Ibn Abī Uṣaybiʿah and his *Best Accounts* to a larger audience.

Throughout the project, we have been able to work with the knowledge that our efforts will be brought into print by Brill and set at a very high standard. We thank Kathy van Vliet-Leigh, Acquisitions Editor for Middle East, Islam & South and Central Asian Studies, and the editors of the *Handbook of Oriental Studies* series for making this possible. Moreover, they have allowed us to have within these volumes all the material that we felt was required for a full presentation of Ibn Abī Uṣaybiʿah's *Best Accounts of the Classes of Physicians*. The open access version of *A Literary History of Medicine* has been underwritten by the Wellcome Trust so as to enable Brill to present the volumes as a single XML online reference work, making it possible for all the volumes to be searched in a single online search. We also wish to thank Pieter te Velde for his meticulous guidance of these volumes through production.

We close this acknowledgment list by again thanking the Wellcome Trust, who have generously supported our project at every stage of the way. Without their continuous support, the project could not have been carried out.

Editorial Policy

The present edition of the *ʿUyūn* represents Ibn Abī Uṣaybiʿah's final Version 3, which he was apparently still adding to at the time of his death in 668/1270. MS A, which has Version 2 compared with Version 3 written in margins and on interleaves, was taken as our most important manuscript and formed the initial basis of our edition. Seven other manuscripts supplied additional important copies of Version 3 as well as complete copies of Version 1 and Version 2. The differences between particular readings in the three versions of the *ʿUyūn* can be found in the critical apparatus. One of the most obvious differences between the three versions is the deletion of the dedication to Amīn al-Dawlah which prefaced Version 1, following Amīn al-Dawlah's execution in 648/1250–1251. For the two subsequent versions of the *ʿUyūn* there is no dedicatee. In our edition, however, we have left the dedication to Amīn al-Dawlah in the Preface but have italicized it and noted that it was omitted in the two subsequent versions. In Ch. 15, which included a biography of Amīn al-Dawlah in all three versions, we have presented the biography of Amīn al-Dawlah (Ch. 15.49) as it appeared in Versions 2 and 3 and placed the quite different text of Version 1 in an addendum. In exceptional instances we have preferred readings from Version 1 or 2 or a reading from the tradition of source texts quoted by Ibn Abī Uṣaybiʿah, which have been collated by us largely for the first time, and in these instances the choices are duly noted in the critical apparatus. We present the text in modern standard orthography, with minimal vowelling for the prose and some more vowelling in the poetry.

Detailed indexes to the edition and translation of the *ʿUyūn* – covering subjects, people, and places, as well as Qur'anic quotations, rhymes and weights and measures – are to be found at the end of Volume 1, following a general index to the essays and introductory material forming Volume 1.

Figures and Tables

Figures

Volume 1

Volume 2-1

Volume 3-1

Volume 3-2

Tables

Volume 1

Abbreviations

Manuscripts

A — Istanbul, Süleymaniye, Şehid Ali Paşa MS 1923. Completed 27 Shaʿbān 773 (4 Mar 1372). Copyist: ʿAbd al-Hādī ibn Abī l-Mufaḍḍal ibn Abī l-Faraj. [complete copy, from an autograph of Version 2 and completed with a draft of Version 3]

B — Oxford, Bodleian Library, MS Pococke 356. Completed 15 Rajab 869 (13 Mar 1465) in Mecca; unnamed copyist. [complete copy, Version 1]

Ga/Gb — Leiden, Universiteitsbibliotheek Or. 59a and Or. 59b (old Cod. 59a. Gol. & Cod. 50b. Gol.). Undated; 16th cent. 2 vols. [Ga is partial copy Version 2; Gb, partial copy Version 3]

Gc — Leiden, Universiteitsbibliotheek, Or. 76 (old Cod. 76. Gol.). Dated 20 Ṣafar 977 (4 Aug 1569). [almost complete copy, Version 2]

H — Istanbul, Topkapı Sarayı Müzesi, Ahmad III Library, MS Taʾrīkh 2859/70. Dated 5 Ṣafar 735 (28 Nov 1329). Copyist: Muḥammad ibn ʿAbd Allāh ibn ʿAbd al-Raḥmān in the great mosque of the city of Hamah. [partial copy, Version 3]

L — London, British Library MS Add. 23364. Dated 10 Rajab 669 (22 Feb 1271); Damascus; copyist: Syrian physician Muḥammad ibn Ibrāhīm Muḥammad al-Suwaydī al-Anṣārī, copied from the copy endowed by IAU to the Maqṣūrah of Ibn ʿUrwah. [partial copy, Version 3]

P — Paris, Bibliothèque nationale de France, MS arabe 5939. Copy completed 19 Muḥarram 690 (22 Jan 1291) in Jayrūn near Damascus by Abū l-Faḍl al-Jarāʾiḥī ibn al-Faraj ibn Naṣr al-ʿAsqalānī; annotated, collated against copy in author's hand. [partial copy, Version 2]

R — London, British Library, MS Add. Rich. 7340. Dated 22 Rajab 1017 (1 Nov 1608) in Isfahan. Copyist: Ibn Muḥammed Shafīʿ Mulla Zayn al-ʿĀbidīn [complete copy, Version 3]

Sa / Sb — London, British Library, MS Add. 25736. Dated 10 Safar 713 (6 Jun 1313); copied by the physician Ibrāhīm al-Khawālifī for his own use. [Sa (first seven folios) is partial copy of Version 2; Sb, partial copy of Version 3]

Earlier Editions

Müller — Ibn Abī Uṣaybiʿah, *ʿUyūn al-anbāʾ fī ṭabaqāt al-aṭibbāʾ*, ed. A. Müller. 2 vols. Cairo: al-Maṭbaʿah al-Wahbiyyah / Königsberg: Selbstverlag, 1882–1884.

Riḍā — Ibn Abī Uṣaybiʿah, *ʿUyūn al-anbāʾ fī ṭabaqāt al-aṭibbāʾ*, ed. Nizār Riḍā. Beirut: Dār Maktabat al-Ḥayāt, 1965.

Najjār Ibn Abī Uṣaybiʿah, *ʿUyūn al-anbāʾ fī ṭabaqāt al-aṭibbāʾ*, ed. ʿĀmir al-Najjār. Vol. 1. Cairo: Dār al-Maʿārif, 1996. Vols. 2–6. Cairo: al-Hayʾah al-Miṣriyyah al-ʿĀmmah lil-Kitāb, 2001–2004.

Other Abbreviations

AI Appendix 1: Ibn al-Nafīs

AII Appendix 2: Additional Marginalia

IAU Ibn Abī Uṣaybiʿah

Ch.xx.xx Chapters within the *ʿUyūn* are designated by numbers 1 through 15, followed by the number of an individual biography within that chapter. A third set of numbers refers to a given subsection of a biography. Subsequent numbers in a given sequence refer to items within numbered lists, such as book-lists or lists of aphorisms.

Cross-References

The three volumes have independent pagination. Where a cross-reference does not specify a volume number, the reference will be found in the same volume.

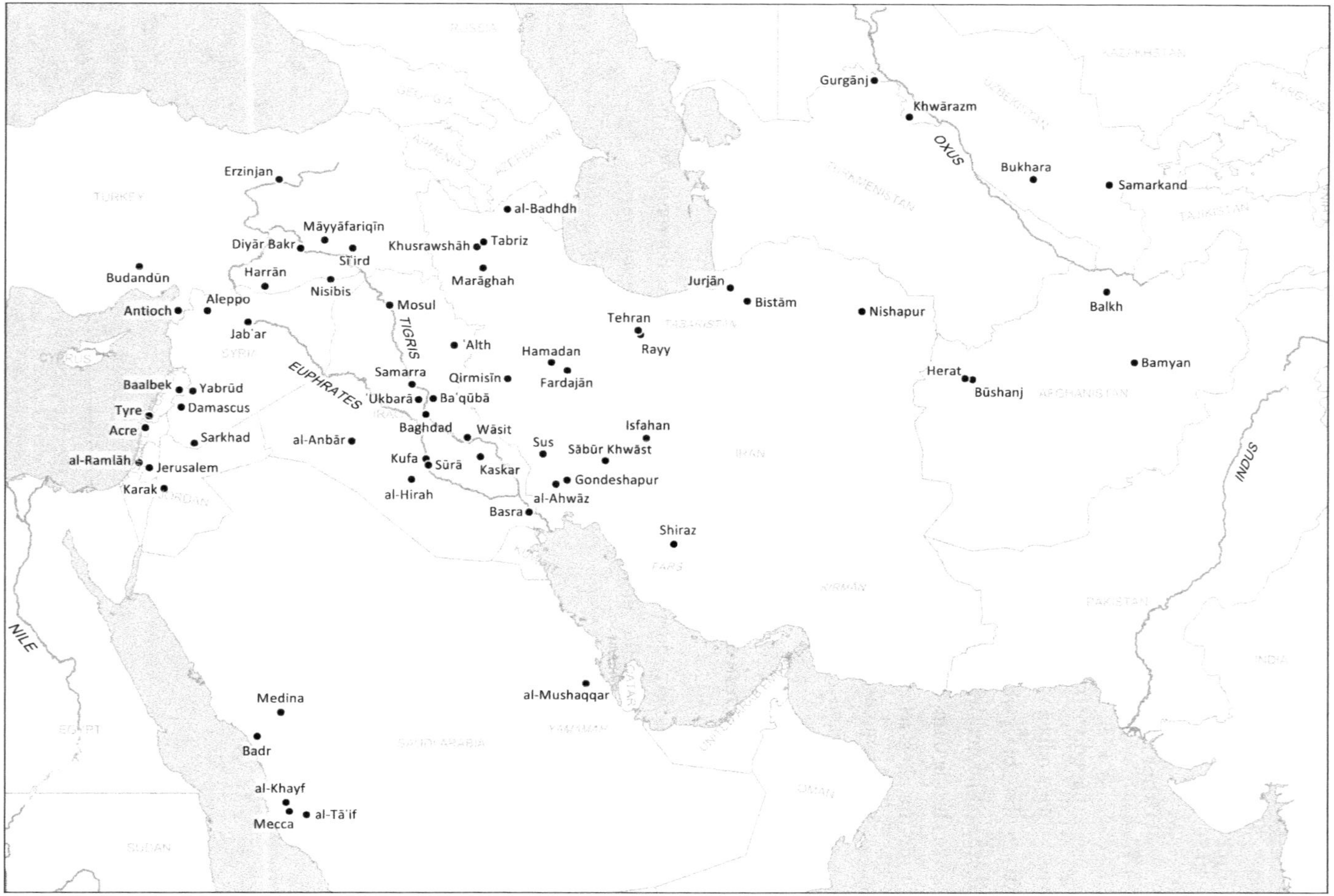

Map of the Islamic World ca. 1200

Map of the Islamic World ca. 1200

CHAPTER 1

Introductory Remarks

In the mid-thirteenth century, a practising physician in Syria named Ibn Abī Uṣaybiʿah set himself the task of recording the history of medicine throughout the known world of his day. His book, *ʿUyūn al-anbāʾ fī ṭabaqāt al-aṭibbāʾ* (The Best Accounts of the Classes of Physicians) covers 1700 years of medical practice, from the mythological beginnings of medicine with Asclepius through Greece, Rome, and India, down to the author's day. Written as much to entertain as to inform, it is not only the earliest comprehensive history of medicine but the most important and ambitious of the medieval period, incorporating accounts of over 432 physicians, their training, practice, and medical compositions, all interlaced with amusing poetry and anecdotes illustrating their life and character.

To undertake a full edition, translation, and study of this complex treatise required a wide range of skills. For the past six years, the Wellcome Trust has generously funded a team of eight scholars to engage in making this remarkable historical source available for the first time in a complete, reliable, and (we hope) readable translation and study. Because this massive treatise is as much a piece of literature as historical chronicle, the project was titled 'A Literary History of Medicine'. The team, otherwise known as the *ALHOM Team*, consisted of Emilie Savage-Smith, Simon Swain and Geert Jan van Gelder, Ignacio Sánchez, N. Peter Joosse, Alasdair Watson, Bruce Inksetter, and Franak Hilloowala, aided by the expert electronic skills of Daniel Burt.

While certain chapters are credited to particular team members, we all in fact read through the efforts of our colleagues and made numerous suggestions to one another. It was truly a team effort, with members bringing various approaches and backgrounds to the project – from love and knowledge of Arabic poetry to Greek and Latin classics to translation techniques or particular interests in the history of medicine.

For nearly three hundred years European scholars have shown interest in learning more of what Ibn Abī Uṣaybiʿah had to say in his early history of medicine. Various transcriptions and translations were undertaken (all of them incomplete), some published and some now residing unread in library vaults.

1 Earlier Translations

Sometime before 1721 the London physician Richard Mead paid for a transcription of two manuscript copies of the *'Uyūn* in the Bodleian Library in Oxford (MS Huntington 171 and MS Pococke 356) to be made by Henry Wild of Norwick (d. 1721), who was known as the 'Arabick Tailor' because of his writing out copies of Arabic manuscripts. Mead also arranged for these transcriptions to be translated into Latin by Sulaymān ibn Yaʿqūb al-Shāmī al-Ṣāliḥānī, known to Londoners as Salomon Negri (d. 1729). The purpose of this undertaking was to provide material to John Freind (d. 1728), who was at that time composing a history of medicine.[1]

Only the biography of Jibrīl ibn Bukhtīshūʿ (Ch. 8.3) was actually used by John Freind, who printed Salomon Negri's Latin translation (*Vita Gabrielis Filii Bachtishuæ, filii Georgii, Ex Arabica Latine reddita a Salomone Negri Damasceno*) as a 26-page Appendix No. 1 in the second volume of his *The History of Physick: From the Time of Galen, to the Beginning of the Sixteenth Century. Chiefly with Regard to Practice. In a Discourse Written to Doctor Mead* (London, Printed for J. Walthoe, jun. over against the Royal-Exchange in Cornhill, 1725–1726).

The complete (?) Latin translation made by Salomon Negri of Henry Wild's transcription was sold as lot 19 at the sale of Richard Mead's library on 8 May 1755 and was purchased by Dr. William Hunter (d. 1812); it is now in the library of the University of Glasgow. As for Henry Wild's transcription of the two Oxford manuscripts, they were bought by Thomas Hunt (then Regius Professor of Hebrew) at the sale of Mead's library and are now also part of the Bodleian's holdings (designated MS Bodl. Or. 422 and 423).

Sometime before 1725, another partial transcription of the Bodleian MS Huntington 171 manuscript, along with an incomplete Latin translation in parallel columns, was made by John Gagnier (d. 1740 or 1741, the Lord Almoner's Professor of Arabic at Oxford). Gagnier's transcription was intended to cover portions omitted by Henry Wild. This transcription and Latin translation remained in the library of Richard Mead until his death, when it was sold (presumably to Thomas Hunt) and is now also part of the Bodleian Library (MS Bodl. Or. 295).[2]

---

1 Savage-Smith, *NCAM-1*, 454–459; Dunlop, 'Arabic Medicine in England', 177. For Mead, see *Oxford Dictionary of National Biography*, art. 'Mead, Richard (1673–1754)' (A. Guerrini); and for Wild, see Dunlop, 'Arabian Tailor'.

2 August Müller in the 1884 Königsberg printing (IAU Müller), on p. xix, of the introduction, mentions the Wild and Gagnier transcriptions of MS Huntington 171, but did not use them.

More European translations of extracts soon followed, no doubt following on from the interest in medieval Arabic medicine generated by John Freind's immensely influential *History of Physick*. Below is a chronological listing:

1746 Johann Jakob Reiske, *Excerpta ex Ibn Abí Uszeibi'ae historia medicorum illustrium*. Unpublished Latin translations of a few extracts made in 1746, now in Copenhagen, Det Kongelige Bibliotek, MS N. 31 (CLXXI). Portions (with Arabic text and Latin translation) were published in 1777 in J.J. Reiske, *Miscellanea medica ex Arabum monimentis*, as part of J.J. Reiske and J.E. Faber, *Opuscula medica ex monimentis Arabum et Ebraeorum* (Halle, 1777), 1–80, Reprinted in Sezgin, *Studies* II, 1–82.

1808 John Mousley, *Abdollatiphi Bagdadensis vita, auctore Ibn Abi Osaiba, e codicibus MSS. Bodleianis descripsit, et Latine vertit* (Oxford: e typographeo Clarendoniano, 1808). Latin translation and edition of the life of ʿAbd al-Laṭīf al-Baghdādī (IAU Ch. 15.40), using the same two Oxford manuscripts transcribed by Henry Wild (MS Pococke 356 and MS Huntington 171)

1833 Friedrich Reinhold Dietz, published the Arabic text with Latin translations of the twelfth chapter of the *ʿUyūn* on physicians of India (based on British Library MS Add. Rich. 7340) in a catalogue of Sanskrit medical manuscripts in the India Office Library; F.R. Dietz, *Analecta Medica ex Libris MSS. Fasciculus Primus, 2: Catalogus codicum Mss. de re medica Sanscritorum Londinensium* (Leipsig, 1833), 111–124. Reprinted in Sezgin, *Studies* II, 83–98.

1840 Pascual de Gayangos, *The History of the Mohammedan Dynasties in Spain* 2 vols. (London: The Oriental Translation Fund, 1840), vol. 1, APPENDIX pp. iii–xxvii. English translations of the lives of Ibn Juljul, Ibn Bājjah, and the Ibn Zuhr family (IAU Ch. 13.36 and 13.59–13.64).

1841 Willian Cureton, 'Extract from the work entitled *ʿUyūn al-anbāʾ fī ṭabaqāt al-aṭibbāʾ*, or, *Fountains of information respecting classes of physicians* by Muwaffik-uddin Abú-'labbás Ahmad Ibn Abú [*sic*] Usaibiâh; with remarks by Horace Hayman Wilson', *Journal of the Royal Asiatic Society*, 6 (1841), 105–119. An English translation of lives of five Indian physicians (IAU Ch. 12.1–12.5). Reprinted in Sezgin, *Studies* III, 1–15.

c. 1847 William Alexander Greenhill, London, Royal College of Physicians, MSS GREEW/264/149, 150, 152, and 153. Unpublished translations and notes made about 1847 from Oxford MS Bodl. 295 (the transcription with parallel Latin translation made by Gagnier), as well as MS

Pococke 356. Covers the the lives ʿAbd Laṭīf al-Baghdādī, Ibn Juljul, and Ibn al-Bayṭār (IAU Ch. 15.40; 13.36; 14.58).

1854 B.R. Sanguinetti, 'Extraits de l' ouvrage arabe d' Ibn Aby Ossaïbiʿah sur l' histoire des médecins, traduction fraņcaise, acompagnée de notes', *Journal Asiatique*, 5th series, 3 (1854), 230–291; 4 (1854), 177–213; 5 (1855), 401–469; 6 (1855), 129–190; 8 (1856), 175–196, 316–353. French translations of extracts from the early chapters on the origins of medicine and on Greek physicians. Reprinted in Sezgin, *Studies* III, 16–309.

1880 August Müller, 'Arabische Quellen zur Geschichte der indischen Medizin', *Zeitschrift der Deutschen Morgenländischen Gesellschaft* 34 (1880), 465–556. An edition, German translation, and extensive commentary on the twelfth chapter of the *ʿUyūn* (on Indian physicians). Reprinted in Sezgin, *Studies* II, 99–190.

1910 Hamed Waly, *Drei Kapitel aus der Ärztegeschichte des Ibn Abī Oṣaibiʿa*, Inaug.-diss., Friedrich-Wilhelms-Universität zu Berlin (Berlin: G. Schade, 1910). German translations al-Ḥārith ibn Kaladah, al-Naḍr ibn al-Ḥārith, Ibn Abī Ramthah al-Tamīmī, ʿAbd al-Malik ibn Abjar al-Kinānī, Ibn Uthāl, and Abū Ḥakam (IAU 7.1–7.6). Reprinted in Sezgin, *Studies* II, 215–259.

1939 Sadik Atallah, *Ibn Abi Useibià – Schrift über die arabischen Ärzte und ihre Kunst. Abschnitt über die Anfänge.* (Inaug.-diss., Ludwig-Maximilians-Universität zu München, 1939). German translations of very short extracts from nineteen biographies of early Islamic physicians. Reprinted in Sezgin, *Studies* II, 261–280.

1940 Ahmed A. Abou-Elgheit, *Gedanken aus Ibn Abi Useibia über die arabischen Ärzte und ihre Kunst* (Inaug.-Diss., Ludwig-Maximilians-Universität zu München, 1940). German translations of very short extracts from thirty-four lives. Reprinted in Sezgin, *Studies* II, 281–308.

1944 Willis Alexander McGill, 'Ibn Abī Uṣaibiʿa's biographies of Ḥusain ibn Isḥāq, Isḥāq ibn Ḥunain, and Ḥubaish ibn al-Ḥasan: Ninth century physicians and translators of Baghdad'. Translator's thesis (M.A.), Kennedy School of Missions, Hartford Seminary Foundation, 1944. Unpublished. Presents in 212 pp. an English translation (from Müller edition) of IAU Ch. 8.29–8.31.

1958 Henri Jahier and Abdelkader Noureddine, *Ibn abi Uçaibiʿa, ʿUyûn al-Anbâʾ fī T'abaqāt al-At'ibbâʾ. Sources d'informations sur les classes des médecins. XIIIe Chapitre: Médecins de l'Occident Musulman*, Publications de la faculté mixte de médecine et de pharmacie d' Alger, 4 (Algiers: Farārīs, 1958). A French translation of the thirteenth chapter of the *ʿUyūn*.

c. 1968 Lothar Kopf, 'Ibn Abu [*sic*] Usaibi'ah: History of Physicians', Bethesda, MD, National Library of Medicine, MS C 294, box 1. A carbon-copy of a typed draft of a translation prepared before 1969. It was commissioned by NLM but never published nor intended for publication. It covers all fifteen chapters, but omits most of the poetry, contains numerous errors and infelicities, and was never completed. It was recently made available on-line, without the (deceased) author's or library's permission, at http://www.tertullian.org/fathers/ibn_abi_usaibia_01.htm.

1981 M.N. Istanbouli, 'The history of Arabic medicine based on the work of Ibn Abi Usaybe'ah, 1203–1270'. Ph. D, Thesis in Social Science, Loughborough University of Technology, 1981. Unpublished, 345 pages, with pp. 147–216 giving a very abbreviated translation/summary of the fifteenth chapter of the *'Uyūn*, employing the Müller edition.

2000 Franak Hilloowala, 'An analysis of Ibn Abi Usaybi'ah's 'Uyūn al-anbā' fī ṭabaqāt al-aṭibbā' (Ahmad ibn al-Qasim Ibn Abi Usaybi'ah, Syria, Egypt)'. Ph.D. dissertation, University of Arizona, 2000. Includes an English translation of the fourteenth chapter of the *'Uyūn*. Unpublished.

In addition to the translations and editions listed above, there are many small portions or paragraphs that have been translated by others within larger studies – too many to mention here. Most of these will be cited in the notes accompanying our translation.

While we hold in great respect the efforts of those before us who have tried to translate this monumental work – and especially the Herculean efforts of Lothar Kopf, who attempted the entire treatise single-handedly – as our project developed, we found it more useful to not employ earlier translations and interpretations. Rather, we have used the new edition, based in part upon manuscripts not available to earlier editors, as the basis for our translations, and employed earlier translations only as later comparative material, if at all.

2 Earlier Editions

The German Arabist August Müller (1848–1892) published a critical edition of the Arabic text of Ibn Abī Usaybi'ah's book in 1884. His was an outstanding achievement but he had every reason to be unhappy with the result. As he writes in his German preface:

> In more than one aspect the form in which the present book sees the light corresponds with neither my original intention nor my subsequent expectations. This is due to the history of its publication, a short account of which, as far as is possible and I hope as far as is necessary, will serve as justification and apology for those shortcomings that to me are most frustrating.[3]

This 'short account' takes some ten pages and is followed by scores of pages with critical remarks, variant readings, and corrections. Müller had been persuaded by his friend Spitta Bey (Wilhelm Spitta, 1853–1883) to have the Arabic text printed in Cairo. To his chagrin, the result was a dog's dinner. The Arabic title page of the first of its two volumes looks impressive, with its ornate border and prolix text in traditional style; it translates as

The Book of
Best Accounts of the Classes of Physicians
Composed by the Eminent Physician and Erudite Scholar
Muwaffaq al-Dīn Abū l-ʿAbbās Aḥmad ibn al-Qāsim ibn Khalīfah
ibn Yūnus as-Saʿdī al-Khazrajī
Known as Ibn Abī Uṣaybiʿah
God rest his Soul
Transcribed and Edited from the Manuscripts Found in Several Libraries
by the Humble Man Needing God's Help and Mercy[4]
Imru' al-Qays ibn al-Ṭaḥḥān
First Edition, at the Printing House of Wahb[5]
in the Year 1299 of the Hijra Corresponding to the Year 1882

One supposes August Müller is responsible for the whimsical 'translation' of his own name; Imru' al-Qays, a famous pre-Islamic name, has few sounds in common with August, but ibn al-Ṭaḥḥān, at least, means 'the Miller's (Müller's) Son'.

For all its deficiencies Müller's edition remained until now the standard edition of the text. The often-used edition by Nizār Riḍā (Beirut, 1965) is wholly dependent on Müller's; it is lightly annotated and by no means a critical edition. The edition by Muḥammad Bāsil ʿUyūn al-Sūd (Beirut, 1998) is similar; it is basically the Riḍā edition with additional indexes. The attempt by ʿĀmir al-Najjār to make a new critical edition (employing seven copies not available to

3 Translation from the German by GJvG.
4 A customary formula of humility.
5 This refers to the publisher, Muṣṭafā Wahbī.

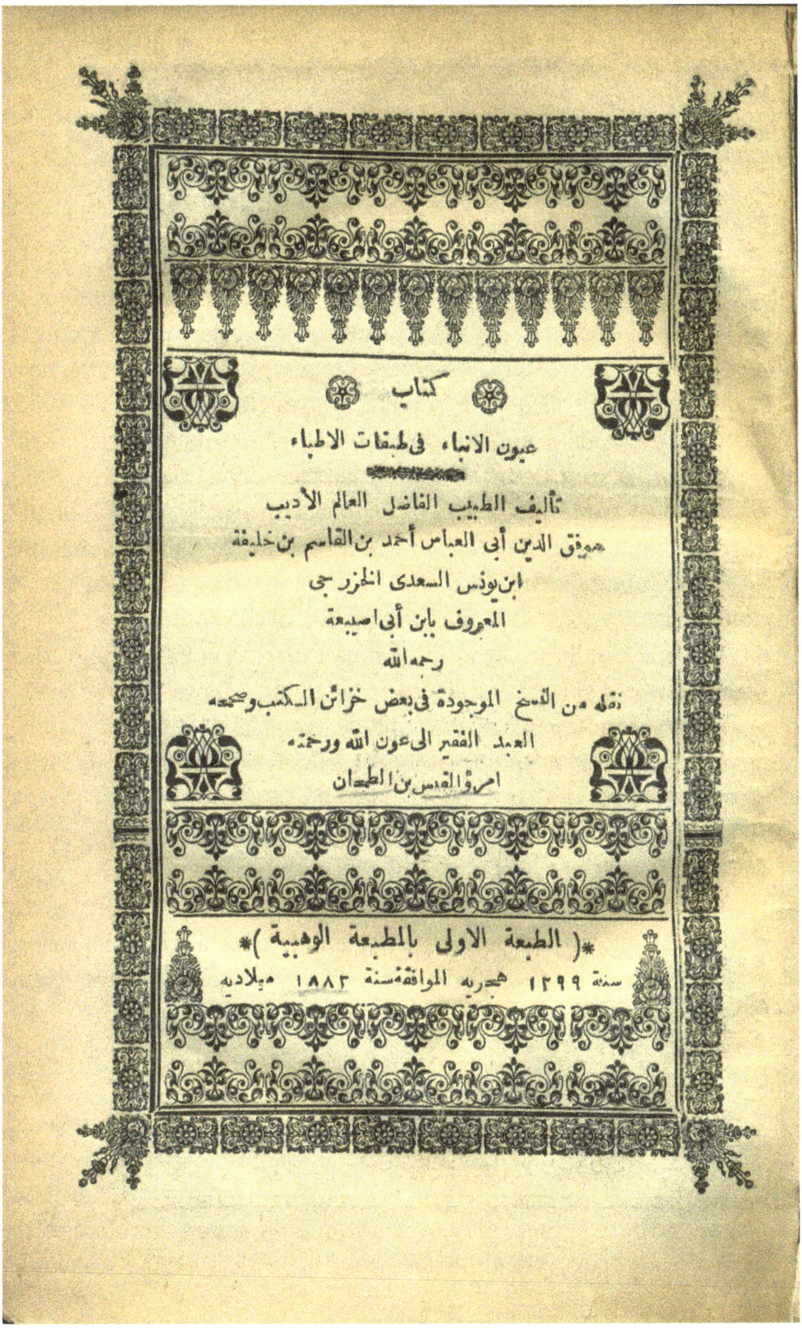

كتاب

عيون الانباء في طبقات الاطباء

تأليف الطبيب الفاضل العالم الأديب

موفق الدين أبي العباس أحمد بن القاسم بن خليفة

ابن يونس السعدي الخزرجي

المعروف بابن أبي اصيبعة

رحمه الله

نقله من النسخ الموجودة في بعض خزائن الكتب وصححه

العبد الفقير الى عون الله ورحمته

امرؤ القيس بن الطحان

(الطبعة الاولى بالمطبعة الوهبية)

سنة ١٢٩٩ هجريه الموافقة لسنة ١٨٨٢ ميلاديه

FIGURE 1.1 Title page from Ibn Abī Uṣaybiʿah, *ʿUyūn al-anbāʾ fī ṭabaqāt al-aṭibbāʾ*, ed. A. Müller. 2 vols. Cairo: al-Maṭbaʿah al-Wahbiyyah/Königsberg: Selbstverlag, 1882

Müller), which appeared in Cairo between 1996 and 2004 in six volumes with extensive introduction, critical apparatus, and indexes, is unfortunately marred by errors, typos, and inaccuracies to such an extent that it cannot be deemed to be a proper scholarly edition.

3 The New Brill Edition and Translation

Ibn Abī Uṣaybiʿah was by no means a reticient author. He often intervenes in the text to give his views ('I say …'). But for the most part he is careful to avoid direct praise of his masterpiece. In the Preface, where we might expect an author to promote his talents, he instead lauds medicine. It is a discipline that aspires to 'what is good' and 'to what is pleasant'. But he does commend his work (the *ʿUyūn*) on one point: 'no one with a thorough knowledge of [the] art has ever written a comprehensive book dealing with physicians through the ages and recounting their history in a coherent fashion'. We see then that Ibn Abī Uṣaybiʿah was proud that he was a practising physician and because he had used his knowledge to write something original. This pat on his own back is found again in a poem of his in Ch. 15: 'It is unique, no one in the past preceded me'.[6] The general worth of the book is something he also allows into the text in the last two chapters and again confirms that he was pleased with his creation, even if modesty attributes the praise of its qualities to the acquaintances who asked for a copy.[7]

The Preface, with the original dedication to Amīn al-Dawlah stripped out in the last two versions of the *ʿUyūn* (following the execution of Amīn al-Dawlah in 1250), essentially restricts itself to an account of subject material and gives a table of contents. Thus we are left to plunge into this great rolling history of persons and stories to revel in the sheer pleasure of its writing and scale. Readers who enjoy *The Best Accounts* from start to finish ought to be reminded of the famous instruction at the start of Apuleius' *Golden Ass*: 'reader, pay attention, for you are going to enjoy yourself!'

To begin to do justice to all the literary qualities and historical contexts of *The Best Accounts* is well beyond the scope of our present work. Rather, we hope that the fruit of our efforts will lead to a renewed focus on Ibn Abī Uṣaybiʿah and his book. We would be delighted to see a renaissance of work on the author with whom we have lived in close proximity for the past several years. The Wellcome

6 Ch. 15.49.6.1.

7 Cf. Muhadhdhab al-Dīn Muḥammad ibn Abī Ḥulayqah at 14.55.2, Amīn al-Dawlah at 15.49.6 (with Ibn Abī Uṣaybiʿah's poem).

Trust's support for open access publishing will surely encourage this by allowing Ibn Abī Uṣaybiʿah to be known anywhere in the world for free. But although we have necessarily avoided comprehensive literary or historical enquiries into the character and context of *The Best Accounts*, we have thought it desirable to include in our introduction a number of essays which serve to highlight key aspects of the book and to prompt readers to make their own investigations of these areas.

We begin with the biography of Ibn Abī Uṣaybiʿah and his career. Following this we discuss the genre and title of his book, setting the author's endeavours in the avowedly literary context of *adab* and high culture. The succeeding essay on the textual transmisison and manuscript tradition of the *ʿUyūn* is particularly important. For when we first investigated the possibility of undertaking a new text and translation of the *ʿUyūn* we had the good fortune to find a most important manuscript witness of the text through the good offices of Dr Peter Starr, Istanbul. This manuscript, Istanbul, Süleymaniye, Şehid Ali Paşa MS 1923, had not been used in any earlier edition or study. It was completed in 773/1372. It turned out to be a copy of a direct copy of an autograph of the second published version of the *ʿUyūn*, Version 2. What makes it special is that the copyist then came across Ibn Abī Uṣaybiʿah's draft of his final published version, Version 3. The intelligent copyist completed his work by recording additions from the draft which he wrote in the margins and on a number of interleaves. So it is that the Şehid Ali Paşa copy gives us precious information about the development of the final text authored by Ibn Abī Uṣaybiʿah and confirmation of how it differed from the previous version. It also shows that Müller was close to solving the puzzle of how the author wrote. In this third essay we give an account of the three versions of the *ʿUyūn* and the manuscripts we have relied on to represent these versions.

We next turn to sources. Ibn Abī Uṣaybiʿah used many of these. He takes care to name most of them scrupulously and to quote them with care. The way in which he deploys sources is of course central to his authorial art, and we have therefore included in the fourth essay below an examination of his methods of selection, quotation, and adaptation. One of the most important of these sources, and undoubtedly the main reason why previous attempts to translate the *ʿUyūn*, have proved vain, is the very large quantity of verse. To tackle this requires a rare technical knowledge and its character and purposes certainly merit the explanatory account of our fifth essay. It seems that our author loved the ordinary poetry of the social and professional milieux he lived in, and the inclusion of some 3600 lines of verse, mainly in the last three chapters and especially Ch. 15, testifies to the expectations and predilections of his audience. This poetry has very little to do with medicine: it is rather the currency of the educated classes, used for polite communication to patrons and friends, for lampoons and scurrilous attacks, for eulogies or congratulations.

An equally distinctive, if less surprising, feature of the work is the amount of material devoted to ancient Greek medicine. The chattering classes of the medieval Islamic world, whatever their faith or interest, were particularly keen to demonstrate knowledge of their intellectual heroes from Greek and Roman antiquity. And since the medicine of the medieval age was built upon the Hippocratic-Galenic legacy, and the study of Galen's core texts continued to be important for all students of medicine and even philosophy, the 'Greek chapters' (Chs 2–6) are expected. We offer an interpretation of how they work in our sixth essay. Our author uses them in part to establish the basic framework for the learning and behaviour expected of the physician through his detailed accounts of legendary and historical Greek doctors before he turns to his own world and the coming of Islam in Ch. 7.

We shall finish our introductory essays with the topic most dear to our author's heart: the practise of medicine. Ibn Abī Uṣaybiʿah was a hospital physician and has left a rich, if unsystematic, view of the ailments and cures of his day alongside glimpses into medical education and attainment. His concentration on the lives of his subjects led him away from a focus on the medical, as opposed to the social, aspects of medicine. Nevertheless, what Ibn Abī Uṣaybiʿah did in his day job was never far from his mind and historians of medicine and science have much to learn from him of orthodoxies and innovations in both theory and action.

There is a large amount of medical terminology in the *ʿUyūn* but rather than go through all of this in the introductory essay we have included references to it in our general subject index along with a myriad of other matters. By means of this index we have tried our best to provide a guide to readers seeking particular items. Naturally readers will wish to find information on what interests them and may have escaped our notice. The open access searchable version of the Brill volumes made available courtesy of the Wellcome Trust will allow them to do this. The indexes of places and people and the smaller indexes of Qur'anic quotations and rhymes and the glossary of weights and measures are intended to be as full as possible. These indexes will be found at the back of Volume 1.

We have described Ibn Abī Uṣaybiʿah's book as a 'literary history' of medicine; and we have tried to produce a translation that is neither too literal nor too free. We have included notes of varying detail to the chapters to help with some of the background. These are extensive overall but we hope our readers will add to or correct the information we give. One particular feature of our annotation needs further comment. Ibn Abī Uṣaybiʿah was extremely fond of books and took care to provide sometimes lengthy book-lists and even alternative book-lists. In these lists, a treatise is now to be taken as lost unless we state otherwise. No consistent attempt has been made to record manuscript copies of works that have yet to receive a modern edition or translation. As far as possible references have been provided to the standard bibliographic sources (such as Ullmann and Sezgin and occasionally the earlier Brockelmann) where indications of manuscripts may be found. Again, not all studies and published editions/translations of works are recorded by us, when they can be found listed in the standard bibliographical sources for a given biographical subject.

It is worth pointing finally to our two appendixes following the translation of the fifteen chapters comprising the *ʿUyūn*. One of Ibn Abī Uṣaybiʿah's great contemporaries, Ibn al-Nafīs, was not included in the *ʿUyūn*, or perhaps purposefully excluded. Biographical information on him was added by copyists to two manuscripts of the book, and for completeness we give this in Appendix 1. In Appendix 2 we edit and translate various marginalia, especially poems, which could not be handled in the footnotes. This material includes a remarkable previously unpublished prayer attributed to Ibn Sīnā asking forgivenness for his love of wine.

The electronic age of rapid communication and the ability to place a document in the 'cloud' so that others could read and comment on it has allowed the team members to reside on separate continents: in the western United States, in Canada, in Germany and Holland, and of course in England. Once a year or more often we have gathered in Oxford to hammer out problems and exchange ideas, and occasionally reverse decisions made at the meeting before. It has been great fun to work on a text that has provided us with so much entertainment as well as a fascinating insight into life in another time and another place, some seven hundred years and more ago. We have found it a book such as the author's colleague and friend, Sadīd al-Dīn ibn Raqīqah, describes in one of his epigrams:

> I stayed at home and took as my companion
> a book that speaks of all kinds of virtues.
> In it, whenever I take it up to leaf through what it contains,
> I have a lush and pretty garden.[8]

8 Ch. 15.46.3.2.

4 Note on Transliteration

Arabic names, words, and phrases have been transliterated according to the system used in the third edition of *The Encyclopaedia of Islam* (*EI Three*), with a few exceptions:

– The common feminine ending known as *tāʾ marbūṭah* is *-ah*, not *-a* (when followed by a genitive it is *-at*).
– Compound names are always given as two words, as they are in Arabic script (Imruʾ al-Qays, ʿAbd Allāh, ʿAbd al-Raḥmān, Hibat Allāh), whereas *EI Three* writes names with Allāh as one word (ʿAbdallāh, Hibatallāh).
– Ibn ('son of') and Bint ('daughter of') are never abbreviated to 'b.' and 'bt.'

Place names and terms current in English appear in their customary forms, such as Mecca, Cairo, Euphrates, Islam, Qur'an. We have used *shaykh* rather than 'sheikh', because the meanings of the latter form in English ('tribal leader; member of a ruling family') are very different from its use by Ibn Abī Uṣaybiʿah ('teacher, authoritative scholar').

CHAPTER 2

Ibn Abī Uṣaybiʿah: His Life and Career

Franak Hilloowala

The book *ʿUyūn al-anbāʾ fī ṭabaqāt al-aṭibbaʾ* (*The Best Accounts of the Classes of Physicians*), by the thirteenth-century Syrian physician Ibn Abī Uṣaybiʿah, is of great importance to scholars today as a key source for the social history of science and medicine in medieval Islam. In addition, the work offers valuable insight into the structure and function of medieval society in the Middle East and forms a major guide to the knowledge of Greek and Indian classical learning available in the Islamic world.

But what do we know of its author, a physician from Damascus? Most of what we know today about his life and career is found in his own personal comments and observations inserted into various biographies of the physicians featured in this book. Thus, it is from the pages of the *ʿUyūn al-anbāʾ* itself that we can attempt to piece together a biographical sketch of Ibn Abī Uṣaybiʿah's life and career.

Muwaffaq al-Dīn Abū l-ʿAbbās Aḥmad ibn al-Qāsim ibn Khalīfah ibn Yūnus al-Khazrajī was commonly known by his nickname (*laqab*), Ibn Abī Uṣaybiʿah. This nickname, he tells us, was also given to his grandfather. It literally means 'the son (or descendant) of the man with a little finger'. Most likely, it indicated that an ancestor of his had either a deformity or other distinctive quality to his little finger.

It is commonly agreed among medieval historians that Ibn Abī Uṣaybiʿah was born in Damascus. The precise date of his birth, however, is uncertain. The date 590/1194, given by some scholars,[1] seems too early because Ibn Abī Uṣaybiʿah informs us that his father, Sadīd al-Dīn al-Qāsim, was born in 575 [= June 1179 to May 1180]. This would have made his father only 15 years of age when Ibn Abī Uṣaybiʿah was born. On the other hand, the historian Ibn Taghrībirdī (d. 874/1470) records in his history, *al-Nujūm al-zāhirah fī mulūk Miṣr wa-l-Qāhirah*, that Ibn Abī Uṣaybiʿah's death occurred in 668, in the month of Jumādā I [= 27 December 1269 to 25 January 1270], when he was over 70 years of age – implying simply that Ibn Abī Uṣaybiʿah was born sometime *before* 598/1201.

1 *GAL*, I, 326 S I, 560.

From the *ʿUyūn al-anbāʾ* we learn that in 590/1194, his father, Sadīd al-Dīn al-Qāsim, was living in Cairo with Ibn Abī Uṣaybiʿah's uncle and Sadīd al-Dīn's brother, Rashīd al-Dīn, and their father. Ibn Abī Uṣaybiʿah reports that his grandfather returned to Syria with his two sons in the year 597/1200–1201, thus making 597 [October 1200–October 1201] a more likely date for Ibn Abī Uṣaybiʿah's birth.

Ibn Abī Uṣaybiʿah composed his famous book toward the end of Ayyubid control in Syria. He and his family, and many of his colleagues, served the ruling Ayyubid princes and governors of the day, experiencing both the benefits of these elite connections and the disadvantages/dangers of political instability and competition among the governing elite. Additionally, Ibn Abī Uṣaybiʿah must have witnessed in 658/1260 the fall to the Mongols of the cities of Damascus and Ṣalkhad,[2] where he was based for the latter part of his life.[3]

The majority of biographical material regarding the life of Ibn Abī Uṣaybiʿah and his family occurs in the fourteenth and fifteenth chapters of the *ʿUyūn al-anbāʾ*, which are concerned with the physicians of Egypt and Syria, both regions where he and members of his family lived and worked as physicians. More specifically, we get biographical information about the careers of his grandfather, father and uncle from the biography of his paternal uncle, Rashīd al-Dīn ʿAlī ibn Khalīfah, found in Ch. 15.51.

Ibn Abī Uṣaybiʿah's grandfather was Khalīfah ibn Yūnus ibn Abī l-Qāsim ibn Khalīfah, of [the tribe of] Khazraj, of the line of Saʿd ibn ʿUbādah,[4] and he also was known as Ibn Abī Uṣaybiʿah. He was a physician in Damascus and served the famous Ayyubid ruler Ṣalāḥ al-Dīn Yūsuf ibn Ayyūb (Saladin),[5] as well as his sons, and successors. When Saladin conquered Egypt in 1171, a group of physicians from Damascus accompanied him. Among them was Ibn Abī Uṣaybiʿah's

2 Ṣalkhad (Ṣarkhad) was an important town and stronghold in southern Syria set in the Ḥawrān region near the border of present-day Jordan. The town contains an important fortress, built between 611/1214 and 645/1247. See Yāqūt al-Rūmī, *Muʿjam al-Buldān*, 3:401. See *EI²* art. 'Ṣalkhad' (M. Meinecke).

3 Ibn Abī Uṣaybiʿah uses the term al-Tatar, referring to the Mongols, and calls the Mongol ruler, Ilkhan. He briefly mentions the Mongol conquests numerous times. See for instance biography 15.32, where he discusses briefly the fall of Aleppo and Damascus, and biography 11.13, his comments on the poem (attributed, obviously wrongly, to Ibn Sina) on the Mongol conquests and the Mamluk victory in 1260.

4 Saʿd ibn ʿUbādah ibn Dulaym of the Banū l-Khazraj (d. c. 14/636), a Companion of the Prophet Muḥammad; see *EI²*, entry 'Saʿd b. ʿUbāda' (W. Montgomery Watt). See also Ch. 15.51.

5 Reg. 532–589/1138–1193, founder of the dynasty of the Ayyūbids. *EI2* art. 'Ṣalāḥ al-Dīn' (D.S. Richards).

grandfather, as well as two oculists, Jamāl al-Dīn ibn Abī l-Ḥawāfir[6] and Shihāb al-Dīn Abū l-Ḥajjāj Yūsuf,[7] both of whom were acquaintances and friends of Ibn Abī Uṣaybiʿah's grandfather in Damascus and both of whom later taught Ibn Abī Uṣaybiʿah's father and uncle during the beginning of their medical training in Cairo.

Ibn Abī Uṣaybiʿah tells us that his grandfather desired to teach his two sons the art of medicine because he, 'was well aware of its noble rank … and held that one who was committed to its truths would be honoured and favoured in this world and given the highest rank in the world to come.'[8] Indeed, the two sons, Sadīd al-Dīn and Rashīd al-Dīn both became successful physicians. They studied medicine in Egypt and Syria with some of the most influential teachers of the time, were attached to the famous hospitals in both Cairo and Damascus, and served the rulers of their day.

Although Ibn Abī Uṣaybiʿah's father was the elder of the two brothers, Sadīd al-Dīn al-Qāsim does not have a separate biography in the *ʿUyūn al-anbāʾ*. Rather, Ibn Abī Uṣaybiʿah discusses his father's accomplishments within the biography of Sadīd al-Dīn's brother, Rashīd al-Dīn ʿAlī ibn Khalīfah. However, Ibn Abī Uṣaybiʿah seems no less impressed with the accomplishments of his father, who was the longer lived of the two brothers.

Our author's father, Sadīd al-Dīn al-Qāsim, was born in 575/1179 in Cairo, where he continued to reside and undertook his early studies of medicine. Sadīd al-Dīn studied with the leading authorities of the day, including the physician and philosopher Maimonides (Mūsā ibn Maymūn).[9] Ibn Abī Uṣaybiʿah's grandfather directed Sadīd al-Dīn to study the science of ophthalmology, sending him to practice under the aforementioned Abū l-Ḥajjāj Yūsuf, who was serving as an oculist at the hospital in Cairo. An interesting detail regarding their place of residence in Cairo is given by Ibn Abī Uṣaybiʿah when he tells us that this hospital was the older one and not the one connected to the fort (presumably Saladin's Nāṣirī hospital established in 1171). He tells us that the older one was situated near the flea markets of lower Cairo and near to where Ibn Abī Uṣaybiʿah's grandfather and the family lived, so that Sadīd al-Dīn was easily able to frequent the lectures at the hospital.[10]

6 For his biography and an anecdote that Ibn Abī Uṣaybiʿah relates about Ibn Abī l-Ḥawāfir and the honour of the art of medicine, see Ch. 14.44.

7 No further information is available on this oculist (*al-kaḥḥāl*) who is mentioned at the start of Ch. 15.51.

8 Ch. 15.51.1.

9 Abū ʿImrān Mūsā ibn Maymūn al-Qurṭubī (Maimonides) (527–606/1135–1208), Jewish theologian, philosopher and physician. For his biography, see Ch. 14.39.

10 Ch. 15.51.1

Sadīd al-Dīn al-Qāsim eventually became a famous oculist in his own right. With time, his skill became known to the courts of the Syrian Ayyubid princes and he entered into their service, attaining a good reputation and many benefits and favours from them. Sadīd al-Dīn served al-Malik al-ʿĀdil[11] while the ruler stayed in Damascus and accompanied him when he travelled. After the latter's death, he served his son, al-Muʿaẓẓam ʿĪsā,[12] and then went into the service of al-Muʿaẓẓam's son, and the house of Ayyub generally in Damascus, attending to all of al-ʿĀdil's sons. He also became a physician in the famous al-Nūrī hospital founded in Damascus by Nūr al-Dīn ibn Zangī (r. 541–569/1146–1175). Ibn Abī Uṣaybiʿah writes that the reputation of his father's skill was such that people came specifically to him to be treated.[13] On Friday night, the twenty second of the month of Rabīʿ II in 649 (= 14 July 1251), he died in Damascus and was buried at the Paradise Gate on the way to Mount Qāsiyūn.[14]

Of the two brothers, Ibn Abī Uṣaybiʿah chose to include a biography in the *ʿUyūn al-anbāʾ* of his uncle, the younger brother, rather than his father, and thus, we can only speculate that he may have been the more successful or well-known of the two.

Ibn Abī Uṣaybiʿah's uncle, Rashīd al-Dīn ʿAlī ibn Khalīfah was born in Aleppo in the year 579/1183 and raised there until he began his education in Cairo. He began his study of medicine under Jamāl al-Dīn ibn Abī l-Ḥawāfir when Abū l-Ḥawāfir was head physician in Egypt during the rule of Saladin. He studied ophthalmology and surgery under Nafīs al-Dīn ibn al-Zubayr,[15] whom al-Malik al-Kāmil ibn al-Malik al-ʿĀdil[16] had appointed chief physician (*riʾāsat al-ṭibb*) in Egypt and who also held a position at the Nāṣirī Hospital.[17] Nafīs al-Dīn ibn al-Zubayr was also one of Ibn Abī Uṣaybiʿah's sources for the later biographies in the *ʿUyūn al-anbāʾ*.

11 Al-Malik al-ʿĀdil Sayf al-Dīn Abū Bakr ibn Ayyūb (d. 619/1218). Ayyubid ruler. Took over the sultanate (569/1199–1200) and dispersed Saladin's sons (i.e., his own nephews). He was recognised by the Abbasid Caliph in 604/1207.

12 Al-Malik al-Muʿaẓẓam ʿĪsā Sharaf al-Dīn, the son of al-Malik al-ʿĀdil I Muḥammad reigned as governor of Damascus from 597–615/1201–1218 and was Sultan from 615–624/1218–1227.

13 Ch. 15.51.3

14 The Gate of Paradise (Bāb al-Farādīs, also known as Bāb al-Imārah) is one of the seven gates of Old Damascus. The gate was given its name because of its proximity to numerous water sources and lush gardens. There were initially eight gates of Old Damascus, but one was destroyed in Ottoman times.

15 For a biography of this physician, see Ch. 14.47.

16 Ayyubid ruler of Egypt from 615/1218 to 635/1238.

17 Built in 1171 by Saladin after he conquered Egypt. See Ragab, *Medieval Islamic Hospital*, 83–89.

When Ibn Abī Uṣaybiʿah's grandfather finally returned to Damascus around 597/1200, Ibn Abī Uṣaybiʿah tells us that his uncle was then 'around 20 years of age' and continued to study with influential teachers in Damascus. The scholar-physician Raḍī al-Dīn al-Raḥbī,[18] and close friend of the family, had heard about Rashīd al-Dīn's academic accomplishments and agreed to take him on as a student. Consequently, Rashīd al-Dīn attended al-Raḥbī's *majlis* and studied medicine under him. While in Damascus, Rashīd al-Dīn was also attached to al-Nūrī hospital where, at that time, Muwaffaq al-Dīn ibn al-Ṣaraf (an otherwise unidentified physician) and the shaykh Muhadhdhab al-Dīn Abū Muḥammad ʿAbd al-Raḥīm ibn ʿAlī ibn Ḥāmid, known as al-Dakhwār (d. 628/1230),[19] were practising. According to Ibn Abī Uṣaybiʿah, Rashīd al-Dīn became a shaykh – that is, teacher – of medicine at twenty-five years of age. He had a *majlis* where he gave instruction in medicine, and Ibn Abī Uṣaybiʿah reports that many of his students became distinguished in the profession of medicine.[20]

In addition to medicine and ophthalmology, Rashīd al-Dīn also studied other subjects such as belles-lettres, astronomy, the art of music, and the Persian language, as well as Arabic and philosophy under the guidance of Muwaffaq al-Dīn ʿAbd al-Laṭīf al-Baghdādī.[21] Rashīd al-Dīn first studied the religious sciences and mathematics under Abū l-Tuqā Ṣāliḥ ibn Aḥmad ibn Ibrāhīm ibn al-Ḥasan ibn Sulaymān al-Qurashī, known as al-Taqī.[22]

Rashīd al-Dīn, like his father and brother, also served at the courts of various Ayyubid princes. Al-Muʿaẓẓam, ruler of Damascus, had heard of Rashīd al-Dīn and summoned him for an audience, praised him and asked that he serve him, but Ibn Abī Uṣaybiʿah's uncle declined asking to remain in Damascus. After serving al-Malik al-Amjad Majd al-Dīn Bahrām Shāh, the son of ʿIzz al-Dīn Farrukh Shāh ibn Shāhān Shāh ibn Ayyūb, the governor of Baalbek who had known Ibn Abī Uṣaybiʿah's grandfather, Rashīd al-Dīn finally entered as a physician the service of al-Muʿaẓẓam until the latter's death. During his service to al-Muʿaẓẓam, at that time ruler of Damascus, he also attended to the ruler's two brothers, al-Malik al-Kāmil ruler of Egypt, and al-Malik al-Ashraf, ruler of the Diyār Bakr, during their visits to al-Muʿaẓẓam. Ibn Abī Uṣaybiʿah reports that during one of these visits, al-Kāmil was so impressed, that he gave Rashīd al-Dīn a robe of honour and five hundred Egyptian dinars for his service.

---

18 For his biography, see Ch. 15.36.

19 He was the founder of a school in Syria devoted entirely to the study of medicine. For his biography, see Ch. 15.50.

20 Ch. 15.51.2.

21 See Ch. 15.40.

22 Additional information on this scholar is not currently available.

In the year 616/1219, Ibn Abī Uṣaybiʿah's uncle was personally summoned by al-Malik al-Ṣāliḥ Ismāʿīl, the son of al-Malik al-ʿĀdil, to the southern Syrian town of Bosra,[23] where a great epidemic was raging. to come and treat his mother and other sufferers at the court; following which he would be allowed to return to Damascus. Although in poor health from his previous activities and travelling to attend to the Ayyubid royal court, Rashīd al-Dīn nonetheless obeyed the summons and successfully treated the Sultan's mother. Shortly afterwards, however, he was stricken with an acute fever that grew worse, even after his return to Damascus. He was unable to recover, despite the care of some of the best physicians in Damascus, and died in the second hour of Monday, the seventeenth of Shaʿbān, of the year 616 [= 28 October 1219] at the age of 38. He too was buried outside the Gate of Paradise as was Ibn Abī Uṣaybiʿah's grandfather and Rashīd al-Dīn's brother, Sadīd al-Dīn.[24]

Given his family background, it is not surprising that Ibn Abī Uṣaybiʿah also pursued a career in medicine. He studied the theoretical aspects of medicine under the tutelage of his father and uncle and with the most influential teachers and medical professionals of his day, learning the practical aspects of medicine at the al-Nūrī hospital in Damascus. His tutors were among some of the most famous of their time. They included Raḍī l-Dīn al-Raḥbī, whom Ibn Abī Uṣaybiʿah refers to as shaykh (teacher) in reference to the many generations who studied the art of medicine under him. There was also the Syrian scholar Ibn al-Bayṭār (d. 646/1248), an authority in the field of botany (*ʿilm al-nabāt*) and famous for his compendium on medicinal substances (*K. al-Jāmiʿ fī l-adwiyah al-mufradah*).[25] Ibn Abī Uṣaybiʿah writes of Ibn al-Bayṭār, 'My first meeting with him was in Damascus in the year 633/1235–1236 when I also observed his easy social qualities, the range of his honourable virtues, the excellence of his disposition, the goodness of his character and the nobleness of his soul, which were beyond description.' He goes on to say of Ibn al-Bayṭār, 'In his company, I inspected many plants in their natural habitats on the outskirts of Damascus. Under his guidance, I also studied his commentary on the names of the medicinal substances in Dioscorides' book, and so I was able to observe at first hand his vast knowledge and his understanding of a great number of subjects.'[26]

23 Bosra (ancient Bostra) is a town in southern Syria that used to be an important stopover on the caravan route to Mecca.

24 Ch. 15.51.8

25 For his biography, see Ch. 14.58.

26 Ch. 14.58.2.

One of the most influential of Ibn Abī Uṣaybiʿah's teachers was the renowned Muhadhdhab al-Dīn ʿAbd al-Raḥīm ibn ʿAlī (d. 628/1230), known as al-Dakhwār,[27] the founder of a school in Syria devoted entirely to the study of medicine. Ibn Abī Uṣaybiʿah also mentions that in addition to studying with al-Dakhwār in Damascus, he had earlier worked with him at the military camp of al-Malik al-ʿĀdil where al-Dakhwār and Ibn Abī Uṣaybiʿah's father served the Sultan.[28]

Around the year 633/1235, Ibn Abī Uṣaybiʿah went to Cairo for the purpose of continuing his medical education. He followed in the footsteps of his grandfather, uncle and father, and trained at the Nāṣirī hospital. Indeed, the professional connections of Ibn Abī Uṣaybiʿah, his family and colleagues, to the hospitals of their day, make Ibn Abī Uṣaybiʿah's *ʿUyūn al-anbāʾ* one of the primary sources we now have for the history of hospitals and their function during the medieval Islamic period.

Like his father, he studied and practiced ophthalmology. Amongst others, he trained under the Jewish physician Ibn Abī l-Bayān (d. 634/1236), author of a medical formulary of compound drugs intended for use in the Nāṣirī hospital.[29] Of his training with Ibn Abī l-Bayān, Ibn Abī Uṣaybiʿah says, 'Whenever we treated the patients at the Nāṣirī hospital in Cairo, I – Ibn Abī Uṣaybiʿah – witnessed the excellence of his achievements with regard to the knowledge and identification of diseases, the recollection of appropriate therapies, and his acquaintance with what Galen had said concerning them – all of which defies description.'[30]

Circumstances, however, seemed to have cut short Ibn Abī Uṣaybiʿah's studies in Cairo. He left after only one year, reportedly in 634/1236–1237, to return to Syria and into the service of ʿIzz al-Dīn Aybak, the governor of Ṣalkhad,[31] a stronghold in southern Syria near the border of present-day Jordan. One might speculate as to why Ibn Abī Uṣaybiʿah chose to cut short the beginning of a seemingly prestigious career in the hospitals of Cairo in order to serve a relatively minor Ayyubid ruler in an area that was not one of the intellectual centres of the Ayyubid Empire, in contrast to those of Cairo or Damascus.[32] Indeed, some of Ibn Abī Uṣaybiʿah's circle of friends and colleagues seemed

27 For his biography, see Ch. 15.50. See also, Ragab, *Medieval Islamic Hospital*, 142–145.

28 Ch. 15.50.3.

29 For his biography, see Ch. 14.53.

30 Ch. 14.43.1.

31 ʿIzz al-Dīn Aybak al-Muʿaẓẓamī ruled Ṣarkhad (or Ṣalkhad as it is better known today) as major-domo of the Ayyubid prince al-Muʿaẓẓam ʿĪsā from 611/1214 until 644/1247; see *EI*² art. 'Ṣalkhad' (M. Meinecke).

32 See ʿĀnūtī, *Ibn Abī Usaybiʿah*; Kurd ʿAlī, *Kunūz al-ajdād*; IAU Wahb.

to have questioned Ibn Abī Uṣaybiʿah's situation in Ṣalkhad as well. A letter from Sharaf al-Dīn ibn al-Raḥbī, a physician in his own right from Damascus and son of Raḍī l-Dīn al-Raḥbī,[33] includes a poem to Ibn Abī Uṣaybiʿah about his situation in Ṣalkhad under the patronage of ʿIzz al-Dīn Aybak al-Muʿaẓẓamī.

Muwaffaq al-Dīn![34] What's this mindlessness of yours,
despite the rank you have earned in knowledge and erudition?
Have you sold your soul for something trifling and paltry?
You sold it cheap, after being serious and assiduous for so long!
You have been staying in a town that mocks its inhabitants;
no sensible person of standing would be content with it.
It is remote from all that is good; barren; nothing is there
except rocks and blazing heat.
You are wasting a life that cannot be replaced:
when a time has passed it will not return.

Sharaf al-Dīn ibn al-Raḥbī goes on:

If where you are now were a place to acquire riches,
it would not compensate for spending your life in hardship.
So how is it what with the little regular pay and its meanness,
and being so far from all virtuous and erudite people?
Come back, then, to the Paradise on Earth:[35] she has come forward
for her beauty to be revealed, in her new clothes,
And do not stay anywhere else if you have earned riches,
for life in any other place is not worth considering.[36]

Ibn Abī Uṣaybiʿah's own poetry, included in correspondence to his colleagues and friends, seems to reiterate the sentiment expressed by Sharaf al-Dīn ibn al-Raḥbī. In a reply to Ibn al-Raḥbī's poem, Ibn Abī Uṣaybiʿah essentially refers to Ṣalkhad as somewhat of a backwater locality:

33 For a biography of this physician see Ch. 15.37.
34 That is, Ibn Abī Uṣaybiʿah.
35 Syria and Damascus are often called *jannat al-dunyā*, 'paradise of (this) world' (e.g. al-Thaʿālibī, *Thimār al-qulūb*, 694).
36 Ch. 15.37.

How can life be enjoyed by someone whom Time has allotted
to people who are firewood![37]
In their ignorance they do not know the worth of a scholar,
which is not surprising in the case of ignorant people.
I came to someone in whose courtyard my merit was wasted. Would the
stupidity of the non-Arabs be aware of the intelligence of the
Arabs?[38]

From an historical and strategic standpoint, Ṣalkhad was not the insignificant backwater that Ibn Abī Uṣaybiʿah and his colleagues mock in their poetry. It was a town and stronghold of strategic and military importance during the medieval period as it was the southernmost advance post of Syria towards the desert lands of the Arabian Peninsula, as well as a junction of important trade routes. During the Ayyubid period, when it was brought under control by Saladin in 1187, the town took on importance as a buffer to the continuous threats from crusades. The town was highly fortified, containing an important fortress, built between 611/1214 and 645/1247 by the Ayyubid dynasty. The peak of the town under the Ayyubids came under al-Muʿaẓẓam, a well-educated ruler who had received a thorough education in belles-lettres and sciences. He became ruler of Damascus in 1198 under his father's (al-ʿĀdil) tutelage and was made ruler of Damascus in 1218. To hold power during his reign, which was characterized by a political struggle with his brothers, al-Kāmil and al-Ashraf Mūsā, he replaced the old guard of his father with his own men. This included the appointment of ʿIzz al-Dīn Aybak al-Muʿaẓẓamī as governor and holding the *iqṭāʿ* (fief) of Ṣalkhad. Aybak held this *iqṭāʿ* from 608–644/1211–1247, when he fell out of favour and was suspected of treason, losing his political standing. It was Aybak who sponsored major building and endowments in Ṣalkhad.

In another poem, this one sent to a Syrian physician named Muwaffaq al-Dīn ʿAbd al-Salām,[39] Ibn Abī Uṣaybiʿah writes about Damascus and his longing to return home:

Perhaps a time that has gone by in Damascus[40]
will return, the abode will be near after separation,
Time, after its tyranny, will grant justice,

37 Presumably alluding to Q al-Jinn 72:15, «*As for the unjust, they are firewood for Hell*».

38 This unkind description of ʿIzz al-Dīn Aybak al-Muʿaẓẓamī contrasts with the very positive entry on him in al-Ṣafadī, *Wāfī*, ix:480–481.

39 For his biography, see Ch. 15.54.

40 On Jilliq as a 'poetic' name of Damascus, see below, Ch. 10.69.3.9.

and I may meet with loved ones.
For I have looked forward for so long to seeing its remains
and have yearned for so long for its inhabitants.[41]

So what then of Ibn Abī Uṣaybiʿah's decision to move to Ṣalkhad after a potentially promising career in Cairo, and his decision to remain there despite his longing to return to Damascus? It does appear that he remained in Ṣalkhad for the remainder of his career and that he died in Ṣalkhad.[42]

Perhaps the answer is to be found in the information that Ibn Abī Uṣaybiʿah presents to us about the careers of his family members. We observe that the family, in its professional capacity as court physicians, was tied to the service of the Ayyubid princes in Syria and Egypt. Ibn Abī Uṣaybiʿah states that his grandfather had served Saladin, following him when he conquered Egypt in 569/1171: 'He had moved to Egypt when al-Malik al-Nāṣir Ṣalāḥ al-Dīn Yūsuf ibn Ayyūb conquered it and was in his service and that of his sons.'[43] In 597/1200–1201 [?] Ibn Abī Uṣaybiʿah's grandfather returned to Syria, the same year that al-ʿĀdil[44] took control of the Ayyubid Sultanate. Ibn Abī Uṣaybiʿah relates that his father, Sadīd al-Dīn had previously served al-ʿĀdil in medical matters and had asked to be excused from the leader's request to follow him on his campaigns in order to remain in Damascus. This request however, was denied and Sadīd al-Dīn finally enrolled in al-ʿĀdil's service on the 15th Dhū l-Ḥijjah of the year 609 [9 May 1213], adding that the sultan and all his sons relied on his father for medical treatment and rewarded him accordingly. Thus Ibn Abī Uṣaybiʿah's father was serving al-ʿĀdil's son al-Ashraf al-Mūsā, emir of Damascus from 626/1229 to 635/1237, when Ibn Abī Uṣaybiʿah was summoned to Ṣalkhad in 634/1236.

Ibn Abī Uṣaybiʿah's uncle also was summoned into service of the Ayyubid princes. Ibn Abī Uṣaybiʿah writes that while Rashīd al-Dīn was serving at the court of al-Malik al-Amjad in Baalbek, al-Muʿaẓẓam took notice of him, and al-Muʿaẓẓam engaged him in his service in 610/1213. Al-Muʿaẓẓam also appointed Rashīd al-Dīn to act as military secretary when al-Muʿaẓẓam was in Damascus, a post that Ibn Abī Uṣaybiʿah's uncle did not want. Ibn Abī Uṣaybiʿah says that the only thing his uncle could do was to obey the order, although he was eventually released from this administrative position when the sultan finally acceded to his request. Ibn Abī Uṣaybiʿah states that this obligation to service eventually cost Rashīd al-Dīn his life, for he fell ill in 616/1219, following a sum-

41 Ch. 15.54.
42 Ibn Kathīr, *al-Bidāyah*, year 668.
43 Ch. 15.51.1.
44 Sultan of Egypt, (r. 596–615/1200–1218).

mons to Bosra to treat members of the Ayyubid ruling family suffering from an epidemic raging there, and Rashīd al-Dīn died shortly thereafter.

Given his family's past service and his father's continuing service to the Ayyubid court, Ibn Abī Uṣaybiʿah's move to Ṣalkhad to serve as a phyician to the Ayyubid sultan's representative there, can be seen as an obligation. Ibn Abī Uṣaybiʿah's family (as was probably the case with many of the physicians and medical professionals found in the pages of the *ʿUyūn al-anbāʾ*) owed their livelihood and good standing to the ruling elite of the day. We see this in another poem in the correspondence with Sharaf al-Dīn ibn al-Raḥbī. In the poem, Ibn Abī Uṣaybiʿah suggests that perhaps one has no power over that which is preordained:

> If my staying among such people was a mistake on my part
> and part of my life went by in hardship,
> Well, my namesake, in the past, stayed among people
> in the land of Naḥlah, complaining of time's vicissitudes.[45]
> These things come pre-ordained;
> nothing in this world happens without a cause.[46]

Whatever Ibn Abī Uṣaybiʿah's reasons for remaining in Ṣalkhad, it was there that he undertook to write the *ʿUyūn al-anbāʾ*, with the encouragement of others, such as the physician Amīn al-Dawlah,[47] who was the vizier for al-Malik al-Ṣāliḥ and keen patron of the sciences. As Ibn Abī Uṣaybiʿah explains: 'He took a keen interest in collecting and studying books, and purchased many outstanding works in all the sciences.'[48] Ibn Abī Uṣaybiʿah also tells us that Amīn al-Dawlah heard about the book he was composing and asked to have a copy made. Ibn Abī Uṣaybiʿah fulfilled the request and composed a panegyrical poem for the vizier that he sent along with the copy of the book.[49]

Ibn Abī Uṣaybiʿah's motivation for writing the *ʿUyūn al-anbāʾ* seems evident from the Preface, where he classifies his subject, the art of medicine (*ṣināʿat*

45 The famous poet al-Mutanabbī (d. 354/965), whose given name was Aḥmad like that of Ibn Abī Uṣaybiʿah, said in a poem: 'My stay in the abode of Naḥlah is just as Christ's dwelling among the Jews' (*Dīwān*, 32, cf. Arberry, *Poems of al-Mutanabbī*, 20–21). Naḥlah (Nahle) being a village just north of Baalbek.

46 Ch. 15.37.

47 For Amīn al-Dawlah's biography, see Ch. 15.49.

48 Ch. 15.49.6.

49 For the dedication to Amīn al-Dawlah and the various versions of the *ʿUyūn al-anbāʾ*, see the essay by Ignacio Sánchez on the manuscript tradition of the *ʿUyūn al-anbāʾ* and also Ch. 15.49 and the brief Addendum to 15.49.

al-ṭibb) as among the noblest of professions. He goes on to state that 'Its excellence is attested in scripture and the provisions of law, to such an extent that knowledge of the human body is deemed to be second in importance only to knowledge of religion.'[50] He states his intentions are to focus on the 'most distinguished of physicians' – an undertaking that, to his knowledge, had not been previously attempted. He also explains that he will use the form of the biographical dictionary (*ṭabaqāt*), not simply to present individual biographies of great men of medicine, but to write a comprehensive, coherent history of physicians throughout the ages from antiquity up to recent times (his own day), seemingly in an unbroken connection with a past tradition.

In addition to his desire to show the physician class as an elite class with a long and respected past, one might argue that his motivations were more self-serving in nature, a desire to make a contribution to his profession. Despite the heavy reliance on past scholarly authorities, Ibn Abī Uṣaybiʿah's use of personal narratives is a significant characteristic of the *ʿUyūn al-anbāʾ*. He often relates his own eye-witness accounts and observations of a person, place or event, and often recounts anecdotes and information transmitted orally through family, friends and colleagues.

His target audience appears to have been the scholarly-elite (his colleagues and teachers) and the ruling-elite – either officials or princely rulers. By highlighting the famous works and accomplishments in medicine of the physicians, and the patrons who have supported the profession, he tells his patrons and rulers that it is necessary to have learned physicians at court, and perhaps, more importantly, they should be highly valued by those who seek their services.

Not only does Ibn Abī Uṣaybiʿah highlight the medical achievements of the physicians included in the *ʿUyūn al-anbāʾ*, he also intentionally includes the other scholarly achievements of these medical professionals. He states that he has also included the accounts of 'wise men and philosophers' who also gave at least some of their attention to the art of medicine. It is notable that most of the physicians in Ibn Abī Uṣaybiʿah's *ʿUyūn al-anbāʾ* did not devote their entire time to medicine. The emphasis is on the polymath who had wide-ranging knowledge, for in the medieval period such an education identified someone as a scholar and elite. Lists of works and achievements of these physician/scholars include fields such as philosophy, astronomy, mathematics, engineering, languages and belles-lettres, and of course, as we have seen, poetry. All these subjects feature prominently in the education of a well-rounded medieval scholar

50 See Preface.

of Ibn Abī Uṣaybiʿah's day. In addition, he praises those scholars who are theological leaders within their communities as well, regardless of their religious affiliation.

Ibn Abī Uṣaybiʿah's *ʿUyūn al-anbāʾ* was not his only literary contribution. As we have seen, Ibn Abī Uṣaybiʿah himself composed poetry, examples of which are found in the *ʿUyūn al-anbāʾ*. He also wrote four other treatises on medicine or related topics that are now lost, but are mentioned by the author himself in the *ʿUyūn al-anbāʾ*:

- *Iṣābāt al-munajjimīn* (*The Predictions of the Astrologers*)
- *al-Tajārib wa-l-fawāʾid* (*Experiences and Useful Lessons*)
- *Ḥikāyāt al-aṭibbāʾ fī ʿilājāt al-adwāʾ* (*Anecdotes of Physicians on the Treatment of Illnesses*)
- *Maʿālim al-umam* (*The Outstanding Personalities of All Nations*)

Ibn Abī Uṣaybiʿah occasionally mentions, within the text *ʿUyūn al-anbāʾ*, various scholars and colleagues who have seen his book and praised it. Thus, we know from Ibn Abī Uṣaybiʿah himself that he continued to work on his biography of physicians, and added to it, probably up to the end of his life. From these personal anecdotes, we see that, for Ibn Abī Uṣaybiʿah, the *ʿUyūn al-anbāʾ* was a lifetime's work in which he took great personal pride. One might even argue, it was a work that he hoped would perhaps help make a name for himself and would serve to elevate the prestige of the medical profession.

It is nice to think that his life's work was not in vain, and that perhaps Ibn Abī Uṣaybiʿah was successful in what he sought to achieve. Today, the *ʿUyūn al-anbāʾ* is one of the most useful sources available for those of us who study the social history of science and medicine. Making this work available to non-Arabic speakers creates a much wider audience of scholars and the inquisitive, who can benefit from Ibn Abī Uṣaybiʿah's efforts and perhaps enjoy the persona of this thirteenth-century physician.

Indeed, we might chose to communicate the importance of the *ʿUyūn al-anbāʾ* in the same manner in which Ibn Abī Uṣaybiʿah and his colleagues did so, by citing a poem – specifically one found in the biography of his friend, colleague and contemporary, ʿIzz al-Dīn ibn al-Suwaydī.[51] In this poem al-Suwaydī expresses to Ibn Abī Uṣaybiʿah (whom he calls Muwaffaq al-Dīn) his admiration for his book on the history of physicians entitled '*The Best Accounts of the Classes of Physicians*' (*K. ʿUyūn al-anbāʾ fī ṭabaqāt al-aṭibbāʾ*):[52]

51 For a biography of this physician, see Ch. 15.57.

52 Ch. 15.57.

Muwaffaq al-Dīn, you have achieved what you desire
and have reached the highest of splendid ranks!
You have provided a fine history of those who have gone,
though their bones have now decayed.[53]
May God single you out with His beneficence
in this world and the next.

53 Compare Q al-Nāziʿāt 79:11, «*when we have become decayed bones*».

CHAPTER 3

ʿUyūn al-anbāʾ fī ṭabaqāt al-aṭibbāʾ: Its Genre and Title

Geert Jan van Gelder

Ibn Abī Uṣaybiʿah's book is the first to offer a 'world history of medicine', but as a work consisting of a large number of biographies it stands in a long tradition in Arabic literature. As its title proclaims, it is part of what is called the genre of *ṭabaqāt*.[1] The singular, *ṭabaqah*, may be translated as 'layer', '(social) class', or 'category', especially a group of people who share a profession or ability. As a genre it started when Islamic scholars, keen to establish the names, dates, and reliability of the transmitters of Hadith from the time of the prophet Muḥammad and his Companions, began to compile lists. An early standard work on the Prophet's Companions is *Kitāb al-ṭabaqāt al-kabīr* (*The Great Book of Classes*) by Ibn Saʿd (d. c. 320/844), and even larger compilations for this category were produced in later times. A contemporary of Ibn Saʿd, Ibn Sallām al-Jumaḥī (d. 231/845) compiled the first work on poets, *Ṭabaqāt fuḥūl al-shuʿarāʾ* (*The Classes of the Master Poets*). Similar works were written on jurisprudents, theologians, Qur'an commentators, ascetics and mystics, and philologists and grammarians. The organisation of these works differs and 'class' is usually not clearly defined; the various sections may be arranged roughly chronologically or geographically. Ibn Saʿd's multi-volume work is mainly chronological; within a 'class' there may be tribal or geographical grouping, and all women are together in the final volume, segregated from men.

Among the predecessors of Ibn Abī Uṣaybiʿah is the Andalusian Ibn Juljul (d. after 384/994), who collected biographies of philosophers and physicians in his *Ṭabaqāt al-aṭibbāʾ wa-l-ḥukamāʾ* (*The Classes of Physicians and Sages*), a work much smaller than that of Ibn Abī Uṣaybiʿah and with a more limited scope. Only the last of its seven 'classes' deals with physicians in the Islamic period. Another predecessor, also from Spain, is Abū l-Qāsim Ṣāʿid ibn Aḥmad, usually called Ṣāʿid al-Andalusī (d. 462/1070). His slim but important volume entitled *Ṭabaqāt al-Umam* (*The Classes of Nations*) deals with the scientists and scholars

1 See *EI*² art. 'Ṭabaḳāt' (Cl. Gilliot); Ibrahim Hafsi, 'Recherches'; al-Qāḍī, 'Biographical Dictionaries'; Young, 'Arabic Biographical Writing'.

DOI:10.1163/9789004410312_004

of various 'civilised' nations, which to him are the Indians, Persians, Chaldaeans, Greeks, *Rūm* (Romans or Byzantines), (pre-Islamic) Egyptians, Arabs, and Jews.

In a number of large and general biographical compilations the idea of 'class' is abandoned in favour of an alphabetical arrangement on the *ism* ('given name'), which is at least straightforward and systematic, but often cumbersome for the user because in very many cases a person is known by another part of his (more rarely her) name. Examples of such general alphabetical works are *Wafayāt al-aʿyān* (*Obits of Prominent People*) by Ibn Khallikān (d. 681/1282) and the massive *al-Wāfī bi-l-Wafayāt* (*The Completion of the Obits*) by al-Ṣafadī (d. 764/1363). These and many other works also list the books, epistles and treatises written by scholars, so that one may call them bio-bibliographical dictionaries. Ibn al-Qifṭī (d. 646/1248) adopted the same alphabetical arrangement for his dictionary on philologists and grammarians, *Inbāh al-ruwāh ʿalā anbāh al-nuḥāh* (*Informing Transmitters of the Reports on Grammarians*) and in his work on philosophers and physicians, *Ikhbār al-ʿulamāʾ bi-akhbār al-ḥukamāʾ* (*Informing Scholars of the Reports on Sages*).[2]

There is something unsatisfactory about a work where the mythical Hermes Trismegistus rubs shoulders with the 4th/10th-century Hilāl ibn Ibrāhīm al-Ṣābiʾ merely on account of the contingencies of the Arabic alphabet, even though this facilitates quick consultation. Ibn Abī Uṣaybiʿah apparently did not want to write a dictionary or encyclopaedia to be used for quick reference. Rather, he intended to write a work that could be read through from beginning to end, a history showing the origins, development, and progress of medical science, so he decided to go back to the system of *ṭabaqāt*, combining Ṣāʿid's category of 'nation' and geography with chronology; the alphabet plays no part in his book. Although the word *ṭabaqāt* appears in the title, the fifteen sections are called *abwāb* (sing. *bāb*), 'chapters'. Apart from the introductory first chapter, each chapter is said to contain the '*ṭabaqāt*' of the relevant section, but this does not mean that these chapters containing the entries on individuals are subdivided: at most there is a rough chronological order. Association also played a part and may distort the chronologic al sequence.[3] It is clear that the word *ṭabaqah* has virtually lost its meaning of 'class' or 'category' and has come to mean something like 'generation'. In fact, the word *ṭabaqāt* in each chapter

2 Edited by Julius Lippert as *Taʾrīkh al-ḥukamāʾ* (*The History of Sages*).

3 One may compare the remark by Ibn al-Nadīm in his *al-Fihrist* (Sayyid), i:450–451, a work known to Ibn Abī Uṣaybiʿah: 'When I mention one famous person, I let him be followed by whoever is close to him and resembles him, even though the latter may be from a later period than the former. This is my method in compiling this book.'

title could be omitted without any loss of meaning. The shortest, Chapter XII, *Fī ṭabaqāt al-aṭibbāʿ alladhīna kānū min al-Hind* (literally, 'On the Classes of Physicians Who Were from India'), contains a mere six biographies and a better translation is simply 'The Physicians of India'. It is impossible, however, for reasons of euphony, to omit the word *ṭabaqāṭ* from the rhyming book title: *ʿUyūn al-anbāʾ fī l-aṭibbāʾ* would sound badly unbalanced. Ibn Abī Uṣaybiʿah could have used another word, such as *tarājim* (biographies), or expanded his title as *ʿUyūn al-anbāʾ fī l-ḥukamāʾ wa-l-aṭibbāʾ*, since many physicians in the book could be called *ḥukamāʾ* (sing. *ḥakīm*, 'sage' or 'philosopher', often also meaning 'physician'). One may suppose he chose *ṭabaqāt* for his book, not only on account of the alliteration (the repetition of *ṭ-b* may have appealed to him), but also in order to emphasise its pedigree in a long and respectable tradition.

The word *aṭibbāʾ* is the plural of *ṭabīb*. Originally meaning 'skilled, expert, knowledgeable', it became the most common word for physician or medical doctor, cognate with *ṭibb*, 'medicine, medical practice'. The feminine is *ṭabībah*; but among the 442 physicians in the book who have individual entries only one is a woman, the rather obscure Zaynab of the Arab tribe of Banū Awd.[4] Cleopatra, listed as Kilāwubaṭrah in a series of names in the section on Socrates,[5] is called an *imraʾah ṭabībah*, 'skilled woman' or 'woman doctor'.

Rhyming titles[6] are extremely common in pre-modern Arabic literature from the 4th/10th century onward and Ibn Abī Uṣaybiʿah's book is no exception. Usually, they consist of two rhyming halves, the first being somewhat flowery and figurative and the second, mostly introduced by *fī* ('in', but here rather 'on, about', or the equivalent of the colon preceding an English subtitle), indicating more clearly the topic of the book. The word *anbāʾ* ('reports', plural of *nabaʾ*) would have readily suggested itself: not only is it used in similar biographical compilations, as in the celebrated work by his contemporary Ibn Khallikān mentioned above, it also provides a rhyme, even a rich rhyme, with *aṭibbāʾ*, 'physicians'. The word *ʿuyūn* is one the plurals of *ʿayn*,[7] which has the basic meanings of 'eye' and often means 'well, source'. Figuratively, it may stand for anything prime, choice, select, or prominent. Thus *ʿUyūn al-anbāʾ* can mean '*The Sources of Reports*' as well as '*Choice Reports*' and it evokes the prac-

4 IAU Ch. 7.10.

5 Ch. 4.1.11.2; in the section on Galen (Ch. 5.1.21.1) her name is spelled Qilāwubaṭr. See also al-Masʿūdī, *Murūj*, ii:27–29, where, appearing as Qilābaṭrah, she is called a 'female sage and philosopher' (*ḥakīmah mutafalsifah*), author of works on medicine.

6 See Ambros, 'Beobachtungen'.

7 Other plurals are *aʿyun* ('eyes'), *aʿyān* ('notables, prominent people'), and *ʿaynāt* (plural of the 18th letter of the Arabic alphabet, called *ʿayn*).

tically synonymous title of a seminal literary anthology, *ʿUyūn al-akhbār*, by the polymath Ibn Qutaybah (d. 276/889). This work is generally considered a proto-type of what is called '*adab* literature', in which *adab* (not only 'good manners', 'proper conduct' but also '*belles-lettres*')[8] stands for the mixture of literary entertainment, moral education, and instruction or information that is characteristic of such works. It is obvious that Ibn Abī Uṣaybiʿah, by his inclusion of so much poetry, entertaining anecdotes, maxims and wise sayings,[9] saw his work as standing in the same tradition. It inspired our decision to translate the title as *A Literary history of Medicine*, with a more literal version, *The Best Accounts of the Classes of Physicians* as a kind of subtitle.[10]

---

8 In modern Arabic, *adab* means 'literature' but has also kept its original sense of 'good manners'. On *adab*, see e.g. *EI*[2] art. 'Adab' (F. Gabrieli); *EAL* 54–56 (H. Kilpatrick); Bonebakker, 'Adab and the Concept of belles-lettres'; Horst, 'Die Entstehung der adab-Literatur'; Khalidi, *Arabic Historical Thought* (see Ch. 3, 'History and *Adab*', 83–130).

9 Maxims and wise sayings (*ḥikam*) are more prominent in *ʿUyūn al-anbāʾ* than in, for instance, the works of Ibn Juljul and Ibn al-Qifṭī. Ibn Abī Uṣaybiʿah may have been inspired by the work of the 5th/11th-century Egyptian author Mubashshir ibn Fātik, *Mukhtār al-ḥikam wa-maḥāsin al-kalām* ('Choice Wise Sayings and Fine Statements'), on the ancient Greek sages.

10 Other translations of the title include: *Fountains of Information respecting the classes of Physicians* (Cureton, 'Extract'); *Sources of Information regarding the Classes of Physicians* (Young, 'Arabic Biographical Writing', 174); and *The Sources of Knowledge about the Generations of Physicians* (Reynolds, *Interpreting the Self*, 60).

CHAPTER 4

The Textual and Manuscript Tradition of Ibn Abī Uṣaybiʿah's *ʿUyūn al-anbāʾ fī ṭabaqāt al-aṭibbāʾ*

Ignacio Sánchez

1 The Edition of Ibn Abī Uṣaybiʿah's Work

The first edition of the *ʿUyūn*, made by the German scholar August Müller and published under the pen-name of Imruʾ al-Qays al-Ṭaḥḥān, saw the light in Cairo in 1882. Much to his despair, this publication was flawed by innumerable mistakes resulting from the stubbornness and lack of diligence of Muṣṭafā Efendī Wahbī, the owner of the Cairene press commissioned to print the books. Müller was forced to sponsor a reprint of the text accompanied with long lists of variant readings and corrections and a detailed introduction describing the difficulties of editing this work and his problems with the publisher.[1] This second edition was printed in Königsberg in 1884.

The Königsberg edition of the *ʿUyūn* also served as the basis for two further editions. The first of them was published in Beirut in 1965 by Nizār Riḍā. Riḍā's work as editor essentially consisted of amending the Arabic text of Müller's edition by incorporating some of the corrections listed in the second edition of 1884 and adding some minimal – and haphazard – annotation; this edition still contains numerous errors, and the index of names is chaotic. Riḍā's edition of 1965 was, in turn, the basis for yet another non-critical edition made by Qāsim Wahhāb, published in Beirut in 1987.

In addition to these two publications that rely directly on Müller's editorial work, we also have a second critical edition made by ʿĀmir al-Najjār, originally published in Cairo in 1996 and reprinted in 2001–2004. Al-Najjār worked with eight manuscripts, some of them unknown to Müller, and took also into consideration Qāsim Wahhāb's edition of 1987. This publication does not have the multitude of typographical and editorial mistakes that marred the text edited by Müller – though some mistakes are noticeable especially in his edition of the poetry – but it is not free from problems. Najjār is not as careful as Müller when

1 The dealings and disagreements of Müller with his publisher are described in detail in the first pages of Müller's introduction to his second edition of the *ʿUyūn al-anbāʾ*: 'Vorwort', v–xvi.

DOI:10.1163/9789004410312_005

it comes to differentiating between the different textual traditions, and he occasionally adopts readings that differ from the majority of the manuscripts without further explanation, or even without citing any manuscript to support them.[2]

A further complication that affects all the editions arises from the peculiarities of the *ʿUyūn* itself, already pointed out by Müller in the long introduction to his 1884 edition. Ibn Abī Uṣaybiʿah suggests in several instances that he wrote different versions of his work, and the extant manuscripts bear testimony to this variety. The identification of the different textual traditions is extremely problematic but is crucial both for the edition and for the evaluation of Ibn Abī Uṣaybiʿah's method.

Textual variations present obvious problems when it comes to producing an authoritative edition. In this regard, Müller's edition cannot be praised enough; he was able to identify different recensions and trace the chronological evolution of the text and, consequently, to select the manuscripts containing the latest version of the work. But this came with a caveat: the great divergences in the contents of the manuscripts and the nature of the additions found in the latest stage of the work cast doubt on the authorship of several passages. Müller, who was aware of the use of a draft copy of the author (*musawwadah*) in some manuscripts,[3] confessed that he was unable to ascertain whether the passages that appear in these copies and are lacking in the other manuscripts were added by Ibn Abī Uṣaybiʿah himself or by a copyist after his death. He had particular doubts about the use of Ibn al-Qifṭī's *Taʾrīkh al-ḥukamāʾ*.[4]

The challenges that a new edition of the text poses to its editors are various: first, one must establish an authoritative text, freed of the many mistakes that are found in previous editions; second, we must identify and separate the different textual versions of the *ʿUyūn al-anbāʾ* and the manuscript traditions associated with each version of the work; finally, the doubts cast on Ibn Abī Uṣaybiʿah's authorship of some passages should be revised in the light of new manuscript evidence, especially regarding the material taken from Ibn al-Qifṭī and the lacunas that Müller considered to be indications of posthumous additions.[5]

---

2 On these editions see Hilloowala, 'An Analysis of Ibn Abi Usaybiʿah', 200–203.

3 Specifically, he believed that the Cambridge, University Library, MS Or. 1461 (**n** in Müller's edition), and the London, British Library, MS Add. 7340 (**d** in Müller's edition) were redacted using Ibn Abī Uṣaybiʿah's *musawwadah* after the death of the author. See Müller, 'Über Text und Sprachgebrauch', 867–877.

4 See Müller, 'Vorwort', xxi, and the discussion below.

5 See Müller, 'Über Text und Sprachgebrauch', 857 f.

2 Müller's Study of the Textual and Manuscript Traditions

The best way to approach the study of the manuscript tradition of the *ʿUyūn al-anbāʾ* is to climb on the shoulders of Müller's gigantic scholarship. Müller's edition was based on an large number of manuscripts that he discussed both in the introduction to the Königsberg edition of 1884,[6] and in a couple of journal articles, later collected by Fuat Sezgin in his *Studies on Ibn Abī Uṣaybiʿa (1270) and his ʿUyūn al-anbāʾ fī ṭabaqāt al-aṭibbāʾ*.[7] In these articles, Müller concluded that Ibn Abī Uṣaybiʿah wrote two recensions of the treatise, and that, in addition, it is possible to identify a third recension that included rearranged materials.[8] Müller also argued that the differences in the copies ascribed to the second version of the text might be the result of additions made after the death of the author, especially the glosses or quotations from Ibn al-Qifṭī's *Taʾrīkh al-ḥukamāʾ*, only accessible to Ibn Abī Uṣaybiʿah in the last years of his life.[9]

The existence of at least two versions of the *ʿUyūn al-anbāʾ* was deduced from internal evidence. The first version – and the manuscripts preserving it – is clearly identifiable from a dedication to the vizier Amīn al-Dawlah that is not present in later versions of the text. As Müller notes, the last date mentioned in the manuscripts containing this dedication is 639/1242.[10] In the manuscripts in which this dedication is not present, there are additions in which Ibn Abī ʿUṣaybiʿah refers to an earlier version of his work on at least three occasions.[11]

The dedication to Amīn al-Dawlah and Ibn Abī Uṣaybiʿah's references, together with a careful collation of many of the manuscripts, drove Müller to identify two textual traditions. Additionally, he pointed out the existence of a third hybrid recension of the text represented by the copy of the work held at

6 See Müller, 'Vorwort'.

7 Specifically, Müller, 'Über Text und Sprachgebrauch', and 'Über Ibn Abi Oçeibiʿa' [reprinted in Sezgin, *Studies* I, 139–263; and Sezgin, *Studies* II, 191–214].

8 Vernet, in his article of the second edition of the *Encyclopaedia of Islam*, referred to these versions as 'a major and a minor' plus a 're-written version after the author's death', see *EI*[2] art. 'Ibn Abī Uṣaybiʿah' (J. Vernet). For the sake of clarity, I will refer to Müller's versions of Ibn Abī Uṣaybiʿah's text as *recensions 1, 2, and 3*. The classification used in our edition and in this study, which does not correspond with Müller's, will be referred to as *Versions 1, 2, and 3*.

9 Müller, 'Über Text und Sprachgebrauch', 858–859, and 882 f.

10 Müller, 'Über Text und Sprachgebrauch', 856. See below the discussion on the dating of the versions pp. 55–56.

11 See below pp. 54–55.

TABLE 4.1 IAU's recensions (Müller)

Recension 1	Recension 2	Recension 3 (amalgamation of previous recensions)
c = Munich, Staatsbibliothek, MS Or. 243–244	d = London, British Library, MS Add. 7340	a = Berlin, Staatsbibliothek, MS Wetzst. 323 and MS Sprenger 312
p = Oxford, Bodleian Library, MS Pococke 356	e = London, British Library, MS Add. 25736	
t = Tübingen, Universitätsbibliothek, MS M. A. VI.14	f = London, British Library, MS Add. 23364	
v = Vienna, Nationalbibliothek, MS 1164 (Mxt. 180)	g = Gotha, Stadtbibliothek, MS 1769	
	i = Paris, Bibliothèque Nationale, MS Supp. Ar. 674	
	k = Leiden, Universiteitsbibliotheek, MS Or. 59b (1st part)	
	l = Leiden, Universiteitsbibliotheek, MS Or. 59b (2nd part)	
	m = Oxford, Bodleian Library, MS Marsh 153	
	n = Cambridge, University Library, MS Or. 1461	
	s = Paris, Bibliothèque Nationale, MS Suppl. Ar. 5939 (Schefer 2116)	

the Staatsbibliothek in Berlin and preserved in two volumes, MS Wetzst. 323 and MS Sprenger 312. This third version, according to Müller, contained abridged biographies resulting from the amalgamation of the two previous versions, i.e., the recension dedicated to Amīn al-Dawlah and the second recension written by Ibn Abī Uṣaybiʿah after 639/1242. Table 4.1 shows Müller's tripartite classification of the manuscripts.[12]

12 See Müller, 'Vorwort', xvii–xxi.

The identification of the manuscripts containing the first recension of the text does not pose many problems since they all share a clear set of characteristics, the most evident of which are the presence of the dedication to Amīn al-Dawlah (lacking in later versions) and the lack of references to events that happened after the year 639/1242. The singularity of the hybrid Berlin copy clearly sets it apart from other versions of the text.[13] But the manuscripts that Müller ascribed to the second recension, on the contrary, show important divergences in need of further elucidation.

For Müller, the numerous textual variants present in the copies with recension 2 might have resulted either from common accidents in the textual transmission or from posthumous editorial interventions, or from both – hence his doubts about Ibn Abī Uṣaybiʿah's authorship of a significant number of passages. Müller believed that the major part of the additions and changes from recension 1 to recension 2 were made during the life of the author. These changes are mainly revisions resulting from personal experiences that enriched the information about the physicians, especially after 639/1242, and from a systematic use of Ibn al-Qifṭī's *Taʾrīkh al-ḥukamāʾ*.[14] More serious doubts were raised, however, by the comments written in the margins of the Cambridge, University Library, MS Or. 1461 (siglum **n** in Müller's edition). This manuscript, which contains a full copy of the *ʿUyūn an-anbāʾ*, reproduces in the margin of f. 231a a marginal note present in the exemplar from which it was copied. In this note, the copyist of the exemplar (used as source for the Cambridge MS) stated that parts of Ibn Abī Uṣaybiʿah's work were not present in the first exemplar he had used, which was a copy in the author's handwriting; and that these additional sections were taken from a second exemplar that was the autograph draft of Ibn Abī Uṣaybiʿah (*musawwadat al-muṣannif*).[15]

The marginalia in the Cambridge manuscript describe the copyist's struggle to reproduce the contents of Ibn Abī Uṣaybiʿah's *musawwadah*, but these marginal comments were not made by the copyist of the Cambridge MS himself, but rather by the copyist of the manuscript from which it was itself copied, as stated in a clarification in the same hand added after the colophon:[16]

---

13 None of the manuscripts used by Müller or Najjār, nor any of those consulted for this study show any relationship with the Berlin MS, on which see below p. 63.

14 Müller, 'Über Text und Sprachgebrauch', 856.

15 See Cambridge, University Library, MS Or.1461 f. 231a; and Müller, 'Über Text und Sprachgebrauch', 867 f. This marginal note is discussed below, see pp. 43 and 47.

16 See Cambridge, University Library, MS Or.1461 f. 326a.

وكتبت هذه النسخة من النسخة التي كتب من نسخة المصنف وقوبل بمسودة المصنف رحمه الله تعالى وغفر ذنوبه

> This copy [i.e., the Cambridge MS] was copied from the exemplar that had been copied from the autograph of the author and [then] collated with his [the author's] draft, may God have mercy on him and pardon his faults.

Müller believed that the majority of the variants that represent the transition from recension 1 to recension 2 correspond with the changes and updates in the *musawwadah* to which the marginalia of the Cambridge MS refers. But these variations are far too many: 'In long sections of our text in all manuscripts' – says Müller – 'every page presents a great number of variants, omissions or arrangements'.

The main question that confronted Müller was to what extent these changes were due to authorial revision or to a posthumous intervention of a copyist. But the annotations reproduced in the Cambridge MS are virtually incomprehensible without the exemplar from which it was copied. Fortunately, this exemplar has been identified as Istanbul MS Şehid 'Ali Paşa 1923, the existence of which was unknown to Müller – or to any other editor. Müller was certainly conscious of the problems that his lack of information presented, so much so that his discussion of the *musawwadah* is to a great extent a succession of unanswerable questions. The first of them is obviously: 'How was the copyist [of the Cambridge MS] able to know the state of the *musawwadah* and to describe it, if he did not have it in front of him?'.[17]

Upon reading the marginal note in the Cambridge MS that refers to the use of both the author's autograph and the author's draft, Müller's first interpretation was that the clean copy (*mubayyaḍah*) was an exemplar of recension 1, not least because the copyist states that it was written for some eminent men (*akābir al-nās*) and the manuscripts with the first recension have the dedication to Amīn al-Dawlah. Consequently, it would be plausible to assume that the *musawwadah* referred to by the copyist would be the last version of the work – in progress towards recension 2 – in the author's hand.

Müller noticed, however, that the identification of the clean autograph (*mubayyaḍah*) with recension 1 was not certain, since some of the biographies present in all manuscripts with recension 1 are lacking in the Cambridge MS.[18]

17 See Müller, 'Über Text und Sprachgebrauch', 869. For the Istanbul Şehid Ali Paşa, MS 1923, see Section IV below.

18 Müller, 'Über Text und Sprachgebrauch', 872.

Müller was unable to provide a clear answer to this issue, although he suggested that these omissions and similar variants might have been the result of accidents in the textual and manuscript transmission of the *mubayyaḍah* (the use of 'ein Doppelgänger' with notable differences), or even had been produced by the intervention of the copyist of the Cambridge MS.[19]

Müller's intuition was, again, correct, but not his explanation. The exemplar to which he referred as the 'clean copy' (*mubayyaḍah*) is not a heavily modified version of the first recension dedicated to Amīn al-Dawlah. As we will discuss below, the author's autograph to which the copyist of the Cambridge MS refers is an intermediate version (Version 2 in our edition), that should be placed between the first version dedicated to Amīn al-Dawlah and the last one represented by the *musawwadah*. This could only be known by directly examining the Istanbul Şehid Ali Paşa, MS 1923 from which the Cambridge MS was copied.

Müller's introduction to the 1884 edition of the *ʿUyūn* is mainly concerned with distinguishing between the different recensions. His article on the text and the language of Ibn Abī Uṣaybiʿah's work, published that same year, examines in detail the chronology of the additions and its implications for assessing the authorship of the variants in recension 2. As mentioned, Müller was unable to make his mind up about the use of Ibn al-Qifṭī's *Taʾrīkh al-ḥukamāʾ*, but he found some instances that suggested a possible posthumous intervention by later editors. The first example is the addition of small glosses, such as that explaining the etymology of the ethnonym 'Sabians' at the beginning of the life of Thābit ibn Qurrah,[20] which is missing in the manuscripts with recension 1; or the gloss commenting on the significance of Monastir's *rābiṭah* in the biography of Ibn al-Jazzār.[21] Müller, however, admitted that he had no way of knowing whether these glosses stemmed from the draft of the author or not. As can be seen in the Istanbul MS Şehid Ali Paşa – discussed below –, these glosses were added by Ibn Abī Uṣaybiʿah himself.

A further example of suspicious additions spotted by Müller is the filling of blank spaces. At the beginning of the biography of al-Suhrawardī, for instance, the manuscripts from both recensions 1 and 2 have left a blank space instead

19 Müller, 'Über Text und Sprachgebrauch', 875–876. The missing biographies of the Cambridge MS correspond, in fact, with biographies written on some interleaves added to the Şehid Ali Paşa manuscript that have not come down to us, and might have been lost when the Cambridge MS was copied.

20 'It is said that the Sabians are related to Ṣāb, that is Ṭāṭ the son of the prophet Idrīs, may salutations be upon him', see Müller, i:215.26–27 [= Ch. 10.3.1].

21 'Monastir – the latter is a place for devout Muslims bound to God (*murābiṭah*), well-known for its blessings (*barakah*) and mentioned in the history books, on the coast of the Mediterranean Sea', see Müller, ii:38.7 (incorrectly written *al-mustanīr*) [= Ch. 13.3.2.1].

of the name of al-Suhrawardī's father. This omission was also recorded by Ibn Khallikān when he quotes from Ibn Abī Uṣaybi'ah in his biography of this scholar, stating that there was a blank space after 'ibn' just where the name of al-Suhrawardī's father should have appeared. Of all the manuscripts consulted by Müller only the Berlin MS (Müller's **a**), which belongs to recension 3, has filled the blank space with *'A-m-r-k-ā'*.[22] This seems to be an obvious example of scribal intervention, but the Berlin MS (recension 3) is by no means representative of the textual tradition of the *'Uyūn al-anbā'*.[23]

In the light of these conjectures, Müller attempted a chronological classification of the manuscripts based on their stemmatic affinities and his interpretation of the role played by the use of the *musawwadah*. He divided the manuscripts into the four types described in Table 4.2.[24]

As will be discussed below, this classification cannot be maintained in view of the direct copy of the *musawwadah* preserved in MS Şehid Ali Paşa 1923. But, for all the difficulties, and despite the doubts about the authorship of some additions, Müller was certainly correct when he identified the possible stages in the transmission of the work, and also when he selected the best manuscripts among those at his disposal. He identified London British Library MS Add 7340 (**d** in Müller's edition, **R** in ours), and Cambridge MS Or. 1461 (Müller's **n**, not used in our edition) as those containing a most complete version of the work, since he believed that they had made use of the *musawwadah* of the author. The third of the core manuscripts for Müller was the London British Library Add. 23364 (Müller's **f**, in our edition **L**), copied from an exemplar endowed by Ibn Abī Uṣaybi'ah to the *maqṣūrah* of Ibn 'Urwah. This selection, as Müller argued, was also supported by indirect later witnesses, whose quotations from the *'Uyūn al-anbā'* reproduce the text as it can be read in these manuscripts rather than the contents of recension 1.[25] On the whole,

22 Müller, 'Über Text und Sprachgebrauch', 858–859. All MSS consulted in our edition have the blank space, except for H, which has 'Amīrak', i.e., the correct name (see Ch. 15.18.1).

23 Müller, 'Über Text und Sprachgebrauch', 859. The relevance of this recension 3 is somehow exaggerated in the article of the *Encyclopaedia of Islam*, where Vernet states: 'From the two redactions a not very careful copyist produced a re-written version after the author's death', see *EI*² art. 'Ibn Abī Uṣaybi'a' (J. Vernet).

24 Müller, 'Über Text und Sprachgebrauch', 885–888. In this classification, Müller does not refer to the Oxford, Bodleian Library, MS Marsh 153 (**m**) and the Gotha, Stadtbibliothek, MS 1769 (**g**), since they contain partial or incomplete copies; for his comments on these MSS see Müller, 'Vorwort', xxiv. He does not discuss at this point the manuscripts with recension 1 either, since they do not present the kind of variants that might result from the use of the *musawwadah*.

25 Müller, 'Vorwort', xx.

TABLE 4.2 Classification of MSS according to the use of the *musawwadah* (Müller)

1	2A	2B	2C
Recension 1: version dedicated to Amīn al-Dawlah.	Recension 2: transcriptions from IAU's *musawwadah* made before the death of the author or shortly thereafter.	Recension 2: lacunose manuscripts, corrected (perhaps by the author) but without further additions.	Recension 2, manuscripts posthumously corrected with IAU's *musawwadah*.
– Munich, Staatsbibliothek, MS Or. 243 – Oxford, Bodleian Library, MS Pococke 356 – Tübingen, Universitätsbibliothek, MS M. A. VI.14 – Vienna, Nationalbibliothek, MS 1164	– London, British Library, MS Add. 25736 – London, British Library, MS Add. 23364 – Paris, Bibliothèque Nationale, MS Supp. Ar. 674 – Leiden, Universiteitsbibliotheek, MS Or. 59b (1st part)	– Berlin, Staatsbibliothek, MS Wetzst. 323 and MS Sprenger 312 (one of the copies used in this MS would have belonged to this group) – Leiden, Universiteitsbibliotheek, MS Or. 59b (2nd part) – Paris, Bibliothèque nationale de France, MS arabe 5939	– London, British Library, MS Add. 7340 – Cambridge, University Library, MS Or. 1461

and taking into consideration the difficulties of the textual transmission of the work, Müller's intuition was usually spot on, and his edition stands out as one of the monuments of 19th-century scholarship. He was wrong, however, about the use of the *musawwadah* and its implications, as will be discussed below.

3 Najjār's Discussion of the Manuscripts

In contrast with Müller, Najjār provides a very succinct description of the manuscripts used in his critical edition of the *ʿUyūn al-anbāʾ* and does not discuss at all the implications of the different textual versions of the text. From his description and the footnotes referring to the biographies wanting in each manuscript, it is possible to infer that four out of the eight copies on which he based his edition belong to the first version of the work, easily identifiable by the presence of the dedication and the absence of the last nine biographies of chapter 15; these manuscripts are Cairo, Dār al-Kutub, MS 2104 *Taʾrīkh Ṭalʿat*; Cairo, Dār al-Kutub, MS 182 *Taʾrīkh*; Cairo, Dār al-Kutub,

MS 1341 *Ta'rīkh Taymūr*; and Tübingen, Universitätsbibliothek, MS M. A. VI.14, the latter also used by Müller.[26] Najjār chose Cairo, Dār al-Kutub, MS 2104 *Ta'rīkh Ṭalʿat*, dated in 1003/1595, as the basis for the edition of the first part of the book, since it was the oldest copy available to him containing this section.[27]

The other four manuscripts used in Najjār's edition are: Cairo, Dār al-Kutub MS 219 *Ta'rīkh*, which is the base manuscript for chapters 8–14 due to its early date (707/1308); Copenhagen, Kongelige Bibliotek, MS 2/1144 (Cod. Ar. CIXXI), which contains chapters 7–15; Istanbul, Ahmad III, MS 2860/71 *Ta'rīkh*, which contains chapters 11–15; and Istanbul, Ahmad III, MS 2859/70, which goes from the end of chapter 12 to the end of the work. In addition to these manuscripts, Najjār also consulted Qāsim Wahhāb's edition of the *ʿUyūn al-anbā'*, printed in Beirut in 1987.

Najjār approached his edition without apparently paying attention to the differences between versions, as if the textual variants were only the product of divergences in the manuscript traditions. In his edition, for example, the first seven chapters are only covered by manuscripts containing Version 1 of the work, without any collation of the manuscripts he used representing later versions, which are used by him only for chapters 8 to 15.

4 Manuscript Şehid Ali Paşa 1923

Most of the questions that troubled Müller in his study were related to the use of the *musawwadah*, which he could only envisage from the vague indications found in the Cambridge MS. This manuscript was copied from Şehid Ali Paşa, MS 1923, a volume of 307 folios held at the Süleymaniye Library in Istanbul. Examination of this manuscript sheds extraordinary light onto the complicated history of the textual transmission of the *ʿUyūn*.

The Şehid Ali Paşa MS originally contained a full copy of an early version of the *ʿUyūn* taken from a manuscript that had been directly copied from the author's autograph. Afterwards, its copyist found and used as an exemplar the autograph draft of the author (*musawwadah*); when he collated his copy with this second manuscript, he found variant readings, with which he corrected the

26 For the description of the manuscripts see Ibn Abī Uṣaybiʿā, *ʿUyūn al-anbā' fī ṭabaqāt al-aṭibbā'*, ed. Najjār, i:101–108.

27 Najjār does not state precisely what he means by 'the beginning of the book (*awwal al-kitāb*)', but in a later commentary he seems to refer to the first eight chapters, see Najjār, i:102.

text, and additional information that was added in the margins and on interleaves pasted to the volume.[28] The revision of works and the coexistence of various versions is not an oddity in the writerly practices of the pre-Modern Islamic lands. Famous works such as Ḥunayn ibn Isḥāq's *Risālah*, for instance, have come down to us in two versions.[29] Autograph *musawwadāt* are not rare either and author's drafts of relevant works have survived and been studied, such as Ibn al-Furāt's *Taʾrīkh al-duwal wa-l-mulūk*;[30] and sometimes even edited, as in the case of the *musawwadah* of al-Maqrīzī's *Khiṭaṭ*.[31]

These two exemplars on which the Şehid Ali Paşa MS is based are the ones mentioned in the colophon of the Cambridge MS known to Müller, but the identification of their contents and their relationship with the versions of the work is only possible thanks to the diligence of the copyist of the Istanbul manuscript, who was careful to mark all the additions and variants. By looking at the distribution of the text in this manuscript, is possible to differentiate between (1) the version represented by the main body text, copied in the first place and (2) a second, extended version consisting of the body text with interlineal or marginal corrections from the variants found in the *musawwadah* together with additions written in the margins and interleaves taken from the author's draft, which might be full biographies, extended paragraphs, or small additions and glosses.

The colophon of the Şehid Ali Paşa MS (see Fig. 4.1) was most likely written before its collation with the *musawwadah*, since it does not make any reference to textual variants and additions:[32]

تم كتاب عيون الأنباء في طبقات الأطباء على يد أقل عبيد الله المشمولين بنعمته والراجين
غفرانه بسعة رحمته المتوسلين إليه بعظم ربوبيته أن يصلى على النبي المرسل رحمة لكافة خليقته
وعلى آله وصحبه ويسلم تسليماً كثيراً عبد الهادي بن أبي المفضل بن أبي الفرج وكان الفراغ
من تعليقه في يوم ٥ ٢٧ ٨ من عام ٧٧٣ للهجرة النبوية على صاحبها أفضل الصلاة والسلام
والتحية والإكرام وذلك بلطف الله وكرمه وتطوله ونعمه له الحمد كثيراً أولاً وأخيراً

28 See Fig. 4.3, where an entire biography is added in margin.

29 See Bergsträsser, 'Neue Materialen.'

30 See Bora, 'A Mamluk Historian's Holograph'.

31 Edited by F. Sayyid; see al-Maqrīzī, *Musawwadat K. al-Mawāʿiẓ*. On the different autographs of this work and their additions, see Witkam, 'Reflections on al-Maqrīzī's Biographical Dictionary'.

32 See Istanbul MS Şehid Ali Paşa 1923, f. 306a.

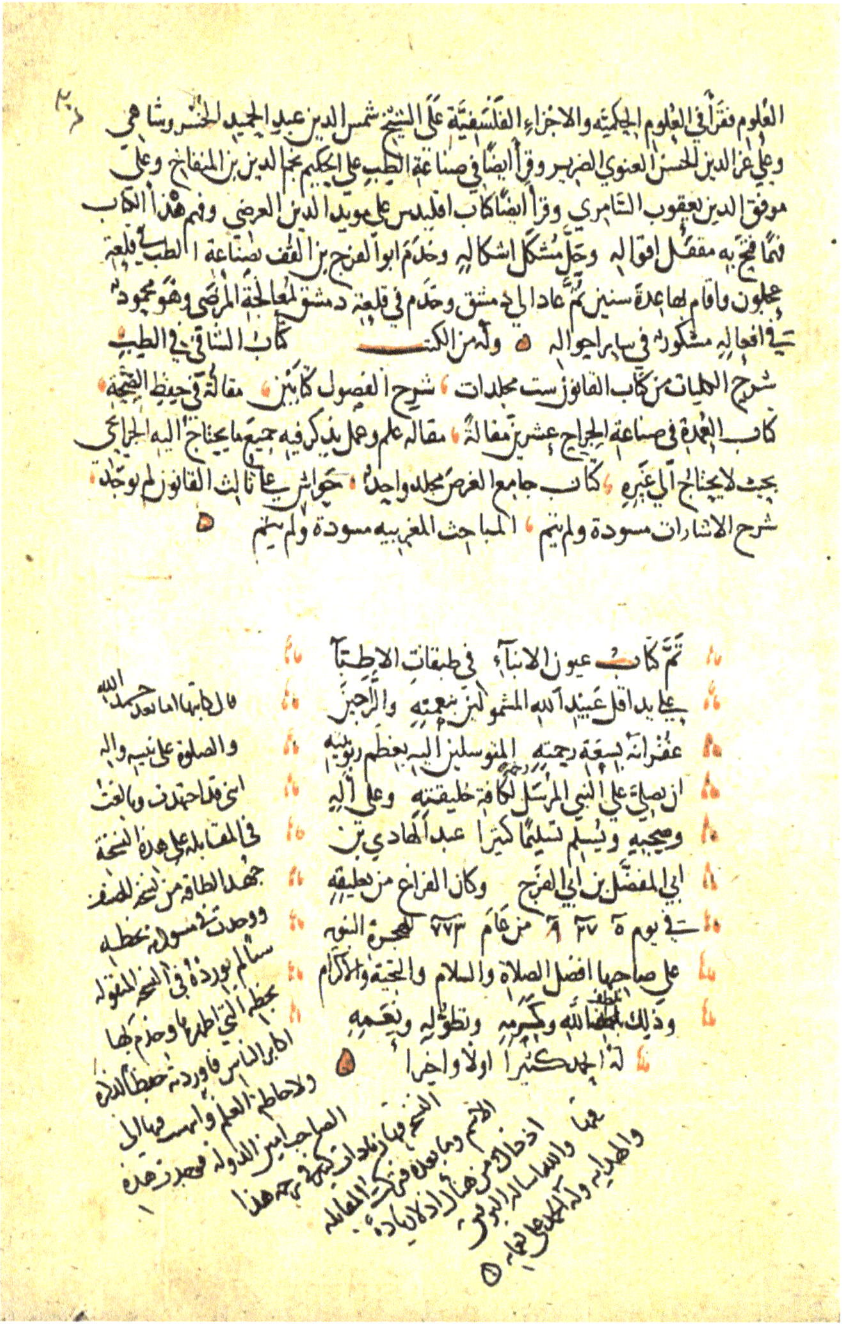

FIGURE 4.1 Istanbul, Süleymaniye, Şehid Ali Paşa MS 1923, fol. 306a showing colophon. Completed 27 Shaʿbān 773 (4 Mar 1372). Copyist: ʿAbd al-Hādī ibn Abī l-Mufaḍḍal ibn Abī l-Faraj. Siglum **A** in edition

The Best Accounts of the Classes of Physicians was completed by the humblest servant among those who enjoy the grace of God, trust in His great mercy for their forgiveness, and who asks Him, with the power of His divinity, to cherish the Prophet sent as mercy to all His creatures, and his people and companions, and to grant them many blessings, ʿAbd al-Hādī ibn Abī l-Mufaḍḍal ibn Abī l-Faraj. It was finished on 27 Shaʿbān 773 of the Hijrah [4 Mar 1372].[33] May its author receive the most perfect blessings, greetings and honours. This was accomplished thanks to God's kindness and magnanimity, and His generous grace; to Him is all praise from the beginning to the end.

The copyist, however, left two explanatory notes expressing his doubts and explaining the process of collation and the problems he had to solve. The first one was written at the beginning of Ch. 14 (see Fig. 4.2):[34]

حاشية قال كاتبها وجدت في مسودة المصنف أسماء لم ترد النسخة المنقول منها والمنقول منها نسخة قد نقلت من نسخة كتبها المصنف بخطه أيضاً وهجس بخاطري أنه أهمل الأسماء من المبيضة إما لأنه أراد الاختصار لكونهم لا فائدة فيهم وإما لغرض ما آخر وأردت أن لا تكون هذه النسخة ناقصة اسماً من الأسماء الواردة المسودة وكنت قد أوردت ما وجدته غير وارد النسخة المنقول منها في الأبواب المتقدمة من هذا الكتاب في هذه النسخة كل اسم حيث أورده المصنف في مسودته خشية أن يكون الناقل من مبيضته أهملهم والآن وقع في نفسي أن أورد جميع الأسماء الواردة المسودة تلو بعضهم بعض في قائمات وأنبه على مواضعهم حسب الطاقة إن استطعت ذلك ومن الله أستمد المعونة على ذلك

The writer of this marginal note said: I found in the draft of the author names that were not in the exemplar copy. And the exemplar is a copy that was transcribed from a copy also written by the author in his own hand. It occurred to me that he omitted names in the fair copy either because he intended to compose an abridgement, since they [i.e. the omitted names] are of no benefit or for some other reason. I did not want this copy to lack a single name from amongst those found in the draft [of the author]. I

33 The date in this colophon, written in black ink actually reads the 9th month, i.e., Ramaḍān, but the number was corrected on top, in red, to 8. The correction is appropriate for otherwise the date of the week (Friday) does not match.

34 See Istanbul MS Şehid Ali Paşa 1923, f. 215b.

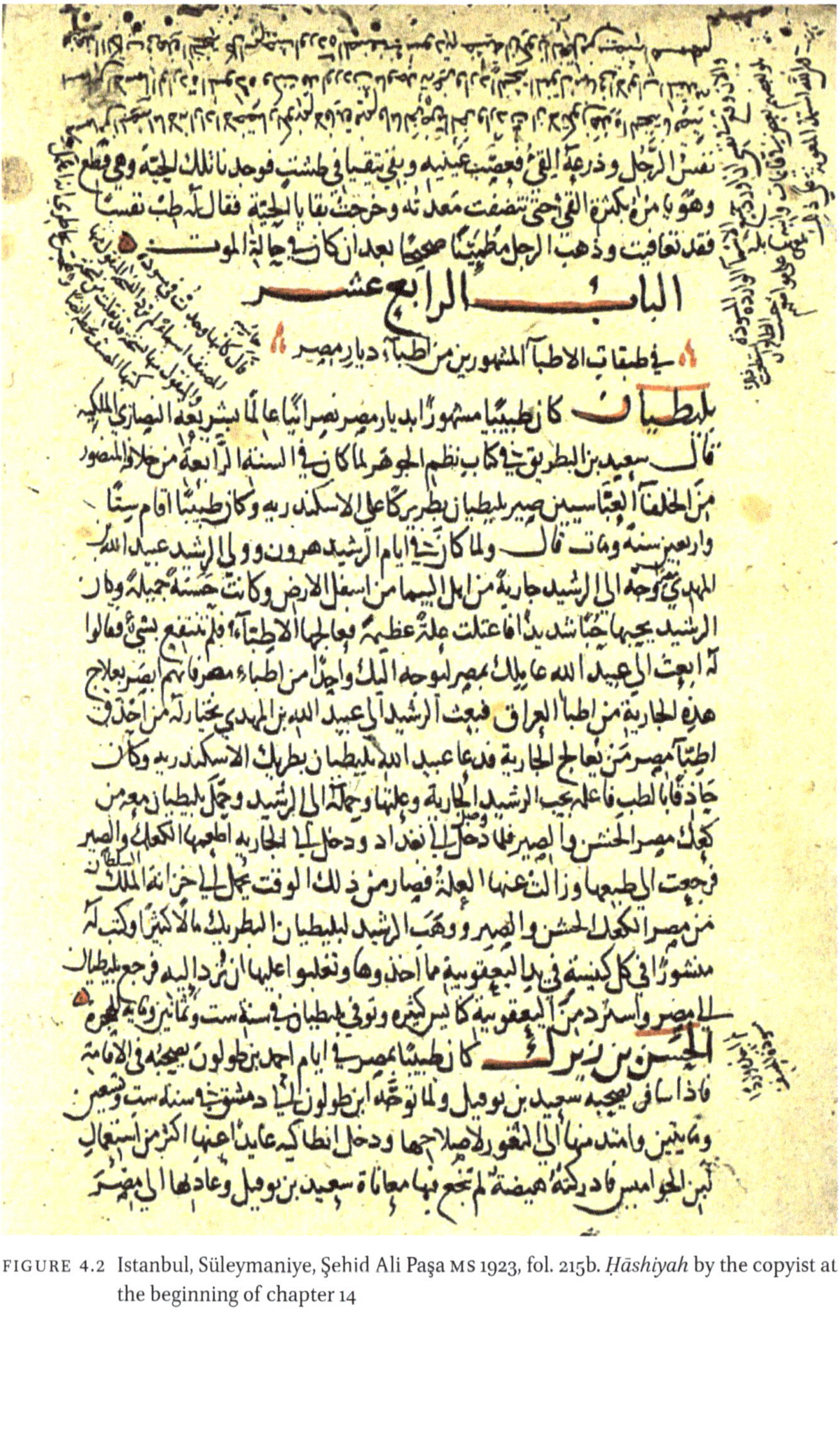

FIGURE 4.2 Istanbul, Süleymaniye, Şehid Ali Paşa MS 1923, fol. 215b. *Ḥāshiyah* by the copyist at the beginning of chapter 14

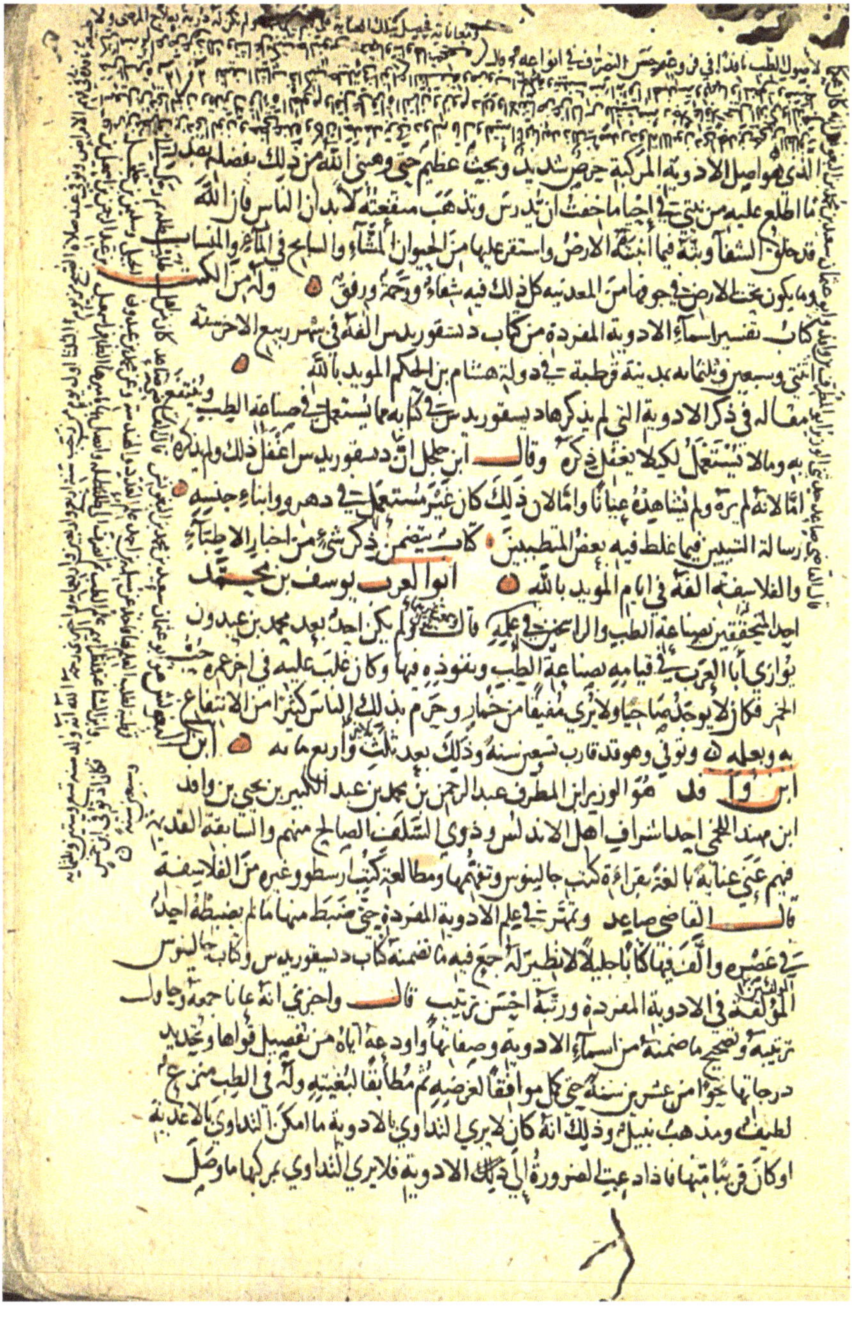

FIGURE 4.3 Istanbul, Süleymaniye, Şehid Ali Paşa MS 1923, fol. 200a, with an entire biography in the margin

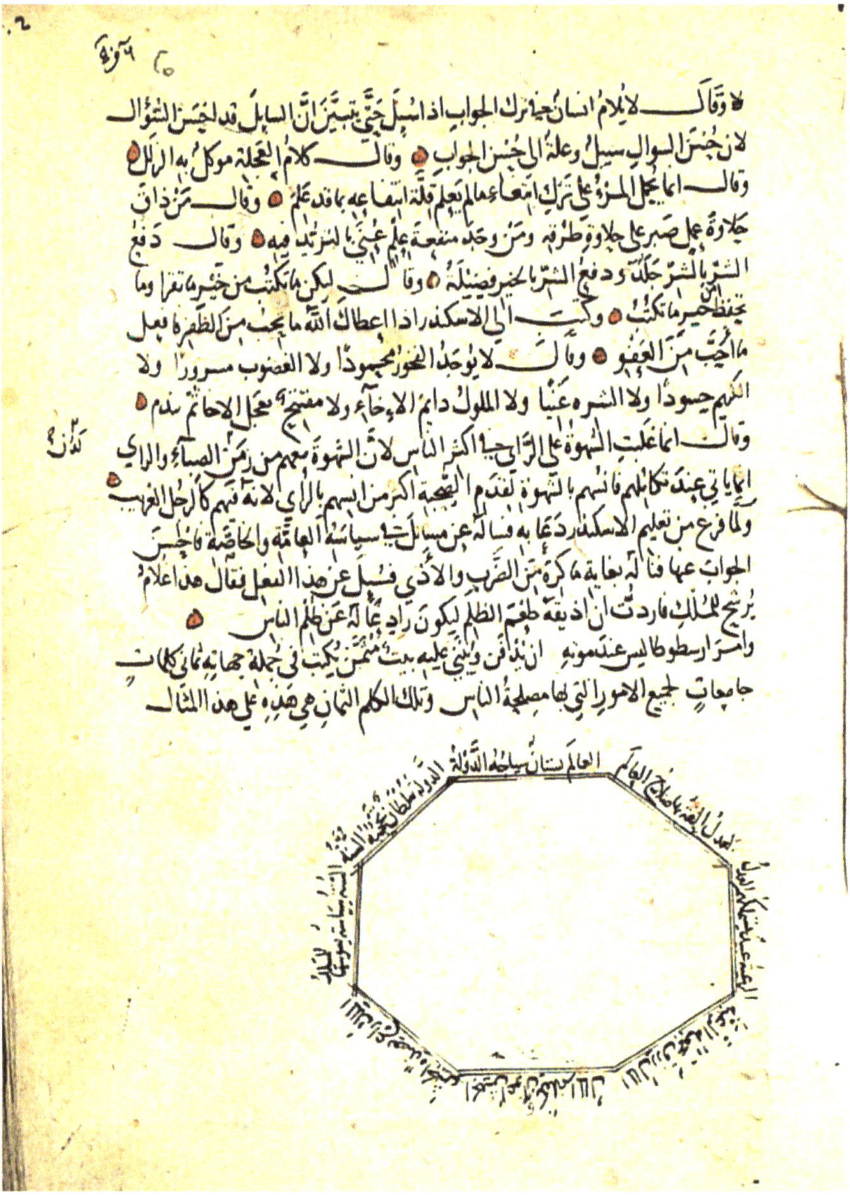

FIGURE 4.4 Istanbul, Süleymaniye, Şehid Ali Paşa MS 1923, fol. 45a showing placement of sayings on the octagonal mausoleum of Aristotle

have included whatever was not present in the exemplar in the preceding chapters of this book in this copy, each name at the point where the author [Ibn Abī Uṣaybiʿah] mentioned it in his draft, out of concern that the copyist of his [Ibn Abī Uṣaybiʿah's] fair copy had omitted them. And now it has occurred to me that I should include all of the names present in the author's draft following one another in lists (*qāʾimāt*) and indicate their positions [in the text] as far as I am capable and able to do so. From God I ask assistance in this.

The second marginal note was copied next to the colophon (see Fig. 4.1):[35]

قال كاتبها أما بعد حمد الله والصلوة على نبيه وآله أنني قد اجتهدت وبالغت في المقابلة على هذه النسخة جهد الطاقة من نسخة المصنف ووجدت في مسودته بخطه شيئاً لم يورده في النسخة المنقولة بخطه التي أظهرها وخدم بها أكابر الناس فأوردته حفظاً لذكره ولإحاطة العلم وانتهيت فيها إلى الصاحب أمين الدولة فوجدت هذه النسخة فيها زيادات كثيرة في ترجمة هذا الاسم وما بعده فتركت المقابلة إذ ذاك من هناك إذ لا زيادة فيها والله أسأله التوفيق والهداية وله الحمد على نعماته

The copyist says: praise be to God and blessing on His prophet and his family. I have meticulously collated this copy with the copy of the author to the best of my efforts. Then I found in the draft of the author, in his handwriting, things that he had not included in the autographed exemplar that he presented and dedicated to important people, so I included this in order to preserve what he reports and to encompass all knowledge. I reached up to [the biography of] al-Ṣāḥib Amīn al-Dawlah [al Ghazāl] and I found that this copy had many additions in the biography under this name and in the following. I abandoned the collation from this point onwards, since there is no extra material.

In the first of these notes the copyist refers to three manuscripts, two of them autographs: on the one hand, the exemplar from which he had made the initial copy (*al-nuskhah al-manqūl minhā*), which had been copied from an autograph (*qad nuqilat min nuskhah katabahā al-muṣannif bi-khaṭṭihi*);[36] on the other

35 See Istanbul MS Şehid Ali Paşa 1923, f. 306a (left margin).
36 The colophon of the Cambridge MS states that it had been directly copied from an autograph, see above pp. 35–36.

hand, the copyist had direct access to a draft copy in the author's handwriting (*musawwadah*) – from this draft he took additional passages and biographies not included in the first manuscript he had used.

The reference to the collation of the manuscript with the *musawwadah* is, however, rather puzzling since the collation marks reveal a procedure that is quite the opposite of the description that can be read in the colophon.

The collation notes found in the manuscript are the following (with reference to the exemplar used for the collation in brackets): f. 120b (*nuskhat al-aṣl lil-muṣannif*); f. 130b (*musawwadah lil-muṣannif*); f. 140b (*nuskhat al-muṣannif*); f. 142a (without reference to exemplar); f. 147a (without reference to exemplar); f. 150a (without reference to exemplar); f. 150b (without reference to exemplar); two in f. 160b (one without reference to exemplar, one stating *nuskhat al-aṣl bi-khaṭṭ al-muṣannif*); f. 167a (*nuskhat al-muṣannif*); f. 170b (without reference to exemplar); f. 180b (*nuskhat al-muṣannif bi-khaṭṭihi*); f. 192b (*nuskhat al-muṣannif bi-khaṭṭihi*); f. 198b (without reference to exemplar); f. 208b (without reference to exemplar); f. 215b (without reference to exemplar); f. 229b (without reference to exemplar); f. 232b (without reference to exemplar); f. 238b (without reference to exemplar); f. 258b (*nuskhat al-muṣannif*); f. 268b (without reference to exemplar); f. 278b (without reference to exemplar); f. 288b (without reference to exemplar).

The last collation note occurs on f. 288b, just before the biography of Amīn al-Dawlah (15.49), as stated in the colophon. But this is not due to the numerous additions that the copyist found afterwards, but rather because the differences between the text taken from the first exemplar and that of the *musawwadah* are minimal after this point, as the lack of marginalia and collation notes show. This is also proved by the correspondence between the last part of the work and the copy of the manuscript personally endowed by Ibn Abī Uṣaybiʿah to the *maqṣurah* of Ibn ʿUrwah and that is now in British Library MS Add. 23364.

It is difficult to find an explanation for the copyist's claim about abandoning the collation after the biography of Amīn al-Dawlah. In this marginal comment he refers to the collation with the copy of the author (*nuskhat al-muṣannif*): this might mean that he stopped collating his copy, once finished, with the first exemplar; or also that he might have had access to the original autograph from which the manuscript he used had been copied. However, the references in the collation notes are inconsistent, and, even though they only refer to the *musawwadat al-muṣannif* in one instance, they stop just when the differences between the main body text and the text of the *musawwadah* are minimal, which suggest that the collation referred to by the copyist is a collation with the author's draft. The most plausible explanation, though speculative, is that he stopped the collation because there were no further additions *after* the biography of Amīn al-Dawlah (Ch. 15.49), as it happens, and that the copyist mistakenly referred to additions after this biography (*mā baʿdahu*) instead of the additions that occur before (*mā qablahu*).

Further marginal notes also refer inconsistently to the exemplars consulted by the copyist. The first of them is a customary reference to the end of a *juz*ʾ in the exemplar (*al-aṣl al-manqūl minhu*) from which the main body text was copied, which occurs in f. 58b.[37] The first mention of a second copy, later referred to by the copyist as *musawwadah*, occurs on f. 102a, right at the end of the biography of ʿAbd Allāh ibn al-Ṭayfūrī (8.10): "I have found an addition in another copy that I have annexed before [this biography]" (*wajadtu ziyādah fī nuskhah ukhrā qad aḍaftuhā qablahā*). This addition is, in fact, a long text complementing the biography of Ibn al-Ṭayfūrī that the copyist added on an interleaf between f. 100 and f. 101. Up to this point – i.e., the first seven chapters and the first ten biographies of chapter 8 – the intervention of the copyist is limited to minor corrections in the margins and occasional marginalia with personal comments introduced by the term *ḥāshiyah*.[38]

The first reference to the second copy as the *musawwadah* occurs shortly afterwards in f. 115b, when marking the end of a *juz*ʾ in a marginal note: 'This is the end of the second *juz*ʾ of the copy of the author in his draft (*hādha ākhir al-juzʾ al-thānī min nuskhat al-muṣannif fī musawwadatihi*)'. A similar marginal note marking the end of the third *juz*ʾ comes on f. 167; however, this time it refers to the copy as *nuskhat al-muṣannif*, not *musawwadah*: 'This is the end of the third *juz*ʾ of the copy of the author (*hādha ākhir al-juzʾ al-thālith min nuskhat al-muṣannif*)'. As happened with the collation notes, these references are inconsistent. There is no reference to a second *juz*ʾ following the mention to the end of the first *juz*ʾ of *al-aṣl al-manqūl minhu* on f. 58b, which is written in the body text and clearly refers to the first exemplar used by the copyist; instead, we find a marginal reference to the end of the second *juz*ʾ in the *musawwadah* on f. 115.

Despite the inconsistency of the references, the explanations of the copyist and the clear separation between the main body text and the additions in margins and interleaves clearly reveal the use of two different versions and,

---

37 It reads: *tamma al-juzʾ al-awwal min tajziʾat al-aṣl al-manqūl minhu min kitāb ʿUyūn al-anbāʾ fī ṭabaqāt al-aṭibbāʾ min nuskhah bi-khaṭṭ al-muṣannif raḥimahu Allāh wa-yatlūhu fī l-juzʾ al-thānī min tajziʾatihi: ʿwa-min alfāẓ Jālīnūs …ʾ.*

38 For instance, marginalia referring to al-Mubashshir ibn Fātik in order to provide a date for Hippocrates (Şehid Ali Paşa, MS 1923, f. 18a); or a note comparing the travels of Galen with the *hijrah* of the Prophet Muḥammad (Şehid Ali Paşa, MS 1923, f. 48a).

most importantly, allow us to identify the exact additions and variants added at a later stage. But which versions are these? Müller's considerations about the Cambridge MS that is a copy of the Şehid Ali Paşa MS are also applicable in this case. The most intuitive would be to assume that they represent the two recensions identified by Müller, especially when the copyist claims that he copied the first text from a work that had been presented to important people (*akābīr al-nās*). But, as Müller had already noticed, even though this seems to refer to the patronage of Amīn al-Dawlah, the dedication is absent from these manuscripts and their contents do not present any characteristic that might point to a first recension.

5 Versions of the Text

Müller tried to find a solution to the conundrum of the different versions of the *ʿUyūn* by postulating the existence of copies that had incorporated parts of the author's draft at different stages. As the classification in Table 4.2 shows, he postulated that some manuscripts were partial transcriptions of the author's *musawwadah* (type 2A), some were lacunose manuscripts corrected during the lifetime of the author (type 2B), and some posthumously corrected using the author's draft (type 2C). A major problem in Müller's approach is the centrality he attributes to the role of the *musawwadah*, all the more so since he did not have access to the direct copy of the author's draft from which the Cambridge MS stems.

But there is a second aspect largely overlooked in his study of the manuscript tradition: the need of a systematic discussion of the number, nature and order of the biographies included in each manuscript of Ibn Abī Uṣaybiʿah's work. Apart from the obvious discussion of the nine last biographies of chapter 15, which are devoted to physicians who were not active before 639/1242 and consequently lacking in the first version, Müller only dealt with the differences in the number of biographies included in each copy as textual variants.

If we focus on the number and distribution of the biographies included in each manuscript, it is possible to find clear patterns regarding their inclusion in the work, the chronology of the different versions of the text, and their relationship. In an important number of manuscripts this corresponds with the distribution of the biographies in the Şehid Ali Paşa MS – that is, all the biographies added in the margins of this copy belong to the last version of the work and the biographies present in the main body text belong to a previous version that *is not* the version dedicated to Amīn al-Dawlah.

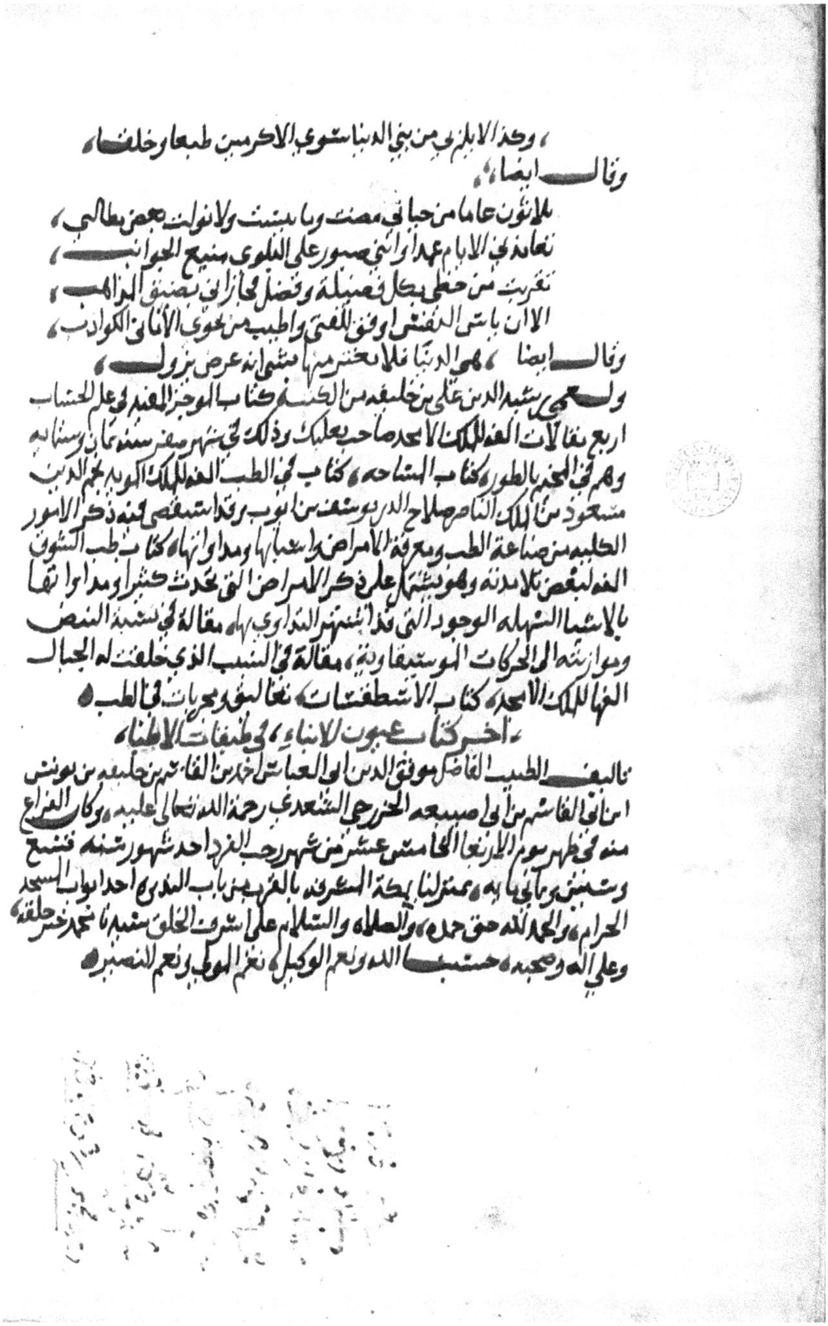

FIGURE 4.5 Oxford, Bodleian Library, MS Pococke 356, fol. 281b, colophon. Completed 15 Rajab 869 (13 Mar 1465) in Mecca. Siglum **B** in edition. Complete copy of Version 1

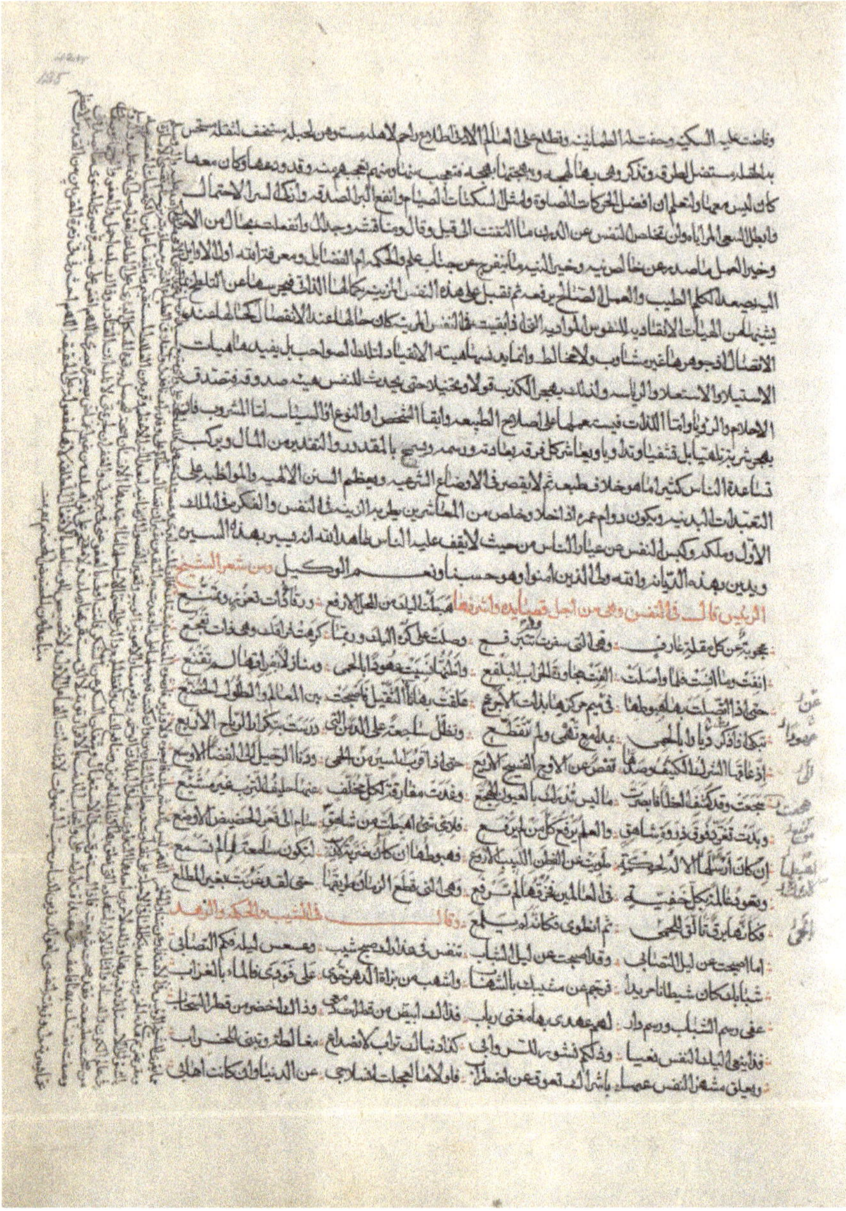

FIGURE 4.6 London, British Library, MS Add. Rich. 7340. fol. 125a, with a prayer attributed to Ibn Sīnā asking for forgiveness for drinking wine, copied into the margin of his biography. Dated 22 Rajab 1017 (1 Nov 1608) in Isfahan. Copyist: Ibn Muḥammed Shafīʿ Mulla Zayn al-ʿĀbidīn. Siglum **R** of edition. Complete copy of Version 3

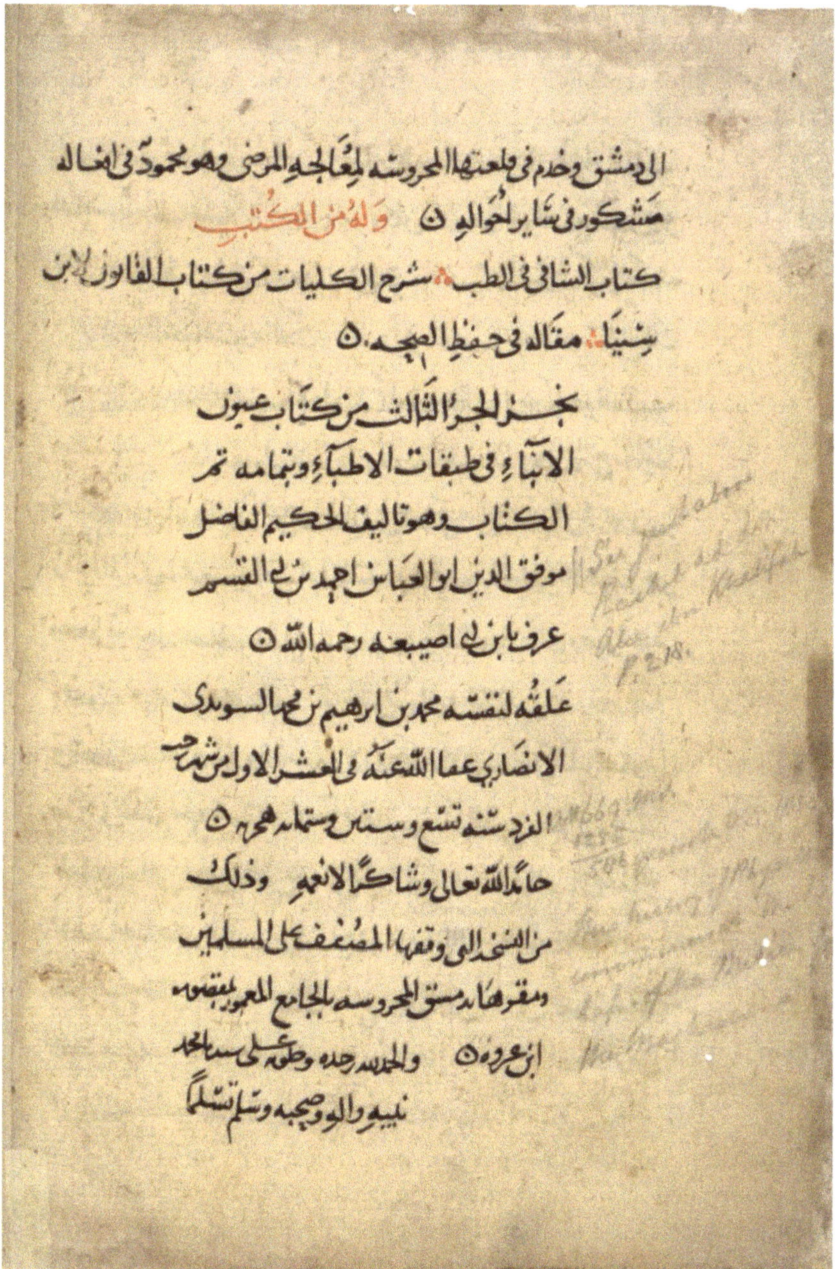
الى دمشق وخدم في قلعتها المحروسه لمعالجه المرضى وهو محمود في افعاله
مشكور في سائر احواله ٥ وله من الكتب
كتاب الشافي في الطب ٥ شرح الكليات من كتاب القانون لابن
سينا ٥ مقاله في حفظ الصحه ٥
نجز الجزء الثالث من كتاب عيون
الانباء في طبقات الاطباء وتمامه تم
الكتاب وهو تاليف الحكيم الفاضل
موفق الدين ابو العباس احمد بن ابي القسم
عرف بابن ابي اصيبعه رحمه الله ٥
علقه لنفسه محمد بن ابرهيم بن محمد السويدي
الانصاري عفا الله عنه في العشر الاول من شهر رجب
الفرد سنه تسع وستين وستمايه هجريه ٥
حامدا لله تعالى وشاكرا لانعمه وذلك
من النسخه التي وقفها المصنف على المسلمين
ومقرها بدمشق المحروسه بالجامع المعمور بمقصوره
ابن عروه ٥ والحمد لله وحده وصلواته على سيدنا محمد
نبيه واله وصحبه وسلم تسليما

FIGURE 4.7 London, British Library MS Add. 23364, fol. 251b, colophon. Dated 10 Rajab 669 (22 Feb 1271), Damascus. Copyist: Syrian physician Muḥammad ibn Ibrāhīm Muḥammad al-Suwaydī al-Anṣārī, copied from the copy endowed by IAU to the Maqṣūrah of Ibn ʿUrwah. Siglum L in edition. Partial copy of Version 3

In view of these findings, it is possible to identify three versions of the text, with the addition of the hybrid version represented by the Berlin MS. The existence of these three versions is also supported by internal evidence provided by the author himself, and by their textual particularities.

5.1 *Version 1*

Ibn Abī Uṣaybiʿah makes explicit reference to the existence of an early version of his work on at least four occasions. The first instance occurs in the biography of Muhadhdhab al-Dīn Abū Saʿīd Muḥammad Ibn Abī Ḥulayqah:

> I – Ibn Abī Uṣaybiʿah – received a letter from Muhadhdhab al-Dīn while he was in the military encampment of al-Manṣūr al-Ẓāhirī in the month of Shawwāl in the year 667 [June 1269] (…) In the letter, he stated that he had found in Cairo a copy of this book that I have written on the classes of physicians and that he had acquired it, making it part of the collection in his library.
>
> Ch. 14.55.2

The second reference is found in the biography of Ibn Abī Uṣaybiʿah's teacher Rafīʿ al-Dīn al-Jīlī:

> I – Ibn Abi Uṣaybiʿah – say: It is curious to note that the judge Rafīʿ al-Dīn went over a copy of this book in my presence, in which I had not included him. He looked through it, but stopped when he had finished [reading] the account of Shihāb al-Dīn al-Suhrawardī. He was much impressed by it and spoke: 'You have included him, but you have omitted others who were greater than he', referring to himself.
>
> Ch. 15.20

The third, more extensive reference occurs in the biography of Amīn al-Dawlah as it appears in later versions of the work:[39]

> When Amīn al-Dawlah – may God have mercy upon him – was occupying the office of vizier in Damascus in the days of al-Malik al-Ṣāliḥ Ismāʿīl, he was a close friend of my father's. One day he said to him, 'Sadīd al-Dīn, I have heard that your son has composed a book on the classes of

39 His biography in Version 1 is, in fact, a eulogy that can be read as a continuation of the dedication; it has been edited and translated as an addendum to biography Ch. 15.49.

> physicians that is unprecedented. All the physicians in my service praise him greatly for his highly valuable book. I have in my library more than twenty thousand volumes, but none in that particular domain. I would like you to send him a letter and ask him to have a copy of that book made for me.' At that time I was in Ṣarkhad, at the court of its ruler, the emir ʿIzz al-Dīn Aybak al-Muʿaẓẓamī, and subject to his orders. Upon receiving my father's letter, I went to Damascus, taking along with me the rough drafts of my book. There, I called upon the illustrious copyist Shams al-Dīn Muḥammad al-Ḥusaynī, who did a lot of copying work for us: his handwriting was excellent, and his mastery of the Arabic language was admirable. I gave him space at our home, where he copied the book in a fairly short time, putting it into four sections, in quarter Baghdādī format.
>
> Having had these bound, I composed a panegyrical poem for the Ṣāḥib Amīn al-Dawlah and sent all these items to him by the hand of the chief judge of Damascus, Rafīʿ al-Dīn al-Jīlī, who was one of the teachers with whom I was on friendly terms and under whom I had studied and read a section of Ibn Sīnā's *Book of Remarks and Admonitions* (*K. al-Ishārāt wa-l-tanbīhāt*). When Amīn al-Dawlah read the book and poem, he was greatly surprised and extremely happy. He sent the judge back with a large sum of money and honorary robes for me, along with many expressions of gratitude. 'It is my desire that you notify me of every new book you write,' he said.
>
> Ch. 15.49.6

A last, brief reference occurs in the biography of Ibn Abī Uṣaybiʿah's friend Ibn al-Suwaydī, who wrote some verses in praise of his book (Ch. 15.57.2).

After the report on the patronage of Amīn al-Dawlah, Ibn Abī Uṣaybiʿah added a *qaṣīdah* that he had written to praise the generosity of the dedicatee of the work. Only eight verses of the poem are quoted in the first version of the work, but Versions 2 and 3 contain the entire *qaṣīdah* and the date of its composition: the beginning (*awāʾil*) of the year 643/1245.[40] This date corresponds with the last year of Amīn al-Dawlah's vizierate in Damascus, and is consistent with the last date mentioned in the manuscripts containing the first version, which is 642/1244–1245. However, the death date of al-Jīlī that Ibn Abī Uṣaybiʿah gives in the biography of this physician included in Versions

40 Müller, ii, 237 [= Ch. 15.49.6.1].

2 and 3 is Dhū l-Ḥijjah of 641/May-June 1244.[41] If we accept the accuracy of this date and take al-Jīlī's death as the *terminus ante quem*, the first version of the work should have been finished and sent to Amīn al-Dawlah no later than 641/1244.

The differences between the first version of the *ʿUyūn al-anbāʾ* and later versions are both qualitative and quantitative. In terms of the number of biographies it contains and their order, Version 1, which corresponds with Müller's recension 1, is defined by the following characteristics:

1. These manuscripts include the following dedication to the vizier Amīn al-Dawlah:[42]

   وخدمت به خزانة المولى الصاحب الوزير العالم العادل الرئيس الكامل سيد الوزراء ملك الحكماء إمام العلماء شمس الشريعة أمين الدولة كمال الدين شرف الملة أبي الحسن بن غزال بن أبي سعيد أدام الله سعادته وبلغه في الدارين إرادته

   [*The Best Accounts of the Classes of Physicians*] is a contribution to the library of the great, learned and just minister, the paragon of masters, lord of viziers, wisest of wise men, chief of learned men, glorious sun of religion, Amīn al-Dawlah Kamāl al-Dīn Sharaf al-Millah Abū l-Ḥasan ibn Ghazāl ibn Abī Saʿīd, may God prolong his happiness and grant him his wishes in this world and the next.
2. The last nine biographies of Ch. 15 (15.52–15.60) are lacking in all the manuscripts containing the dedication, since these physicians were not active at the time of the composition of this first version before 641/1244.
3. In manuscripts with Version 1 the biographies of Ḥakam al-Dimashqī (7.7), ʿĪsā ibn Ḥakam (7.8), and al-Tamīmī (14.14) are included in Ch. 15 rather than in Ch. 7 and Ch. 14, as in the manuscripts of Versions 2 and 3.
4. These manuscripts contain only 350 biographies, 84 less than those present in the last version of the work. With minor exceptions that can be considered accidents of transmission, the missing biographies are the same in all of the manuscripts and reveal a clear pattern.
5. The text in Chs. 1–7, with exception of the dedication and minor variants, is the same in all versions.

41 Müller, ii, 172 [= Ch. 15.20].

42 This reproduces the text of Oxford, Bodleian Library, MS Pocock 365. The dedication is present in all the manuscripts of Version 1 consulted for this study (see table below).

Version 1 shows also its own textual characteristics. The most obvious divergences result from the lapse of time – at least twenty-four years – that passed between the preparation of the first and the later versions. Chapters 1–7 containing biographies of ancient physicians and philosophers do not show much alteration, but the biographies of physicians who lived in the Islamic period, especially those contemporary with Ibn Abī Uṣaybiʿah, were updated with new information personally acquired by the author himself, or with information taken from newly available sources.

An illustrative example of the way in which Ibn Abī Uṣaybiʿah updated the biographies of his contemporaries can be found in the life of Sadīd al-Dīn Abū Manṣūr (Ch. 15.44).

Version 1 (from MS B):[43]

وللحكيم يعقوب ولد فاضل في صناعة الطب وهو سديد الدين أبو منصور بن يعقوب استخدمه الملك الناصر خلد الله تعالى أيامه وهو الآن في صحبته بالكرك

> The physician Yaʿqūb has a son who earned an outstanding reputation in the art of medicine. His name is Sadīd al-Dīn Abū Manṣūr ibn Yaʿqūb. He served al-Malik al-Nāṣir, may God make his days eternal. He now enjoys this ruler's company in al-Karak.

Versions 2 and 3:

هو الحكيم الأجل العالم أبو منصور بن الحكيم موفق الدين يعقوب بن (٢٨٢أ) سقلاب من أفاضل الأطباء وأعيان العلماء متميز في علم صناعة الطب وعملها متقن افصولها وجملها اشتغل على والده وعلى غيره بصناعة الطب وقرأ أيضاً بالكرك على الإمام شمس الدين الخسروشاهي كثيراً من العلوم الحكمية وخدم الحكيم سديد الدين أبو منصور الملك الناصر صلاح الدين داود ابن الملك المعظم عيسى بن أبي بكر بن أيوب وأقام في صحبته بالكرك وكان مكيناً عنده معتمداً عليه في صناعة الطب ثم أتى أبو منصور إلى دمشق وتوفي بها

> Sadīd al-Dīn Abū Manṣūr is the revered and learned physician Abū Manṣūr, son of the physician Muwaffaq al-Dīn Yaʿqūb ibn Siqlāb. He was an

43 Text edited from Oxford, Bodleian Library, MS Pococke 365, f. 262. See Fig. 4.5 with its colophon.

> outstanding physician and an eminent scholar, distinguished in the theory and practice of the art of medicine and a master in the particulars and universals of medicine. He studied the art of medicine under his father and others. In al-Karak, he also studied many of the natural sciences under the learned authority Shams al-Dīn al-Khusrawshāhī.

A remarkable case where the differences go beyond the mere addition of biographical and bibliographical information is the entry on al-Ṣāḥib Amīn al-Dawlah, the dedicatee of the first version. This biography (Ch. 15.49) occupies only a few paragraphs in the manuscripts containing the first version of the work and is essentially an encomium of the dedicatee that seems to continue the eulogies of the dedication. It consists of a long passage with dithyrambic epithets, eight verses written by Ibn Abī Uṣaybiʿah in honour of his patron, and a list of Amīn al-Dawlah's books. In later versions of the work Amīn al-Dawlah's biography still contains some encomiastic passages, but it is substantially longer[44] and of a completely different tenor. The biography in Versions 2 and 3 provides factual information about Amīn al-Dawlah's life, including negative aspects that Ibn Abī Uṣaybiʿah would have not written about his dedicatee, such as the mention of Amīn al-Dawlah's Samaritan origins and his conversion to Islam, and a story about his venality and enrichment at the expense of the Damascene citizens. It also contains an updated list of his works. In addition to that, Ibn Abī Uṣaybiʿah turns the eulogy of Amīn al-Dawlah into a celebration of his own work by introducing the story about the dedication of the first version of the *ʿUyūn* and reproducing his poem in praise of Amīn al-Dawlah in its entirety (16 verses in contrast with the 8 verses of the first version).

The use of additional information taken from sources that might not have been available when he wrote the first version of the text is noticeable in many cases, especially in updated book-lists. In some biographies Ibn Abī Uṣaybiʿah also resorted to very specific works to enrich his reports; for instance, the biography of the Fatimid physician Isḥāq ibn Sulaymān al-Isrāʾīlī (Ch. 13.2) is complemented in later versions with a quotation from Ibn al-Jazzār's *Kitāb al-Dawlah*,[45] and in the biography of Abū l-Ṣalt (Ch. 13.58) some of the poems included in later versions and missing in Version 1 were taken from the *Rasāʾil* of Ibn al-Ṣayrafī.[46] However, the most relevant example of the use of new

44 In Şehid Ali Paşa, MS 1923, for instance, it consists of two folios, ff. 289a–291a.
45 See Oxford, Bodleian Library, MS Pococke 356 f. 187b, where this quote is missing.
46 See Oxford, Bodleian Library, MS Pococke 356 f. 195a, where this quote is missing.

sources are the quotations from Ibn al-Qifṭī's *Taʾrīkh al-ḥukamāʾ*, which, as Müller already noticed, was not used in the first version of the *ʿUyūn*.[47]

5.2 *Version 2*

The second version corresponds with the main body text of the Şehid Ali Paşa MS and has important quantitative and qualitative differences with regards to Versions 1 and 3. The manuscripts containing Version 2 were ascribed by Müller to recension 2 of the text, and some of them fall within Müller's type 2A, which he considered to be incomplete transcriptions of the *musawwadah*. Several characteristics allow us to identify them:

1. The dedication to Amīn al-Dawlah is missing in all of them.
2. With minor exceptions that can be considered particular accidents of transmission, the manuscripts of this group include 72 new biographies that were not present in the manuscripts of Version 1, including the last 9 biographies of Ch. 15.
3. Likewise, the manuscripts of this group lack 57 biographies that were present in the manuscripts of Version 1. These missing biographies are the same in all copies.
4. In all the manuscripts of Version 2, the biography of ʿImād al-Dīn al-Dunaysirī (Ch. 15.58) occurs before that of ʿIzz al-Dīn ibn al-Suwaydī (Ch. 15.57).
5. Overall, the manuscripts of this version usually have a total number of 375 biographies: 15 biographies more than the manuscripts of Version 1 and 69 biographies less than Version 3 of the work. With minor exceptions that can be considered accidents of transmission, the additional and missing biographies are the same in all of these manuscripts and reveal a clear pattern.[48]
6. The text of Chs. 1–7, with exception of the dedication and minor variants, is the same in all versions. Differences of content in comparison with Version 3, which contains a large number of additions, affect biographies from Ch. 8.10 to Ch. 15.49.

Textual differences between Version 1 and Version 2 are essentially the result of updated information, as shown by the biography of Sadīd al-Dīn Abū Manṣūr (Ch. 15.44) discussed above. By contrast, the differences between Version 2 and Version 3 are exclusively quantitative: all the text present in Version 2 is kept in Version 3 with minor alterations, but a fair amount of new materials is added

47 See the discussion of Ibn al-Qifṭī in the essay on Ibn Abī Uṣaybiʿah's sources, pp. 83-84.

48 See the list with these biographies in the appendix to this chapter.

to Version 3 and the biographies missing in Version 2 and present in Version 1 are recovered. This is why the layout of Şehid Ali Paşa, MS 1923 is so relevant for distinguishing between these two stages: all the text of Version 2 – in the main body of the copy – is kept in Version 3, which consists of the main-body text plus the additions in margins and interleaves.

The formal and textual differences between Versions 2 and 3 do not reveal anything regarding the date and the circumstances of their composition. Müller suggested that some manuscripts that belong to Version 2 might have been imperfect copies of the *musawwadah* of the author – i.e., an abridgement that stems from the complete text of Version 3. The copyist of the Şehid Ali Paşa MS also envisages the possibility of an abridgement, but he traces it back to the author since the copy of the Version 2 he used stemmed from an autograph.

Both explanations assume that Version 3 predated Version 2, but a close study of the additions shows that Version 2 is in fact an intermediate version, probably a copy of an incomplete draft that nonetheless generated a textual tradition today represented by approximately a third of all the manuscripts at our disposal. First, the presence and absence of biographies in Version 2 does not seem to follow any logic and cannot be explained by assuming that this is an abridgement. The only explanation one can adduce for the elimination of biographies that were present in Version 1 is that they were, in general, short biographies that needed to be reworked to be later added in a revised form, but the text of the major share of these short biographies that is missing in Version 2 reappears in Version 3 with the same wording they had in Version 1 – the last biographies of Ch. 13 (the Maghrib and al-Andalus) are a good example of that. Second, a fair number of the additions found in Version 3 with regard to Version 2 are small glosses that clarify the extant text. The absence of these glosses in Version 2 can by no means be interpreted as an abridgement. In this regard the Istanbul Ahmed III MS 2859/70 is especially illustrative: it contains a copy of Ch. 13–15 from Version 3, but it lacks the small biography of Jābir ibn Manṣūr al-Sukkarī (Ch. 15.4), which occupies a couple of lines and was missing in Version 2, and a few minor glosses added in Version 3 especially to Ch. 13. For instance, it lacks the gloss of Isḥāq ibn ʿImrān's nickname 'Instant Poison':

> because, for all his knowledge, he was frequently unsuccessful in his treatment, and those who made use of his services died, whence his nickname;
>
> Ch. 13.1.1

or a comment about the city of Monastir:

> the latter is a place for devout Muslims bound to God (*murābiṭah*), well-known for its blessings (*barakah*) and mentioned in the history books, on the coast of the Mediterranean Sea.[49]
>
> Ch. 13.3.2.1

Unfortunately, MS Istanbul Ahmed III MS 2859/70 only has the last three chapters of the work, but the absence of the short biography in Ch. 15.4 and the missing glosses suggest that the text of Version 2 was reworked and updated to Version 3 and that these additions might have not been present in the author's draft when the Ahmed III MS 2859/70 or the exemplar from which it derives was copied.

The dating of Version 2 is also difficult to pin down. Both Version 2 and Version 3 include the anecdotes of Abū Ḥulayqah's, al-Jīlī's, and Amīn al-Dawlah's biographies, which refer to the composition of the Version 1, but no reference is made at any point to the existence of two further versions. The latest date in the manuscripts with Versions 2 and 3 is the same: 667/1269, that is, one year – or less – before the death of the author. This date cannot be considered a later addition to the manuscripts of Version 2 since it occurs three times and in three different biographies: Muhadhdhab al-Dīn Abū Saʿīd Muḥammad Ibn Abī Ḥulayqah (Ch. 14.55), Sharaf al-Dīn ibn Raḥbī (15.36), and ʿImād al-Dīn al-Dunaysirī (Ch. 15.57). This means that two different versions of the *ʿUyūn* were copied and somehow distributed in the same year. We cannot know, however whether the distribution of Version 2 was authorised by its author – which seems unlikely – or was only a discarded draft that was copied at some point and subsequently gave rise to an independent textual tradition.

According to the testimony of the copyist of the Şehid Ali Paşa MS, both Version 2 and Version 3 stem from autographs. The colophon of the Paris, Bibliothèque Nationale MS ar. 5939, which is a copy of Version 2, also states that this copy was collated with an autograph. The reasons for preparing two rather similar versions are unclear and, if we consider that Version 2 lacks biographies that were present in Version 1, illogical. A possible explanation would be that the author might have worked on different drafts from which different fair copies were made, but this is only speculation. In any case, there is no doubt that Version 2 comprised a differentiated textual tradition represented by a large

49 Examples of these glosses or very short additions missing in Ahmed III MS 2859/70 can be found in: 13.1.1; 13.2.1; 13.3.2.1; 13.3.2.3; 13.3.4 no. 6; 13.3.4 no. 27; 13.22; 13.24.1; 13.24.2; 13.24.3; 13.26; 13.36.2.1; 13.58.2.1; 15.5; 15.12.

number of manuscripts, nor that, as we will discuss below, the contents of both versions came from the pen of Ibn Abī Uṣaybiʿah.

5.3 *Version 3*

As with the previous version, there is no evidence to date accurately the completion of Version 3 beyond the fact that it was finished sometime between 667/1269 and the death of the author in 668/1270.

The differences with regard to the second version are essentially quantitative. The manuscripts of this group were also ascribed by Müller to the recension 2 of the text, most of them falling into Müller's types 2B and 2C. The most relevant characteristics of Version 3 are:

1. The dedication to Amīn al-Dawlah is missing in all these manuscripts.
2. The biographies included in these manuscripts correspond, with minor exceptions that can be considered accidents of transmission, to those included in our and Müller's edition – i.e., the last version of the work. Complete copies have a total number of 434 biographies.
3. In all the manuscripts of this version, unlike those of Version 2, the biography of ʿIzz al-Dīn ibn al-Suwaydī (Ch. 15.57) occurs before that of ʿImād al-Dīn al-Dunaysirī (Ch. 15.58).
4. All the text from Version 2 is kept in Version 3 with only minor variants.
5. The text of Chs. 1–7, with exception of the dedication and minor variants, is the same in all versions. Differences of content in comparison with Version 2, which lacks the information added to Version 3, affect biographies from Ch. 8.10 to Ch. 15.49.

The text of Version 3 corresponds with the entire text of the Şehid Ali Paşa MS – i.e., the main body text together with the additions from the *musawwadah* placed in the margins and on interleaves. The authorship of these additions is confirmed by the oldest manuscript of the work that has come down to us, also with Version 3, which is a direct copy of the author's fair copy given as a *waqf*, British Library MS Add. 23364 copied in Damascus in 1271.[50]

Some minor textual differences found in manuscripts containing Version 3 deserve attention, such as the glosses missing in the Ahmed III MS 2859/70 commented above, which might point to final minor additions written in the *musawwadah* once a fair copy of it had been made. In general, the text within

50 This is stated in the colophon, see London, British Library, MS Add. 23364, f. 251b; see Fig. 4.7. The *waqf* copy of this work is also mentioned in Ibn Kathīr, *al-Bidāyah wa-l-nihāyah*, xiii:299. The institution referred to as *maqṣūrah* in the colophon is the library of the *Dār al-Ḥadīth* founded by the Shāfiʿī scholar Sharaf al-Dīn Muḥammad Ibn ʿUrwah al-Mawṣilī in the Umayyad Mosque of Damascus; see al-Nuʿaymī, *al-Dāris*, i:62.

the manuscripts of this version shows the common variants of any textual tradition. Version 3 of the *ʿUyūn al-anbāʾ* is the last and most complete version of the text.

5.4 *Hybrid Versions*

A last question that merits discussion is the existence of manuscripts showing a significant degree of editorial intervention. The Şehid Ali Paşa MS is, in fact, a hybrid of Versions 2 and 3, but thanks to the meticulousness of the copyist both versions are clearly separated. The copyist of the Leiden MS Universiteitsbibliotheek Or. 59b (Gb in our edition) was also puzzled by the existence of two different versions when he tried to collate his text but, unlike the copyist of the Şehid Ali Paşa MS, the text he had already copied belonged to Version 3 and the second text he had at his disposal to Version 2 – i.e., it was a shorter version. He did not leave any explanation of his method, but he diligently added, right after the end of his copy of Version 3, the text of biographies in Chs. 14.31–15.60 as he found them in Version 2.

The most notable of these hybrid versions is the Berlin MS preserved in two volumes: Wetzst. II.323 and Spreng, 312. Müller discussed this manuscript at length and adopted some of its variants in his edition. If we examine the number and distribution of biographies in this version, it reveals itself as an amalgamation of Version 3 with previous versions: it contains all the biographies of Version 3, but their order sometimes follows that of Version 1. For instance, Ḥakam al-Dimashqī (Ch. 7.7) and ʿĪsā ibn Ḥakam (Ch. 7.8) occur at the beginning of chapter 15 instead of in chapter 7. Changes in the arrangement are also noticeable in biographies in Chs. 13.25, 13.27, 13.28, 13.29, 13.38, 13.40, 13.42, 13.43, 13.44, 13.45, 15.15, 15.16, 15.19, 15.20, which are missing in Version 2, but included in this copy in places that do not correspond with the order they occupy in Version 3. Biographies 13.30, 13.31, 13.32, 13.33, and 13.34, missing in Version 2, are also lacking in this manuscript. None of the manuscripts consulted in this study is related to the Berlin MS, and it seems that it did not generate any textual tradition.

6 Classification of Manuscripts

If we exclude the hybrid Berlin MS, the manuscripts considered in this study can be divided as follows in terms of the version they contain:[51]

51 Not all these MSS have been used in the edition, those that have been used are referred to with lower-case sigla for Müller's edition, and capitalized sigla for ours.

Version 1	Version 2	Version 3
– Tübingen, Universitätsbibliothek, MS M. A. VI.14 (**t**) – Oxford, Bodleian Library, MS Pococke 365 (**p**/**B**) – Oxford, Bodleian Library, MS Hunt 171 – Istanbul, Murad Molla, MS 1464 – Istanbul, Fatih, MS 4438 – Istanbul, Hamidiye, MS 1028 – Istanbul, Nurusmaniye, MS 3560 – Istanbul, Yeni Çami, MS 891	– Leiden, Universiteitsbibliotheek, MS Or. 76 (**Gc**) – Istanbul, Damad Ibrahim Paşa, MS 935 – Oxford, Bodleian Library, MS Marsh 153 – Istanbul, Ahmad III, MS 2860 – Leiden, Universiteitsbibliotheek, MS Or. 59a (**l**/**Ga**) – Oxford, Bodleian Library, MS Marsh 419 – London, Wellcome Institute, MS Arabic 432 – London, Wellcome Institute, MS Arabic 433 – Istanbul, Köprülü, MS 1104 – Gotha, Stadtbibliothek, MS 1769 (**g**)	– Istanbul, Şehid Ali Paşa, MS 1923 (unknown to Müller/**A**) – Cambridge, University Library, MS Or. 1461 (**n**) – London British Library Add. 23364 (**f**/**L**) – London, British Library, MS Add. 25736 (**e**/**S**) – London, British Library, MS Add. 7340 (**d**/**R**) – Leiden, Universiteitsbibliotheek, MS Or. 59b (addenda from V2) (**k**/**Gb**) – Paris, Bibliothèque Nationale, MS Ar. 5939 (**s**/**P**) – Istanbul, Ahmad III, MS 2859/70 (**H**) – Damascus, Maktabat al-Asad al-Waṭaniyyah, MS 148 Ṭ M (former 4883) [fairly abridged version]

7 Problems of Authorship

The analysis of the Şehid Ali Paşa MS also allows us to answer many of the questions raised by Müller about the authorship of some of the additions incorporated to what he called the second recension, since all the text included in the MSS of Version 3 appears in the marginal additions copied from the *musawwadah* with a few minor exceptions easily identifiable as scribal interventions.

The most obvious of these apocryphal additions are the death dates added to the biographies of physicians that outlived the author, deceased in 668/1270. This can be found in two of the manuscripts used in our edition, the Ahmad III MS 2859/70 (**H** in our edition) and British Library Add. 7340 (**R** in our edition), which add death dates to biographies in Chs. 15.52 (Badr al-Dīn Ibn Qāḍī

Baʿlabakk, d. 670/1271), 15.53 (Shams al-Dīn Muḥammad al-Kullī, d. 675/1276), 15.57 (ʿIzz al-Dīn ibn al-Suwaydī, d. 690/1291), 15.58 (ʿImād al-Dīn al-Dunaysirī, d. 685/1286), 15.60 (Ibn al-Quff, d. 685/1286).

A close examination of the copy of the *musawwadah* and the rest of the manuscripts also shows that Ibn Abī Uṣaybiʿah left some blank spaces in the place of names and dates, probably in the hope of eventually finding the missing information.[52] As Müller noted in his study, the addition of the missing names and dates might reveal a later intervention. The example adduced by Müller was the name of al-Suhrawardī's father, added posthumously to Ibn Abī Uṣaybiʿah's work in the Berlin MS and, among the manuscripts used in our edition, also in the Ahmed III MS 2859/70 (not used by Müller). Overall, there are twenty-one small blanks left to be completed in the *ʿUyūn al-anbāʾ*. Apart from the addition of the name of al-Suhrawardī's father, we have found only one other instance in which a blank space is filled in one of the manuscripts, MS Ahmed III 2859/70, which gives the death date of Abū Marwān ibn Abī l-ʿAlāʾ ibn Zuhr, missing in the rest of our copies. These additions are isolated cases and do not cast any doubt on the authorship of the contents added to Versions 2 and 3.

It is also worth noting that the copyist of the British Library MS 7340 added two marginal notes after the long book-lists in the biographies of al-Kindī and al-Sarakhsī referring to blank pages: 'The author (*al-muṣannif*) left here an additional blank page, perhaps with the hope of adding other works that he had not mentioned' (Ch. 10.1.14);[53] and 'The author (*al-muṣannif*) left a blank space of approximately a page' (10.2.3.4).[54] This copyist does not give any reference concerning the exemplar used in his copy, but this was a common practice attested in bibliographical works, such as Ibn al-Nadīm's *Fihrist*, and there is no evidence of the addition of further titles in any manuscript.

Some editorial work by Ibn Abī Uṣaybiʿah is also noticeable if we compare the *musawwadah* with copies containing Version 3. In some cases, it is possible to trace it back directly to the author, as the following marginal note in the Şehid Ali Paşa MS (f. 232a) shows:

52 Specifically, these blanks occur in Chs. 10.2.3.2 (date); 10.45 (name); 10.50 (date); 10.63.3 (date); 10.77.2 (date); 11.19.3 (date); 11.24 (place); 13.40 (name); 13.61.3.2 (date); 13.62.3 (date); 14.8 (name); 14.31.6 (date); 14.47.2 (date, filled in MS H); 14.50 (date, filled in Version 1); 15.1.4 (date); 15.9 (date); 15.11.2 (date); 15.13 (date);15.18.1 (name, filled in MS H and Berlin MS); 15.23.4.2 (date); 15.46.4 no. 3 (date).

53 See London, British Library Add. 7340 (R in our edition, d in Müller's), f. 85a.

54 See London, British Library Add. 7340 (R in our edition, d in Müller's), f. 86a.

قال كاتبها ورأيت في مسودة المصنف أبياتاً يهجى بها ابن جميع المذكور ثم أنه ضرب عليها واضرب عنها وكأنه إنما فعل ذلك لما تقدم من كلامه مع الصاحب جمال الدين بن المطروح ولم يورد الأبيات ما في المبيضة قال في مسودته

> The copyist [of this note] says: I saw in the draft copy of the author some verses satirizing the aforementioned Ibn Jumayʿ, which he had crossed out and abandoned; he did so after discussing this with al-Ṣāḥib Jamāl al-Dīn ibn al-Maṭrūḥ and were not to be included in the fair copy. He said [that] in his draft copy.

These expunged verses do not appear in any of the consulted copies, but three other poems from the *musawwadah* that had disappeared from Version 2 were copied in the margin of the manuscript.[55] These poems do not appear in the copy of the *waqf*-manuscript or in other copy containing Version 3, but there is no doubt that they come from Ibn Abī Uṣaybiʿah's pen since they are present in Version 1.[56]

A second example of editorial work is the suppression of titles from the book-list of the biography of Ibn al-Quff (Ch. 15.60.2), the last biography of the work. In the copy of the *musawwadah* we find five additional titles, of which three seem to refer to lost works or works in progress of which the author was aware (no. 8 is 'not extant', no. 9 is a draft, and no. 10 was left unfinished). These titles are not present in any other copy of Versions 2 and 3 with the exception of the Cambridge MS (Müller's **n**), which is a direct copy of the Şehid Ali Paşa MS, and the amalgamated version of the Berlin MS (Müller's **a**).[57] Since these titles are not in the copy of the manuscript endowed by the author to the *maqṣūrah* of Ibn ʿUrwah (Müller's **f**, in our edition **L**), it is likely that they are the result of direct authorial intervention by Ibn Abī Uṣaybiʿah, who would have supervised the production of the *waqf*-copy himself.

If we exclude the occasional posthumous interventions discussed above, among all the variants and additions present in the different versions of the manuscripts there is only one instance that raises doubts about Ibn Abī Uṣaybiʿah's authorship. In the biography of Zayn al-Dīn al-Ḥāfiẓī (Ch. 15.32) the text

55 See Şehid Ali Paşa MS, f. 232a.

56 These poems have been added to our edition.

57 Following Müller, and for the sake of completion, we decided to keep these titles as part of the main text of the work.

containing the report of the al-Malik al-Nāṣir's defeat at the hands of the sultan Aybak is clearly corrupt in the the Şehid Ali Paşa Manuscript (MS **A** in our edition) copied from the *musawwadah*, the copy of the *waqf*-manuscript, and other copies. All these manuscripts have lacunas in the report and wrongly identify the sultan Quṭuz as the opponent of the Ayyubid prince.[58] There are, however, two manuscripts with Version 3 that have a slightly more extended and correct version of the story, naming Aybak instead of Quṭuz: MS Ahmed III 2859/70 (our **H**), and British Library Add. 7340 (Müller's **d**, in our edition **R**).[59] The fact that the copy of the *waqf*-manuscript has preserved the corrupt version suggests that the text in MSS **H** and **R** is the result of a later editorial intervention. However, a close examination of the fragment reveals scribal mistakes, such as homoioteleuton, which seems clear in the text omitted after *ilā*, and the abrupt ending:

Long version (MSS **H** and **R**), with variants and additions in bold letters:

وهرب الملك الناصر يوسف من دمشق إلى **جهة زيزاء وملك النواحي وما زال الحافظي به إلى أن سلمه إلى هولاكو فبقي عنده إلى أن تسلمه على ما هو مشهور وكان الملك الناصر لما ورد دمشق في سنة ثمان وأربعين وستمائة** إلى مصر وقصد أن يملكها فخرجت عساكر مصر وملكها يومئذ **الملك المعز المعروف بأيبك التركماني** فكسر الملك الناصر وتفرقت عساكره وزال ملكه **عن مصر بعد أن ملكها وعاد إلى الشام وبقي حاكماً عليه إلى سنة ثمان وخمسين وستمائة فطرق البلاد العدو على ما هو مشهور**

Short version (rest of MSS):

وهرب الملك الناصر يوسف من دمشق إلى مصر وقصد أن يملكها فخرجت عساكر مصر وملكها يومئذ المظفر سيف الدين قطز فكسر الملك الناصر وتفرقت عساكره وزال ملكه

In this case, it is likely that the version in **H** and **R** might be a posthumous correction since in the Şehid Ali Paşa MS this passage is part of the main body text, i.e., the text common to Versions 2 and 3, and the copyist did not register any variant in the *musawwadah*. But in view of the obvious corruption of the passage and the absence to references to any source as the origin of this informa-

58 Al-Nāṣir Yūsuf was defeated by Aybak's forces in the battle of al-Ṣāliḥiyyah in 1250.

59 Müller kept in his edition the corrupt version of the story, without referring to variants in the 'Lesarten'. In our edition we have followed the text of MSS H and R.

tion, we should not exclude the possibility that it might have been corrected through the intervention of the author, or from the collation of his notes.

The British Library MS Add. 7340 has yet another addition at the end of the biography of Jamāl al-Dīn ibn al-Raḥbī (Ch. 15.38), with five verses that are not present in the rest of the manuscripts from any version and might be the result of a posthumous edition.[60]

Finally, it is worth noting that in at least two manuscripts the biography of Ibn al-Nafīs was added at the end of the work.[61] Of course, this should not be ascribed to Ibn Abī Uṣaybiʿah.

Appendix

1 *Biographies Missing in Version 1 Compared with Version 2 and Version 3*

Biographies missing in Version 1 and present in Version 2 [72 biographies]	Biographies missing in Version 1 and present in Version 3 [83 biographies]
Chapter 8	
8.7 Khaṣīb al-Naṣrānī	8.7 Khaṣīb al-Naṣrānī
8.17 Isrāʾīl ibn Sahl	8.17 Isrāʾīl ibn Sahl
8.32 Yūḥannā ibn Bukhtīshūʿ	8.32 Yūḥannā ibn Bukhtīshūʿ
8.33 Bukhtīshūʿ ibn Yūḥannā	8.33 Bukhtīshūʿ ibn Yūḥannā
8.35 ʿĪsā ibn Yaḥyā	8.35 ʿĪsā ibn Yaḥyā
Chapter 9	
9.12 Al-Ḥajjāj ibn Maṭar	9.12 Al-Ḥajjāj ibn Maṭar
9.13 Ibn Nāʿimah	9.13 Ibn Nāʿimah
9.14 Zarūyā ibn Mānaḥūh al-Nāʿimī al-Ḥimṣī	9.14 Zarūyā ibn Mānaḥūh al-Nāʿimī al-Ḥimṣī
9.15 Hilāl ibn Abī Hilāl al-Ḥimṣī	9.15 Hilāl ibn Abī Hilāl al-Ḥimṣī
9.16 Pethion the Translator	9.16 Pethion the Translator
9.17 Abū Naṣr ibn Nārī ibn Ayyūb	9.17 Abū Naṣr ibn Nārī ibn Ayyūb
9.18 Basīl al-Muṭrān	9.18 Basīl al-Muṭrān
9.20 Mūsā ibn Khālid	9.20 Mūsā ibn Khālid
9.21 Usṭāth	9.21 Usṭāth
9.22 Khayrūn ibn Rābiṭah	9.22 Khayrūn ibn Rābiṭah
9.23 Theodore the Syncellus	9.23 Theodore the Syncellus

60 See London, British Library MS Add. 7340, f. 185b.

61 See Appendix 1 following the fifteen chapters comprising the *ʿUyūn*.

(*cont.*)

Biographies missing in Version 1 and present in Version 2 [72 biographies]	**Biographies missing in Version 1 and present in Version 3 [83 biographies]**
Chapter 10	
10.6 Ibrāhīm ibn Sinān	10.6 Ibrāhīm ibn Sinān
10.7 Ibrāhīm ibn Zahrūn al-Ḥarrānī	10.7 Ibrāhīm ibn Zahrūn al-Ḥarrānī
10.11 Abū ʿUthmān Saʿīd ibn Ghālib	10.11 Abū ʿUthmān Saʿīd ibn Ghālib
10.12 ʿAbdūs	10.12 ʿAbdūs
10.14 Daylam	10.14 Daylam
10.15 Dāwūd ibn Daylam	10.15 Dāwūd ibn Daylam
10.26 ʿĪsā the physician of al-Qāhir	10.26 ʿĪsā the physician of al-Qāhir
10.28 Also missing in Version 2	10.28 Isḥāq ibn Shalīṭā
10.35 Abū l-Faraj ibn Abī Saʿīd	10.35 Abū l-Faraj ibn Abī Saʿīd
10.39 Al-Faḍl ibn Jarīr al-Takrītī	10.39 Al-Faḍl ibn Jarīr al-Takrītī
10.40 Abū Naṣr Yaḥyā ibn Jarīr	10.40 Abū Naṣr Yaḥyā ibn Jarīr
10.41 Ibn Dīnār	10.41 Ibn Dīnār
10.47 Also missing in Version 2	10.47 Muḥammad ibn Thawāb al-Mawṣilī
10.48 Aḥmad ibn Muḥammad al-Baladī	10.48 Aḥmad ibn Muḥammad al-Baladī
10.51 Ibn Shibl al-Baghdādī	10.51 Ibn Shibl al-Baghdādī
10.57 Isḥāq ibn ʿAlī al-Ruhāwī	10.57 Isḥāq ibn ʿAlī al-Ruhāwī
10.60 Abū l-Khaṭṭāb Muḥammad ibn Muḥammad	10.60 Abū l-Khaṭṭāb Muḥammad ibn Muḥammad
10.72 Saʿīd ibn Uthrudī	10.72 Saʿīd ibn Uthrudī
10.73 Al-Ḥasan ibn Uthrudī	10.73 Al-Ḥasan ibn Uthrudī
10.78 Abū l-Ḥusayn Ṣāʿid ibn Hibat Allāh	10.78 Abū l-Ḥusayn Ṣāʿid ibn Hibat Allāh
Chapter 11	
11. 3 Rabban al-Ṭabarī	11. 3 Rabban al-Ṭabarī
11.9 Abū l-Faraj ibn Hindū	11.9 Abū l-Faraj ibn Hindū
11.10 Also missing in Version 2	11.10 Al-Fasawī
Chapter 12	
12.2 Ṣanjahal	12.2 Ṣanjahal
12.3 Shānāq	12.3 Shānāq
12.4 Also missing in Version 2	12.4 Jawdar
12.5 Mankah al-Hindī	12.5 Mankah al-Hindī
Chapter 13	
13.56 Khalaf al-Zahrāwī	13.56 Khalaf al-Zahrāwī
13.57 Ibn Bakkalārish	13.57 Ibn Bakkalārish

(*cont.*)

Biographies missing in Version 1 and present in Version 2 [72 biographies]	Biographies missing in Version 1 and present in Version 3 [83 biographies]
Chapter 14	
14.1 Baliṭīyān	14.1 Baliṭīyān
14.2 Also missing in Version 2	14.2 Ibrāhīm ibn ʿĪsā
14.3 Al-Ḥasan ibn Zīrak	14.3 Al-Ḥasan ibn Zīrak
14.4 Saʿīd ibn Tawfīl	14.4 Saʿīd ibn Tawfīl
14.12 Also missing in Version 2	14.12 ʿĪsā ibn al-Biṭrīq
14.13 Aʿyan ibn Aʿyan	14.13 Aʿyan ibn Aʿyan
14.15 Also missing in Version 2	14.15 Sahlān
14.18 Al-Ḥaqīr al-Nāfiʿ	14.18 Al-Ḥaqīr al-Nāfiʿ
14.45 Fatḥ al-Dīn ibn Jamāl al-Dīn ibn Abī l-Ḥawāfir	14.45 Fatḥ al-Dīn ibn Jamāl al-Dīn ibn Abī l-Ḥawāfir
14.46 Shihāb al-Dīn ibn Fatḥ al-Dīn ibn Abī l-Ḥawāfir	14.46 Shihāb al-Dīn ibn Fatḥ al-Dīn ibn Abī l-Ḥawāfir
14.48 Afḍal al-Dīn al-Khūnajī	14.48 Afḍal al-Dīn al-Khūnajī
14.49 Abū Sulaymān Dāwūd ibn Abī l-Munā	14.49 Abū Sulaymān Dāwūd ibn Abī l-Munā
14.52 Also missing in Version 2	14.52 Abū Naṣr ibn Abī Sulaymān
14.53 Also missing in Version 2	14.53 Abū l-Faḍl ibn Abī Sulaymān
14.54 Rashīd al-Dīn Abū Ḥulayqah	14.54 Rashīd al-Dīn Abū Ḥulayqah
14.55 Muhadhdhab al-Dīn ibn Abī Ḥulayqah	14.55 Muhadhdhab al-Dīn ibn Abī Ḥulayqah
14.56 Rashīd al-Dīn Abū Saʿīd	14.56 Rashīd al-Dīn Abū Saʿīd
Chapter 15	
15.4 Also missing in Version 2	15.4 Jābir ibn Mansūr al-Sukkarī
15.12 Also missing in Version 2	15.12 Abū l-Faḍl ibn Abī l-Waqqār
15.15 Sukkarah al-Ḥalabī	15.15 Sukkarah al-Ḥalabī
15.17 Ibn al-Ṣalāḥ	15.17 Ibn al-Ṣalāḥ
15.19 Shams al-Dīn al-Khuwayyī	15.19 Shams al-Dīn al-Khuwayyī
15.20 Rafīʿ al-Dīn al-Jīlī	15.20 Rafīʿ al-Dīn al-Jīlī
15.21 Shams al-Dīn al-Khusrawshāhī	15.21 Shams al-Dīn al-Khusrawshāhī
15.22 Sayf al-Dīn al-Āmidī	15.22 Sayf al-Dīn al-Āmidī
15.25 Al-Sharīf al-Kaḥḥāl	15.25 Al-Sharīf al-Kaḥḥāl
15.31 Najm al-Dīn ibn al-Lubūdī	15.31 Najm al-Dīn ibn al-Lubūdī
15.32 Zayn al-Dīn al-Ḥāfiẓī	15.32 Zayn al-Dīn al-Ḥāfiẓī
15.38 Jamāl al-Dīn ibn al-Raḥbī	15.38 Jamāl al-Dīn ibn al-Raḥbī
15.52 Badr al-Dīn ibn Qāḍī Baʿlabakk	15.52 Badr al-Dīn ibn Qāḍī Baʿlabakk
15.53 Shams al-Dīn Muḥammad al-Kullī	15.53 Shams al-Dīn Muḥammad al-Kullī
15.54 Muwaffaq al-Dīn ʿAbd al-Salām	15.54 Muwaffaq al-Dīn ʿAbd al-Salām
15.55 Muwaffaq al-Dīn al-Minfākh	15.55 Muwaffaq al-Dīn al-Minfākh
15.56 Najm al-Dīn ibn al-Minfākh	15.56 Najm al-Dīn ibn al-Minfākh
15.57 ʿIzz al-Dīn ibn al-Suwaydī	15.57 ʿIzz al-Dīn ibn al-Suwaydī
15.58 ʿImād al-Dīn al-Dunaysirī	15.58 ʿImād al-Dīn al-Dunaysirī
15.59 Muwaqqaf al-Dīn Yaʿqūb al-Sāmirī	15.59 Muwaqqaf al-Dīn Yaʿqūb al-Sāmirī
15.60 Abū l-Faraj ibn al-Quff	15.60 Abū l-Faraj ibn al-Quff

2 *Biographies Missing in Version 2 Compared with Versions 1 and 3*

Biographies added to Version 2 and missing in Version 1 [72 biographies]	Biographies missing in Version 2 and present in Version 1 [57 biographies]	Biographies missing in Version 2 and present in Version 3 [69 biographies]
	Chapter 8	
8.7 Khaṣīb 8.17 Isrāʾīl ibn Sahl 8.32 Yūḥannā ibn Bukhtīshūʿ 8.33 Bukhtīshūʿ ibn Yūḥannā 8.35 ʿĪsā ibn Yaḥyā ibn Ibrāhīm	8.10 ʿAbd Allāh al-Ṭayfūrī 8.11 Zakariyyā ibn al-Ṭayfūrī 8.23 Ibrāhīm ibn Ayyūb al-Abrash 8.24 Jibrīl, 'the oculist of al-Maʾmūn'	8.10 ʿAbd Allāh al-Ṭayfūrī 8.11 Zakariyyā ibn al-Ṭayfūrī 8.23 Ibrāhīm ibn Ayyūb al-Abrash 8.24 Jibrīl, 'the oculist of al-Maʾmūn'
	Chapter 9	
9.12 Al-Ḥajjāj ibn Maṭar 9.13 Ibn Nāʿimah 9.14 Zarūyā ibn Mānaḥūh al-Nāʿimī al-Ḥimṣī 9.15 Hilāl ibn Abī Hilāl al-Ḥimṣī 9.16 Pethion the Translator 9.17 Abū Naṣr ibn Nārī ibn Ayyūb 9.18 Basīl al-Muṭrān 9.20 Mūsā ibn Khālid the Translator 9.21 Usṭāth 9.22 Khayrūn ibn Rābiṭah 9.23 Theodore the Syncellus	9.38 Abū l-Ḥasan ʿAlī ibn Ibrāhīm ibn Baks	9.38 Abū l-Ḥasan ʿAlī ibn Ibrāhīm ibn Baks
	Chapter 10	
10.6 Ibrāhīm ibn Sinān ibn Thābit ibn Qurrah 10.7 Ibrāhīm ibn Zahrūn al-Ḥarrānī 10.11 Abū ʿUthmān Saʿīd ibn Ghālib 10.12 ʿAbdūs 10.14 Daylam 10.15 Dāwūd ibn Daylam 10.26 ʿĪsā, the Physician of al-Qāhir 10.35 Abū l-Faraj ibn Abī Saʿīd al-Yamāmī 10.39 al-Faḍl ibn Jarīr al-Takrītī	10.6 Ibrāhīm ibn Sinān ibn Thābit ibn Qurrah 10.47 Also missing in Version 1	10.47 Muḥammad ibn Thawāb al-Mawṣilī

(*cont.*)

Biographies added to Version 2 and missing in Version 1 [72 biographies]	Biographies missing in Version 2 and present in Version 1 [57 biographies]	Biographies missing in Version 2 and present in Version 3 [69 biographies]
10.40 Abū Naṣr Yaḥyā ibn Jarīr al-Takrītī 10.41 Ibn Dīnār 10.48 Aḥmad ibn Muḥammad al-Baladī 10.51 Ibn Shibl al-Baghdādī 10.57 Isḥāq ibn ʿAlī al-Ruhāwī 10.60 Abū l-Khaṭṭāb 10.72 Saʿīd ibn ʿAlī ibn Hibat Allāh ibn Uthrudī 10.73 Abū ʿAlī al-Ḥasan ibn ʿAlī ibn Uthrudī 10.78 Abū l-Ḥusayn Ṣāʿid ibn Hibat Allāh ibn al-Muʾammil		
	Chapter 11	
11. 3 Rabban al-Ṭabarī 11.9 Abū l-Faraj ibn Hindū	11.10 Also missing in Version 1	11.10 al-Ḥasan al-Fasawī
	Chapter 12	
12.2 Ṣanjahal 12.3 Shānāq (Cāṇakya) 12.5 Mankah al-Hindī (Māṇikya or Maṅkha)	12.4 Also missing in Version 1	12.4 Jawdar
	Chapter 13	
13.56 Khalaf al-Zahrāwī 13.57 Ibn Bakkalārish	13.11 Aḥmad ibn Khamīs ibn ʿĀmir ibn Duminj 13.25 ʿUmar ibn Ḥafṣ ibn Barīq 13.27 Muḥammad ibn Tamlīkh 13.28 Abū l-Walīd ibn al-Kattānī 13.29 Abū ʿAbd Allāh ibn al-Kattānī 13.30 Aḥmad ibn Ḥakīm ibn Ḥafṣūn 13.31 Abū Bakr Aḥmad ibn Jābir 13.32 Abū ʿAbd Allāh al-Malik al-Thaqafī 13.33 Hārūn ibn Mūsā al-Ushūnī	13.11 Aḥmad ibn Khamīs ibn ʿĀmir ibn Duminj 13.25 ʿUmar ibn Ḥafṣ ibn Barīq 13.27 Muḥammad ibn Tamlīkh 13.28 Abū l-Walīd ibn al-Kattānī 13.29 Abū ʿAbd Allāh ibn al-Kattānī 13.30 Aḥmad ibn Ḥakīm ibn Ḥafṣūn 13.31 Abū Bakr Aḥmad ibn Jābir 13.32 Abū ʿAbd Allāh al-Malik al-Thaqafī 13.33 Hārūn ibn Mūsā al-Ushūnī

(*cont.*)

Biographies added to Version 2 and missing in Version 1 [72 biographies]	Biographies missing in Version 2 and present in Version 1 [57 biographies]	Biographies missing in Version 2 and present in Version 3 [69 biographies]
	13.34 Muḥammad ibn ʿAbdūn al-Jabalī al-ʿIdwī	13.34 Muḥammad ibn ʿAbdūn al-Jabalī al-ʿIdwī
	13.38 Ibn Baghūnish	13.38 Ibn Baghūnish
	13.40 Al-Rumaylī	13.40 Al-Rumaylī
	13.42 Ibn al-Nabbāsh	13.42 Ibn al-Nabbāsh
	13.43 Abū Jaʿfar ibn Khamīs al-Ṭulayṭulī	13.43 Abū Jaʿfar ibn Khamīs al-Ṭulayṭulī
	13.44 Abū l-Ḥasan ʿAbd al-Raḥmān ibn Khalaf ibn ʿAsākir al-Dārimī	13.44 Abū l-Ḥasan ʿAbd al-Raḥmān ibn Khalaf ibn ʿAsākir al-Dārimī
	13.45 Ibn al-Khayyāṭ	13.45 Ibn al-Khayyāṭ
	13.48 Isḥāq ibn Qasṭār	13.48 Isḥāq ibn Qasṭār
	13.49 Ḥasdāy ibn Isḥāq	13.49 Ḥasdāy ibn Isḥāq
	13.50 Abū l-Faḍl Ḥasdāy ibn Yūsuf ibn Ḥasdāy	13.50 Abū l-Faḍl Ḥasdāy ibn Yūsuf ibn Ḥasdāy
	13.69 Abū ʿAbd Allāh ibn Yazīd	13.69 Abū ʿAbd Allāh ibn Yazīd
	13.70 Abū Marwān ibn Qablāl	13.70 Abū Marwān ibn Qablāl
	13.71 Abū Isḥāq Ibrāhīm al-Dānī	13.71 Abū Isḥāq Ibrāhīm al-Dānī
	13.72 Abū Yaḥyā Qāsim al-Ishbīlī	13.72 Abū Yaḥyā Qāsim al-Ishbīlī
	13.73 Abū l-Ḥakam ibn Ghalindū	13.73 Abū l-Ḥakam ibn Ghalindū
	13.74 Abū Jaʿfar Aḥmad ibn Ḥassān	13.74 Abū Jaʿfar Aḥmad ibn Ḥassān
	13.75 Abū l-ʿAlāʾ ibn Abī Jaʿfar Aḥmad ibn Ḥassān	13.75 Abū l-ʿAlāʾ ibn Abī Jaʿfar Aḥmad ibn Ḥassān
	13.76 Abū Muḥammad al-Shadhūnī	13.76 Abū Muḥammad al-Shadhūnī
	13.77 Al-Maṣdūm	13.77 Al Maṣdūm
	13.78 ʿAbd al-ʿAzīz ibn Maslamah al-Bājī	13.78 ʿAbd al-ʿAzīz ibn Maslamah al-Bājī
	13.81 Abū ʿAbd Allāh al-Nadrūmī	13.81 Abū ʿAbd Allāh al-Nadrūmī
	13.82 Abū Jaʿfar Aḥmad ibn Sābiq	13.82 Abū Jaʿfar Aḥmad ibn Sābiq
	13.83 Ibn al-Ḥallāʾ al-Mursī	13.83 Ibn al-Ḥallāʾ al-Mursī
	13.84 Abū Isḥāq ibn Ṭumlūs	13.84 Abū Isḥāq ibn Ṭumlūs
	13.85 Abū Jaʿfar al-Dhahabī	13.85 Abū Jaʿfar al-Dhahabī
	13.87 Abū l-ʿAbbās al-Kanbarānī	13.87 Abū l-ʿAbbās al-Kanbarānī
	Chapter 14	
14.1 Politianus (Balīṭīyān)	14.2 Also missing in Version 1	14.2 Ibrāhīm ibn ʿĪsā
14.3 al-Ḥasan ibn Zīrak	14.5 Khalaf al-Ṭūlūnī	14.5 Khalaf al-Ṭūlūnī
14.4 Saʿīd ibn Tawfīl	14.6 Nasṭās ibn Jurayj	14.6 Nasṭās ibn Jurayj
14.13 Aʿyan ibn Aʿyan	14.7 Isḥāq ibn Ibrāhīm ibn Nasṭās	14.7 Isḥāq ibn Ibrāhīm ibn Nasṭās

(*cont.*)

Biographies added to Version 2 and missing in Version 1 [72 biographies]	Biographies missing in Version 2 and present in Version 1 [57 biographies]	Biographies missing in Version 2 and present in Version 3 [69 biographies]
14.18 Al-Ḥaqīr al-Nāfiʿ	14.9 Mūsā ibn al-ʿĀzār al-Isrāʾīlī	14.9 Mūsā ibn al-ʿĀzār al-Isrāʾīlī
14.45 Fatḥ al-Dīn ibn Jamāl al-Dīn ibn al-Ḥawāfir	14.10 Yūsuf al-Naṣrānī	14.10 Yūsuf al-Naṣrānī
14.46 Shihāb al-Dīn ibn Fatḥ al-Dīn ibn Abī l-Ḥawāfir	14.12 Also missing in Version 1	14.12 ʿĪsā ibn al-Biṭrīq
14.48 Afḍal al-Dīn al-Khūnajī	14.15 Also missing in Version 1	14.15 Sahlān
14.49 Abū Sulaymān Dāwūd ibn Abī l-Munā ibn Abī Fānah	14.16 Abū l-Fatḥ Manṣūr ibn Sahlān ibn Muqashshir	14.16 Abū l-Fatḥ Manṣūr ibn Sahlān ibn Muqashshir
14.54 Rashīd al-Dīn Abū Ḥulayqah	14.19 Abū Bishr, the Physician of the ʿAẓīmiyyah	14.19 Abū Bishr, the Physician of the ʿAẓīmiyyah
14.55 Muhadhdhab al-Dīn Abū Saʿīd Muḥammad ibn Abī Ḥulayqah	14.20 Ibn Maʿshar the Physician	14.20 Ibn Maʿshar the Physician
14.56 Rashīd al-Dīn Abū Saʿīd	14.24 Isḥāq ibn Yūnus	14.24 Isḥāq ibn Yūnus
	14.28 Mubārak ibn Salāmah ibn Raḥmūn	14.28 Mubārak ibn Salāmah ibn Raḥmūn
	14.35 Al-Raʾīs Hibat Allāh	14.35 Al-Raʾīs Hibat Allāh
	14.37 Abū l-Barakāt ibn al-Quḍāʿī	14.37 Abū l-Barakāt ibn al-Quḍāʿī
	14.38 Abū l-Maʿālī ibn Tammām	14.38 Abū l-Maʿālī ibn Tammām
	14.40 Ibrāhīm ibn al-Raʾīs Mūsā	14.40 Ibrāhīm ibn al-Raʾīs Mūsā
	14.41 Abū al-Barakāt ibn Shaʿyā	14.41 Abū al-Barakāt ibn Shaʿyā
	14.50 Abū Saʿīd ibn Abī Sulaymān	14.50 Abū Saʿīd ibn Abī Sulaymān
	14.52 Also missing in Version 1	14.52 Abū Naṣr ibn Abī Sulaymān
	14.53 Also missing in Version 1	14.53 Abū al-Faḍl ibn Abī Sulaymān
	14.57 Asʿad al-Dīn ibn Abī l-Ḥasan	14.57 Asʿad al-Dīn ibn Abī l-Ḥasan
	Chapter 15	
15.15 Sukkarah al-Ḥalabī	15.4 Also missing in Version 1	15.4 Jābir ibn Mansūr al-Sukkarī
15.17 Ibn al-Ṣalāḥ	15.7 Jābir ibn Mawhūb	15.7 Jābir ibn Mawhūb
15.19 Shams al-Dīn al-Khuwayyī	15.12 Also missing in Version 1	15.12 Abū l-Faḍl ibn Abī l-Waqqār
15.20 Rafīʿ al-Dīn al-Jīlī	15.14 Abū Zakariyyā Yaḥyā al-Bayāsī	15.14 Abū Zakariyyā Yaḥyā al-Bayāsī
15.21 Shams al-Dīn al-Khusrawshāhī	15.57 Also missing in Version 1	15.57 ʿIzz al-Dīn ibn al-Suwaydī
15.22 Sayf al-Dīn al-Āmidī	15.58 Also missing in Version 1	15.58 ʿImad al-Dīn ibn al-Suwaydī
15.25 al-Sharīf al-Kaḥḥāl (Burhān al-Dīn Abū l-Faḍl Sulaymān)		
15.31 Najm al-Dīn ibn al-Lubūdī		
15.32 Zayn al-Dīn al-Ḥāfiẓī		
15.38 Jamāl al-Dīn ibn al-Raḥbī		
15.52 Badr al-Dīn ibn Qāḍī Baʿlabakk		
15.53 Shams al-Dīn Muḥammad al-Kullī		

(*cont.*)

Biographies added to Version 2 and missing in Version 1 [72 biographies]	Biographies missing in Version 2 and present in Version 1 [57 biographies]	Biographies missing in Version 2 and present in Version 3 [69 biographies]
15.54 Muwaffaq al-Dīn ʿAbd al-Salām		
15.55 Muwaffaq al-Dīn al-Minfākh		
15.56 Najm al-Dīn ibn al-Minfākh		
15.57 ʿImād al-Dīn al-Dunaysirī		
15.58 ʿIzz al-Dīn ibn al-Suwaydī		
15.59 Muwaqqaf al-Dīn Yaʿqūb al-Sāmirī		
15.60 Abū l-Faraj ibn al-Quff		

CHAPTER 5

Written Sources and the Art of Compilation in Ibn Abī Uṣaybiʿah's *ʿUyūn al-anbāʾ fī ṭabaqāt al-aṭibbāʾ*

Ignacio Sánchez

Introduction

The scholarly appreciation of Ibn Abī Uṣaybiʿah is based, to a great extent, on the large number of sources used in the composition of the *ʿUyūn al-anbāʾ fī ṭabaqāt al-aṭibbāʾ*. Ibn Abī Uṣaybiʿah was certainly not the first author to compile a bio-bibliographical work of this kind. His debt with previous scholars such as Ṣāʿid al-Andalusī and Ibn Juljul is repeatedly and explicitly acknowledged, and also with his contemporary Ibn al-Qifṭī, who is quoted in later versions of the work. The amount of references and direct quotations found in the *ʿUyūn* is, however, unmatched by any of these cognate works. Moreover, some of the quoted or paraphrased sources, especially those used in the first chapters, have not come down to us in any other form, either in their original language or in translation, making of Ibn Abī Uṣaybiʿah the last witness of a fair number of lost works that offer a unique glimpse into the Greek and Islamic intellectual history.

Ibn Abī Uṣaybiʿah also parts ways from previous compilers when it comes to record the activities of physicians and men of science. He treats these individuals as authors, paying attention not only to their scientific prowess, but also to their literary achievements, especially in the field of poetry. A prolific poet himself, Ibn Abī Uṣaybiʿah seems to have envisaged an ideal image of the physician that combines the virtues of the *adīb* with those of the philosopher-physician of the Galenic tradition.[1] The *ʿUyūn* is, in this regard, a literary history of medicine.[2]

1 Ibn Abī Uṣaybiʿah was a prolific poetry writer according to Ibn Taghrībirdī (*al-Nujūm al-Zāhirah*, vii:229); although no *dīwān* by him has come down to us, some of his poems are included in the *ʿUyūn*. On poetry in the *ʿUyūn* see van Gelder's essay below, pp. 123–141.

2 The belletrist nature of this work has not received much appreciation in scholarship. Vernet, in his article of the second edition of the *Encyclopaedia of Islam*, acknowledged the value of Ibn Abī Uṣaybiʿah's biographies despite 'some long series of verses which have nothing to do with the main theme'; see *EI*² art. 'Ibn Abī Uṣaybiʿah' (J. Vernet).

© IGNACIO SÁNCHEZ, 2020 | DOI:10.1163/9789004410312_006

Encyclopaedic Works and Compilations

For many scholars, the appearance of compilations based on a large number of sources like the *ʿUyūn* and the loss of many of the works they quote are the two sides of a same coin. A rather widespread opinion, best represented by Ch. Pellat's article in the second edition of the *Encyclopaedia of Islam*, maintains that it was the anxiety prompted by the threat of foreign invasions that moved Arab authors to produce compilations of an 'encyclopaedic nature' from the 8th/13th century:

> All these works and many others besides, even if it is difficult to regard them as encyclopaedias since they only contain one specific section of information, were to become the instruments of a new form of encyclopaedia born of the vicissitudes of history, particularly of the fear of seeing the disappearance of the vast mass of knowledge accumulated over the centuries and of the concern to salvage at least a part from the irreparable catastrophe represented by the Mongol invasions and the fall of Baghdād in 656/1258.[3]

The river Tigris running black with ink after the Mongol invasion of Baghdad is not only a literary commonplace, but also a scholarly trope. For all the importance of factors such as invasions and wars, the reasons behind the loss of many of the books that were available to Ibn Abī Uṣaybiʿah are varied and uncertain. As Konrad Hirschler has argued, evidence for the destruction of libraries is flimsy, and the fate of books and the libraries that hold them depended to a great extent on the preferences of readers and collectors.[4]

On the other hand, both the notion of 'encyclopaedic work' and the social – and deterministic – explanation of the appearance of compilations has been challenged in recent years.[5] Ibn Abī Uṣaybiʿah reports on the destruction caused by the Mongol armies in several instances, but his words do not suggest any existential angst of the kind postulated by Pellat. The introduction to

3 *EI*² art. 'Mawsūʿa' (Ch. Pellat); the works to which he refers are Yāqūt's *Muʿjam al-buldān* and *Muʿjam al-udabāʾ*, Ibn Khallikān's *Wafayāt*, and the works of Ibn al-Qifṭī.

4 On the destruction of libraries see Hirschler, *Written Word*, 127–130.

5 On the relationship between the Mongol invasions and the rise of encyclopedism, see Muhanna, *The World in a Book*, 1–19. For an overview of the debates on encyclopedism, see Weaver, 'What wasn't an encyclopaedia?'

the *ʿUyūn*, for all its brevity, offers some clues for discerning the author's motivations that are better understood if we consider that the first version of this work was written and distributed before the appearance of Ibn al-Qifṭī's *Taʾrīkh al-ḥukamāʾ* or at least before Ibn Abī Uṣaybiʿah knew of it. There were certainly earlier works on medicine and physicians, but Ibn Abī Uṣaybiʿah claims to have written the first comprehensive history of medicine:

> To my knowledge, however, none of these masters of the art of medicine and no one with a thorough knowledge of that art has ever written a comprehensive book dealing with physicians through the ages and recounting their history in a coherent fashion.
>
> Preface

The personal ambition behind this enterprise becomes obvious when Ibn Abī Uṣaybiʿah indulges himself in describing the positive reception of his book and reproduces the encomiastic words of some of his contemporaries, including those of the dedicatee and patron of the first version of the work, Amīn al-Dawlah.[6]

But Ibn Abī Uṣaybiʿah's aim was not only to satisfy his ego. The *ʿUyūn* is also a sincere encomium of medicine and a defence of the role that the physician plays in society. This is perfectly illustrated by the way in which the author explains why his father and uncle learned the art of medicine:

> My grandfather had in mind to teach them both the art of medicine, because he was well aware of its noble rank and the people's great need for physicians, and held that one who was committed to its truths would be honoured and favoured in this world and be given the highest rank in the world to come.
>
> 15.51.1

The preface to the *ʿUyūn* also introduces a clear theological stance that is further developed in the first chapter of the work: medicine is a gift that God bestowed on his subjects millennials ago and that has evolved in the course of time. Physicians are instruments of God's will and vindicating their role throughout history is a way of fulfilling God's purpose:

> I hope by this means to enable the reader to glean some understanding of the learning that Almighty God enabled these men to acquire and

6 See below p. 112.

> the copious natural talent and intelligence that He bestowed upon them. Many of them, though they lived long ago and their days are past, stand in much the same relation to us as a teacher does to his pupil, or a recipient of kindness to his benefactor, owing to the advantages that we derive from their written works and the benefits found in the material they gathered together in their books.
>
> Preface

The plan of Ibn Abī Uṣaybiʿah's work, his selection of materials, his vindication of the figure of Galen as a paragon of wisdom, and his strategies of compilation are to a great extent governed by this idea which, in the 7th/13th century, was a rather contentious stance.[7]

Books and Libraries

Ibn Abī Uṣaybiʿah was able to write such a comprehensive work because he had access to a wide range of sources. References to the origin of the books he used are, in general, scarce. But it is possible to find occasional glimpses into his quest for information when he consults autographs of fellow physicians,[8] and also in the occasional references to libraries and reading practices that appear in some biographies.

Damascus was a city renowned for the quantity and the quality of its *waqf* libraries.[9] We cannot know how many of them were frequented by Ibn Abī Uṣaybiʿah, but in the *ʿUyūn* he provides some information about specific med-

7 The revealed nature of sciences and the relationship between prophecy and medicine had been the subject on many debates since the 3rd/9th century, probably best represented by the polemic between Abū Ḥātim al-Rāzī and Abū Bakr al-Rāzī, on which see Abū Ḥātim al-Rāzī, *Proofs of Prophecy*. On this polemic see also Griffel & Hachmeier, 'Prophets as Physicians of the Soul'.

8 References to autographs appear in quotes introduced by formulas such as *naqaltu* or *wajadtu bi-khaṭṭihi*, concretely on paragraphs: 1.7; 5.1.14; 8.26.16; 8.29.5; 8.29.21; 8.30.5; 10.8.2; 10.13.2; 10.13.4; 10.13.5; 10.23.2; 10.36; 10.37.5 no. 40; 10.38.4; 10.38.6; 10.64.3; 10.66.4; 11.5.15; 11.5.21; 11.21; 13.2.3, no. 1; 14.5; 14.21; 14.22.3.2; 14.22.5.1; 14.25.3; 14.25.7; 14.26.3; 14.31.8; 14.32.4; 15.1.1.2.; 15.5.18; 15.8.4; 15.11.2; 15.17; 15.23.4.1; 15.23.5 nos. 1, 2; 15.40.1; 15.40.2; 15.40.7; 15.40.8; 15.47.1; 15.49.6.2.

9 Still a century later, the Egyptian Ibn Nubātah (d. 768/1366) said in his introduction to his commentary to Ibn Zaydūn's *Epistle*: 'I knew of books in a number of the endowed libraries in Damascus that are helpful for the researcher and a useful aide-mémoire for the new generations of intelligent men.' (*Sarḥ al-ʿuyūn*, 14). On the Damascene libraries, see the introduction to Hirschler, *Medieval Damascus*, 1–17.

ical collections, such as that of his teacher al-Dakhwār, who founded a medical *madrasah*. About al-Dakhwār's books and lessons Ibn Abī Uṣaybiʿah says:

> He never taught anyone unless there was a copy of that book at his disposal for the student to read. He examined and collated it, and if there was an error in the copy that the pupil was reading, he would have it corrected. The copies that the shaykh Muhadhdhab al-Dīn's pupils perused were known for their faultlessness, and most of them were in his own handwriting. His medical and philological books, such as the *K. al-Ṣiḥāḥ* by al-Jawharī, the *Mujmal* by Ibn Fāris, and the *K. al-Nabāt* by Abū Ḥanīfah al-Dīnawarī, were constantly within reach.
>
> 15.50.5

A second large collection of books accessible for students of medicine in Damascus and mentioned in the *ʿUyūn* was the library of Ibn al-Muṭrān, purchased after his death by ʿImrān al-Isrāʾīlī, one of Ibn Abī Uṣaybiʿah's teachers (15.23.4.1).

In addition to these libraries, Damascene bookshops may have held very old materials, if we believe the author's claim about Ḥunayn ibn Isḥāq's autographs:

> I have found large numbers of these works [i.e. Ḥunayn's books], and have purchased a good many of them. They are written in *muwallad* Kūfic script in the handwriting of al-Azraq, Ḥunayn's secretary. The letters are written very large, with broad strokes, and the lines are widely spaced. The paper is very heavy, being three or four times as thick as the paper manufactured nowadays, while the sheets are trimmed to about one third the size of a sheet of Baghdādī paper. Ḥunayn had his works published in that fashion to make them bulkier and increase their weight, inasmuch as he was paid weight for weight in silver dirhams. It is thus clear that he used that particular type of paper deliberately. Small wonder, then, that the manuscripts have lasted so well for so long.
>
> 8.29.21

During his stay in Egypt, Ibn Abī Uṣaybiʿah was also able to use what was left of the library of al-Mubashshir ibn Fātik, as stated in the biography of this author, where he claims to have consulted many books 'in his handwriting concerned with the writings of the Ancients' (14.23.1). Also from Egypt came the books that Saʿīd ibn Hishām al-Ḥajarī Ibn Mulsāqah copied from the personal library of the bibliophile Ifrāʾīm ibn al-Zaffān, which was acquired by the Ayyubid al-

Afḍal so that these books would not abandon Egypt: 'This is why I have come across a great number of medical books – says Ibn Abī Uṣaybiʿah –, as well as volumes on other topics, bearing the name of Ifrāʾīm as well as the honorific titles of al-Afḍal.' (14.26.2)[10]

Another library associated with Ibn Abī Uṣaybiʿah is that of the *Maqṣūrah* of Ibn ʿUrwah, a *Dār al-ḥadīth* to which he bequeathed a copy of the *ʿUyūn*.[11] Abū l-Faḍl ibn ʿAbd al-Karīm al-Muhandis, a physician included by Ibn Abī Uṣaybiʿah in this work (15.33), also donated his books to this institution. We do not know, however, whether it might have held a collection of medical works.

Apart from these brief remarks, nothing is known about other libraries that Ibn Abī Uṣaybiʿah may have used in his stays in Damascus, Aleppo and Cairo.

Sources Employed in the *ʿUyūn*

The sources used by Ibn Abī Uṣaybiʿah cover a wide variety of disciplines and genres, which range from historical chronicles to medical and philosophical treatises, *adab* compilations, and poetic *dīwāns*. Many of them, as happens with the *ʿUyūn* itself, are of an eclectic nature and defy a clear classification, but they may be broadly subsumed under the following rubrics.

1 Universal and Local Histories, Chronographies, and Geographical Works

Ibn Abī Uṣaybiʿah's use of universal histories and chronicles is almost exclusively restricted to the chapters on the pre-Islamic period, where they are employed to discuss chronologies concerning the lifetime of the legendary inventors of medicine and the first physicians. Most of these sources are of Christian origin, such as Orosius, *Historia adversus paganos* (2.1.2), which was not consulted directly, but rather quoted from Ibn Juljul's *Ṭabaqāt al-aṭibbāʾ*; an unidenti-

10 On the Fatimid and Ayyubid libraries, see Behrens-Abouseif, *The Book in Mamluk Egypt and Syria*, 7–15.

11 This is indicated in the colophon of MS L, which is a direct copy of this exemplar; see Fig. 4.7 on p. 53. On the *Maqṣūrah* of Ibn ʿUrwah (*Dār al-ḥadīth al-ʿurwiyyah*), see al-Nuʿaymī, *al-Dāris*, i:61–67.

fied *Akhbār al-jabābirah* (*History of the Tyrants*) (2.1.1), or the chronicle of Eusebius of Caesarea cited as part of the discussion of the dates of Galen's lifetime (5.1.8.1).[12]

Chronologies, annals and dynastic histories by classical Islamic historians are also rare, and they are usually employed as a source of anecdotes. In Ch. 7, focused on the origins of Islam, Ibn Abī Uṣaybiʿah quotes from al-Wāqidī (7.5.6), al-Ṭabarī (7.5.6), and from the *K. Akhbār al-Ḥajjāj* attributed to certain Ibrāhīm ibn al-Qāsim (7.9.1–7.9.4);[13] an anecdote from al-Ṭabarī's *Taʾrīkh al-rusul wa-l-mulūk* is also quoted in 10.12.2.

Similarly, only a few identifiable works of local history are quoted in the *ʿUyūn al-anbāʾ*: Ibn al-Yasaʿ's *al-Muʿrib ʿan maḥāsin ahl al-Maghrib* is used in the chapter on North Africa and al-Andalus (13.52, 13.61.2.2); Ibn al-Dāyah's *Sīrat Aḥmad ibn Ṭūlūn* is used in the first biographies of Ch. 14, dealing with the Ṭūlūnid period (14.3, 14.4.1–14.4.4); and Yaḥyā ibn Saʿīd ibn Yaḥyā, *K. Taʾrīkh al-Dhayl*, which is essentially a history of Egypt, is also used to retrieve information about two Christian physicians (14.10–14.11).

Ibn Abī Uṣaybiʿah only refers to geographical works on three occasions; surprisingly, two of these quotes seem to be misattributions. The famous report of Aristotle's tomb in Sicily by Ibn Ḥawqal, who visited the island and speaks in first person in his *K. Ṣūrat al-arḍ*, is ascribed to al-Masʿūdī's *al-Masālik wa-l-mamālik* (4.6.3.3). This *al-Masālik wa-l-mamālik* attributed to al-Masʿūdī is also quoted when discussing the location of the tomb of Galen (5.1.21.1), but the report comes from al-Iṣṭakhrī's *al-Masālik wa-l-mamālik*. Al-Masʿūdī's *al-Tanbīh wa-l-ishrāf* is also briefly paraphrased in the biography of Aristotle to provide information that his father, Nicomachus, was a Pythagorean (4.6.1).

II Histories of Science, Medicine and Philosophy

a *General Biographical Works*

General histories of medicine, science and philosophy are not only the most often quoted works, but also the genre that informs the structure of the *ʿUyūn* and Ibn Abī Uṣaybiʿah's narrative.

Despite the title, the biographical entries are not organised in formal *ṭabaqāt*, i.e., generations of scholars; rather, they combine geographical and chronographic criteria. In this regard, the *ʿUyūn* clearly follows the model laid by

12 The quotation is apparently embedded in ʿUbayd Allāh ibn Jibrīl's enquiry into the matter.

13 Both quotations have not been found in the extant works of these authors and seem to be secondary quotes taken from a Shiʿite source.

Ibn Juljul in his *Ṭabaqāt al-aṭibbāʾ*, which stands out as the most influential of all the works consulted by Ibn Abī Uṣaybiʿah and the most extensively quoted and paraphrased source. Overall, there are more than fifty passages with paraphrases and quotations from this work.[14] Special mention should be made of Ch. 13, devoted to the physicians of North Africa and al-Andalus, since it is to a great extent a reproduction of Ibn Juljul's model, and the text of the biographies of the *Ṭabaqāt al-aṭibbāʾ* is quoted or paraphrased and sometimes complemented with later sources, especially Ṣāʿid al-Andalusī's *Ṭabaqāt al-umam*. This latter work, though used especially for its information about Andalusī scholars, is also employed to write the history of pre-Galenic physicians and philosophers (4.1.11.4, 4.2, 4.3.1, 4.4.1, 4.6.4.1), and to describe the works of al-Kindī (10.1.1–10.1.3, 10.1.8).

The third great biographical work used in the *ʿUyūn*, together with Ibn Juljul and Ṣāʿid al-Andalusī, is Ibn al-Qifṭī's *Taʾrīkh al-ḥukamāʾ*, but Ibn Abī Uṣaybiʿah's relationship with Ibn al-Qifṭī requires detailed explanation. It is important to emphasise that, even though Ibn al-Qifṭī and Ibn Abī Uṣaybiʿah share a large number of sources, the *Taʾrīkh al-ḥukamāʾ* is only quoted nine times, especially in Ch. 14 and 15, and only in later versions of the work. The first version of the *ʿUyūn* was written and distributed no later than 641/1244; we do not know whether Ibn al-Qifṭī's work had been disseminated by then, but Ibn Abī Uṣaybiʿah seems not to have been aware of it.

As Müller already noted, there are no references to the *Taʾrīkh al-ḥukamāʾ* in the first version of the *ʿUyūn*, but they appear in later versions.[15] Direct quotes from Ibn al-Qifṭī's work can be found in five biographies (10.77.3, 11.3, 11.21, 14.14.3, 14.22.3.1–14.22.3.2). Additionally, there are three references to Ibn al-Qifṭī that are not present in the extant version of the *Taʾrīkh al-ḥukamāʾ* (13.3.4 no. 5, 15.23.2–15.23.2.1, 15.36.1.1).

The text that has come down to us is not the original work by Ibn al-Qifṭī, but an epitome written by al-Zawzanī in 647/1249, one year after the death of the author.[16] Some scholars have argued that the quotations of Ibn Abī Uṣaybiʿah that are not present in al-Zawzanī's recension may have been part of the lost, original version of the *Ṭabaqāt al-ḥukamāʾ*. For instance, in Version 3 of the *ʿUyūn* Ibn Abī Uṣaybiʿah added the following gloss to the list of the works of Ibn al-Jazzār:

14 For the quotations and paraphrases see the Appendix to this chapter.

15 See the discussion in the essay on the manuscript tradition pp. 31–75.

16 See *EI*[2] art. 'Ibn al-Ḳifṭī' (A. Dietrich).

> Jamāl al-Dīn al-Qiftī reports (*wa-ḥakā*) that he saw in Qifṭ a large book written by him [i.e. Ibn al-Jazzār] on medicine entitled *Nourishment for the Sedentary* (*Qūt al-muqīm*), which comprised twenty volumes.
>
> 13.3.4 no. 5

This statement is not in the extant text of the *Taʾrīkh al-ḥukamāʾ*. In the introduction to his edition of al-Zawzanī's recension, Lippert claimed to be absolutely confident that this fragment stemmed from Ibn al-Qifṭī's original version, because there were not many sources available for Ibn al-Jazzār.[17] Lippert also believed that the original version of the *Taʾrīkh al-ḥukamāʾ* was the source of another short quotation that Ibn Abī Uṣaybiah attributes to Ibn al-Qifṭī in the biography of al-Raḥbī and which is missing in al-Zawzānī's recension: 'Al-Ṣāḥib Jamāl al-Dīn Abū l-Ḥasan 'Alī ibn Yūsuf ibn Ibrāhīm al-Qifṭī has told me that the physician al-Raḥbī closely followed the basic principles that are necessary for the preservation of health.' (15.36.1.1).[18]

Lippert may be right when he attributes these discrepancies to the editorial work of al-Zawzānī, but we should not exclude two other possibilities: the quotes may either be secondary quotations that Ibn Abī Uṣaybiʿah took from an unidentified source in which Ibn al-Qifṭī is used, or they may be personal communications. Ibn Abī Uṣaybiʿah and Ibn al-Qifṭī were contemporaries and at the very least they shared a common acquaintance, since al-Raḥbī was one of Ibn Abī Uṣaybiʿah's teachers. It is not unlikely that they could have met in Cairo or Aleppo. Likewise, the gloss to Ibn al-Jazzār's title is a personal report (*raʾaytu*), and another unidentified quotation in the biography of Ibn al-Muṭrān, also ascribed to Ibn al-Qifṭī, ends with a personal report in first person in which Ibn al-Qifṭī states that he saw (*wa-raʾaytu*) Ibn al-Muṭrān's wife in Aleppo after the year 600/1203 (15.23.2.1).

b *Chronicles and Histories of Physicians*

For the biographies of Greek scholars, Ibn Abī Uṣaybiʿah primarily draws on the work of Yaḥyā al-Naḥwī, an obscure figure often – wrongly – identified by Arab historians with John Philoponos of Alexandria (ca. 490–575), the famous Monophysite scholar and commentator of Aristotle.[19] Yaḥyā al-Naḥwī has his own and rather mythical biography in the *ʿUyūn*, taken from ʿUbayd Allāh ibn

17 Ibn al-Qifṭī, *Taʾrīkh al-ḥukamāʾ*, 12 (Einleitung).

18 Ibn al-Qifṭī, *Taʾrīkh al-ḥukamāʾ*, 12 (Einleitung). This statement, in almost the exact wording and without any reference to Ibn al-Qifṭī, occurs also in Barhebraeus, *Mukhtaṣar*, 217.

19 This identification was already dismissed by Meyerhof, who suggested that this Yaḥyā al-Naḥwī was an Alexandrian physician who lived between the 5th or 6th centuries, either Palladios, Gessios or Sergios; see Meyerhof, 'Joannes Grammatikos (Philiponos)'.

Jibrīl's *Manāqib al-aṭibbāʾ* (6.1.3). This portrayal of Yaḥyā al-Naḥwī appears to represent three individuals: the aforementioned John Philoponos, an otherwise unknown Alexandrian medical author who wrote Galenic commentaries and summaries and, according to this report, established the canon of the sixteen Galenic books,[20] and a very shadowy figure who was the author of a chronology. Nothing is known about this chronology of physicians, which Ibn Abī Uṣaybiʿah makes good use of apparently through Isḥāq ibn Ḥunayn's *History of Physicians*.[21] The extant fragments are quoted in Ch. 2, dealing with Asclepius, the mythical origins of medicine and the role of Hippocrates as saviour of the art in the island of Kos (2.1.5), and especially Ch. 3 on the successors of Hippocrates. Ch. 3 is almost entirely based on Yaḥyā al-Naḥwī's history and consists of a list of names and dates in what appears to be a combination of real scholars and fictional figures. It is not unlikely that Ibn Abī Uṣaybiʿah might have also consulted the *Taʾrīkh al-aṭibbāʾ* of Isḥāq ibn Ḥunayn to write these sections instead of the chronology of Yaḥyā al-Naḥwī, which is not listed among his books.[22] Isḥāq ibn Ḥunayn's work is also largely based on Yaḥyā al-Naḥwī and seems to be the source of some passages on the life of Dioscorides quoted by Ibn Abī Uṣaybiʿah (4.1.11.1).

Yaḥyā al-Naḥwī's *Taʾrīkh* is one of the most important classical sources in terms of biographical and chronographic information. Additionally, Ibn Abī Uṣaybiʿah quotes from other Greek works of different genres which also contain historical and biographical data, such as Ps.-Plato's *K. al-Nawāmīs* (2.1.2) and Plato's *Republic* (2.1.4) used in the biography of Asclepius.

Two other classical works dealing with the history of philosophy are quoted at length in the first chapters of the *ʿUyūn*. The first one is Porphyry's history of philosophers, entitled in Arabic *K. Akhbār al-falāsifah wa-qiṣaṣihim wa-ārāʾihim*; this is the most important source for the extended biography of Pythagoras (4.3.3, 4.3.4.1–4.3.4.4). The second work is an epistle on the life of Aristotle addressed to certain Gallus (*K. Ilā Ghalus fī sīrat Arisṭūṭālis*) and written by a biographer known in Arabic as Ptolemy al-Gharīb (fl. ca.300), i.e. 'Ptolemy the Stranger' as opposed to the familiar Claudius Ptolemy. This is in

20 On these commentaries see Ullmann, *Medizin*, 89–90. A further mystification found in other sources seems to stem from Ibn al-Nadīm's *Fihrist*, in which Yaḥyā al-Naḥwī is said to have lived until the conquest of Egypt and met ʿAmr ibn al-ʿĀṣ; this might be a confusion with the Chalcedonian patriarch of Egypt Cyrus, who negotiated with the Arab conquerors. See Ibn al-Nadīm, *Fihrist* (Sayyid), 2/1:178–179.

21 Quotations from Yaḥyā al-Naḥwī can be found in early histories of medicine and science such as Ibn Juljul's *Ṭabaqāt al-aṭibbāʾ*, or al-Sijistānī's *Ṣiwān al-ḥikmah*.

22 On Yaḥyā al-Naḥwī and Ibn Abī Uṣaybiʿah's use of his chronology, see Swain's essay pp. 150–154.

fact a brief biography accompanied by a detailed and long list with Aristotle's books that Ibn Abī Uṣaybiʿah quotes in its entirety (biography: 4.6.2.1–4.6.2.3, 4.6.6.1; book-list in 4.6.13.1). A further work that relies on classical texts with biographical information about the proponents of the three medical schools and is used by Ibn Abī Uṣaybiʿah is Ibn Riḍwān's *Tafsīr K. al-Firaq li-Jālīnus* (*Commentary on Galen's On Sects*) (1.7).

Among the authors who lived in the Islamic period, the most relevant work for the chapters on Greek medicine and philosophy is al-Mubashshir ibn Fātik's *Mukhtār al-ḥikam wa-maḥāsin al-kalim* (*The Choicest Maxims and Best Sayings*). The only surviving work of this author, the *Mukhtār al-ḥikam* contains a series of biographies of Greek sages which combines historical information and gnomic literature.[23] Al-Mubashshir ibn Fātik is used to provide historical information – sometimes of legendary nature – about Asclepius (2.1.3), Hippocrates (4.1.7), Pythagoras (4.3.2, 4.3.4.1–4.3.4.4), Socrates (4.4.2.1–4.4.2.5), Plato (4.5.2), Aristotle (4.6.3.1–4.6.3.2, 4.6.3.4), and Galen (5.1.21.1, 5.1.23, 5.1.25–5.1.26, 5.1.31). The longest extracts from this work, however, are the sections of aphorisms and witty sayings included in the biographies of these eminent Greek philosophers and physicians (4.1.8.4, 4.3.5, 4.4.4, 4.5.3, 4.6.11, 5.1.35), usually complemented with maxims taken from the *Ādāb* or *Nawādir al-falāsifah* attributed to Ḥunayn ibn Isḥāq (4.1.7, 4.1.8.2, 4.4.3, 4.6.3.5, 4.6.7.1–4.6.9, 5.1.3.4).[24]

In the chapters on the Greek tradition several miscellaneous sources written by Arab authors stand out. Abū Maʿshar's *K. al-Ulūf*, now lost, is quoted several times as part of the discussion on the origins of medicine (2.1.1, 2.1.3, 2.1.4), and also once in the biography of the Indian physician Kankah al-Hindī (12.1). Ibn al-Muṭrān's *Garden of physicians* (*Bustān al-aṭibbāʾ*), quoted on several occasions (1.2, 1.3.1), is one of the most relevant sources in the polemic on the origins of medicine in Ch. 1, where Ibn Abī Uṣaybiʿah resorts to it in order to confirm the statements made in the Ps.-Galenic *Commentary on the Hippocratic Oath.*

Special attention is due to the quotations of a work attributed to Abū Sulaymān al-Sijistānī and referred to as *Taʿālīq* (*Annotations*). Al-Sijistānī is the alleged author of an important work on the Greek philosophical and scientific tradition, the *Ṣiwān al-ḥikmah*, which has only survived in an abridged version

23 See *EI*[2] art. 'al-Mubashshir b. Fātik' (F. Rosenthal). Ibn Abī Uṣaybiʿah, who had access to what was left of al-Mubashshir ibn Fātiq's personal library, also quotes him from another unidentified source in 10.22.3.

24 In many ways, the works of al-Mubashshir ibn Fātik and Ḥunayn ibn Isḥāq can be considered works of *adab*. The discussion of this concept and the role of *adab* in Ibn Abī Uṣaybiʿah's narrative discussed below should also be taken into consideration for these works.

that, as Dimitri Gutas has suggested,[25] might stem from an intermediary.[26] The ascription of this work to al-Sijistānī has been discussed in scholarship, since all the documentary basis for this attribution is a single remark by Ẓahīr al-Dīn al-Bayhaqī. Gimaret and Wadad al-Qadi cast doubt on al-Sijistānī's authorship, and the latter suggested that the author of the *Ṣiwān al-ḥikmah* was Abū l-Qāsim, a disciple of the philosopher Abū l-Ḥasan al-ʿĀmirī.[27] Kraemer, in turn, claims that the *Ṣiwān al-ḥikmah* was a compendium compiled by al-Sijistānī's students.[28]

Ibn Abī Uṣaybiʿah is relevant to this discussion. He quotes four times from a work by al-Sijistānī, which he refers to as *Taʿālīq* (*Annotations*) (2.1.1, 4.6.3.5, 6.1.2, 15.1.3.1). A further reference to al-Sijistānī occurs when describing the famous dream in which Aristotle appeared to al-Maʾmūn. Ibn Abī Uṣaybiʿah states that he quotes from an autograph of certain Ḥasan ibn al-ʿAbbās al-Ṣanādīqī, and that al-Sijistānī was, in fact, transmitting the report of Yaḥyā ibn ʿAdī (8.29.5);[29] information on Yaḥyā ibn ʿAdī is also taken from al-Sijistānī's *Taʿālīq* in Ch. 15 (15.1.3.1), and it is not implausible that the story of al-Maʾmūn's dream may have been also part of this work. As Dunlop noted, none of the quotations of the *ʿUyūn* can be found in the text of the *Ṣiwān al-ḥikmah* preserved in the abridged version that has come down to us,[30] and it seems that these quotations are the only extant evidence of al-Sijistānī's lost *Taʿālīq fī l-ḥikmah*.

Histories of medicine and physicians are the most important source for the chapters focused on the Islamic period. The entries on the Syriac Christian physicians from the school of Gundeshapur are largely based on a history written by a Nestorian scholar and translation named Pethion. Parts of this work have survived in the chronicles by Elias of Nisibis and Barhebraeus, with passages close to those quoted by Ibn Abī Uṣaybiʿah concerning Jurjīs ibn Bukhtīshūʿ for events which occurred in the years AD 765 and 768.[31] Ibn Abī Uṣaybiʿah refers to Pethion as 'the translator' and has a brief entry on him on Ch. 9. His work, never referred to by title,[32] is quoted at length in Ch. 8, especially with

25 See Gutas, 'Ṣiwān al-Ḥikmah', 646.
26 See Dunlop, *Ṣiwān al-ḥikmah*.
27 Gimaret, 'Sur un passage', 154–155; al-Qadi, *'Kitāb Ṣiwān al-Ḥikmah'*.
28 See the detailed discussion in Kraemer, *Philosophy in the Renaissance of Islam*, 119–135.
29 A report on Yaḥyā ibn ʿAdī is also quoted from al-Sijistānī's *Taʿālīq* in 15.1.3.1.
30 See Dunlop, *Ṣiwān al-ḥikmah*, xiv.
31 Elias of Nisibis, *Opus Chronologicum*, i:178–180; Barhebraus, *Chronicum Syriacum*, 125–126. For further bibliographical references see: http://www.late-antique-historiography.ugent.be/database/works/97 [consulted: 25/02/2019].
32 ʿUbayd Allāh ibn Bukhtīshūʿ refers to this work as *akhbāruhu*, i.e. Pethion's news or notices, see ʿUbayd Allāh ibn Bukhtīshūʿ, *On Apparent Death*, 152 (§ 81).

regard to the Bukhtīshūʿ family (8.1.1–8.1.2, 8.2, 8.3.2–8.3.5, 8.3.20–8.3.21, 8.4.2, 8.4.8, 8.25.1).

The most important source for the chapter on Syriac physicians is Ibn al-Dāyah's *Akhbār al-aṭibbāʾ*. This work has not come down to us and all we know from it are the quotations in the *ʿUyūn* and Ibn al-Qifṭī's *Taʾrīkh al-ḥukamāʾ*.[33] Ibn Abī Uṣaybiʿah never mentions the title of the work and introduces these quotations with the formula *qāla* Yūsuf ibn Ibrāhīm. A secretary to Ibrāhīm ibn al-Mahdī, the musician and half-brother of the caliphs al-Hādī and Hārūn al-Rashīd, Ibn al-Dāyah was a privileged witness to the history of the ʿAbbasid courts of Baghdad and Samarra in the 3rd/9th century, in which a large number of Syriac physicians worked. He is quoted at length in Ch. 8, where his work is used as one of the main sources for fourteen biographies.[34] Ibn Abī Uṣaybiʿah also reproduces a long anecdote from Ibn al-Dāyah's *Akhbār al-aṭibbāʾ* in the biography of Galen, narrating an expedition of Ibrāhīm ibn al-Mahdī to visit the house of the physician at Smyrna (5.1.16.1).

Two other major works written by physicians largely focused on medical stories and anecdotes stand out as the most important sources for Chs 8 and 10. The most extensively used is ʿUbayd Allāh ibn Jibrīl ibn ʿUbayd Allāh ibn Bukhtīshūʿ's *Manāqib al-aṭibbaʾ* (*The Merits of Physicians*). As a member of a famous family of physicians, ʿUbayd Allāh ibn Jibrīl could benefit from the recollections and reports of his relatives and their personal archives. This work, written in 423/1032, has not come down to us, but the excerpts quoted in the *ʿUyūn* suggest that ʿUbayd Allāh had a good knowledge of classical sources and was a competent historian. Ibn Abī Uṣaybiʿah shows admiration for his scholarship and largely relies on him to shed light onto the contradictory reports about Galen's lifetime (5.1.7, 5.1.8.1–5.1.8.2, 5.1.11–5.1.13);[35] ʿUbayd Allāh is also the main source for the life of Yaḥyā al-Naḥwī (6.1.3). Most of the quotations from the *Manāqib al-aṭibbāʾ*, however, occur on chapters 8, 10, and also 11.[36] The anecdotes quoted from this work describe the lavish life of the famous physicians of the ʿAbbasid court are similar to those found in *adab* works.

A second source largely used in chapters 8 and 10 is Thābit ibn Sinān ibn Thābit ibn Qurrah's *Taʾrīkh*, which according to the description that Ibn Abī Uṣaybiʿah provides in the biography of this physician is a 'chronicle he wrote in which he mentions the occurrences and events which took place during the

33 See *EI*² art. 'Ibn al-Dāya' (F. Rosenthal).

34 See Appendix for exact references.

35 Ibn Abī Uṣaybiʿah seems to have expanded his quotes from ʿUbayd Allāh with other sources; see the footnotes to these sections.

36 See the Appendix for exact references.

time of al-Muqtadir bi-Allāh until the time of al-Ṭāʾiʿ li-Allāh'[37] (10.5.1). There are large quotations from this chronicle in the biographies of the members of the Banū Qurrah family, including autobiographical reports in the entry of Thābit Ibn Sinān (10.3, 10.4, 10.5, 10.6). Ibn Abī Uṣaybiʿah also employs information from this work in three other biographies of this chapter (10.16, 10.26, 10.28).

There is a third medical work that, despite its title, seems to have been conceived of as a collection of anecdotes and is quoted by Ibn Abī Uṣaybiʿah in four biographies of Ch. 10, sometimes with long extracts: Ibn Buṭlān's *M. Fī ʿillat naql al-aṭibbāʾ al-maharah tadbīr akthar al-amrāḍ allatī kānat tuʿālaj qadīman bi-l-adwiyah al-ḥārrah ilā al-tadbīr al-mubarrad* (*An essay on the reason the master physicians changed the regimen for most of the diseases which were, of old, treated with hot medicines to a cold regime*). This work is also lost.

c Particular Histories and Autobiographies

Together with these comprehensive historical and biographical works, Ibn Abī Uṣaybiʿah occasionally used other texts focused on particular topics or individuals in order to provide additional information or to discuss specific matters.

Ch. 9 seems to be entirely based on an unacknowledged history of translators, most likely of Syrian origin, if we judge by the rendition of foreign names. The parallels between the *ʿUyūn* and the *K. al-Fihrist*, which also fails to provide references to any source, suggest that Ibn Abī Uṣaybiʿah and Ibn al-Nadīm relied on the same work.

Other works that can be listed under this rubric are Abū Saʿīd Zāhid al-ʿUlamāʾ's *K. fī l-Bīmāristānāt*, a lost history of hospitals from which Ibn Abī Uṣaybiʿah quotes on one occasion (11.5.2); or the book on medical errors from which he quotes in Ch. 10: Abū Saʿīd al-Ḥasan ibn Aḥmad ibn ʿAlī's *K. Warṭat al-ajillāʾ min hafwat al-aṭibbāʾ* (*The Perils of Patricians due to the Errors of Physicians*) (10.13.5).

It is also worth noting that several biographies rely on autobiographical reports. The most relevant example is the entry on Ibn Sīnā, which is largely based on the author's autobiography and the continuation by his pupil Abū ʿUbayd al-Jūzjānī (11.13.2.1–11.13.3.19). Autobiographical works are also quoted in the biographies of Ḥunayn ibn Isḥāq (8.29.14–18.29.21),[38] Ibn al-Haytham (14.22.4.1–14.22.4.4), Ibn Riḍwān (14.25.1–14.25.2), and ʿAbd al-Laṭīf al-Baghdādī (15.40.2–15.40.4).

37 Al-Muqtadir ruled in 295–320/908–932, and al-Ṭāʾiʿ li-Allāh – or al-Ṭāʾiʿ li-Amr Allāh – in 363–381/974–91. Thābit ibn Sinān died in 363/973–974.

38 This work, in all certainty apocryphal, bears the title *R. fīmā aṣābahū min al-miḥan wa-l-shadāʾid* (*On his trials and tribulations*).

III Bibliographical Works

One of the most important traits of Ibn Abī Uṣaybiʿah's scholarship is his diligent search for titles. The *'Uyūn* is not only a precious source of biographical information but probably the richest bibliographical source for the history of medicine. Unlike the quotations used to provide biographical information, the origin of the book-lists is rarely acknowledged and, with a few exceptions, it is impossible to ascertain their origin.

The most important bibliographical source is Ibn al-Nadīm's *K. al-Fihrist*, which is used to provide historical information in at least twenty occasions[39] and is likely one of the unacknowledged sources behind many book-lists.

Ibn Abī Uṣaybiʿah also had access to an important number of catalogues. The *'Uyūn* not only bears witness to the preservation of book-lists of Greek origin, but also to the continuation of this practice among physicians and philosophers in the Islamic period. Among the classical bibliographies we can mention the long list of works of Pythagoras, taken from Porphyry's *History*, which reproduces the list compiled by the philosopher Archytas (4.3.6.1) and the philosopher Plutarch (4.3.6.2). The list of Aristotle's books is copied from the epistle of Ptolemy al-Gharīb (4.6.13.1), and in the biography of Galen Ibn Abī Uṣaybiʿah quotes from two books of Galen, namely *On My Own Books* (5.1.9–5.1.10.2, 5.1.28) and *Marātib qirā'at kutubihi* (*The Order for Reading his Books*) (5.1.3); and also from two books of Ḥunayn ibn Isḥāq: the *Risālah* (5.1.37) and the *M. Fī dhikr al-kutub allatī lam yadhkurhā Jālīnūs fī fihrist kutubihi* (*On the Books not listed by Galen in the Catalogue of his Works*) (5.1.38).[40] The list of books in the biography of Hippocrates seems to come also from classical sources, since it was most likely taken from the book-list that Ibn al-Riḍwān had translated from Greek by Yaḥyā ibn Saʿīd and included in his *K. al-Taṭarruq ilā al-saʿādah bi-l-ṭibb*[41] (4.1.9.2).

The book-lists of authors who lived in Islamic times are sometimes taken from autobiographical works. Such seems to be the case of Ibn Sīnā's book-list included by al-Jūzjānī in the biography of his teacher (11.13.3.2), later complemented by Ibn Abī Uṣaybiʿah himself (11.13.8). The autobiography of Ibn al-Haytham also includes a long list of books (14.22.5.1).

---

39 See the Appendix for exact references.

40 The latter is not acknowledged by Ibn Abī Uṣaybiʿah; on this work by Ḥunayn ibn Isḥāq see the footnote to Ch. 5.1.38.

41 See Ibn Riḍwān, *Weg zur Glückseligkeit*, 15–16.

In some cases, it is possible to point to lists collected by other scholars as the likely source of Ibn Abī Uṣaybiʿah. Thus, he may have relied on Ibn al-Imām, the editor of the works of Ibn Bājjah, for the long list of books in the biography of the philosopher (13.59.4), and also used an extant book-list in his biography of Ibn Rushd, similar to the one to be found in the Escorial Library (13.66.6).[42] In some cases, the existence of a previous catalogue can be ascertained when we collate the text of the *ʿUyūn* with cognate sources. For instance, the similarities and slight differences of the extended lists of al-Kindī's books in Ibn Abī Uṣaybiʿah, Ibn al-Nadīm, and Ibn al-Qifṭī suggests that they used a common, unacknowledged source.

a *Adab and Poetry*

The Galenic contention that the good physician should be an expert in both medicine and philosophy is implicit throughout Ibn Abī Uṣaybiʿah's narrative. He not only includes biographies of philosophers in his work, but also quotes long excerpts from philosophical works, and lists books of philosophy next to those of other sciences. To this Greek ideal, Ibn Abī Uṣaybiʿah adds a third notion: *adab*. In Islamic times this concept encompassed both the idea of good behaviour and manners, and the possession of a vast general and literary culture.[43] *Adab* can also be broadly – and vaguely – understood as a literary genre of miscellaneous works that provide both instruction and entertainment. This definition is, of course, arbitrary and some of the works listed under the rubric 'histories', such as ʿUbayd Allāh ibn Jibrīl's *Manāqib al-aṭibbaʾ*, the *Nawādir al-falāsifah* (or *Adab al-falāsifah*) attributed to Ḥunayn ibn Isḥāq, and al-Mubashshir ibn al-Fātik's *Mukhtār al-ḥikam* would fit in this category well. The works discussed below as *adab* are only those that do not deal exclusively with medical or scientific topics, but the following considerations may also be applied to works of a similar nature listed under other rubrics.

In terms of his use of literary motifs and anecdotes Ibn Abī Uṣaybiʿah is more indebted to ʿUbayd Allāh ibn Jibrīl's *Manāqib al-aṭibbāʾ* than to the compilations of Ṣāʿid al-Andalusī and Ibn Juljul. The accomplished physician portrayed in the *ʿUyūn* is also an *adīb*. This image is conveyed by means of two narrative strategies: on the one hand, physicians are presented as paradigmatic courtiers, taking part in the social activities of the court and enjoying the favour of the caliphs, especially in the golden age of the ʿAbbasid period that Ibn Abī

42 El Escorial MS Árabe 884 ff. 82b–83a, discussed in Cruz Hernández, *Abū l-Walīd Muḥammad Ibn Rušd*, 359–369.

43 For an overview see *EI Three* art. 'Adab' (J. Hämeen-Anttila et al.).

Uṣaybiʿah seems to regard with nostalgia; on the other hand, physicians appear as actors of the cultural community: they are poets, fluent in the art of epistle-writing (*inshāʾ*), and are able to reply with witticisms and provide advice in the form of sapiential maxims.

Adab literature and poetry is used as a source of factual information and anecdotes, but is also instrumental in conveying the ideal of what a physician should be. A good example to illustrate this point is the biography of Jibrīl ibn Bukhtīshūʿ which ends with two poems of praise composed in his honour: the first by Abū Nuwās probably taken from Ibn al-Buṭlān's *Daʿwat al-aṭibbāʾ*, the second by the caliph al-Maʾmūn, quoted from al-Iṣfahānī's *K. al-Aghānī* (8.3.23). Thus Ibn Bukhtīshūʿ appears not only as a man of science, but as a central member of the caliphal court celebrated by rulers and poets.

Rich literary works such al-Iṣfahānī's *K. al-Aghānī* are used for different purposes, not only to quote poetry as in the aforementioned example and, possibly, as an unacknowledged source of many other poems throughout the book. Ibn Abī Uṣaybiʿah employs the *K. al-Aghānī* also as a source of historical information and anecdotes in the chapter on the beginning of Islam and the first Muslim physicians (7.2.3, 7.2.4, 7.5.1–7.5.5). It is also likely that another major *adab* compilation, Ibn ʿAbd Rabbih's *al-ʿIqd al-Farīd*, may have been used in this chapter without acknowledgment.

Most of the quotations from *adab* works appear in the biographies of the famous physicians who work at the ʿAbbasid courts of Baghdad and Samarra. For instance, in the biography of Yuḥannā ibn Māsawayh, Ibn Abī Uṣaybiʿah quotes from al-Khālidī's *K. al-Tuḥaf wa-l-hadāyā* (*Book of Gifts and Precious Things*) (8.26.15), from a work by Jirāb al-Dawlah – likely the lost *Tarwīḥ al-arwāḥ wa-miftāḥ al-surūr wa-l-afrāḥ* – (8.26.16), and from al-Ṣūlī's *K. al-Awrāq* (8.26.17). The biography of Ḥunayn ibn Isḥāq contains quotations from al-Sarakhsī's *K. al-Lahw wa-l-malāhī* (8.29.11), and Ibn al-Dāyah's *R. fī l-mukāfaʾah* (8.29.13). In Ch. 10 Ibn Abī Uṣaybʿah also quotes from Ibn al-Dāyah's *K. Ḥusn al-ʿuqbā* (10.1.7), Ibn Qutaybah's *K. Farāʾid al-durr* (10.1.10), al-Tanūkhī's *K. al-Faraj baʿd al-shiddah*, quoted in one biography (10.62.3) and later in Ch. 11 (11.5.1–11.5.12), and al-ʿAntarī's *al-Nawr al-mujtanā* (10.69.3.1).

It is important to note that many of the anecdotes quoted from ʿUbayd Allah ibn Jibrīl's *Manāqib al-aṭibbāʾ* are of the same tenor as those found in these works. More importantly, Ibn Abī Uṣaybiʿah often adopts this tone of narrative to write the biographies of his contemporaries, which are based on his own experiences and do not rely on written sources. Perhaps the most conspicuous example is the recreation of the 'skill-test motif' whereby a physician advances in his career after being put to the test by his employer, usually a ruler. Ibn Abī Uṣaybiʿah incorporates this motif in several instances, especially for teachers and members of his family, as in the biography of al-Dakhwār (15.50.1), or the stories about his father included in the biography of his uncle (15.51.3).

Another defining trait of the *ʿUyūn* is the large number of verses that its author quotes, many of them their own.[44] When quoting poetry, Ibn Abī Uṣaybiʿah only refers to a written source on a few instances. Together with the aforementioned *K. al-Aghānī*, we also find a reference to al-Maʿarrī's *Kitāb Istaghfir wa-staghfirī* (5.1.33), ʿImād al-Dīn al-Iṣbahānī's *Kharīdah* (15.13), or al-Jilyānī's *al-Tuḥfah al-jawhariyyah* (15.8.4).

b Particular Works Written by Physicians and Philosophers

Ibn Abī Uṣaybiʿah sometimes quotes from the works composed by the physicians and philosophers about whom he is writing to provide additional biographical information or discuss the data given in other sources. There are occasional references to medicines and treatments, or philosophical matters, but these texts are mainly used as historical sources.

The most relevant of all the authors quoted is, needless to say, Galen. In many ways, Galen is the alpha and omega of the *ʿUyūn*. The first quotation of the work comes from the *Commentary of Hippocrates' Oath*, which was believed to be a Galenic work (1.1); the last quote of Ch. 15 is from Galen's *Examination of the Best Physician*, brought up by Ibn Abī Uṣaybiʿah to extol the skills and virtue of his uncle (15.51.3).

In his biography of Galen, Ibn Abī Uṣaybiʿah expressed his intention of writing a monograph on this physician: 'My intention is to write a monograph setting out all noteworthy material of this sort which I have found throughout his books and in others – if God the Exalted wills it' (5.1.27). We do not know if this work was ever written, but Ibn Abī Uṣaybiʿah's admiration for Galen is visible in many ways throughout his work.[45] It is no wonder that Galen is far and away the most quoted author, but the quotations that we find in the *ʿUyūn* are not the usual ones that appear in most medical texts and reveal a profound knowledge of the Galenic corpus, thoroughly read by Ibn Abī Uṣaybiʿah in his search for biographical and bibliographical information.

The works of Galen – or attributed to Galen – used in the *ʿUyūn* are the following:

- *K. Akhlāq al-nafs* (*Character Traits*) (4.1.3.4; 5.1.6)
- *Bīnaks*, i.e., *Pinax* (*On my own Books*) (2.1.2; 5.1.9–5.1.10.2; 5.1.28; 5.1.37 no. 21)

44 On poetry, see van Gelder's essay in this volume.

45 On Galen, see Swain's essay in this volume.

- *K. Fī l-adwiyah al-musʾhilah* (*Laxative Drugs*) (1.2)
- *K. Fī l-amrāḍ al-ʿasirat al-burʾ* (*Diseases That Are Difficult to Cure*) (5.1.19)
- *K. Fī anna al-akhyār min al-nās qad yantafiʿūna bi-aʿdāʾihim* (*The Best People Can Derive Benefit From their Enemies*) (5.1.30)
- *K. Fī ārāʾ Abuqrāṭ wa-Aflāṭūn* (*On the Opinions of Hippocrates and Plato*) (4.5.4 no. 5)
- *K. Fī aṣḥāb al-ḥiyal* (*On the Methodist Sect*) (5.1.19)
- *K. Fī l-faṣd* (*On Bloodletting*) (1.6)
- *K. Fī l-ḥathth ʿalā taʿallum ṣināʿat al-ṭibb* (*Exhortation to Study the Art of Medicine*) (2.1.2)
- *K. Fī l-kaymūs al-jayyid wa-l-radīʾ* (*Good and Bad Juices*) (5.1.24.1–5.1.24.2)
- *Ḥīlat al-burʾ* (*Method of Healing*) (1.6; 2.1.2)
- *Marātib qirāʾat kutubihi* (*The Order for Reading his Books*) (5.1.3)
- *K. Miḥnat al-ṭabīb al-fāḍil* (*Examination of the Best Physician*) (5.1.18.1–5.1.18.3; 5.1.29.1–5.1.29.3; 15.51.3)
- *Nawādir taqdimat al-maʿrifah* (*Remarkable Stories of Prognosis*) (5.1.28)
- *K. Qāṭājānis* (*Composition of Drugs by Types*) (5.1.20)
- Ps.-Galen, *K. Fī l-ḥuqan* (*On Enemas*) (1.11)
- Ps.-Galen, *Tafsīr K. al-Aymān li-Abuqrāṭ* (*Commentary of Hippocrates' Oath*) (1.1; 1.6; 2.1.6.1–2.1.6.4; 4.1.1)

In Ch. 1 and Ch. 2 the Ps.-Galenic *Commentary on Hippocrates' Oath* is the most important source to discuss the origins of medicine. This work was central in the polemics concerning the revealed nature of science and in the defence of medicine against theologians.[46] This commentary was often adduced by physicians as proof of Galen's acknowledgement of God and His role as creator, together with the Christianized translation of the *Usefulness of the Parts of the Body*, which is not quoted by Ibn Abī Uṣaybiʿah.[47] The Ps.-Galenic *Commentary on Hippocrates' Oath* is quoted at length in these first chapters as part of the discussion on the origins of medicine; the other Galenic quotations are used to support claims made in this text of by other physicians. For instance, *Fī l-adwiyah al-musʾhilah* (1.2) and the *Ḥīlat al-burʾ* (1.6) are quoted to corroborate claims defending the importance of experience in the development of medicine, whilst the *K. Fī l-faṣd* (1.6) is quoted to prove that Galen believed that dreams may have been used by God to reveal medical knowledge.

46 On this see Rosenthal, 'Defense of Medicine'.

47 On 'Christian Galenism' see, Schwarb, 'Early Kalām', especially 113–120.

In Chs. 4–5 Ibn Abī Uṣaybiʿah shows a great ability to identify the autobiographical accounts scattered in the works of Galen and the quotes from the Galenic corpus are essentially used to provide historical, biographical and bibliographical information.

In addition to Galen, other medical works are occasionally quoted, also as part of the discussion on the origins of medicine or as sources of historical information for the pre-Islamic period, for instance Ibn al-Muṭrān's *Firdaws al-ḥikmah* (1.2, 1.3.2), Oribasius, *al-Kunnāsh al-kabīr* (1.7), the anonymous *K. al-Diryāq* (1.8), al-Rāzī's *K. al-Khawāṣṣ* (1.10), or Abū Maʿshar's *K. al-Ulūf* (2.1.1, 2.1.3, 2.1.4).

In the rest of the chapters, particular medical or philosophical works are mostly used, as with the Galenic corpus, for the autobiographical accounts they might contain, or to illustrate particular aspects of a physician's work mentioned in his biography. The short entry on the life of Dioscorides reproduces an autobiographical remark from the introduction to the *Materia Medica* (4.1.11.1), and Plato's *Apology* is quoted in the biography of Socrates to show that he had never had problems with the law before his conviction (4.4.3). Al-Ruhāwī's *Adab al-ṭabīb*, in spite of the diverse topics it addresses, is only quoted in Ch. 8 as source of biographical information; and the same can be said about Ibn Hindū's *K. Miftāḥ al-ṭibb* (6.4, 11.8.5) and Abū l-Khaṭṭāb's *K. al-Shāmil fī l-ṭibb* (10.37.4, 10.58.2, 11.8.3). On the other hand, the entry on the life of Hippocrates is complemented with quotations from his *Oath* (4.1.3.1), his *Medical Law* (*Nāmūs*) (4.1.3.2), and his *Testament* (*al-Waṣiyyah*) (4.1.3.3), which set the standards of behaviour for all physicians and work with the long extracts from Galen's practice in Ch. 5 to show the 'best physician'. The biography of Aristotle has a long quotation from al-Fārābī's *Iḥṣāʾ al-ʿulūm* explaining the contents of the *Organon* (4.6.5.1).

Ibn Abī Uṣaybiʿah is certainly interested in conveying messages that go beyond the scope of a given biography. This is especially relevant in Ch. 7, where the topic of the superiority of prophecy over other kinds of knowledge is brought up after the biography of the son of al-Ḥārith ibn Kaladah, misidentified by Ibn Abī Uṣaybiʿah as al-Naḍr ibn al-Ḥārith ibn ʿAlqamah ibn Kaladah, a bitter enemy of Muḥammad. Interestingly enough, Ibn Abī Uṣaybiʿah does not quote from a Muslim treatise, but from the Ps.-Platonic *K. al-Nawāmīs* (7.2.2), which was also used in the discussion on the origins of medicine in Ch. 2 (2.1.2).

c *Documents and Personal Letters*

Although the *ʿUyūn* is essentially based on narrative sources and oral reports, Ibn Abī Uṣaybiʿah may have had access to at least one documentary source and to the personal correspondence of some physicians.

The biography of Jibrīl ibn Bukhtīshūʿ includes a long inventory of the goods and properties of the physician taken, according to Ibn Abī Uṣaybiʿah, from Ibn Bukhtīshūʿ's personal account book: 'An account book was found in Bukhtīshūʿ ibn Jibrīl's cabinet containing records written by the secretary of the elder Jibrīl ibn Bukhtīshūʿ, with corrections in Jibrīl's handwriting' (8.3.22). This may well be a secondary quote from an unacknowledged source, perhaps ʿUbayd Allāh ibn Bukhtīshūʿ's *Manāqib al-aṭibbāʾ*; but it is worth noting that Ibn Abī Uṣaybiʿah decided to reproduce the entire document, describing in excruciating detail the wealth of the physician.

Ibn Abī Uṣaybiʿah also refers to epistolary sources that may have been drawn from compilations. For instance, in the biography of Amīn al-Dawlah ibn al-Tilmīdh, he quotes from letters sent by Amīn al-Dawlah to his son (10.64.15), and to Hibat Allāh ibn al-Faḍl ibn Ṣāʿid (10.64.19.17), al-ʿAzīz Abū Naṣr ibn Muḥammad ibn Ḥāmid (10.64.19.19), Ibn Aflaḥ (10.64.19.19), and Muwaffaq al-Dīn Abū Ṭāhir al-Ḥusayn ibn Muḥammad (10.64.19.19). He also quotes from letters addressed to Amīn al-Dawlah by Ibn al-Sharīf al-Jalīl (10.64.17.1) and Ibn Aflaḥ (10.64.19.18). All these epistolary interchanges contained poetry, as also does a letter written by certain secretary named Muwaffaq al-Dīn Hibat Allāh Abū l-Qāsim ibn ʿAbd al-Wahhāb ibn Muḥammad ibn ʿAlī, which includes an epigram reproduced in the biography of Amīn al-Dawlah ibn Ghazāl (15.49.6.2) that, according to Ibn Abī Uṣaybiʿah, was copied from the handwriting of the author – we cannot say whether this was a letter or an edited volume of poems.

Ibn Abī Uṣaybiʿah also had access to the correspondence that his father and uncle had with fellow physicians and explicit use of this family archive is acknowledged in the biography of ʿAbd al-Laṭīf al-Baghdādī, where he reproduces part of a letter that this physician sent to his father (15.40.7).

d *Oral Sources*

A proper study of oral sources in the *ʿUyūn* would require a separate essay; the following are only a few general remarks to provide some contextualisation.

The chapters on the pre-Islamic period and the beginning of Islam are almost entirely based on written sources, with very rare remarks based on personal communications, such as that of certain Jamāl al-Dīn al-Siʿirdī referred to in Ch. 1 who had told Ibn Abī Uṣaybiʿah about a plant similar to the one mentioned by Galen. Chs. 8, 10, 11, and 12 have only a few minor comments based on oral communications.

It is in Ch. 13 where personal reports begin to play an important role. This chapter can be divided into two sections: biographies 13.1–13.66 follow to a great extent the *ṭabaqāt* of Ibn Juljul and are based on written sources; biographies 13.67–13.88, in contrast, were composed thanks to the oral information provided by an Andalusī traveller that Ibn Abī Uṣaybiʿah met in Damascus in 643/1237.[48]

A simple look at the Appendix with the list of quotes and paraphrases suffices to show that Chs. 14–15 are for their most part based on personal experiences and oral reports, since a large number of the physicians included in these chapters were contemporaries of Ibn Abī Uṣaybiʿah. These two last chapters abound in poetic quotation, but the scarcity of quotations of written sources contrasts with the rest of the work.

Method and Treatment of Sources

Ibn Abī Uṣaybiʿah has traditionally been regarded as an uncritical compilator. This contention stems from Müller's study on the sources of the *ʿUyūn al-anbāʾ*, specifically from a paragraph praising the intellectual honesty of its author. Müller claims that Ibn Abī Uṣaybiʿah's references and his methodical identification of sources is entirely worthy of praise.[49] This diligence, however, is but the result of lacklustre scholarship. According to Müller, Ibn Abī Uṣaybiʿah uncritically copies everything he finds with subservient mentality,[50] and that he shows lacunae in the fields of history and geography, and poor acquaintance with astronomy and mathematics.[51]

Müller's analysis of the *ʿUyūn* was biased as a result of 19th-century Orientalist obsession with originality and decadence. This led him not only to overlook the active role of Ibn Uṣaybiʿah as compiler and reject any originality in his work, but also to dismiss the medical thought of his fellow physicians altogether ('Medical research was dead since Ibn Sīnā').[52] Müller's opinion was echoed by Vernet in an article of the influential second edition of the *Encyclopaedia of Islam*, where he also presents the *ʿUyūn* as a mere collection of texts taken from previous sources:

48 See 13.63.2 and the corresponding footnote.

49 Müller, 'Über Ibn Abī Oçeibiʿa', 270. The value of the *ʿUyūn* as repository of quotations from ancient texts had already be praised by scholars since its discovery; for an overview see Waly, *Drei Kapitel*, 1–19.

50 'Er schreibt seine Vorlagen, wie sie sind, gut oder schlecht, mit gradezu sklavischer Traue ab, ohne auch nur nächstliegenden Verbesserungen vorzunehmen'; see Müller, 'Über Ibn Abī Oçeibiʿa', 271.

51 Müller, 'Über Ibn Abī Oçeibiʿa', 271.

52 Müller, 'Über Ibn Abī Oçeibiʿa', 263.

> He based his work on the bibliographical productions of his predecessors (Ibn Djuldjul for example), and a comparison between their texts and that of Ibn Abī Uṣaybiʿa shows how he either copies them, very often literally, or summarizes them, and how this mass of raw material was amplified by successive additions.[53]

Vernet also denounced the biases of Ibn Abī Uṣaybiʿah (the physician Ibn al-Nafīs was not included in his book), and the author's indulgence in poetic quotations that have nothing to do with the main topic of the book.

These negative opinions have been contested by other scholars, who present the *ʿUyūn* in a more positive light. Ullmann noted that this work was not conceived of as a history of a particular scientific discipline, but rather as a collection of biographies of relevant personalities in the field of medicine. Ibn Abī Uṣaybiʿah's acquaintance with fellow physicians, says Ullmann, allowed him to write an extraordinary cultural and historical portrait of his time.[54] In his *Ärztliches Leben und Denken im arabischen Mittelalter*, Bürgel also calls attention to Ibn Abī Uṣaybiʿah's role in the selection and critique of sources, rejecting Müller's claim that this author copied uncritically.[55] Specifically, Bürgel points to Ibn Abī Uṣaybiʿah's use of different sources to write his biography of Ḥunayn ibn Isḥāq and his discussion of the lifetimes of al-Rāzī and of Galen.[56]

The projection of concepts such as 'originality' and 'criticism' onto a 13th century work is anachronistic. A proper contextualization of the author's method should rest on his engagement with the scholarly practices proper to his time,[57] and on the discussion of the medieval understanding of originality, borrowings, and 'intellectual theft',[58] which falls beyond the scope of this essay. A further problem is that the scholarship on the *ʿUyūn* has traditionally read this work against the background of medical literature or 'proper' histories of medicine, as if the literary contents of the text were but a superfluous embellishment. The notion of *adab*, as discussed, was central in Ibn Abī Uṣaybiʿah's conception of the ideal physician. He is very careful in his selection of anecdotes and poetry, and discusses the false attribution of verses in several

---

53 See *EI*[2] art. 'Ibn Abī Uṣaybiʿa' (J. Vernet).

54 Ullmann, *Medizin*, 232.

55 'Müller tut aber unserem Autor unrecht, wenn er ihm vorwirft, seine Vorlagen völlig unkritisch kopiert zu haben'; see Bürgel, *Ärztliches Leben*, xx–xxi.

56 Bürgel, *Ärztliches Leben*, xxi.

57 In general, see Rosenthal, *Technique and approach*.

58 See *EI*[2] art. 'Sariḳa' (W.P. Heinrichs). Medieval depictions of plagiarism might be found in Ullmann, *Der verstohlene Blick*, 174–183.

instances, but this has not been regarded as 'criticism' since it does not belong to the field of medicine.

The following sections aim at discussing the methodology followed by Ibn Abī Uṣaybiʿah when dealing with the sources that we have surveyed above. But it should be emphasised that the *ʿUyūn* is a 'literary history of medicine' and the author's aims and methods can only be properly understood if we approach this work as such.

a *Structure of the Entries and Combination of Sources*

The first chapter of the *ʿUyūn* is a rather contentious survey of the different points of view on the origins of medicine. This chapter may have been partly inspired by Galen's work on sects and its Arab commentaries,[59] but its structure is similar to the *maqālāt* works that convey different, sometimes discordant opinions from various scholars. It ends with the exposition of the author's own opinions.

Chapters 2 and 3 consist mainly of long lists of names due to the poor knowledge of the disciples of Hippocrates and other pre-Galenic physicians, but the rest of the chapters contain biographies of scholars. All these biographical entries present a rather similar structure: they begin with a customary eulogizing paragraph – in which the source of information is usually unacknowledged – followed by basic biographical information taken on most occasions from medical histories (Pethion, Yaḥyā al-Naḥwī, etc.), or from comprehensive bio-bibliographical works (Ibn Juljul, Ṣāʿid al-Andalusī, Ibn al-Nadīm). Subsequently, the author may complement the information with quotes – or paraphrases – taken from selected sources; these additions often consist of anecdotes that bear witness to the skills of the physician, his success at the court, or, more rarely, his piety. In the case of contemporaries or contemporaries of his teachers Ibn Abī Uṣaybiʿah relies on oral reports and personal recollections.

These core biographical sections are normally followed by examples that illustrate the physician's worth in the field of *adab*, understood in the broad sense of both literary education and good manners and morals; these examples may include poetry, witticisms, or – especially in the biographies of philosophers – aphorisms. The entries end with a list of books, which are sometimes glossed although the source of the bibliographical information is rarely acknowledged.

---

59 Ibn Riḍwān's commentary on Galen's *K. al-Firaq* is explicitly mentioned (1.7). This work has not come down to us, but Ibn Riḍwān deals with the story of medicine and its mythical prophetic origins in the unpublished *R. Fī Sharaf al-ṭibb*.

A paradigmatic example could be the biography of Abū l-Ṣalt Umayyah (13.58). It begins with a eulogizing paragraph providing basic biographical information for which no source is acknowledged (13.58.1), complemented with the story of Abū l-Ṣalt's imprisonment in Alexandria as related to Ibn Abī Uṣaybiʿah by Sadīd al-Dīn al-Manṭiqī (13.58.2.1–13.58.2.2). Then, and as illustration of Abū l-Ṣalt's literary skills, Ibn Abī Uṣaybiʿah copies part of an epistolary interchange between Abū l-Ṣalt and Ibn al-Ṣayrafī, written in rhymed prosed and with epigrams (13.58.3.1–13.58.3.2); and poems written by Abū l-Ṣalt on various topics (13.58.4.1–13.58.4.9). The biography ends with the list of his works (13.58.5).

The use of sources in the different versions of the *ʿUyūn* also deserves comment. The additional information incorporated to Versions 2 and 3 of the work does not seem to be related to the availability of sources with one exception: Ibn al-Qifṭī. As discussed, the first version of the *ʿUyūn* was published before *Taʾrīkh al-aṭibbaʾ* (or at least before it was known to our author) and Ibn Abī Uṣaybiʿah was only able to use Ibn al-Qifṭī in later versions. Versions 2 and 3 also have additional verses, and additional items in the book-lists; some of these additions may come from sources that were not previously available for Ibn Abī Uṣaybiʿah, but this cannot be known owing to the scarce references provided for these particular sections.

b *Quotation and Paraphrase*

As Müller claimed rather dismissively, Ibn Abī Uṣaybiʿah copied what he saw, reproducing the texts as he encountered them without correcting the Arabic. In terms of the fidelity to the original text, the quotations of the *ʿUyūn* are worthy of praise and have preserved traces of middle Arabic that other would have been polished by other scholars.[60]

Another trait that sets Ibn Abī Uṣaybiʿah apart from other compilators is the special attention he paid to acknowledging the origin of his information. This awareness is sometimes reflected in his comments about other physicians. When glossing on the *Comprehensive Book* (*K. al-Ḥāwī*), for instance, he states:

> In it al-Rāzī gathers everything about diseases and their cures which he found to be dispersed among all other medical books from the Ancients and those who came after them until his own time and he attributes every quote to its source.
>
> 11.5.25 no. 1

60 On the language of the *ʿUyūn*, see Müller, 'Text'.

A further reference to the acknowledgement of sources comes up in the biography of Muwaffaq al-Dīn Yaʿqūb ibn Siqlāb, who used to quote Galen at all times:

> Whenever he was asked a question on some medical problem or a certain passage, whether difficult or otherwise, he would simply reply by saying 'Galen says' and quoting some of Galen's utterances. For this he was greatly admired. Sometimes, when quoting some of Galen's sayings, he even indicated such and such page of a certain chapter of Galen's works, referring to the copy in his possession, for he had studied that copy so many times that he had become wholly accustomed to it.
>
> 15.43.1

A similar statement is also made about Ibn al-Bayṭār:

> he never mentioned a drug without also citing in which chapter it is to be found in the book of Dioscorides or Galen, and even under which numbered item it appears amongst all the drugs mentioned in that chapter.
>
> 15.58.2

Referring to chapters, as Ibn al-Bayṭār did, was a rather common practice in medical scholarship; not so the mention of item numbers. Ibn Siqlāb's references to the pages of his own copies of Galen is a curious rarity, impossible to implement in the pre-print era beyond the limits of one's personal library. In a time were texts circulated in manuscript copies and particular passages were difficult to find, quotations were the best way of acknowledging the source material.[61] References to other works were rudimentary, but, rudimentary as it may have been, Ibn Abī Uṣaybiʿah had a method.

c *Acknowledged Quotations*

In general, direct quotations in which the source is explicitly mentioned are far more numerous than paraphrases. Almost all of them are introduced with the formula *qāla fulān*, although it is also possible to occasionally find *ḥakā fulān* or *ḥaddatha fulān*, which is a formula most commonly used for oral sources. In most biographical entries, these quotations are juxtaposed without any personal intervention.

61 See Rosenthal, *Technique and Approach*, 37–40.

Ibn Abī Uṣaybiʿāh usually provides the name of the authority and sometimes also the title of the work from which he is taking his information, but he is not consistent. If we take a look at the references to Ibn al-Nadīm we find that both name and title are given in most of the quotations (4.6.3.5, 6.1.2, 8.29.6, 10.1.6, 10.1.9, 10.4.2, 10.22.2, 15.5.9), but Ibn Abī Uṣaybiʿah may simply refer to Ibn al-Nadīm (8.26.4, 10.1.12, 10.44.3, 11.4, three times in 15.5.9).

In other cases, probably assuming the familiarity of his audience with these works, Ibn Abī Uṣaybiʿah does not provide the titles of the books he quotes. For instance, Ibn al-Dāyah (d. 265/878), the author of an often-quoted history of physicians is referred to as Yusūf ibn Ibrāhīm (8.3.6, 8.3.11, 8.8.2, 8.8.3, 8.9, 8.26.14, 8.26), and once as 'Yūsuf ibn Ibrāhīm the astrologer, known as Ibn al-Dāyah' (8.3.10). The title of his work, *Akhbār al-aṭibbāʾ*, is never mentioned. This is also the case of the quotations from 'Ubayd Allāh ibn Jibrīl's *Manāqib al-aṭibbāʾ* and those from Yaḥyā al-Naḥwī's history, which are introduced with the name of the authors only.

The quotations from the biographical works of Ṣāʿid al-Andalusī and Ibn Juljul deserve special mention. These authors are most of the times acknowledged with their names only, although the titles of their works are occasionally mentioned.[62]

d Second-Hand Quotations

There are several instances of second-hand or secondary quotations taken from compilations in which Ibn Abī Uṣaybiʿah does not acknowledge the compiler, but rather quotes an utterance as if came directly from the original source. In Ch. 1, for instance, there is a long passage directly attributed to Andromachus, which is in fact a secondary quote taken from the *K. Diryāq*, itself unacknowledged (1.8). Ch. 2 has a quotation from Orosius' *Historia adversus paganos*, a work surely unknown to Ibn Abī Uṣaybiʿah, introduced with *ʿalā mā ḥakā Hurūsiyus ṣāḥib al-qiṣaṣ* (2.1.2). This, like a following quote by Galen, was taken from not from Orosius' or Galen's works, but from Ibn Juljul's *Ṭabaqāt al-aṭibbāʾ*, which is also unacknowledged.

Some of these secondary quotations can be traced back to Ibn al-Nadīm's *K. al-Fihrist*, unacknowledged in these cases. In Ch. 2 Ibn Abī Uṣaybiʿah quotes from Thābit ibn Qurrah using the formula *'wa-min khaṭṭ Thābit ibn Qurrah al-Ḥarrānī lammā dhakara al-Baqāriṭah qāla'*. This quote does not come from an autograph consulted by Ibn Abī Uṣaybiʿah, who always uses a verb in the first person in these cases (*wajadtu, raʾaytu, naqaltu*), but most likely from Ibn al-

62 See the Appendix to this chapter for precise references.

Nadīm's *K. al-Fihrist*, in which the words of Thābit are introduced by a similar rubric (*min khaṭṭ Thābit fī l-Baqāriṭah*).[63] A quote from some ancient history book in which Hippocrates is said to be contemporary of Kay Bahman, is also an unacknowledged second-hand quote from the *K. al-Fihrist* (4.1.5).

Since many of the works used by Ibn Abī Uṣaybiʿah have not come down to us, it is impossible to pin down all these secondary quotations, but some quotes from 3rd/9th and 4th/10th century individuals introduced by *qāla* without further references are probably second-hand quotes taken from an unacknowledged source, perhaps Yūsuf ibn Ibrāhīm Ibn al-Dāyah's lost history of physicians. This seems to be the case of the quotations attributed to Maymūn ibn Hārūn (8.3.13), Isḥāq ibn Ḥunayn (8.20.3), an unidentified Abū ʿAlī al-Qiyānī (8.4.12, 8.14, 8.20.12), Ibrāhīm ibn al-ʿAbbās ibn Ṭūmār al-Hāshimī (8.32.2–8.32.3), and Abū ʿAlī ibn Makīkhā (10.8.10).

e Beginning and Ending Quotations

Medieval Arab authors are usually careful to mark the beginning of a quotation. The commonest formula to introduce these passages, also employed by Ibn Abī Uṣaybiʿah on most occasions, is *qāla/ḥakā/ḥaddatha fulān* which often appears rubricated or overlined in red in manuscripts. When the relevance of a source or the value of the manuscript consulted needs to be emphasised, Ibn Abī Uṣaybiʿah resorts to paraphrases with a verb in first person, such as *wajadtu* or *naqaltu*. This is a common formula to introduce quotations from autographs (*naqaltu min khaṭṭi-hi*).[64]

Marking the end of a quotation is an entirely different matter, and a major problem for editors. End of quotes are only occasionally indicated; they are sometimes marked with textual formulae, such as *intahā* or *qāla*, sometimes with rubrics in manuscripts.[65] Ibn Abī Uṣaybiʿah is, in this regard, also rather methodical. On occasion, he may explicitly mark the end of the quote with the expression *hādhā mā dhakarahu* or *hādha ākhar mā dhakarahu* (1.8, 5.1.13, 8.29.3, 11.13.4). But most biographical entries consist of juxtaposed quotations from sources and poetry in which the beginning of a new quote (*wa-qāla …*) also marks the end of the previous one. When he glosses a quote or continues with his own narrative after a quotation, this transition is marked with verbs in the first person, usually *aqūlu* ('I say'), or periphrasis with *naqaltu* ('I copied'), *wajadtu* ('I found'), etc.[66]

---

63 See Ibn al-Nadīm, *Fihrist* (Sayyid), 2/1:288.
64 See the references to autographs above in note 8.
65 See Rosenthal, *Technique and Approach*, 39.
66 On the use of these first-person formulae, see below p. 110.

f Unacknowledged Borrowings

The majority of the quotations in the *ʿUyūn* are acknowledged. There is also a number of unacknowledged quotes and paraphrases that, although not identified with a specific reference, are contiguous or almost contiguous with acknowledged quotations. The most obvious example is offered by the chapter on al-Andalus, the first part of which is almost entirely based on Ibn Juljul and Ṣāʿid al-Andalusī.

Most biographies of Ch. 13 begin with a paragraph that paraphrases one of these authors without acknowledgement and continues with an acknowledged quote. The use of Ibn Juljul and Ṣāʿid al-Andalusī in biographies that only have a few lines is not always acknowledged, especially if they have been quoted before; for instance, biographies 13.41–13.49 are a slightly paraphrased quotation from Ṣāʿid al-Andalusī, who is not mentioned in this section; and biographies 13.30–13.33 are also an unacknowledged borrowing from Ibn Juljul, to whom the brief biography 14.5, in fact a paraphrase from the *Ṭabaqāt al-aṭibbāʾ*, is also not attributed.

Another work occasionally paraphrased without acknowledgement is the *K. al-Aghānī*, but these paraphrases occur in proximity to acknowledged quotations from the work in Chs. 7 and 8.[67] Ibn al-Nadīm's *K. al-Fihrist*, despite being one of the most often quoted texts, is used without acknowledgement as the source of the biographies of Oribasius and his followers (5.2.1), of biographies 8.36–8.38, and of some passages in the biography of Abū l-Ḥasan Thābit ibn Qurrah (10.3.2.1–10.3.2.3, 10.3.12).

This practice suggests that Ibn Abī Uṣaybiʿah was particularly scrupulous in marking direct quotations and, usually, also isolated paraphrases; in contrast, the paraphrases that appear next to acknowledged quotations were not always identified. The lack of references in these cases cannot be interpreted as an attempt to conceal the origin of the information.

There is one exception to this rule. In Ch. 14 Ibn Abī Uṣaybiʿah quotes from a work that is neither acknowledged in this section nor mentioned in any other part of the *ʿUyūn al-anbāʾ*: Aḥmad ibn Yūsuf ibn Ibrāhīm Ibn al-Dāyah's *Sīrat Ibn Ṭūlūn* (14.3, 14.4.1–14.4.4). An unidentified work by Aḥmad Ibn al-Dāyah is quoted in 7.6, also his *R. fī l-mukāfaʾah* (8.29.13), and his *K. Ḥusn al-ʿuqbā* (10.1.7).

The origin of book-lists is only acknowledged when they are part of larger quotations from bio-bibliographical sources, such as Porphyry, Ptolemy al-Gharīb, al-Jūzjānī, or Ibn al-Haytham.

67 See the Appendix for details.

g References and Cross-References

Giving exact references to texts in manuscripts was impossible, but medieval authors divided their texts into units that helped locate quotations. Most of the works were divided into *maqālāt*, 'books' or 'sections', which in turn were divided into chapters, *abwāb*. Referring to a particular *maqālah* and a *bāb* to indicate the origin of quotes was a common practice in medical literature, especially for references to the Galenic and the Hippocratic corpus. Other extended works such as Ibn al-Nadīm's *K. al-Fihrist* also used these divisions.

Glimpses of this kind of scholarly awareness can be perceived in Ibn Abī Uṣaybiʿah's treatment of book-lists, where he often notes the number of *maqālāt* of every book, and in a few quotations, especially from Galenic works. For instance, he quotes from Galen's *K. al-Akhlāq* referring to the first *maqālah* (4.1.3.4, 5.1.6), from the first *maqālah* of Galen's *Commentary on the Prognosis* (4.1.9.2, 52), from the first *maqālah* of Porphyry's *History of Philosophers* (4.3.3), from the second *maqālah* of Galen's *On the Opinions of Hippocrates and Plato* (4.5.4); he locates a reference to Galen's first visit to Rome 'at the very start of the first section (*maqālah*) of the *Anatomical Procedures*' (5.1.8.2), and reproduces a quote from Ibn al-Muṭrān with a reference to the fourth *maqālah* of Galen's *Anatomy According to the Views of Hippocrates* (5.1.14). These are, however, exceptions. Ibn Abī Uṣaybiʿah's references to the quoted works do not follow the medical practice. His aim is not to help the reader to find a particular text in a referenced work, but rather to quote texts by giving credit to their authors.

Referrals to other books and cross-references to sections of the *ʿUyūn* are also rare. The *ʿUyūn* consists, in its major part, of a series of biographies that can be read independently. There are no explicit cross-references to other biographies and chapters of the work, even when the author makes remarks about previous discussions. For instance, Ibn Abī Uṣaybiʿah refers to the claim that Hippocrates was 'the first to consign the art of medicine to written works' when he points out that the report of Yaḥyā al-Naḥwī contradicts that (3.5), but he does not refer the reader to Ch. 2, where this was discussed.

References to other works are also exceptional and circumscribed to catalogues and to Ibn Abī Uṣaybiʿah's own books. At the very end of the biography of Galen, Ibn Abī Uṣaybiʿah states: 'Should anyone wish to refer to their titles or to Galen's aim in each of them, he may consult this book [i.e. his *Pinax*]' (5.1.40). In the biography of al-Rāzī, Ibn Abī Uṣaybiʿah also refers the reader to a book of his entitled *Anecdotes of Physicians in Treating Illnesses* (*K. Ḥikāyāt al-aṭibbāʾ fī ʿilājāt al-adwāʾ*), in which he has collected further anecdotes about this physician (11.5.13); he does the same in the biography of al-Badīʿ al-Aṣṭurlābī, in this case referring to another book of his that has not come down to us either, entitled *The Correct Predictions of the Astrologers* (*K. Iṣābāt al-munajjimīn*) (10.67.3)

h Editing Quotes

Since most of the works quoted by Ibn Abī Uṣaybiʿah are lost, it is impossible to assess the accuracy of all his quotations and to detect possible editions. The collation of his text with extant works shows that he had a deep respect for his sources. There are only a couple of instances in which quotations seem to have been edited.

The first of them is a quote from ʿUbayd Allāh ibn Jībrīl about the origins of medicine that contains a citation from Galen (in 5.1.9–10). It is likely that Ibn Abī Usaybiʿah has edited the quotation from the *Akhbār al-aṭibbāʾ* by amplifying the citation of Galen. A second case occurs in Ch. 10, where a quote from Ibn al-Nadīm's *K. al-Fihrist* seems to have been expanded in Version 3 of the *ʿUyūn* using Ibn al-Qifṭī's *Taʾrīkh al-ḥukamāʾ*:

> The scribe Ibn al-Nadīm of Baghdad says in *The Catalogue* (*K. al-Fihrist*)[68] that the Caliph al-Qāhir bi-Allāh urged Sinān ibn Thābit ibn Qurrah to convert to Islam whereupon he fled to Khorasan for fear of the Caliph. Subsequently, however, he did convert, and returned to Baghdad a Muslim, where he died.
>
> 10.4.2

The version in Ibn al-Qifṭī, found also in Ibn Abī Uṣaybiʿah's *musawwadah* and in Ms R, reports the caliph request and adds:

> but he [i.e. Sinān] strongly refused. Then al-Qāhir threatened him and Sinān, fearing him and the severity of his strength, entered Islam and settled for a time. Then he witnessed something from al-Qāhir that frightened him and he fled to Khorasan.[69]

Ibn Abī Uṣaybiʿah also adds supplementary information to book-lists. These sections often come from unacknowledged catalogues and the additions from the author's hand can only be noticed when he makes them explicit, usually with a verb in first person. The most relevant examples concern the book-lists taken from classical catalogues which are supplemented by a further list of

68 Ibn al-Nadīm, *Fihrist* (Sayyid), 2/1:313.

69 Ibn al-Qifṭī, *Taʾrīkh al-ḥukamāʾ*, 190.

works known to Ibn Abī Uṣaybiʿah and not included in the *pinax*. For instance, in the list of Aristotle's works, taken from Ptolemy al-Gharīb, Ibn Abī Uṣaybiʿa appends a supplementary list with works not mentioned by Ptolemy that he had personally seen, most of them apocrypha (4.6.13.3).

i *Critique of Sources*

Scholarly appreciation for Ibn Abī Uṣaybiʿah rests on the assumption that he was, above all, a trustworthy and methodical amanuensis who did not engage in textual or source criticism. It is true that most of the biographies consist of a collection of information from varied sources that the author rarely discusses. But this contention overlooks two main facts. First, the *ʿUyūn* is a 'literary history of medicine' aimed, among other things, at conveying an ideal image of the physician in which *adab* plays a crucial role; it cannot be compared undiscriminatingly with biographical works more focused on the history of science, such as Ibn Juljul's and Ṣāʿid al-Andalusī's *ṭabaqāt*, let alone with scientific works in which scholarly practices are driven by different aims and rely on different methodologies. Second, the critical approach of Ibn Abī Uṣaybiʿah is evident in his treatment of some sources, notably classical works dealing with pre-Islamic medicine, and also poetry.

There are several instances in the *ʿUyūn* in which discordant reports are weighed against each other, especially when the reports are contradictory. The first occurrence of contradictory reports is in the biography of Asclepius (2.1.3) in which Ibn Abī Uṣaybiʿah points out that al-Mubashshir ibn Fātik claims once that Asclepius was a pupil of Hermes, whilst on another occasion he says that:

> Asclepius was a man of the antediluvian age, and that he was a pupil of Agathodaemon the Egyptian. Agathodaemon figured among the Greek and Egyptian prophets. The name means 'the fortunate'.[70]

The claim that Asclepius brought the art of medicine to the Greeks is also contrasted with a statement from Abū Maʿshar's *K. al-Ulūf*:

> This Asclepius was not the first practitioner of the art of medicine to be deified. Furthermore, he was not the originator of that art: he learned it from his predecessors, walking the path that they had trodden.
>
> 2.1.3

70 A Greco-Egyptian god who is represented in Arabic tradition as one of the ancient Egyptian sages. On his importance as an authority in the occult sciences, see *EI*² art. 'Aghāthūdhīmūn' (M. Plessner).

This statement is followed by other reports on the life of Asclepius – sometimes contradictory – from Yaḥyā al-Naḥwī, Thābit ibn Qurrah, and the Ps.-Galenic *Commentary on Hippocrates' Oath* (2.1.5). In this case, Ibn Abī Uṣaybiʿah does not express his own opinion, he simply collects reports that might convey discordant opinions.

The same procedure is followed when discussing the lifetime of Aristotle:

> The *shaykh* Abū Sulaymān Muḥammad ibn Ṭāhir ibn Bahrām al-Manṭiqī[71] in his *Annotations* (*Taʿālīq*) states that, 'Theophrastus was appointed by Aristotle, who lived sixty-one years. As for Plato, he lived a long life'. Ibn al-Nadīm al-Baghdādī, in turn, says in his *Catalogue* (*K. al-Fihrist*) that 'Aristotle died when he was sixty-six years old'.[72] And Isḥāq [ibn Ḥunayn] claimed that Aristotle lived sixty-seven years.
>
> 4.6.3.5

On other occasions, Ibn Abī Uṣaybiʿah intervenes in the discussion casting doubt on the quality of the information. For instance, after quoting the anecdote about Polemon's physiognomical assessment of Hippocrates, he adds: 'I [i.e., Ibn Abī Uṣaybiʿah] say: This story has also been told about the philosopher Socrates and his students.' (4.1.6).

The most relevant example is perhaps his engagement with Yaḥyā al-Naḥwī's *Ta'rīkh*. This work, also used extensively by Isḥāq ibn Ḥunayn, was one of the few comprehensive sources for the history of Greek medicine available for Arab historians, at least in the time of Ibn Abī Uṣaybiʿah. But its information proved to be often contradictory when compared with other sources, some of them judged to be more reliable, such as Galen. The first reference in which the contradictions between Yaḥyā al-Naḥwī and other sources are explicitly noted occurs in the section on the disciples of Hippocrates. The legendary reports on Hippocrates' life discussed in Ch. 2 present him as the saviour of the art of medicine and the first physician who composed medical books. Yaḥyā al-Naḥwī's account is contradictory:

> I – Ibn Abī Uṣaybiʿah – say: John the Grammarian's account of these works, assuming it is accurate and that they really were composed as he asserts, is a refutation of those who allege that Hippocrates was the first

71 Al-Sijistānī al-Manṭiqī is the author of the *Ṣiwān al-ḥikmah*. See his biography at Ch. 11.7.

72 See Ibn al-Nadīm, *Fihrist*, ed. Tajaddud, 309; ed. Sayyid, ii:159.

> to consign the art of medicine to written works, because the individuals who composed them were much earlier than Hippocrates.
>
> 3.5

The differences between Yaḥyā al-Naḥwī and other sources are also raised in the section on the disciples of Hippocrates, fourteen according to Yaḥyā, twelve according to other sources (4.1.10.1). The most critical remarks against this chronographer, however, occur in the biography of Galen, where his information is contrasted with the many autobiographical accounts scattered in the Galenic corpus. In Ibn Abī Uṣaybiʿah's own words,

> The length of Galen's life was 'eighty-seven years. He was a youth and a student for seventeen years and a scholar and a teacher for seventy years'. So runs the account of John the Grammarian. In the same way I have based myself on John's reports for the divisions in the lives of the previously mentioned great master physicians down to my own time with respect to their studying and their teaching.[73] This matter requires careful scrutiny for information on it cannot be restricted, as has been indicated. Indeed, logic dictates that some of it is inconceivable and this includes his information that Galen was a youth and a student for seventeen years and a scholar and a teacher for seventy years. The only way to check what he says is to use the reports of Galen himself. Following statements by someone like Galen about himself is preferable to following the words of someone else about him.
>
> 5.1.3

The sections of poetry also contain occasional examples of source criticism, especially concerning the authorship of the poems. For instance, Ibn Abī Uṣaybiʿah notices that a poem by Ibn Shibl al-Baghdādī of philosophical tenor is sometimes wrongly attributed to Ibn Sīnā (10.51.2); in the entry on Muḥammad ibn al-Mujallī al-ʿAntarī he discusses how a poem has been also attributed to Ibn Sīnā and Ibn Buṭlān (10.69.3.1). In the biography of Ibn Sīnā, Ibn Abī Uṣaybiʿah also expresses some doubts about the attribution to Ibn Sīnā of a poem foreseeing the advent of the Mongols; perhaps due to his own interest in astrology he concludes: 'God knows best if al-Shaykh al-Raʾīs really composed this poem or someone else, but it occurred to me that I should quote the poem here, whether it is by Ibn Sīnā or someone else.' (11.13.7.9).

---

73 Cf. Rosenthal, 'Isḥāq ibn Ḥunayn', p. 66/76.

j First-Person Interventions

In addition to the aforementioned cases in which Ibn Abī Uṣaybiʿah, explicitly or implicitly, pointed out to contradictions in the sources, there are other explicit authorial interventions that deserve attention.

1 Introductory Formulae with Verbs in First Person

These formulae are often used to mark the end of a quotation and the beginning of a new one. On most occasions, however, they introduce personal remarks about the information previously given, or emphasise the contradictory or complementary nature of two different pieces of information.

The verb *aqūlu* sometimes marks the end of citation or the beginning of a new quote, but it might have other uses. The first chapter begins with a statement in first person introduced by *aqūlu* and contains several sections in which the author speaks using the first singular or plural person. In the rest of the chapters the formula *aqūlu* may be used to mark the author's glosses to citations, names, and titles; and to introduce personal opinions and experiences, such as his discussion of the contradictory reports on Galen's lifetime (5.1.4).

The formulae *naqaltu* and, less often, *wajadtu* may also mark a transition often a citation, but they are especially used when Ibn Abī Uṣaybiʿah emphasizes that his source was an autograph.[74] This is the usual formula to introduce unidentified sources, referred to as *baʿḍ al-tawārīkh* (8.3.9, 8.4.3, 8.12.2), *baʿḍ al-kutub* (8.4.13), *taʾrīkh mukhtaṣar qadīm* (5.1.8.2), or *baʿḍ tawārīkh al-naṣārā* (6.1.3).

2 Personal Opinions

The expression of personal opinions referring to sources is extremely rare. But there is one instance in which the voice of the author let itself be clearly heard and guides his selection of sources. In Ch. 1, Ibn Abī Uṣaybiʿah surveys the different thesis concerning the origins of medicine (1.4), but he himself holds the opinion that medicine is God's gift to His creation, since it has been directly conveyed by prophets (1.5), revealed in dreams (1.6, 1.7), or might stem from divinely inspired instincts that are especially observable in animals (1.11); but he also acknowledges that large part of it has been acquired by mere chance (1.8, 1.9), or empirical observation (1.10) and has evolved over time.

The veracity of reports is discussed in a couple of instances by appealing to personal experiences. A story told by Galen in *On Laxative Drugs* about a plant that produces nose-bleeding is confirmed with a personal report about a plant

74 See occurrences above footnote 8.

with similar effects, although Ibn Abī Uṣaybiʿah is unable to say whether the drug mentioned by Galen is the same one about which he has been informed (1.3.1). A second example is the claim that al-Rāzī had been a moneychanger before becoming a physician. Ibn Abī Uṣaybiʿah proves the accuracy of this report stating:

> It is said that early in his life al-Rāzī was a moneychanger. This is supported by an old manuscript I have seen of al-Rāzī's book *al-Manṣūrī*, which has pages missing at the end and is badly deteriorated with age. Its title is in the same script as the rest of it and is as follows: '*The Manṣūrī Compendium, composed by Muḥammad ibn Zakariyyā al-Rāzī the moneychanger* (*Ṣayrafī*). The owner of the manuscript said that it was in the handwriting of al-Rāzī.'
>
> 15.5.18

3 Biases and Censorship

For many scholars, the first name that comes to mind when the *'Uyūn* is mentioned is a physician about whom Ibn Abī Uṣaybiʿah does not say a single word: Ibn al-Nafīs. One of the flaws denounced by Vernet in his article in the *Encyclopaedia of Islam* was precisely 'the one-sidedness of the choice of subjects: he provides no mention of persons such as Ibn Nafīs, who, like him, was a pupil of Ibn [*sic*] al-Dakhwār (d. circa 628/1230), but whom he disliked'.[75]

This omission was also noted by medieval scholars and the biographies of Ibn al-Nafīs taken from other sources have been sometimes added in the margins of the manuscripts of the *'Uyūn*.[76] We do not really know Ibn Abī Uṣaybiʿah's reason for not including Ibn al-Nafīs in his collection of biographies. For all the relevance of Ibn al-Nafīs, this omission only does not justify Vernet's contention. But there are other biases and omissions that deserve attention.

The arbitrariness behind the selection of materials in this kind of collection is always difficult to assess. It depends to a great extent on the availability of sources, which in the case of the *'Uyūn* is especially relevant in some chapters. For instance, the chapter on al-Andalus essentially relies, in terms of written sources, on Ibn Juljul's *Ṭabaqāt al-aṭibbāʾ*. As noted by Balty-Guesdon, Ibn Juljul only pays attention to eastern scholars active during the pre-Buyid period and excludes eminent Andalusī physicians who served under Abū ʿĀmir al-Manṣūr (d. 392/1002) such as ʿArīb ibn Saʿīd (d. ca. 369/980).[77] Ibn Abī Uṣaybiʿah inher-

75 *EI*² art. 'Ibn Abī Uṣaybiʿa' (J. Vernet).
76 See Appendix 1.
77 Balty-Guesdon, 'Les *Ṭabaqāt al-Aṭibbā*''.

its Ibn Juljul biases. Moreover, the last biographies of this chapter are based on the oral reports of Abū Marwān al-Balkhī, a scholar from Seville, and almost all these physicians are from or have affiliation with this city. This chapter is affected by a twofold bias that should not be attributed to Ibn Abī Uṣaybiʿah.

Other omissions point to different constraints. The biographies of Amīn al-Dawlah ibn Ghazāl (15.49) and Rafīʿ al-Dīn al-Jīlī (15.20) are especially relevant in this regard. The first version of the *ʿUyūn*, dedicated to Amīn al-Dawlah, did not have an entry on al-Jīlī, even though he was a notable physician and personal friend of Ibn Abī Uṣaybiʿah. His biography, added to Versions 2 and 3 of the work, has a Pirandellian anecdote in which this physician complained about not having been included in Ibn Abī Uṣaybiʿah's book. We also learn from this biography that al-Jīlī had earned the enmity of the vizier Amīn al-Dawlah, who had him thrown down a cliff. Ibn Abī Uṣaybiʿah's decision of not including a biography of al-Jīlī in the version of his work dedicated to Amīn al-Dawlah was most likely an act of prudence. A comparison of the treatment of Amīn al-Dawlah in the first, dedicated version and in later versions of the work supports this hypothesis: whilst the biography of the dedication is, in fact, a eulogy,[78] the biography in Versions 2 and 3, written after the death of Amīn al-Dawlah and in which the dedication disappears, includes negative information. This is an obvious case of self-censorship that may also have influenced the way in which other biographies were written and the identity of the physicians included in different versions of the work.

If we turn to the kind of information conveyed in the *ʿUyūn*, it is evident that Ibn Abī Uṣaybiʿah tried to protect the image of physicians, and he occasionally dismisses the veracity of sources that might damage the reputation of his admired figures. In the biography of Galen, for instance, he quotes from an unknown source in which the death of Galen was attributed to a diarrhoea that he misdiagnosed and was unable to cure. Ibn Abī Uṣaybiʿah claims that this is a fictional, fabricated story (*hādhihi al-ḥikāyah aḥsabuhā muftaʿalah ʿan Jālīnus*) (5.1.22). Likewise, in his biography of his admired al-Rāzī Ibn Abī Uṣaybiʿah never mentions the physician's polemical opinions on prophecy and the reports about al-Rāzī's unbelief are strongly contested in a gloss to a work entitled 'On publicizing his claims about the vices of the Saints' (*K. fīmā yarūmuhu min iẓhār mā yaddaʿī min ʿuyūb al-awliyāʾ*):

> I – Ibn Abī Uṣaybiʿah – say that this book, if he did indeed compose it (and God only knows), may have been composed by an evil opponent of

78 This biography has been edited and translated as an addendum to Ch. 15.49.

> al-Rāzī and attributed to him so that whoever comes to see it or hear of it will form a bad opinion of al-Rāzī. Al-Rāzī, however, is far above having any dealings with such a subject or compiling a work on this matter. Even some of those who criticize al-Rāzī – or rather excommunicate him – such as ʿAlī ibn Riḍwān[79] of Egypt and others name this book *al-Rāzī's book on the false-miracles of the Prophets* (*K. al-Rāzī fī makhārīq al-anbiyāʾ*).
>
> 11.5.23 no. 166

A further example is his rejection of Ṣāʿid al-Andalusī's critiques against the works of al-Kindī:

> I – Ibn Abī Uṣaybiʿah – would say that what Judge Ṣāʿid has said about al-Kindī is unjust and does not detract from al-Kindī's knowledge, nor should it prevent the people from studying his books and deriving benefit from them.
>
> 10.1.8

Conversely, the immorality of Yūḥannā ibn Masawayh is explained by appealing to his lack of religion:

> I say that Ibn Māsawayh acted in that way because he was deficient in virtue, in religion, and fidelity. He was not a Muslim, but he did not even hold to his own religion, as is evident from the stories given above related by Yūsuf ibn Ibrāhīm. No sensible person should give credit to, nor a prudent person rely upon, someone who does not have a religion which he follows and to which he is bound.
>
> 8.26.18

Despite this argument based on the poor religiosity of Ibn Māsawayh, Ibn Abī Uṣaybiʿah is also conspicuously scrupulous about theological debates. He quotes hadiths transmitted by physicians who also excelled in this field but avoids engaging on the discussion of prophetic medicine[80] and does not quote any theological source in his discussion of prophecy. His defence of medicine as a revealed science in Ch. 1 is based on the works of physicians, especially, the Ps.–Galenic *Commentary of Hippocrates' Oath*; and, more relevantly,

79 See his entry Ch. 14.25.

80 Prophetic medicine was already an unavoidable topic in the time of Ibn Abī Uṣaybiʿah; on the works on *al-ṭibb al-nabawī*, see Ragab, *Piety and Patienthood*, 46–94.

his defence of the superiority of prophecy over knowledge relies on the Ps.–Platonic *K. al-Nawāmīs* (7.2.2).

Appendix: Identified Written Sources

This appendix lists all the sources employed by Ibn Abī Uṣaybiʿah in constructing the *ʿUyūn al-anbāʾ*. Some of the sources are acknowledged and some not. The authors are listed alphabetically, with chapter and section references. Those quotations that are acknowledged will be indicated by (a.q.), with acknowledged paraphrases indicated as (a.p.); (a.s.q.) indicates acknowledged secondary quote. Unacknowledged quotations will be marked (u.q.) and unacknowledged paraphrases (u.p.). Poetry is not included unless there is explicit reference to a written source.

ʿAbd al-Laṭīf al-Baghdādī, aphorisms copied from various works – 15.40.8 (a.q.)
ʿAbd al-Laṭīf al-Baghdādī, autobiography (*sīratuhu*) – 15.40.2–15.40.4 (a.q.)
ʿAbd al-Laṭīf al-Baghdādī, *Kitāb al-Ifādah wa-l-iʿtibār fī l-umūr al-mushāhadah wa-l-ḥawādith al-muʿāyanah bi-arḍ Miṣr* (*Information and Details about Events Witnessed and Incidents Observed in the Land of Egypt*) – 15.40.5 (acknowledged used of the book to provide biographical information); 15.40.6 (a.q.)
ʿAbd al-Laṭīf al-Baghdādī, letter sent to Ibn Abī Uṣaybiʿah's father – 15.40.7 (a.q.)
ʿAbd al-Laṭīf al-Baghdādī, unidentified source – 10.64.3 (a.q.); 10.66.4 (a.q.); 11.21 (a.q.)
ʿAbd al-Malik ibn Zuhr, *K. al-Taysīr fī l-mudāwāh wa-l-tadbīr* (*Facilitation of Treatment and Regimen*) – 1.7 (a.q.)
Abū ʿAlī ibn Makīkhā, unidentified source – 10.8.10 (a.q., likely a secondary quote)
Abū ʿAlī al-Qiyānī, personal report – 8.4.12 (a.s.q. taken from unknown source); 8.14 (a.s.q.; no reference to source); 8.20.12 (a.s.q.; no reference to source); 8.29.10 (a.q.; no reference to source)
Abū Bakr al-Khālidī and Abū ʿUthmān al-Khālidī, *K. al-Tuḥaf wa-l-hadāyā* (*Book of Gifts and Precious Things*) – 8.26.15 (a.q.)
Abū l-Barakāt al-Baghdādī, *al-Muʿtabar* (*Lessons in Wisdom*) – 1.11 (a.p.)
Abū l-Faḍl ibn ʿAbd al-Karīm al-Muhandis, *M. Fī ruʾyat al-hilāl*, (*Treatise on the sighting of the new moon*) – 15.33 (a.q.)
Abū l-Faraj al-Iṣfahānī, *K. al-Aghānī* (*Book of Songs*) – 7.2.3 (u.q. likely); 7.2.4 (a.q.); 7.5.1–7.5.5 (a.q., a.p.); 8.7 (a.p.)
Abū Ḥafṣ al-Kirmānī, *Akhbār al-Barāmikah* (*History of the Barmakids*) – 12.5 (a.q.)
Abū l-Ḥasan ʿAlī ibn ʿAbd al-ʿAzīz ibn al-Imām, prologue to the collected works of Ibn Bājjah – 13.59.2.1–13.59.2 (a.q.)
Abū Isḥāq Ibrāhīm al-Ṣābiʾ, unidentified source – 10.3.7 (a.q.)
Abū Jaʿfar Aḥmad ibn Muḥammad ibn Abī l-Ashʿath, *K. al-Adwiyah al-mufradah* (*On simple drugs*) – 10.46.4 no. 1 (a.q.)

Abū Jaʿfar Aḥmad ibn Muḥammad ibn Abī l-Ashʿath, *K. al-Ghādhī wa-l-mughtadhī* (*On Food and Nutrition*) – 15.3.1.3 (a.q.)

Abū l-Khaṭṭāb Muḥammad ibn Muḥammad ibn Abī Ṭālib, *K. al-Shāmil fī l-ṭibb* (*The Comprehensive Book of Medicine*) – 10.37.4 (a.q.); 10.58.2 (a.p.); 11.8.3 (a.q.)

Abū Manṣūr al-Thaʿālibī, *Tatimmat al-yatīmah* (*A Completion of the Unique Book of its Time*) – 11.9.1 (a.q.)

Abū Maʿshar, *Kitāb al-ulūf* (*Book of Thousands*) – 2.1.1 (a.q.); 2.1.3 (a.q.); 2.1.4 (a.q.); 12.2 (a.q.)

Abū Muḥammad Badr ibn Abī l-Iṣbaʿ, personal reports – 8.4.6–8.4.7 (a.s.q. from unknown source)

Abū l-Qāsim ʿAlī ibn Sulaymān ibn al-Ṣayrafī, likely *Qānūn dīwān al-rasāʾil* (*Rules of the Chancellery*) – 13.58.3.1–13.58.3.4 (a.q.)

Abū Sahl al-Masīḥī, *K. Fī iẓhār ḥikmat Allāh fī khalq al-insān* (*On the Manifestation of God's Wisdom in the Creation of Man*) – 11.12 (a.q.)

Abū Saʿīd al-Ḥasan ibn Aḥmad ibn ʿAlī, *K. Warṭat al-ajillāʾ min hafwat al-aṭibbāʾ* (*The Perils of Patricians due to the Errors of Physicians*) – 10.13.5 (a.q.)

Abū Saʿīd Shādhān ibn Baḥr, *K. al-Mudhākarāt* (*Book of Memoranda*) – 10.1.5 (a.q.)

Abū Saʿīd Zāhid al-ʿUlamāʾ, *K. Fī l-bīmāristānāt* (*On Hospitals*) – 11.5.2 (a.q.)

Abū l-Ṣalt Umayyah ibn Abī l-Ṣalt, *al-Risālah al-Miṣriyyah* (*Egyptian Epistle*) – 14.27.2–14.27.3 (a.q.)

Abū Sulaymān al-Sijistānī, *Taʿālīq* (*Annotations*) – 2.1.1 (a.q.); 4.6.3.5 (a.q.); 6.1.2 (a.q.); 15.1.3.1 (a.q.)

Abū Sulaymān al-Sijistānī, unidentified source – 8.29.5 (a.s.q. taken from Ibn al-Nadīm, who quotes from certain Ḥasan ibn al-ʿAbbās al-Ṣanādīqī)

Abū ʿUbayd al-Jūzjānī, biography of Ibn Sīnā – 11.13.2.1–11.13.3.19 (a.q.)

Abū ʿUbayd al-Qāsim ibn Sallām al-Baghdādī, *K. al-amthāl* (*Book of Proverbs*) – 7.5.5 (a.q.)

Abū Yaḥyā al-Yasaʿ ibn ʿĪsā ibn Ḥazm ibn al-Yasaʿ in *The Book that Declares the Good Qualities of the People of the West* (*al-Muʿrib ʿan maḥāsin ahl al-Maghrib*) – 13.40 (a.p.); 13.52 (a.p.); 13.61.2.2 (a.q.)

Aḥmad ibn al-Ṭayyib al-Sarakhsī, *K. al-Lahw wa-l-malāhī* (*Amusement and Entertainment*) – 8.29.11 (a.q.)

Aḥmad ibn Yaḥyā al-Makkī, *K. al-Mujarrad fī l-aghānī* (*Book of Songs Only*) – 8.3.23 (a.q.; misattributed to Abū l-Faraj al-Iṣbahānī)

Aḥmad ibn Yūsuf ibn Ibrāhīm ibn al-Dāyah, unidentified source – 7.6 (a.q.)

Aḥmad ibn Yūsuf ibn Ibrāhīm ibn al-Dāyah, *R. fī l-mukāfaʾah* (*Epistle on Recompense*) – 8.29.13 (a.p.)

Aḥmad ibn Yūsuf ibn Ibrāhīm ibn al-Dāyah, *K. Ḥusn al-ʿuqbā* (*Book of Fortunate Outcome*) – 10.1.7 (a.q.)

Aḥmad ibn Yūsuf ibn Ibrāhīm Ibn al-Dāyah, *Sīrat Aḥmad ibn Ṭūlūn* (*History of Ibn Ṭūlūn*) – 14.3 (u.p.); 14.4.1–14.4.4 (u.p.)

ʿAlī ibn Yūsuf ibn Abī l-Maʿālī Saʿd ibn ʿAlī al-Ḥaẓīrī, *K. Zīnat al-dahr* (*The Adornment of the Age*) – 10.65.3 (a.q.)
Amīn al-Dawlah ibn Ghazāl, letter to the vizier Burhān al-Dīn – 15.49.6.2 (a.q.)
Amīn al-Dawlah ibn al-Tilmīdh, letter to al-ʿAzīz Abū Naṣr ibn Muḥammad ibn Ḥāmid – 10.64.19.19 (a.q.)
Amīn al-Dawlah ibn al-Tilmīdh, letter to Ibn Aflaḥ – 10.64.19.19 (a.q.)
Amīn al-Dawlah ibn al-Tilmīdh, letter to Muwaffaq al-Dīn Abū Ṭāhir al-Ḥusayn ibn Muḥammad – 10.64.19.1 (a.q.)
Amīn al-Dawlah ibn al-Tilmīdh, letter to his son – 10.64.1 (a.q.)
Amīn al-Dawlah ibn al-Tilmīdh, letter to Jamāl al-Ruʾasāʾ Abū l-Fatḥ Hibat Allāh ibn al-Faḍl ibn Ṣāʿid – 10.64.19.17 (a.q.)
Amīn al-Dīn Abū Zakariyyā Yaḥyā ibn Ismāʿīl al-Bayyāsī, unidentified source – 15.17 (a.q.)
Anonymous, *K. al-Diryāq* (*Book of Antidotes*) – 1.8 (s.q. attributed to Andromachus the Younger; the *K. al-Diryāq* is not acknowledged).
al-ʿAntarī, *al-Nawr al-mujtanā* (*The Blossom Plucked*) – 10.69.3.1 (reference to the work to confirm authorship of poem)
al-Bayhaqī, *Mashārib al-tajārib wa-ghawārib al-gharāʾib* (*Draughts of Experiences and Waves of Wonders*) – 5.1.5 (a.q.)
al-Bīrūnī, *K. al-jamāhir fī maʿrifat al-jawāhir* (*Collected Information on Precious Stones*) – 8.4.11 (a.q.)
Bulmuẓaffar ibn Muʿarrif, unidentified source – 11.5.15 (a.q.); 11.5.21 (a.q.)
Dioscorides, *Hayūlā ʿilāj al-ṭibb* (*De materia medica*) – 1.11 (a.p.); 4.1.11.1 (a.q.)
Eusebius of Caesarea, unidentified source – 5.1.8.1 (a.s.p., likely taken from ʿUbayd Allāh ibn Jībrīl)
Fakhr al-Dīn al-Rāzī, *Waṣiyyah* (*Testament*) – 11.19.6.1 (a.q.)
al-Fārābī, *duʿāʾ* (prayer) – 15.1.4 (a.q.)
al-Fārābī, *Iḥṣāʾ al-ʿulūm* (*Enumeration of Sciences*) – 4.6.5.1 (a.q., no reference to title)
al-Fārābī, unidentified source – 15.1.2 (a.q.)
Galen, *Akhlāq al-nafs* (*Character Traits*) – 4.1.3.4 (a.q.); 5.1.6 (a.q.)
Galen, *Bīnaks*, i.e., *Pinax* (*On my own Books*) – 2.1.2 (a.q.); 5.1.9–5.1.10.2 (a.q.); 5.1.28 (a.p.); 5.1.37 no. 21 (a.q., title not given)
Galen, *K. Fī l-adwiyah al-mus'hilah* (*Laxative drugs*) – 1.2 (a.p.).
Galen, *K. Fī l-amrāḍ al-ʿasirat al-burʾ* (*Diseases That Are Difficult to Cure*) – 5.1.19 (a.q.)
Galen, *K. Fī anna al-akhyār min al-nās qad yantafiʿūna bi-aʿdāʾihim* (*The Best People Can Derive Benefit From their Enemies*) – 5.1.30 (a.q.)
Galen, *K. Fī ārāʾ Abuqrāṭ wa-Aflāṭūn* (*On the Opinions of Hippocrates and Plato*) – 4.5.4 no. 5 (a.q.)
Galen, *K. Fī aṣḥāb al-ḥiyal* (*On the Methodist Sect*) – 5.1.19 (reference to confirm information of Galen's *Kitāb fī l-amrāḍ al-ʿasirat al-burʾ*).

Galen, *K. Fī l-faṣd* (*On Bloodletting*) – 1.6 (a.p.).

Galen, *K. Fī l-ḥathth ʿalā taʿallum ṣināʿat al-ṭibb* (*Exhortation to Study the Art of Medicine*) – 2.1.2 (a.s.q. quote taken from Ibn Juljul's *Ṭabaqāt*; Ibn Juljul is not acknowledged)

Galen, *K. Fī l-kaymūs al-jayyid wa-l-radīʾ* (*Good and Bad Juices*) – 5.1.24.1–5.1.24.2 (a.q.)

Galen, *Ḥīlat al-burʾ* (*Method of Healing*) – 1.6 (a.p.); 2.1.2 (a.p.)

Galen, *Marātib qirāʾat kutubihi* (*The Order for Reading his Books*) – 5.1.3 (a.q.)

Galen, *K. Miḥnat al-ṭabīb al-fāḍīl* (*Examination of the Best Physician*) – 5.1.18.1–5.1.18.3 (a.q.); 5.1.29.1–5.1.29.3 (a.q.); 15.51.3 (a.q.)

Galen, *Nawādir taqdimat al-maʿrifah* (*Remarkable Stories of Prognosis*) – 5.1.28 (a.q.)

Galen, *K. Qāṭājānis* (*Composition of Drugs by Types*) – 5.1.20 (a. p.)

Galen, unidentified work (*fī mawāḍiʿ kathīrah*) – 2.1.2 (a.s.q. taken from Ibn Juljul's *Ṭabaqat*; Ibn Juljul is not acknowledged).

Ps.-Galen, *K. fī l-ḥuqan* (*On Enemas*) – 1.11 (secondary quote–Herodotus – with acknowledgement of the source).

Ps.-Galen, *Tafsīr K. al-Aymān li-Abuqrāṭ* (*Commentary of Hippocrates' Book of Oaths*) – 1.1 (a.p.); 1.6 (a.q.); 2.1.6.1–2.1.6.4 (a.q. includes glosses by Ḥunayn ibn Isḥāq, the translator of this text); 4.1.1 (a.p.)

Ḥakīm al-Zamān ʿAbd al-Munʿim al-Jilyānī, *qaṣīdah* entitled *al-Tuḥfah al-jawhariyyah* (*The Jewelled Precious Gift*) – 15.8.4 (a.q.)

al-Ḥasan ibn ʿAbd Allāh ibn Saʿīd al-ʿAskarī, *K. al-Ḥikam wa-l-amthāl* (*The Book of Aphorisms and Proverbs*) – 10.1.13 (a.q.)

Ḥayṣa-Bayṣa, letters – 10.68.1.3–10.68.1.4 (a.q.)

Hilāl al-Ṣābiʾ, unidentified source – 10.4.9 (a.q.); 10.8.7–10.8.9 (a.q.)

Hippocrates, *K. al-Aymān* (*Oath*) – 4.1.3.1 (a.q.)

Hippocrates, *Nāmūs al-ṭibb* (*Medical Law*) – 4.1.3.2 (a.q.)

Hippocrates, *al-Waṣiyyah* (*Testament*), also known as *Tartīb al-ṭibb* (*Etiquette of Medicine*) – 4.1.3.3 (a.q.)

Ḥunayn ibn Isḥāq, *K. al-ʿAshr maqālāt fī l-ʿayn* (*Ten Treatises on the Eye*) – 8.29.22 no. 2 (a.q.)

Ḥunayn ibn Isḥāq, *M. Fī dhikr al-kutub allatī lam yadhkurhā Jālīnus fī fihrist kutubihi* (*On the Books not listed by Galen in the Catalogue of his Works*) – 5.1.38 (a.q.; no reference to title)

Ḥunayn ibn Isḥāq, *Nawādir al-falāsifah wa-l-ḥukamāʾ* (*Anecdotes of the Philosophers and Wise Men*), also known as *Adab al-falāsifah* – 4.1.7 (a.q.); 4.1.8.2 (a.q.); 4.4.3 (a.q.); 4.6.3.5 (a.q.); 4.6.7.1–4.6.9 (a.q.); 5.1.34 (a.q.)

Ḥunayn ibn Isḥāq, *Risālah* (*Letter*) – 5.1.37 (a.q., sometimes heavily paraphrased)

Ḥunayn ibn Isḥāq, *R. Fīmā aṣābahū min al-miḥan wa-l-shadāʾid* (*On his trials and tribulations*) – 8.29.14–8.29.21

Ḥunayn ibn Isḥāq, unidentified work – 4.1.11.1 (a.q.); 5.1.39 no. 167 (a.q.); 8.31 (a.q.)

Ibn ʿAbd Rabbih, *al-ʿIqd al-farīd* (*The Unique Necklace*) – 7.1.2–7.1.4 (u.p. likely)
Ibn Abī Ṣādiq al-Nīsābūrī, *Sharḥ K. al-Masāʾil* (*Commentary of* [*Ḥunayn's*] *Questions*) – 8.29.22 no. 1 (a.q.)
Ibn Abī Ṣādiq al-Nīsābūrī, *Tafsīr K. Manāfiʿ al-aʿḍāʾ li-Jālīnus* (*Commentary on Galen's On the Usefulness of the Parts*) – 11.17 (a.q.)
Ibn Bakhtawayh, *K. al-Muqaddimāt* (*Prolegomena*) – 5.1.22 (a.q.); 10.1.12 (a.q.)
Ibn Buṭlān, *Daʿwat al-aṭibbāʾ* (*The Physician's Dinner-Party*) – 8.30.5 (a.q.); 10.38.6 no. 10 (a.q.)
Ibn Buṭlān, *M. Fī ʿillat naql al-aṭibbāʾ al-maharah tadbīr akthar al-amrāḍ allatī kānat tuʿālaj qadīman bi-l-adwiyah al-ḥārrah ilā al-tadbīr al-mubarrad* (*An essay on the reason the master physicians changed the regimen for most of the diseases which were, of old, treated with hot medicines*) – 10.8.2 (a.q.); 10.13.2–10.13.4 (a.q.); 10.23.2 (a.q.); 10.36 (a.q.)
Ibn Buṭlān, *Maqālah ilā ʿAlī ibn Riḍwān* (*Essay Addressed to ʿAlī ibn Riḍwān*) – 14.25.5 (a.q.)
Ibn Buṭlān, *Waqʿat al-aṭibbāʾ* (*The Battle of the Physicians*) – 10.38.2 (u.q.); 10.38.3 (a.q.)
Ibn Buṭlān, unidentified source – 6.1.1 (a.p.); 8.26.16 (a.q.); 10.38.4 (a.q.); 14.25.3 (a.p.)
Ibn Ḥawqal, *K. Ṣūrat al-arḍ* (*The Shape of the Earth*) – 4.6.3.3 (a.q., wrongly attributed to al-Masʿūdī's *al-Masālik wa-l-mamālik*).
Ibn al-Haytham, autobiography and list of own books from unidentified source – 14.22.4.1–14.22.4.4 (a.q.); 14.22.5.1 (a.q.)
Ibn Hindū, *Miftāḥ al-ṭibb wa-minhāj al-ṭullāb* (*Key to Medicine and Guide for Students*) – 6.4 (a.q.); 11.8.5 (a.q.)
Ibn al-Jazzār, *K. Akhbār al-dawlah* (*History of the* [*Fatimid*] *Dynasty*) – 13.2.2.1–13.2.2.2 (a.q.)
Ibn Juljul, *Ṭabaqāt al-aṭibbāʾ* (*The Classes of Physicians*) – 2.2 (a.q.); 4.1.6 (a.q.); 4.5.1 (a.q.); 4.6.1 (a.q.); 4.6.12 (u.p. likely); 5.1.15 (a.q.); 5.1.17 (a.q.); 5.1.21.2 (u.q. introduced by *qāla ghayruhu*); 7.1.1 (u.p.); 7.1.6 (a.q.); 7.3 (u.q. slightly paraphrased); 8.19.2 (a.p.); 8.26.3 (a.q.); 8.29.9 (a.q.); 8.29.12 (a.q.); 10.1.4 (a.q.); 10.9.1 (a.q.); 10.44.1 (a.q.); 11.5.7 (a.q.); 13.1.2 (a.q.); 13.2.1 (u.p.); 13.3.1 (u.p.); 13.3.2.1–13.3.2.3 (a.q.); 13.12 (u.q.); 13.13 (u.p.); 13.14 (u.p.); 13.15 (u.p.); 13.16 (u.p.); 13.17 (a.q.); 13.18 (a.q.); 13.19.1–13.19.2.1 (u.q.); 13.19.3 (a.q.); 13.20 (u.p.); 13.21.2.1–13.21.2.2 (a.q.); 13.22 (u.p.); 13.23 (u.q.); 13.24.1 (u.q.); 13.24.2–13.24.3 (a.q.); 13.25 (u.p.); 13.26 (u.p.); 13.28 (u.p.); 13.30–13.33 (u.p.); 14.5 (u.q. slightly paraphrased)
Ibn Juljul, *K. Tafsīr asmāʾ al-adwiyah al-mufradah min kitāb Diyusqūrīdis* (*Explanation of the names of the simple drugs in the book of Dioscorides*) – 13.36.2.1–13.36.2.4 (a.q.)
Ibn Jumayʿ al-Miṣrī, *Taṣrīḥ al-maknūn fī tanqīḥ al-Qānūn* (*Making Explicit what is Concealed: On Examining the Canon*) – 13.61.2.1 (a.q.)
Ibn al-Muṭrān, *Firdaws al-ḥikmah* (*Paradise of Wisdom*) – 1.2 (a.q.); 1.3.2 (a.q.)
Ibn al-Muṭrān, unidentified source – 5.1.14 (a.q.)

Ibn al-Nadīm, *K. al-Fihrist* (*Catalogue*) – 2.1.5 (u.s.q); 4.1.5 (u.q.); 4.6.3.5 (a.q.); 4.7 (u.q.); 4.8.1–4.8.2 (u.q.); 5.2.1 (u.q.); 5.2.2 (u.q.); 6.1.2 (a.q.); 8.26.4 (a.q.); 8.29.6 (a.q.); 8.36–8.38 (u.q.); 10.1.6 (a.q.); 10.1.9 (a.q.); 10.3.12 (u.q.); 10.4.2 (a.p.); 10.22.2 (a.q.); 10.44.2 (a.q.); 11.1 (u.q.); 11.4 (a.q); 11.5.9 (a.q.)

Ibn al-Qifṭī, *Taʾrīkh al-ḥukamāʾ* (*History of Learned Men*) – 10.77.3 (a.q.); 11.3 (a.q.); 11.21 (a.q.); 14.14.3 (a.q.); 14.14.6 (a.q.); 14.22.3.1–14.22.3.2 (a.q.)

Ibn al-Qifṭī, unidentified source – 13.3.4 no. 5 (a.q.; maybe oral report); 15.23.2–15.23.2.1 (a.q.); 15.36.1.1 (a.q.)

Ibn Qutaybah, *K. Farāʾid al-durr* (*The Unique Pearls*) – 10.1.10 (a.q.)

Ibn Riḍwān, autobiography – 14.25.1–14.25.2 (a.q.)

Ibn Riḍwān, *Fawāʾid* – 4.1.9.1 (u.p. in several instances)

Ibn Riḍwān, *K. Ḥall shukūk al-Rāzī ʿalā Jālīnūs* (*Resolution al-Rāzī's Doubts regarding Galen*) – 11.8.4 (a.q.)

Ibn Riḍwān, *al-Kitāb al-nāfiʿ fī kayfiyyat taʿlīm ṣināʿat al-ṭibb* (*Useful Book on How to Study the Art of Medicine*) – 6.3.1–6.3.4 (a.q.)

Ibn Riḍwān, *Tafsīr K. al-Firaq li-Jālīnus* (*Commentary of Galen's On Sects*) – 1.7 (a.q.)

Ibn Riḍwān, *al-Taṭarruq ilā l-saʿādah bi-l-ṭibb* (*Attaining Happiness Through Medicine*) – 4.1.9.2 (u.p.; list of books seems to be largely based on Ibn Riḍwān)

Ibn Riḍwān, sayings from unidentified work – 14.25.7–14.25.8 (a.q.)

Ibn Riḍwān, unidentified work – 4.1.2 (a.p.)

Ibn Rushd, *K. al-Kulliyyāt* (*Book of Generalities*) – 13.66.2 (a.q.)

Ibn Sīnā, *M. Fī naqḍ Risālat Ibn al-Ṭayyib fī l-quwā l-ṭabīʿiyyah* (*Refutation of Ibn al-Ṭayyib's On the Natural Faculties*) – 10.37.2 (a.q., no reference to title)

Ibrāhīm ibn al-ʿAbbās ibn Ṭūmār al-Hāshimī, unidentified source – 8.32.2–8.32.3 (a.q., likely secondary quote taken from an unacknowledged source)

Ibrāhīm ibn ʿAlī al-Ḥuṣrī, *Nūr al-ṭarf wa-nawr al-ẓarf* (*The Light of the Eye and the Blossom of Wit*) – 8.4.5 (a.q.; probable confusion with al-Ḥuṣrī's *Zahr al-ādāb wa-thamar al-albāb*)

Ibrāhīm ibn al-Qāsim *K. Akhbār al-Ḥajjāj* (*Reports of al-Ḥajjāj*) – 7.9.1–7.9.4 (a.q./a.p.)

ʿImād al-Dīn Abū ʿAbd Allāh Muḥammad ibn Ḥāmid al-Iṣbahānī, *al-Kharīdah* (*Book of the Unbored Pearl*) – 15.13 (a.q.)

Isḥāq ibn Ḥunayn, *Taʾrīkh al-aṭibbāʾ* – 4.1.11.1 (u.q., likely); 4.1.11.2 (u.p., likely); 4.1.11.3 (a.p.); 4.6.3.5 (a.q.); 5.1.2 (u. p.); 5.1.4 (a.p.); 5.2.1 (u.q.)

Isḥāq ibn Ḥunayn, personal report – 8.20.3 (a.s.q.; no reference to source, likely to be Ibn al-Dāyah)

Isḥāq ibn Ḥunayn, unidentified source – 8.30.2 (a.q.)

al-Iṣṭakhrī, *al-Masālik wa-l-mamālik* (*Routes and Realms*) – 5.1.21.1 (a.p., wrongly attributed to al-Masʿūdī's *al-Masālik wa-l-mamālik*)

Jamāl al-Mulk Abū l-Qāsim ʿAlī ibn Aflaḥ, letter to Amīn al-Dawlah ibn al-Tilmīdh and Amīn al-Dawlah's reply – 10.64.19.18 (a.q.)

al-Jawharī, *al-Ṣiḥāḥ fī l-lughah* (*The Correct [Forms] in Language*) – 7.1.1 (a.p.)

Jibrīl ibn Bukhtīshūʿ, first person report – 8.3.8 (a.s.q. taken from certain history book *baʿḍ al-tawārīkh*)

Jibrīl ibn Bukhtīshūʿ, personal account book – 8.3.22 (a.p.)

Jirāb al-Dawlah, unidentified source, likely the lost *Tarwīḥ al-arwāḥ wa-miftāḥ al-surūr wa-l-afrāḥ* – 8.26.16 (a.q.)

al-Jurjānī, *K. al-Kināyāt* (*Book of Metonyms*) – 10.3.8 (a.q.)

al-Khaṭīb al-Baghdādī, *Taʾrīkh Baghdād* (*History of Baghdad*) – 8.26.17 (a.q.; no reference to title)

al-Maʿarrī, *Kitāb al-Istighfār* (*Asking for Pardon*), i.e. *Kitāb Istaghfir wa-staghfirī* – 5.1.33 (a.q.)

al-Masʿūdī, *al-Tanbīh wa-l-ishrāf* (*The Book of Notification and Verification*) – 4.6.1 (a.q.); 5.1.6 (a.s.q. taken from Ṣāʿid al-Andalusī's *Ṭabaqāt al-umam*; Ṣāʿid al-Andalusī is not acknowledged)

Maymūn ibn Hārūn, first person report (a.s.q.; probably from Ibn al-Dāyah's *Akhbār al-aṭibbāʾ*)

al-Mubashshir ibn Fātik, *Mukhtār al-ḥikam wa-maḥāsin al-kalim* (*The Choicest Maxims and Best Sayings*) – 1.5 (a.p.); 2.1.3 (a.q.); 2.1.7 (a.q.); 4.1.7 (a.q.); 4.1.8.4 (a.q.); 4.3.2 (u.q.); 4.3.4.1–4.3.4.4 (a.q.); 4.3.5 (a.q.); 4.4.2.1–4.4.2.5 (a.q.); 4.4.4 (a.q.); 4.5.2–4.5.3 (a.q.); 4.6.3.1–4.6.3.2 (a.q.); 4.6.3.4 (a.q.); 4.6.11 (a.q.); 5.1.21.1 (a.q.); 5.1.23 (a.q.); 5.1.25–5.1.26 (a. p.); 5.1.31 (a.q.); 5.1.35 (a.q.)

al-Mubashshir ibn Fātik, unidentified source – 10.22.3 (a.q.)

Muḥammad ibn Aḥmad ibn Ṣāliḥ al-ʿAbdī, unidentified source – 13.61.3.1 (a.q.)

Muwaffaq al-Dīn Hibat Allāh Abū l-Qāsim ibn ʿAbd al-Wahhāb ibn Muḥammad ibn ʿAlī, poem copied from an autograph – 15.49.6.2 (a.q.)

Najm al-Dīn ibn al-Lubūdī, epistle – 15.31.1 (a.q.)

Oribasius, *al-Kunnāsh al-kabīr* (*Great Handbook*) – 1.7 (a.q.).

Orosius, *Tawārīkh* (*Historia adversus paganos*) – 2.1.2 (a.s.q. taken from Ibn Juljul's *Ṭabaqāt*; Ibn Juljul is not acknowledged)

Pethion, *Taʾrīkh* (*History*) – 8.1.1–8.1.2 (a.q.); 8.2 (a.q.); 8.3.2–8.3.5 (a.q.); 8.3.20–8.3.21 (a.q.); 8.4.2 (a.q.); 8.4.8 (a.q.); 8.25.1 (a.q.) [IAU never gives the title]

Plato, *K. al-Siyāsah* (*Republic*) – 2.1.2 (a.q.)

Plato, *Iḥtijāj Suqrāṭ ʿalā ahl Athīniyah* (*Apology of Socrates to the People of Athens*) – 4.4.3 (a.q.)

Ps.-Plato, *K. al-Nawāmīs* (*Laws*) – 2.1.2 (a.s.p. taken from Ibn Juljul's *Ṭabaqāt* – Ibn Juljul is not acknowledged); 7.2.2 (a.q.).

Porphyry, *K. Akhbār al-falāsifah wa-qiṣaṣihim wa-ārāʾihim* (*Book of the History of Philosophers, their Stories and Opinions*) – 4.3.3 (a.q.) 4.3.6.1–4.3.6.2 (a.q.)

Ptolemy al-Gharīb, *K. Ilā Ghalus fī sīrat Arisṭūṭālis* (*Epistle to Gallus on the Life of Aristotle*) – 4.6.2.1–4.6.2.3 (a.q.); 4.6.6.1 (a.q.); 4.6.13.1 (a.q. with book-list) al-Quṭb al-Miṣrī, *Sharḥ al-kulliyyāt min K. al-Qānūn li-l-shaykh al-raʾīs Ibn Sīnā* (*Commentary on the 'Generalities' of the shaykh al-raʾīs Ibn Sīnā's Canon of Medicine*) – 11.20 (a.q.)

al-Rāzī, Muhammad ibn Zakariyyā, *K. al-Khawāṣṣ* (*On Occult Properties*) – 1.10 (a.p.)

al-Rāzī, Muhammad ibn Zakariyyā, *al-Kitāb al-Ḥāwī* (*Comprehensive Book*) – 5.1.32 (a.p.); 12.3 (acknowledged as bibliographical source)

al-Rāzī, Muhammad ibn Zakariyyā, *al-Kitāb al-Manṣūrī* (*The Manṣūrī Compedium*) – 11.5.18 (acknowledged reference to full book title taken from autograph)

al-Ruhāwī, *Adab al-ṭabīb* (*Practical Ethics of the Physician*) – 8.3.8 (a.q.); 8.4.10 (a.q.); 8.8.1 (a.q.); 8.12.1–8.12.2 (a.q.); 8.20.2 (a.q.); 8.23.1 (a.q.); 8.24 (a.q.); 8.25.2–8.25.6 (a.q.); 8.26.2 (a.q.)

Ṣadaqah al-Sāmirī, aphorisms and poems copied from autograph – 15.47.1–15.47.2.1 (a.q.)

Ṣāʿid al-Andalusī, *K. Ṭabaqāt al-umam* (*The Categories of the Nations*) – 4.1.11.4 (a.p./a.q.); 4.2 (a.q.); 4.3.1 (a.q.); 4.4.1 (a.q.); 4.6.4.1 (a.q.); 10.1.1–10.1.3 (a.p.); 10.1.8 (a.q.); 11.5.8 (a.q.); 13.4 (u.p.); 13.5.1 (a.q.); 13.6.1.1–13.6.1.2 (a.q.); 13.7 (u.p.); 13.8 (u.p.); 13.9 (a.q.); 13.10 (u.p.); 13.11 (u.p.); 13.21.1 (u.p.); 13.24.1 (u.q.); 13.27.2 (a.q.); 13.29.2 (a.q.); 13.34.1 (u.p.); 13.34.2(a.q.); 13.37 (a.q.); 13.38.1–13.38.2 (a.q.); 13.39.2 (a.q.); 13.41–13.49 (u.q./u.p.); 13.60.2 (a.q.); 15.1.3.2 (a.q.)

Ṣāʿid ibn Bishr, *M. Fī maraḍ al-marāqqiyyā* (*On Hypocondria*) – 10.13.6 (a.q.)

Saʿīd ibn al-Biṭrīq, *Naẓm al-jawhar* (*The String of Jewels*) – 14.1 (a.q.)

al-Sayyid al-Naqīb al-Kāmil ibn al-Sharīf al-Jalīl, letter to Amīn al-Dawlah ibn al-Tilmīdh – 10.64.17.1 (a.q.)

Shānāq (Cāṇakya), *Muntakhal al-jawhar* (*Sifted Jewels*) – 12.3 (a.q.)

al-Ṣūlī, *K. al-Awrāq* (*Book of Folios*) – 8.26.17 (a.q.)

al-Ṭabarī, *Taʾrīkh* (*History*) – 7.5.6 (a.q.); 10.12.2 (a.q.)

al-Tanūkhī, *al-Faraj baʿd al-shiddah* (*The Book of Relief after Hardship*) – 10.62.3 (reference to the work to confirm information); 11.5.11–11.5.12 (a.q.)

Thābit ibn Qurrah, unidentified work – 2.1.5 (a.s.q., taken from Ibn al-Nadīm's *Fihrist*; Ibn al-Nadīm is not acknowledged)

Thābit ibn Sinān ibn Thābit, unidentified source – 8.4.9 (a.q., probably a.s.q. from unknown source); 8.23.2 (a.p.; unidentified source, could be his *Taʾrīkh*)

Thābit ibn Sinān ibn Thābit ibn Qurrah, *Taʾrīkh* (*History*) – 10.3.6 (a.q.); 10.3.9 (a.q.); 10.3.11 (a.q.); 10.4.4.1–10.4.4.8 (a.q.); 10.5.2 (a.q.); 10.5.3–10.5.4 (a.q.); 10.10.5 2 (a.q.); 10.16 (a.q.); 10.26 (a.q.) [title given only in 10.4.4.1; 10.5.3; 10.26]

al-Ṭurṭūshī, *Sirāj al-mulūk* (*The Lamp of Kings*) – 15.3.1.5 (a.p.)

ʿUbayd Allāh ibn Jibrīl ibn ʿUbayd Allāh ibn Bukhtīshūʿ, *Manāqib al-aṭibbaʾ* (*The Merits of Physicians*) – 5.1.7 (a.q.); 5.1.8.1–5.1.8.2 (a.q.); 5.1.11–5.1.13 (a.q.); 6.1.3 (a.q.); 8.5.2–8.5.4 (a.q.); 8.29.7 (a.q.); 8.39 (a.q.); 10.8.3–10.8.6 (a.q.); 10.17 (a.q.); 10.27–10.30 (a.q.); 10.44.4 (a.q.); 10.46.2 (a.q.); 11.5.6 (a.q.); 11.5.22 (a.q.); 11.12 (a.q.); 14.20 (a.q.); 15.2 (a.q.) [title given only in 6.1.3; 8.29.7]

Unidentified author, *Akhbār al-jabābirah* (*History of the Tyrants*) – 2.1.2 (a.p.)
Unidentified chronicle (*baʿḍ al-tawārīkh*) – 10.14.2.1–10.14.2.2 (a.q.)
Unidentified reports translated from the Greek (*baʿd al-mawāḍiʿ al-manqūlah min al-yunānī*) – 4.1.1 (a.q./a.p.)
Unidentified source – 8.4.3–8.4.4 (a.q.; *baʿḍ al-tawārīkh*); 8.4.13 (a.p. *baʿḍ al-kutub*); 8.5.5 (a.p./a.q. probably from ʿUbayd Allāh ibn Jibrīl's *Manāqib al-aṭibbaʾ*; introduced by *mā jarā lahu*)
Unidentified source of Syriac origin – Ch. 9 (source also used in Ibn al-Nadīm's *K. al-Fihrist*).
Unnamed learned scholar (*baʿḍ al-mashāyikh*) – 15.1.1.2 (a.q.)
al-Wāqidī, *Taʾrīkh* (*History*) – 7.5.6 (a.q.)
al-Wāthiq bi-Allāh, *K. al-Bustān* (*The Garden*) – 7.1.5 (a.p.; nothing is known about this work)
Yaḥyā al-Naḥwī, *Taʾrīkh* (*Chronology*) – 2.1.5 (a.p.); 3.1 (u.p.); 3.2 (a.q.); 3.3 (u.p.); 3.4 (u.p.); 3.5 (u.p./a.q.); 3.6 (u.p.); 4.1.10.1–4.1.10.2 (a.p.); 4.1.10.4–4.1.10.5 (u.p.); 5.1.3 (a.p.); 5.1.4 (a.p.) [IAU never gives the title]
Yaḥyā ibn Saʿīd ibn Yaḥyā, *K. Taʾrīkh al-Dhayl* (*Supplement to History*) – 14.10–14.11 (a.q.)
Yūsuf ibn Ibrāhīm ibn al-Dāyah, *Akhbār al-aṭibbāʾ* (*Accounts of Physicians*) – 5.1.16.1–5.1.16.5 (a.q.); 7.7.2 (a.q.); 7.8.2–7.8.3 (a.q.); 8.3.6–8.3.7 (a.q.); 8.3.10–8.3.12 (a.q.); 8.3.14–8.3.19 (a.q.); 8.8.2–8.8.6 (a.q.); 8.9 (a.q.); 8.10.2–8.10.4 (a.q.); 8.11 (a.q.); 8.13.2–8.13.4 (a.q.); 8.18.1–8.18.4 (a.q.); 8.19.3–8.19.5 (a.q.); 8.20.4–8.20.11 (a.q.); 8.21.2–8.21.4 (a.q.); 8.25.7–8.25.8 (a.q.); 8.26.5–8.26.14 (a.q.); 8.27.2–8.27.4 (a.q.); 8.29.2–8.29.3 (a.q.); 12.6 (a.q.) [IAU never gives the title]

CHAPTER 6

Poetry in *ʿUyūn al-anbāʾ*

Geert Jan van Gelder

Poetry is a very prominent feature of Ibn Abī Uṣaybiʿah's book. Its presence was obviously important to the author and to his intended readership. It was only in modern times that western scholars showed explicit signs of impatience with it.[1] In his lecture 'Ueber Ibn Abi Oçeibiʿa und seine Geschichte der Aerzte' August Müller, the editor of the first scholarly edition, mentions that the author, in addition to his medical studies,

> unfortunately devoted himself so persistently to *adab* and poetry that the *qaṣīdah*s and shorter poems composed by himself and others drove to well-justified despair the otherwise quite patient copyist of the excellent model of the Brit. Mus. manuscript Add. 7340.[2]

Müller speaks of the 'polite letters and poetic dedications' that the author 'included all too conscientiously in his book'.[3] Juan Vernet, in his entry on the author in the second edition of *The Encyclopaedia of Islam*, says somewhat scathingly that the book contains 'some long series of verses which have nothing to do with the main theme'. It is true, much of the poetry is not at all about medicine and medical topics. Worse, perhaps, much of it can only be called mediocre in quality. This did apparently not bother the author. To him the large quantity of poetry was an essential part of his book, and it is obvious that he included several entries on physicians because they composed poetry, rather than being important doctors.

The work, in the version edited here, contains some 3,600 lines of verse; the exact number depends on how one counts the lines, unequal in length, of strophic poems. They are very unevenly distributed. There is no poetry in chapters 1, 2, 3, 5, and 12, which deal with non-Arabs. Chapter 4, on the ancient Greeks, surprisingly has one Arabic line attributed to Socrates. The chapter on

1 Implicit indications of the same attitude from earlier periods may be deduced from the fact that some manuscripts of the book omit much of the poetry, as the following quotation from Müller illustrates.
2 In Müller, 'Über Ibn Abī Oçeibiʿa', 262 (my translation).
3 Ibid., 263.

 | DOI:10.1163/9789004410312_007

Iraq has many lines (965) and by far the most are found in the chapter on Syria (1,672). They are composed by some one hundred different poets, whether mentioned by name or quoted anonymously.

1 Forms

There are many epigrams and short pieces, but also a good number of longer pieces or *qaṣīdahs*,[4] such as an ode of 92 lines[5] by Abū l-Qāsim Hibat Allāh ibn al-Faḍl (better known as Ibn al-Qaṭṭān, although not named thus in the book) on a *kātib al-inshāʾ* (chancery scribe) in Baghdad;[6] a pseudo-prophetic ode by pseudo-Ibn Sīnā on the Mongol conquests of 52 lines;[7] or an ode on Saladin of 82 lines by al-Jilyānī.[8] Some humorous poems by Abū l-Ḥakam al-Maghribī are also long: one about a 'domestic scandal',[9] another on a badly made pair of shoes.[10] The great majority of poems are in standard form, with monorhyme (*aaaaaa* …) and in one of the traditional metres, which are always quantitative, not unlike Classical Greek, Latin, and Sanskrit versification. There are also nineteen poems that employ the *dūbayt* form (with the rhyme scheme *aaba* or *aaaa*) and metre. Originally Persian (the hybrid Persian-Arabic word means 'two-liner' or 'distich'), it is in Persian more usually called *rubāʿiyyah*, 'quatrain' – paradoxically with an Arabic word – counting the rhyming units. Chapter 13, on Spain, contains a few strophic poems (*muwashshaḥāt*, singular *muwashshaḥah*), including some famous ones. With their short lines and many rhymes they are often set to music and, indeed, some are still being performed even today.

4 Many Arabists use *qaṣīdah* for a 'polythematic' poem of some length, often a panegyric ode introduced with a lyrical introduction on love, nature, or a gnomic passage. In premodern critical usage, however, the term is used for any longer poem, to be distinguished (not always clearly) from a shorter piece or epigram, called *qiṭʿah* or *maqṭūʿah*, misleading terms because, literally 'piece' or 'cut-off', they wrongly suggest that any short piece is a fragment of a longer poem.

5 'Line' or 'verse' is *bayt* in Arabic; its length is variable, dependent on the metre chosen by the poet. A typical *bayt* is so long, by English standards (up to thirty syllables, mostly divided between two hemistichs of equal or nearly equal length), that in our translations it almost always takes two lines.

6 Ch. 10.68.2.2 ('O woman who left me').

7 Ch. 11.13.7.9 ('Beware, dear son').

8 Ch. 15.11.2.1 ('A vigorous, astute man's comfort').

9 Ch. 15.8.7 ('Any domestic scandal'); see van Gelder, 'The Joking Doctor'.

10 Ch. 15.17 ('My plight is bewildering').

Virtually all the poetry is in Classical Arabic (*fuṣḥā*), though occasionally with 'Middle Arabic', 'vernacular' elements. The closing strophe (*kharjah*) of a *muwashshaḥah* may also contain vernacular Arabic,[11] as does an isolated line in non-standard metre, sung by Abū l-Ḥakam al-Maghribī:[12]

> Bee-hunter, here's a job for you:
> Come on, go out early, get some honey!

Some poems are composed in a rather shaky Arabic, particularly those by Najm al-Dīn al-Lubūdī.[13] He simply does not know his grammar, using the moods erratically and doing other strange things. Although Ibn Abī Uṣaybiʿah's own prose style is not free from colloquialisms and linguistic oddities, as already pointed out at length by August Müller,[14] his poems (eleven pieces with a total of 197 lines) are relatively free of them.

2 Themes: Medical Poems

Classifying poems by their form is much easier than trying to classify them by theme. Almost all the traditional modes or themes of Arabic poetry – praise, invective, elegy, vaunting, wisdom, religion, mysticism, love, wine, didactic verse, pornography, riddles, etc. – can be combined, either in a natural way, such as elegy and praise, elegy and wisdom, wisdom and religion, love and wine, or in a conventional manner such as the many panegyric odes that open with a lyrical or amorous passage, a convention so common in Arabic that it has become natural. In what follows the main themes and subjects of the poems in *ʿUyūn al-anbāʾ* will be briefly discussed.

In pre-modern Arabic one finds countless didactic poems on every conceivable subject: versified knowledge, meant to be memorised, and without literary pretensions. Most of them use the simple *rajaz* metre and paired rhyme (*aabbccdd* …, which allows for long poems). Among them are medical poems, such as the famous medical poem by Ibn Sīnā,[15] or his poem on the causes of fevers,[16]

11 See Ch. 13.63.8.2 (the last line of a *muwashshaḥah* by Abū Bakr ibn Zuhr).
12 Ch. 15.8.1.
13 Ch. 15.31.2–5.
14 Müller, 'Ueber Text und Sprachgebrauch'.
15 Avicenna, *Poème de la médicine*.
16 Ibn Sīnā, *Urjūzah fī asbāb al-ḥummayāt*.

and many others.[17] One might have expected to find similar poems in Ibn Abī Uṣaybiʿah's book, but they are not there. He does refer to the practice: of his contemporary Sadīd al-Dīn ibn Raqīqah he says,[18] 'As for verse in *rajaz* metre, I have never seen any physician in his time who was quicker in composing it than he. He could take any medical work and render it in the *rajaz* metre in an instant, staying loyal to its contents and doing justice to the beauty of its words.' But he does not quote any examples. After all, his book is not intended as a medical manual. There are, however, some poems with general medical advice, such as one by the same Ibn Raqīqah that begins:[19]

Beware of eating your fill, shun it!
  Digest one kind of food before eating another.
Do not have sex often, for by doing it
  continually one invites illness.
Don't drink water straight after eating
  and you will be safe from great harm,
Nor on an empty stomach and being hungry,
  unless you have a light snack with it …

Or Ibn al-Ṣāʾigh al-ʿAntarī's advice to his son, beginning:[20]

My dear son, memorize my admonition and act by it,
  for all medicine is gathered in the text of my speech:
Before all medication for a sick one, be concerned
  with preserving his strength from day to day.
Existing health is preserved with likes,
  but in opposites lies the cure of every sickness.
Have as little sexual intercourse as you can, for it is
  the water of life that is poured into wombs …

There are several short epigrams in this vein. It also happens that medical matters are discussed in poetic exchanges between physician and patient. The vizier Abū Ṭālib al-ʿAlawī sent a few lines to Ibn Sīnā, complaining of pustules on his forehead, and the latter replied with a poem recommending, among

17 See e.g. Mattock, 'The Medical Muse' (on an anonymous *urjūzah* on the symptoms of impending death).
18 Ch. 15.46.1.
19 Ch. 15.46.3.6.
20 Ch. 10.69.3.1.

other things, a purge, the application of leeches, and abstention from meat.[21] The famous Syrian knight and poet, Usāmah ibn Munqidh complained about his bad knees in a poem addressed to Muhadhdhab al-Dīn ibn al-Naqqāsh and asked for a balm,[22] which was duly sent. The vizier al-Qāsim ibn ʿUbayd Allah politely asked the physician-translator Isḥāq ibn Ḥunayn, in a humorous epigram, about the latter's bowel movements after having taken a purgative:[23]

> Let me know me how you were last night
> and how you felt,
> And how often the she-camel took you
> to the Empty Abode.

He alludes to the ancient euphemism taken from Bedouin practice of going a short distance into the desert for relieving oneself. The answer came promptly and politely:

> I was fine, happy,
> and relaxed in body and spirit.
> As for travelling, the she-camel,
> and the Empty Quarter,
> My respect for you made me forget,
> O goal of my hopes!

Self-mockery is common. The poet-physician Hibat Allāh ibn al-Faḍl, known as Ibn al-Qaṭṭān, describes his less successful visits to the privy.[24] Abū l-Ḥakam al-Maghribī, 'the joking doctor' pretends that his poetry has curative properties:[25]

> … So tell the people who think my medical skill will give them relief
> that it will benefit them if it is mixed with poetry.

---

21 Ch. 11.13.7.6 ('May God cure and banish the complaint …').

22 Ch. 15.13 ('My knees are at the service of al-Muhadhdhab …').

23 Ch. 8.30.5.

24 Ch. 10.68.2.6 ('Often have I gone to the privy …').

25 Ch. 15.8.13.

3 Poems on Physicians

Ibn Abī Uṣaybiʿah quotes an epigram by the great poet and prose writer Abū l-ʿAlāʾ al-Maʿarrī (d. 449/1057) in praise of the physicians of the past, beginning:[26]

> Blessings on that man, Galen,
> and the companions of Hippocrates …

Hippocrates (Abuqrāṭ or Buqrāt) and Galen (Jālīnūs) are very often mentioned. Ibn al-Budhūkh composed a poem in praise of Galen's books.[27] Isḥāq ibn Ḥunayn does so too, listing them alongside Aristotle and some others in a vaunting poem boasting of his medical expertise.[28] In poems in praise of physicians they are said to be 'the successor of Hippocrates in our age';[29] 'He has surpassed Hippocrates in knowledge and wisdom';[30] 'If Galen were alive in his era he would learn and revise under him'.[31] A nephew of the Andalusian anthologist Ibn ʿAbd Rabbih sent his uncle a poem saying that he had taken Hippocrates and Galen as his 'companions' during his phlebotomy, when the uncle had not responded to an invitation to be present at the procedure. Ibn ʿAbd Rabbih replied with a poem justifying himself by pointing out his nephew's miserliness: 'You found that Hippocrates and Galen would not eat and burden a host with expenses.'[32]

That even the great ancient doctors could not avoid death is a common motif in wisdom poetry and elegies: 'Hippocrates was not saved from death by his medical skill, … Galen could not avoid a natural death' (in an elegy by Yūsuf ibn Hibat Allāh on the physician Ibn Jumayʿ)[33] and an anonymous poem found on the flyleaf of manuscript B of *ʿUyūn al-anbāʾ* points out that Hippocrates, Plato, Aristotle, Galen and Ibn Sīnā all died of various diseases: the only true medicine is God's word. The same motif was used in an anonymous lampoon on Ibn Sīnā, alluding to two of his major works on philosophy:[34]

---

26 Ch. 5.1.33.
27 Ch. 15.10 ('How noble, books by Galen!').
28 Ch. 8.30.4 ('I am the son of those …').
29 Ch. 14.51.3.
30 Ch. 15.37.4 (line 3 of Ibn Abī Uṣaybiʿah's poem beginning 'My Lord, O Sharaf al-Dīn').
31 Ch. 15.45.3 (line 20 of poem beginning 'Her phantom came at night').
32 Ch. 13.24.3.
33 Ch. 14.32.4 (line 11 of poem beginning 'O my eye, let flow the tears').
34 Ch. 11:13.5.

This Avicenna, mankind's foe,
died wretchedly of constipation.
He was not cured by his own *Cure*
nor salvaged by his own *Salvation*.

4 Praise Poems, Elegies, and Vaunting Poetry

The most prestigious kinds of poetry were panegyrics (*madīḥ*) and elegies (*rithāʾ*): praising the living or the dead. Both genres are well represented in the book, especially in the entries on the author's contemporaries. Although many of them are full of clever conceits, striking imagery, learned allusions, and ingenious wordplay, they are not among the more appealing poems to a modern readership, especially when much of their rhetorical brilliance is lost in translation. The customary hyperbole, the unrelenting reference to 'lofty qualities' (*al-ʿulā*), 'excellence' or 'eminence' (*faḍl*), the endless series of words for generosity, munificence, benefaction, liberality, beneficence, bounteousness, etc., and the constant need of copious annotation: all this can make for tediousness. To the author and his colleagues, however, they were relevant, serving as tokens of friendship and esteem, an essential part of polite social intercourse between educated equals, between patron and dependent, or between subject and ruler. An anecdote told about Ibn al-Muṭrān tells how a poem could function as a way to introduce oneself to a potential medical patron.[35]

The panegyrics include, naturally, poems on physicians but there are also eulogies on others, such as odes on rulers: one by the polymath Fakhr al-Dīn al-Rāzī on Khwārazm Shāh, congratulating him on defeating an enemy;[36] a lengthy ode on Saladin by al-Jilyānī;[37] an ode on a high official in Baghdad by Ibn al-Qaṭṭān.[38] The opening of the last-mentioned poem, 'O woman who left me and did not care!', illustrates the ancient and very common convention of opening an ode with a lyrical or elegiac introduction about past love.

Among the elegies are a poem by Yaḥyā ibn ʿAlī ibn al-Munajjim on the death of the important scholar Thābit ibn Qurrah;[39] a poem by Ibn al-Shibl al-Baghdādī on his brother Aḥmad;[40] a short piece by the author on the Ayyubid

35 Ch. 15.23.3.
36 Ch. 11.19.6.2 ('Religion's pavilion has been extended …').
37 Ch. 15.11.2.1 ('A vigorous, astute man's comfort lies in embarking boldly …').
38 Ch. 10.68.2.2 ('O woman who left me').
39 Ch. 10.3.11 ('Ah, everything but God is mortal …').
40 Ch. 10.51.3 ('Extreme sorrow and joy is an ending …').

ruler al-Malik al-Ṣāliḥ Najm al-Dīn Ayyūb;[41] and another short poem by Sadīd al-Dīn Ibn Raqīqah on the death of a son.[42] Perhaps not surprisingly the last-mentioned piece seems to the modern reader more heart-felt than the others; but true feelings are not incompatible with clever poetic conceits: the poem ends as follows:

> You were perfect; then fateful death came to you. Likewise,
> an eclipse may come to a moon when it is full.

Akin to praise poetry is the ancient and very important genre of *fakhr*, boasting or vaunting poetry. Above, mention was made of Isḥāq ibn Ḥunayn's poem in which he claims intellectual descent from the ancient Greek physicians and philosophers. Ibn al-Ṣā'igh al-ʿAntarī, a poet-physician with a predilection for philosophical and esoteric themes, describes himself in high-flown, almost blasphemous terms:[43]

> … My mind is a niche, my soul is a glass
> that shines with the burning lamp of understanding,
> And my light is from the divine Light that is always
> cast on to my essence, without being poured out,
> And my oil is from the olive tree with its sweet oil,
> that is exalted above description in East or West …

This is an unmistakable allusion to the famous Qur'anic 'light verse' in which God's light is described in similar words. Poets commonly boast of their own poetic skills. The poet-physician Ibn Hindū describes how, after having abandoned poetry, he resumed it 'while rhymes slipped from my tongue, like a torrent sliding hurriedly from a hill', surpassing the masters of the past.[44] Some poems by Ibn Sīnā are a blend of boasting, complaint, and wisdom poetry (on which see below). In one of them he claims:[45]

> With which glorious feat could anyone be compared with me?
> With which noble deed could the nations imitate me? …

---

41 Ch. 14.56.3 ('Beware your time as much as you are able …').
42 Ch. 15.463.10 ('Dear son, you have left in my breast …').
43 Ch. 10.69.3.2.
44 Ch. 11.9.2.5.
45 Ch. 11.13.7.3 ('O encampment').

As for eloquence, ask me as someone experienced in it:
    I have been the tongue of yore in the mouth of Time …
The maiden of the sciences of truth was unadorned
    until understanding and the pen, with my exposition, unveiled her.

This poem together with the two that follow it is characterised by an archaicising diction and style, full of rare words and expressions. They may well be the three poems mentioned in an anecdote, earlier in the entry on him.[46] Rebuked by a grammarian, Abū Manṣūr al-Jabbān, for his deficient knowledge of Arabic, Ibn Sīnā takes a few years to study the Arabic lexicon and ancient poetry. Then he composes three poems, together with three prose epistles, claiming to have found them somewhere amongst the Bedouins. Abū Manṣūr, asked to explain their difficulties, falters, then realises he has been fooled.

5 Invective and Lampoon (*hijāʾ*)

The epigram on Ibn Sīnā quoted above is one of many poems, mostly short, that mock, ridicule, abuse, vilify, or satirise others. Ibn Buṭlān mocked his colleague, ʿAlī ibn Riḍwān:[47]

When his face showed itself to the midwives
    they turned on their heels in regret,
Saying (but lowering their voices for decency's sake),
    'Ah, if only we had left him in the womb!'

Salāmah ibn Raḥmūn is lampooned for his lack of medical skill in in epigram by a certain Jirjis al-Faylasūf ('George the Philosopher'), with the following punchline:[48]

… Three things enter at the same time:
    his face, a bier, and the man who washes the corpse.

A similar attack is made on the poet-physician Abū l-Ḥakam al-Maghribī:[49]

46 Ch. 11.13.3.13.
47 Ch. 10.38.3; see also below, under 'Translating poetry'.
48 Ch. 14.27.3.
49 Ch. 15.8.2.

> … Whenever he visits a patient in the morning
> he composes an elegy for him the same day.

Ibn Jumayʿ is mocked by a fellow physician:[50]

> … He cannot determine the urine of a sick man
> in the glass, even when he rolls it on the tongue.
> And the strangest of all is that he takes
> a fee for killing his patient, from the next of kin.

Arabic lampoons make frequent use of obscenity and scatology; there is relatively little of this in the book. An example of obscenity and scatology combined is this epigram by Ibn Hindū on an unknown person:[51]

> How strange, this Emir's constipation!
> How did he get so sick?
> Yet he gets a daily enema,
> administered by a prick.

A special kind of lampoon is self-mockery. An example by Abū l-Ḥakam al-Maghribī was quoted above; he made several more, such as epigrams describing how he got his facial scar by falling on his face when drunk.[52]

6 Wisdom Poetry, Complaint

The word *ḥakīm*, literally, 'wise',[53] very often serves as a substantive noun meaning 'sage, physician, or philosopher' – the senses are often hard to distinguish and they regularly overlap. It is not surprising, therefore, that our physicians pose as sages and philosophers, churning out prose maxims and saws as well as gnomic poems and epigrams in verse, of the kind called *ḥikmah*. Ibn Uṣaybiʿah quotes very many of them: countless short epigrams and some longer pieces. Gnomic lines and passages are also very common in other poetic genres, not-

---

50 Ch. 14.32.4.
51 Ch. 11.9.2.5.
52 Ch. 15.8.3 ('I fell on my face …' and 'Wine has left on my cheek …').
53 It should not be confused with *ḥākim*, 'ruler, sovereign', or *ḥakam*, 'arbiter' (both from the same prolific Semitic root, found in Hebrew e.g. in *ḥāḵām* 'sage, wise', which through Yiddish became *goochem* ['xooxəm], 'clever', in colloquial Dutch).

ably elegy and panegyric. Among the longer wisdom poems are notable compositions by Ibn al-Shibl[54] and Ibn Sīnā, such as his famous poem on the descent of the soul as a bird,[55] or poems on grey hair and old age.[56] Complaint (*shakwā*) about the times is a common ingredient, as in another poem by Ibn Sīnā.[57]

Many one-liners, some by famous poets, others anonymous, may be quoted as proverbs, to comment on an event or circumstance. On his death-bed, the Umayyad caliph ʿAbd al-Malik utters a line to his son: 'Many a person asking about us wants us to die, | male or female, while their tears are flowing in streams!'.[58] He is presumably quoting, but the poet is unknown. Ibn Abī Uṣaybiʿah, criticising a colleague for his ignorance, remarks: 'As the poet says: "He tucks up his robe to wade into deep waters | but the waves engulf him on the shore"'.[59] The poet, as our author probably knew, was the great al-Mutanabbī (d. 354/965), many of whose lines are eminently quotable as proverbs. On an elegiac note on the passing of time, he quotes, again without attribution, a line by Abū Tammām (d. 231/846): 'Then those years passed and those who lived in them, | and it was as if years and people were dreams.'[60]

7 Love Poetry (*ghazal*)

An ancient form of love poetry called *nasīb*, usually of the elegiac kind, reminiscing about past love, may open a longer poem, introducing other themes. Examples are found in some of the *qaṣīdah*s or odes quoted by Ibn Abī Uṣaybiʿah. Independent poems about love, whether sad and serious or frivolous, hetero- or homoerotic, are called *ghazal*. There are many such poems, mostly short. The beloved's sex is not always given explicitly, but very many poems are about young boys. Many love epigrams by Amīn al-Dawlah ibn al-Tilmīdh are quoted. Here is one on a woman, by Amīn al-Dawlah ibn al-Tilmīdh:[61]

---

54 Ch. 10.51.2 ('By your Lord! O revolving celestial sphere …').
55 Ch. 11.13.7.1 ('There descended to you from the highest place …').
56 Ch. 11.13.7.2 ('Have you not woken up from the night of childishness …?') and 11.13.7.4 ('Here is grey hair!').
57 Ch. 11.13.7.3 ('O encampment that has been made unrecognizable by events and antiquity!').
58 Ch. 7.6.
59 Ch. 14.27.2. For other such lines by al-Mutanabbī, see Ch. 15.31.1 ('And when I saw …'); 15.40.4 ('A wound cannot hurt a dead man'); 15.52.1.2 ('This is a prayer …').
60 Ch. 15.50.4.
61 Ch. 10.64.19.8.

I embraced her when night's dark veils were hanging down;
then I became aware of the coolness of her jewels, shortly before dawn.
So I kept them warm, fearing that they might awake her,
while taking care not to melt her necklace with my hot breath.

And another is on a boy, by al-Badīʿ al-Asṭurlābī, one of countless Arabic poems on boys with incipient beards that herald the imminent ending of a love relationship:[62]

Did pens stumble when writing the line of his cheek-down,
stretching the letters, the mole being the spot where they tripped?
Or did the line become a circle when the dot became
the centre of that orbit?
And, since his saliva is wine, are his teeth then
the pearly bubbles, neatly strung by the wine?[63]

8 Bacchic Verse

Ibn al-Ṣāʾigh al-ʿAntarī, whose daring self-praise was quoted before, also composed some less controversial religious verse, such as his piece on the virtues of the Shariah.[64] Nevertheless, he is one of the innumerable poets who extolled wine, asking for 'A choice old wine, like the sun in colour and luminescence, purer than imaginings', to be poured 'from the right hand of a fair white antelope, one of the Turks, like a full moon'.[65] Ibn al-Shibl compares wine in a glass to the spirit in a human body: both glass and body seem lighter with it.[66] Amīn al-Dawlah ibn al-Tilmīdh mentions the two glasses he spent his life with: the wine glass and the inkwell.[67] Abū Nuwās, the foremost Bacchic poet in Arabic, says that he asked the Christian doctor Jibrīl ibn Bukhtīshūʿ about the proper

62 Ch. 10.67.4.1. For other poems on cheek-down, see Ch. 10.56 ('O you whose incipient beard …'); 10.67.4.1 ('They said to me …', 'One with a shapely appearance …'); 11.9.2.2 ('A downy beard …', 'They said, "The lover's heart …"', 'Now there is a true testimony …', 'You, whose face …', 'It is bad enough …'); 13.58.4.8 ('Youthful passion …'); 15.46.3.8 ('A slender youth …'); 15.58.1 ('When your dark cheek-down …'); 15.58.2 ('Yes, my heart …').

63 The bubbles, caused by mixing the wine with water, are a common motif in Bacchic poetry.

64 Ch. 10.69.3.13 ('The Shariah, with its salvation …').

65 Ch. 10.69.3.9 ('Shaʿbān has come …').

66 Ch. 10.51.5 ('The glasses, that came to us empty …').

67 Ch. 10.64.19.3 ('With two glasses …').

quantity of wine that one should drink; the reply was four pints, since human constitution consists of four humours.[68]

Although drinking wine is forbidden to Muslims, wine poetry is very common in pre-modern Arabic literature and references to drinking are rife. Physicians recommend it. Ibn Hindū says that wine (here called *sharāb*, 'drink') can restore a ruin (*kharāb*).[69] Ibn al-Ṣāʾigh, in a 'wisdom poem' addressed to his son, tells him to divide his lifetime in three: seeking wisdom in the first part of his life, and looking forward to death in the last; but 'Earn money in the second, eat and drink wine, and do not desire wickedness'.[70] He describes a night with 'a bright-faced young gazelle' who prevented a hangover by taking pomegranate juice straight after drinking wine. He composed, however, also two epigrams explaining that he has abandoned drinking wine: not because the Shariah forbids it, but because it causes 'reason to rust' and it disagrees with his nature and character.[71] Shihāb al-Dīn al-Suhrawardī advises drinking wine in case the wine promised in the Hereafter turns out to be a false promise (since the Qur'an itself mentions it, this sounds distinctly heretical).[72] Ibn Raqīqah, in a 'medical advice' poem, says, 'abstain from drinking wine', but continues with '… Balance the mixing of your wine with water', as if he knew that his first recommendation would not be heeded.[73] Altogether, there are some forty poem in which wine is enjoyed or described in glowing terms, not counting the very frequent lines in which wine is used in erotic and other comparisons. Many poems combine Bacchic and amorous elements, such as when the wine-pourer is at the same time the poet's beloved:

> … Pour me wine, and you will cure the anguish of a heart
> that at night, since you left, is a companion of worries.[74]

Abū l-Ḥakam al-Maghribī imagines that the red wine is pressed from the Ganymede's rosy cheeks:

> … It is passed round by someone with Babel's glances,[75]
> with a mouth delicious to kiss, sweet of teeth.

68 Ch. 8.3.23 ('I asked my friend Abū ʿĪsā …').
69 Ch. 11.9.2.3 ('The army clergyman …').
70 Ch. 10.69.3.4 ('Divide your lifetime …').
71 Ch. 10.69.3.13 ('Since the fire …', 'I have abandoned …').
72 Ch. 15.18.2 ('Enjoy your blessings …').
73 Ch. 15.46.3.6 ('Beware of eating your fill …').
74 Ch. 10.56, by Saʿīd ibn ʿAbd al-ʿAzīz al-Nīlī.
75 i.e., with bewitching eyes.

He who is delighted by the wine's beauty would say,
'Has this wine been procured from his cheeks?'[76]

These lines are part of a longer, panegyric poem on a patron, which shows how genres and themes may freely be combined.

9 'Social Verse'

Some poems can be described as 'social verse', exchanged between friends and colleagues (in Arabic the genre is sometimes called *ikhwāniyyāt*, 'brotherly poems'). Sharaf al-Dīn ibn al-Raḥbī, writing from Damascus. sends a poem with friendly reproaches and praise of Damascus to Ibn Abī Uṣaybiʿah who is in Cairo:[77]

... Come back, then, to the Paradise on Earth: she has come forward
for her beauty to be revealed, in her new clothes,
And do not stay anywhere else if you have earned riches,
for life in any other place is not worth considering.

Ibn Abī Uṣaybiʿāh responds with another poem, saying how much he misses his friend and complaining about the stupidity of his Mamluk patrons:[78]

... In their ignorance they do not know the worth of a scholar,
which is not surprising in the case of ignorant people.
I came to someone in whose courtyard my merit was wasted. Would the
stupidity of the non-Arabs be aware of the intelligence of the Arabs?

He himself described Damascus in glowing terms in another poem, addressed to his colleague Muwaffaq al-Dīn ʿAbd al-Salām.[79]

---

76 Ch. 15.8.6 ('Love has called you ...').

77 Ch. 15.37.4 (lines 11–12 of the poem beginning 'Muwaffaq al-Dīn! What's this mindlessness of yours ...').

78 Ch. 15.37.4 (lines 21–22 of the poem beginning 'My lord, O Sharaf al-Dīn ...').

79 Ch. 15.54 ('Perhaps a time that has gone by in Damascus ...').

10 Ekphrastic Poems and Riddles

There is hardly any poem that does not contain a description of a place, wine, a person, an animal, nature, and so forth. A special genre is the short epigram describing an object, ideally in a witty manner, with a poetic conceit, a striking simile or metaphor, or a play on words. Ibn Abī Uṣaybiʿah quotes epigrams on a bowl with apples,[80] on marigold (or anemones),[81] on a horse,[82] on fleas (who seem to 'know more about veins than Hippocrates'),[83] on an astrolabe,[84] and on a brazier.[85] Several of these poems, among them the three last-mentioned, make a point of not naming the object explicitly, which turns them into riddles. Similar riddling epigrams are those on fish,[86] a cloud,[87] a needle,[88] on the shadow,[89] and on a doormat.[90] The solution is normally provided in introductory words such as 'He said, by way of a riddle on a needle'. A special kind of riddle is that on names. Often they are difficult to translate without extensive annotation. In a riddle on the name ʿUthmān it is said that if one-fifth is taken away eight remain. The name is written, in Arabic with five letters; if the first is taken away the four remaining letters can be read as *thamān^{in}*, 'eight'.[91] Other name riddles make use of the numerical values of Arabic letters;[92] the poem by Ibn Khalīfah on the name Abū l-Karam is a particularly forbidding example ('I give my life in ransom for a man half of whose name is the square root of Q, and whose fifth is L plus Y plus K. // The sixth of its letters multiplied by its half and a fourth part are like the eight nice ones …').[93]

80 Ch. 10.69.3.7.
81 Ch. 11.9.2.4.
82 Ch. 13.58.4.6.
83 Ch. 13.58.4.7.
84 Ch. 13.58.4.7.
85 Ch. 13.58.4.7.
86 Ch. 10.64.19.2 ('They have donned cuirasses …').
87 Ch. 10.64.19.9 ('One that comes over us …').
88 Ch. 10.64.19.9 ('One that earns a living …') and 10.65.3 ('One with a wide-open mouth in her foot …').
89 Ch. 10.64.19.9 ('A thing of bodies yet not itself embodied …').
90 Ch. 10.64.19.10 ('I spread my cheek for guests …').
91 Ch. 15.58.5 ('I asked all people …').
92 In the Latin alphabet this only works with c, d, i, l. m. v, x, but in Arabic every letter has a numerical value.
93 Ch. 15.51.10.3.

11 Humorous Poems

Wit and humour are often found especially in epigrammatic poetry, whether invective, satirical, or merely descriptive. Abū l-Ḥakam, already quoted several times above, a wit whom I dubbed 'the Joking Doctor', composed two long poems included in the book. One is a description of a disastrous party given by him, entitled *Maʿarrat al-bayt*, freely translated as 'The Domestic Scandal';[94] it is one of the relatively few classical Arabic poems that have a title. The other one is a poem put in the mouth of a friend and colleague, a long complaint on a badly made pair of shoes, in which numerous technical terms from the fields of logic and geometry are employed.[95]

Using such technical terms in poetry is to be expected from scientists and physicians. Al-Badīʿ al-Aṣṭurlābī, physician, philosopher-theologian, and maker of astrolabes, describes a comely youth in Euclidean terms: 'His cheek-down is an equator and his mole | a point on it, and his cheek a triangle.'[96] Speaking of his love for several people, he proclaims: 'My heart is the centre and they are to it | a circumference, and my passions are radii to it.'[97] Ṣadaqah al-Sāmirī uses the technical vocabulary of the syllogism in a scurrilous epigram on a triangular sexual encounter between a slave, his mistress, and his master;[98] in another piece his employs the well-worn procedure of using grammatical terms in an obscene sense.[99]

A common form of wit is punning, now frowned upon as unworthy of true poetry but immensely popular with Ibn Abī Uṣaybiʿah's contemporaries. Not all of them are meant to be funny: they are part and parcel of Arabic literary diction and style. The book includes countless examples by major and minor poets. Many epigrams are based on a play of words, as when Abū l-Ṣalt Umayyah makes 'my tears in streams' (*admuʿī l-hummaʿī*) rhyme with 'when they are with me' (*idh hum maʿī*).[100] Sadīd al-Dīn ibn Raqīqah composed a piece of four lines with nearly identical rhyme-words: *in jādū* ('if they are generous'), *injādū* ('support'), *anjādū* ('courageous men'), and *anjādū* ('highlands').[101] Obviously, such word-play is normally impossible to translate, but the reader who

94 Ch. 15.8.7 ('Any domestic scandal …').
95 Ch. 15.17 ('My plight is bewildering …').
96 Ch. 10.67.4.1 ('One with a shapely appearance …').
97 Ch. 10.67.4.3 ('My heart is divided …').
98 Ch. 15.47.2.1 ('Durrī, his mistress and his master').
99 Ch. 15.47.2.2 ('Ibn Qusaym, now you pretend to know grammar').
100 Ch. 13.58.4.6 ('I thought of their absence …').
101 Ch. 15.46.3.3 ('Let it be your nature …').

does not know Arabic ought to be aware of the punning, especially if it is the *raison d'être* of an epigram; in such cases the relevant words have been added in transliteration.

12 A 'Prophetic' Poem Ascribed to Ibn Sīnā

Unique in its genre, among the poetry in the book, is a strange poem of 52 lines ascribed to Ibn Sīnā that seems to predict the destructive Mongol campaigns in the 7th/13th century in considerable detail.[102] Ibn Abī Uṣaybiʿah has some doubts about the attribution, but this still earned him derision by a later author, al-Ṣafadī, for leaving open the possibility that Ibn Sīnā composed it. Al-Ṣafadī's reasons for rejecting it are twofold, logical and literary: Ibn Sīnā could not possibly have foretold such details, nor would he have composed such doggerel. One can only agree with al-Ṣafadī.

13 Attributions and Corrections

In the countless Arabic anthologies, biographies, histories, and many other genres that contain poetry, the poems are mostly attributed to their poets. As is clear from the above, however, it also happens often that lines or short poems are quoted anonymously: 'As someone said …', 'as the poet says …'. It is not always clear if the one who quotes is aware of the identity of the poet. One presumes that Ibn Abī Uṣaybiʿah knew that the often-quoted line, 'When Death plunges its talons in, | you will find that every amulet is of no avail' is from a famous poem by Abū Dhuʾayb al-Hudhalī (d. c. 28/649),[103] but one cannot be sure. I have tried to identify the poets of every anonymous piece but have not succeeded in every case; some lines and short pieces have always been transmitted anonymously and others have been ascribed in the sources to two or more poets, the truth remaining hidden. What is certain, however, is that Ibn Abī Uṣaybiʿah's attributions are sometimes wrong, due either to his sources or to a faulty memory. Whenever I discovered such a misattribution I have provided details in a note. He ascribes, for instance, an epigram to Abū l-Ṣalt Umayyah that is in fact by the much earlier Ibn al-Muʿtazz,[104] and another piece to al-

102 Ch. 11.13.7.9 ('Beware dear son, the tenth conjunction …').
103 Ch. 15.50.5.
104 Ch. 13:58.4.9 ('You, unique in flirtation …').

Ṣāḥib Amīn al-Dawlah that is by Abū Firās.[105] In numerous other cases it is difficult to decide between the attribution by our author and different ones in other sources.

It is also obvious that the text of poems contains errors. Some lines do not scan correctly or have faulty vocalisations, which may be due to a copyist, such as that of MS A, who adds very many vowels but whose knowledge of prosody leaves something to be desired. Other lines are clearly misremembered or wrongly copied. Many of such errors become evident when one reverts to other, mostly older, sources or to the collected verse (*dīwān*) of the poet in question, if available. This poses a dilemma to the editor and translator: should Ibn Abī Uṣaybiʿah's text be followed, or ought the correct, original version, as intended by the poet, be respected? Generally, I have opted for the latter course, always pointing out the divergence in notes. In some cases it is impossible to determine which is the original version, in which case Ibn Abī Uṣaybiʿah's version could be kept, except when it is clearly faulty. In notes at the beginning of each poem I have provided earlier (and sometimes later) sources, though not exhaustively in the case of lines and poems that are frequently quoted, nor have I given the many variants found in such sources, except when these could shed some light on the interpretation.

14 Translating Poetry

There are two basic approaches to translating pre-modern Arabic poetry: either a more or less faithful rendering in prose, without metre or rhyme, or a re-creation that results in a kind of modern English poetry, perhaps with some kind of metre (or even rhyme, even though that is unfashionable in many high-brow circles today). For our translations we have opted for the former, staying close to the original, yet trying to avoid sounding like a crib, but not afraid of some exoticising elements. Unusual idioms and allusions to matters not known to the average English reader are not glossed over or omitted; rather, they are explained in notes. As said above, untranslatable word-play can be shown by providing some transliterations in the text or in a note.

In a 'free' translation a translator is free to cheat, skirting over problems and omitting obscurities – of which there is an abundance in classical Arabic poetry. In our procedure this is as impossible as it is undesirable: scholarly integrity demands confessions of ignorance. The frequency of parenthesised

105 Ch. 15:49.6.2 ('Speak to this glorious lord …').

question marks and notes about problematical lines in the poems quoted in the book will eloquently attest to this.

As it cannot be denied that (as an eminent Arabist said to me once) much of the poetry in *ʿUyūn al-anbāʾ* is mediocre, attempting to craft high-quality, poetic English equivalents would be a waste of effort and would distort the nature of the originals. There is one genre, however, where I have occasionally given in to the temptation of producing 'poetry' or at least verse: it is light verse, lampoons or humorous poems, which in English call for metre and rhyme. Thus, after giving the more or less straightforward prose rendering of a mocking epigram by Ibn Buṭlān on his colleague Ibn Riḍwān, quoted above ('When his face showed itself to the midwives …'),[106] I decided to add a rhymed alternative version:

> As soon as the midwives saw his face
> they were distraught and left the room,
> And whispered (to avoid disgrace),
> 'We should have left him in the womb.'

The same has been done with an obscene lampoon by Ibn Hindū (quoted above) and the anonymous epigram on Ibn Sīnā's death (also quoted above).[107] Of the other long poem by Abū l-Ḥakam, about a rowdy party, also mentioned there, I have only given a rhymed version, copiously annotated. It is a revision of an earlier publication and it seemed undesirable to add a version in plodding prose.

---

106 Ch. 10.38.3.

107 For a rhymed version of the long poem by Abū l-Ḥakam al-Maghribī about the badly made shoes, mentioned above under 'Humourous Poems', see the selected and shortened translations of IAU in *Ibn Abi Usaybiʿah; Anecdotes and Antidotes. A Medieval Arabic History of Physicians. A new translation* (Oxford World's Classics).

CHAPTER 7

The Greek Chapters and Galen

Simon Swain

1 The Focus on Galen

Ibn Abī Uṣaybiʿah's grand sweep of physicians starts with ancient Greece. In this he is a man of his time, since knowledge of the ancients and their wisdom was an essential part of membership of the chattering classes. In medicine in particular the legacy of the ancients was all around, in practice, in theory, in anecdotes and examples of behaviour, and no history of physicians wanted to ignored the Greek origins of the profession. Following the ancients themselves, Ibn Abī Uṣaybiʿah traced the origins to the Greek semi-god Asclepius, who for Arabic authors was essentially an historical figure living before The Flood. From the first and second chapters concerning the invention of medicine and the role of Asclepius in it, Ibn Abī Uṣaybiʿah advances steadily and comprehensively towards late-antique Alexandria and the sources of medicine for his own culture and times (Ch. 6). In so doing he had spent nearly a quarter of his book on Greco-Roman antiquity, and well over a quarter of this material is itself focussed on the dominant figure of Galen (Ch. 5). The Arabic Galen – Jālīnūs – not only commands the first six chapters of the *ʿUyūn* but maintains a strong presence in the rest of the work. He is one of Ibn Abī Uṣaybiʿah's heroes, central to his own training as a physician, the last of the eight 'great master physicians' beginning with Asclepius, and in memorable words 'the Seal of physicians' (*khātam al-aṭibbāʾ*, Ch. 5.1.2, cf. 3.2).[1] Chs. 3–4 on the heirs of Asclepius, on Hippocrates and the leading ancient philosophers, prepare the way for this greatest physician-philosopher in Ch. 5, while Ch. 6 rounds off his influence and legacy in late antiquity and heralds the Islamic era where the Hippocratic-Galenic system was naturalized and remained the basis of theory and practice. The dominance of this professional system based on the legacy of antiquity is reinforced by Ibn Abī Uṣaybiʿah throughout the *ʿUyūn*, and he takes virtually no notice of alternatives, such as

1 Cf. Qur'an 33:40 Muḥammad as the 'Seal of Prophets' (*khātam al-nabiyyīn* rather than *khātam al-anbiyā*). See also Hershkovits & Hadromi-Allouche, 'Divine Doctors', 51.

'prophetic medicine', magic or astrology.[2] It is no surprise that Galen receives the longest individual biography.

Was Ibn Abī Uṣaybiʿah right to concentrate on him so much? When we look in broad terms at the development of Islamic medicine in the centuries up to his time, there is a progression from the age of translations from Greek in the period of the 2nd–4th (8th–10th) centuries, when Galen and other Greek authors had no peers,[3] to the period of the independent authority of canonical writers in Arabic from Abū Bakr al-Rāzī (d. 313/925) and Ibn Sīnā (d. 428/1037) onward. There arose a small but significant literature of criticism of Galenic thought, beginning with al-Rāzī's *Doubts about Galen*,[4] which focussed mainly on philosophical considerations or whether the physician really had to be a philosopher at all.[5] The celebrated outline of medicine in the *Kitāb al-Kulliyyāt* (*Generalities*)[6] of Ibn Rushd (d. 595/1198) followed much of the Galenic programme but also criticized Galen's logic and threw a purer Aristotelianism against him. Yet Ibn Rushd shows the tension involved in replacing and criticizing while also wishing to follow and honour: he was a close reader of Galen and wrote summaries of key Galenic works, albeit these once again show independence of thought.[7] There were medical advances too, no doubt helped by the more systematic training and observation available in Islamic hospital medicine, that showed Galen had not known everything.[8] But a physician still had to study, or claim to have studied, a good deal of the Galenic corpus in order to be taken seriously. If Ibn Abī Uṣaybiʿah's teacher, al-Dakhwār, ever heard 'the words of Galen ... concerning diseases and their treatment, or the fundamentals of medicine (*wa-l-uṣūl al-ṭibbiyyah*),' he would leap up and shout, 'That is medicine!' (*hādha huwa al-ṭibb*, Ch. 15.50.3). Al-Dakhwār (d. 628/1231) worked at the 'Great (Nūrī) Hospital' at Damascus and Ibn Abī Uṣaybiʿah records that it was there that he himself 'began to study the works of Galen' with him (*ib.*). Neither the brilliant systematization introduced by Ibn Sīnā nor the great

2 On these, Pormann & Savage-Smith, *Medieval Islamic Medicine*, ch. 5, and Savage-Smith, 'The Practice of Medicine as seen through the *'Uyūn al-anbā'*', below.

3 Age of the translations: see Gutas, *Greek Thought, Arabic Culture*, Endress, 'Die wissenschaftliche Literatur'; and for Galen in particular Ḥunayn ibn Isḥāq, *Risālah*, *Neue Materialen*, and *Galen Translations*.

4 Pines, 'Rāzī critique'; Pormann, 'Philosophical Topics', 22–32.

5 E.g. Ibn Hindū, *Miftāḥ* (Tibi), 35–36, (Manṣūri), 81–83.

6 Ch. 13.66.6 no. 4.

7 Cf. Ullmann, *Medizin*, 167. For modern editions of his summaries, see the notes to titles 16–23 of Ibn Rushd's works at Ibn Abī Uṣaybiʿah 13.66.6.

8 See Savage-Smith, 'The Practice of Medicine as seen through the *'Uyūn al-anbā'*', below.

Islamic invention of the teaching hospital seriously weakened the practical and emotional tie to Galen, and this is exactly the situation reflected in the *'Uyūn*.[9]

Ibn Rushd's summaries of Galen were part of a large production of such works that kept Galen available and useful in new forms, through epitomes, paraphrases, commentaries and other forms of interpretation. Most of these works were a good deal shorter than the original source texts and this made them easier to learn and easier to teach. A physician of the 10th century, Aḥmad ibn Abī l-Ash'ath (d. ca. 360/970), had the bright idea of introducing proper divisions into the main Galenic texts:

> It was Aḥmad who gave divisions to each one of the sixteen books of Galen by sentences, chapters, and sections and divided them such as none before him had done and thus provided a great aid to those who study the books of the eminent Galen. For it is now easy to find all that one seeks therein and his divisions are as landmarks which point to what one wishes to read and by which every section of the book, its contents, and purpose may be known.
>
> 10.46.1

But many others preferred to reduce the bulk of their reading. The production of adaptations in various compendious forms such as questions and answers or summaries had begun in late antiquity (e.g. the *Alexandrian Summaries*, see below), but is a very marked feature of the Islamic period. These 'small texts' show that practitioners had neither the time nor – perhaps – the education to immerse themselves in the full Galen but needed accessible and practical treatises to use in the real world. Ibn Riḍwān (Ch. 14.25; d. 453/1061), an Egyptian devoted to the books of the ancients and an important source for Ibn Abī Uṣaybi'ah, made a series of 'useful extracts' (*fawā'id*) from Hippocrates, Philagrius, Plato, Porphyry and especially Galen (Ch. 14.25.9 nos. 43–50, 95–96) – while taking care also to condemn lazy students who relied on shortcuts alone.[10] He and Ibn Zuhr *père* (Ch. 13.61; d. 525/1130–1131) objected to books expressing 'doubts' about Galen and penned a surviving refutation of Rhazes.[11]

9 On the rareness of criticism of Galen, see Bürgel, *Ärztliches Leben*, 412–416, with interesting comments from Maimonides, and his study of Ibn Rushd, Bürgel, *Averroes, 'Contra Galenum'*.

10 *M. fī Sharaf al-ṭibb*, MS Istanbul, Süleymaniye, Hekimoğlu Ali Paşa 691, fol. 120^{r} 7–9 'summaries and commentaries … are the main reason why the best qualities of medicine have become extinct: students waste many years of their lives studying them.'

11 Cf. Ch. 14.25.9 no. 51; Ch. 13.61.4 no. 4.

Both original thinkers and writers, they show that Galen's popularity and utility was not at risk from modern methods of teaching and learning.

By the 13th century the large number of high quality, standard works that reduced the importance of the original Galen while advancing knowledge of the Art took their place alongside a Galen now available in new forms. Ibn Abī Uṣaybiʿah acknowledges that the Alexandrians made a significant contribution to the invention of summary literature (Ch. 6.5.1). He does not share reservations about laziness nor mention the negative, anti-Christian allegations about the decline of medicine before its rescue by the caliphs that were played into accounts of the production of summary literature.[12] The only low points of medicine for Ibn Abī Uṣaybiʿah occurred before its revival by Hippocrates and its second rescue by Galen, both of whom in different ways emphatically re-established the highest standards.[13] Once Galen had trounced his opponents, there was no looking back. Ibn Abī Uṣaybiʿah's focus on Galen reflects for sure the currency of Galen. There is no nostalgia for a lost golden age. In Ch. 15 of the *ʿUyūn*, for example, he shows full well his belief in the quality of the physicians of his own day and the vitality of medicine in his generation. Overall the *ʿUyūn* is a celebration of a progress that shows no sign of stopping. Galen's function as the 'seal' of medicine is entirely compatible with this, hence Ibn Abī Uṣaybiʿah's desire to share with his readers the large amount of bio-bibliographical information the Greek had left behind. The core features of the Galenic system, except in so far as they appear in the contents of certain books, were not of interest in this literary history, nor did his readers need instruction in them. In accordance with what he states in his Preface about the professional and moral excellences of the discipline's stars, Ibn Abī Uṣaybiʿah's task in the 'Greek' chapters was to present model physicians, stories and legends about them, and of course useful bibliographies of their writings. For this purpose, the best examples were provided by Hippocrates and Galen alongside the five 'most distinguished philosophers among the Greeks'. These form the subject matter of Chs. 1–4.

---

12 See Gutas, 'Alexandria'. Cf. at the end of this chapter on Ibn Abjar al-Kinānī.

13 Ch. 4.1.1 Hippocrates, 5.1.2 on Galen's confrontation with *al-aṭibbāʾ al-sūfisṭāʾiyyūn*, meaning the Methodists (also 4.1.10.4).

2 The Origin and First Appearance of the Art of Medicine (Chapter One)

To reach Galen, Ibn Abī Uṣaybiʿah first had to traverse history from Asclepius; and to write history, he needed a source. Finding one for the very start of medicine was not a problem. For in a work that Arabic writers regarded as authentic but modern scholars do not, Galen himself had given information about the beginnings of medicine and in particular the nature of its originator, Asclepius. The work in question is a commentary on the famous *Oath* of the Hippocratic corpus. The *Oath* itself is a short piece that takes its place alongside several 'deontological' writings on the obligations and self-presentation of the physician to his colleagues and publics.[14] Ibn Abī Uṣaybiʿah makes pointed use of it in Ch. 4 when he wishes to outline the ideal Hippocratic. Islamic physicians built on this literature with their own crafted presentations of the physician's conduct.[15] The commentary is in fact unlikely to be by Galen, for it is never mentioned in the Greek tradition and its content is unlike any of the genuine Hippocratic commentaries.[16] But it was written in the name of Galen[17] and absorbed into the corpus like a number of other works, and was thus available in the ninth century for translation by Ḥunayn ibn Isḥāq.[18] The beginning of the *Oath* ('I swear by Apollo the Physician, Asclepius, Health, Panacea, and all the gods and goddesses …') is formulaic – for the work is not religious – but it gave Galen the Commentator the cue to discuss in detail both the role of Asclepius, which Ibn Abī Uṣaybiʿah draws on for Ch. 2, and the origins of medicine, which he uses from the outset in Ch. 1.

Ibn Abī Uṣaybiʿah was fortunate in having another of his heroes, the great translator and expert on Galen, Ḥunayn ibn Isḥāq (Ch. 8.29), as a guide to the difficult material of the commentary. Although Ḥunayn inserts explanations into his Galenic translations from time to time,[19] the extensive notes to

---

14 It may be earlier than some of these writings, which are Hellenistic (i.e. third to first century BC), and is dated to the 'late fifth or early fourth century' by Craik, *Hippocratic Corpus*, 149.

15 The best known text is *Practical Ethics of the Physician* (*K. Adab al-ṭabīb*) by Isḥāq ibn ʿAlī al-Ruhāwī (Ch. 10.57; mid to late 3rd/9th century).

16 For these, see Ch. 5.1.37 nos. 87–102, 122 [= 185], cf. 155, 156, 157 [= 184].

17 Cf. the fragment cited at Ch. 1.6: 'Pergamum, which is my native city'. Galen famously complains at the start of *On My Own Books* of a supposititious work found in a bookshop at Rome and inscribed, "The Physician, by Galen". It is not out of the question, however, that the commentary on the *Oath* is genuine.

18 Ḥunayn's biography is at Ch. 8.29.

19 Vagelpohl, 'Translator's Workshop'.

the commentary on the *Oath* reveal his and his readers' vivid interest in the topic. When Ibn Abī Uṣaybiʿah says, 'Let us begin by reviewing Galen's account, together with material we have adduced, in an effort to pin down these various divergent views', i.e. about the genesis of medicine (Ch. 1.1), he makes it clear that the structure of the chapter is his;[20] but in his review of Galen's meaning he practically merges with Ḥunayn. Nor are Galen and Ḥunayn easy to distinguish as Ibn Abī Uṣaybiʿah sets out the basic questions and positions that follow: medicine was, he says, either eternal or created; if created, it was either created when man was or it was invented later; if invented later (the majority view), it was inspired by God or was the discovery of man.

Galen/Ḥunayn deals with human invention first so as to emphasize, it seems, the subsequent remarks on the role of the divine ('Among those who hold that the art of medicine comes from God …'), which inspired men through dreams and also inspired them to gain 'practical experience', an emphasis underscored by direct quotation from Galen that 'God … created the art of medicine and inspired man with it'. The scene set, Ibn Abī Uṣaybiʿah turns to a trusted source, Saladin's famous physician, Ibn al-Muṭrān, to bolster the arguments for human ingenuity. Ibn al-Muṭrān concludes a long discussion by stating that, although Galen and Plato have indicated that divine inspiration was behind medicine and Asclepius, it is nevertheless human ingenuity that counts and to deny this 'is an error' (1.2). After a short aside Ibn al-Muṭrān is quoted again on the contribution of experience and reasoning and we are left with Ibn al-Muṭrān's conclusion that Galen in his book on the invention of the arts[21] 'in essence says nothing more than I have' (1.3.2).

The Galen who was presented to the Abbasid translators was the product of a number of Christian filters, as we see on many pages of Ch. 5 dealing with his chronology. This did not worry Ibn Abī Uṣaybiʿah, even if he was conscious of it. It is noteworthy too that the main arguments in Ch. 1 are given over to Galen through the medium of two Christian commentators, Ḥunayn and Ibn al-Muṭrān.[22] With their help Ibn Abī Uṣaybiʿah is in a position to let Galen propose both divine inspiration and human ingenuity as the constituents of medical knowledge (1.4 'the art of medicine must have been, for the

20 Cf. Brentjes, 'Narratives'; in general and still the authoritative discussion, Rosenthal, 'An Ancient Commentary'.

21 *De constitutione artium*, in three books, of which the book on medicine survives as *De constitutione artis medicae*.

22 In the biography of Ibn al-Muṭrān Ibn Abī Uṣaybiʿah reports Ibn al-Muṭrān's conversion to Islam from al-Qifṭī in a passage not preserved in the surviving version of the *Taʾrīkh* (Ch. 15.23.2).

most part, broadly consistent with the above discussion', i.e. divine inspiration *and* human experimentation), and he is now free to go into details. He makes a quick tour of various signs of divine origin (1.5), then turns to the promising topic of dreams (1.6 'in second place'), starting of course with remarks acknowledging inspiration from dreams in Galen's *On Bloodletting*,[23] *Method of Healing*, and *Commentary on the Oath*, topped up (1.7) by citations of Oribasius (d. ca. 390/400), a commentary on Galen by Ibn Riḍwān, and the *Facilitation* by Ibn Zuhr *fils* (Avenzoar). Following this we are offered a long account (1.8 'in the third place') illustrating the role of 'serendipity' (*al-muṣādafah*), which is allegedly taken from one of Galen's own sources, Andromachus the Younger, confirmed (1.9) by a story told by al-Tanūkhī and an anecdote about the pharmacologist, Philo of Tarsus, whose famous analgesic was known solely through Galen's account of it in *On the Composition of Drugs by Places*. Fourthly (1.10) Ibn Abī Uṣaybiʿah describes imitation of animals as a source of medical knowledge, citing Rhazes and Galen (*On Enemas*).[24] Finally (1.11 'in fifth place') he discusses divine inspiration in animals, citing among others Dioscorides. He concludes (1.12) that, because man stands 'at the apex of the animal kingdom', he must have been divinely inspired and thenceforward applied experimentation and reasoning, with the result that he transformed his instincts into the rules and principles of medicine.

This flagship chapter is a beautiful essay on Ibn Abī Uṣaybiʿah's profession, its antiquity, and its centrality to human life. It is dominated by Greek sources or sources imbued with Greek perspectives, and with their help the question of the origin of medicine announced at the start as 'difficult' – so said Galen – is elegantly resolved to the neat position where the deity inspires and man gains experience. Ibn Abī Uṣaybiʿah steps aside from more polemical elements to the discussion of the role of medicine, which claimed it opposed God's will.[25] For him medicine was quite in tune with the divine. This position was the perfect basis for moving on in Ch. 2 to Asclepius, a very special type of human being who was, we are told, most people's idea of 'the first known physician' and a good empiricist to boot (2.1.1).

---

23 I.e. the *De curandi ratione per venae sectionem*.

24 *De clysteribus et colica*, a lost pseudo-Galenic work. Cf. Ch. 5.1.39 for Ibn Abī Uṣaybiʿah's introduction to this and other such writings.

25 Rosenthal, 'Defense of Medicine', citing for the defence ʿAbd al-Wadūd (cf. *ʿUyūn* 10.66.3), Ibn Hindū (biography at 11.9) and the Christian physician Ibn al-Quff (15.60, the last biography).

3 Physicians Who Perceived the Rudiments of the Art of Medicine and Initiated the Practice of That Art (Chapter Two)

Ch. 2 has similarities with Ch. 1, for almost all of the second half is direct quotation from Galen/Ḥunayn. The first half depends closely on works by two Muslim writers, Abū Maʿshar's famous *Book of Thousands* and al-Mubashshir ibn Fātik's *Choicest Maxims and Best Sayings*, both of which themselves draw on ancient 'Hermetic' literature. For al-Mubashshir, an author with deep classical interests, Asclepius is the pupil of Hermes the First, who features as the fount of religion and science. In Abū Maʿshar Asclepius becomes the pupil of a third Hermes. The tradition of no fewer than three Hermes-figures in Arabic, as we see it in Abū Maʿshar, appears to rest on a conflation of the pagan 'Hermetica' with the Christian chronological traditions that lie behind his historically informed astrology and made him so useful for those intent on understanding ancient times.[26] Again it is Greek tradition that dominates Ibn Abī Uṣaybiʿah's presentation and he draws on several other real or inauthentic (Ps.-Plato, *K. al-Nawāmīs*) Greco-Roman sources, not always accurately or from originals, including Galen,[27] to confirm the divinity of Asclepius. The first half of the chapter ends appropriately with a quotation from a pagan, the Sabian scholar Thābit ibn Qurrah (via Ibn al-Nadīm's *Fihrist*, as often in Ibn Abī Uṣaybiʿah), which links Asclepius to his descendant Hippocrates and the revival of medicine (2.1.5).

In the second half of Ch. 2 Ibn Abī Uṣaybiʿah reverts to Galen's commentary on the *Oath*. In his *Risālah* Ḥunayn specifies the addition to his translation of 'annotations (*sharḥ*) I made on passages judged difficult (*mustaṣʿabah*)'.[28] Behind this judgement one hears the voices of pupils (including his son Isḥāq, cf. below) and friends requesting explanations of Galen's lengthy description of a Greek god. In the passage as we have it, Ḥunayn battles to bring the descriptions back to medical science. Dryness (an allusion to the main ancient etymology of Asclepius[29]) was easy to deal with: 'an individual dies only when there is a preponderance of dryness and cold'. Or take the dull gloss of the teacher on Galen's assertion that Asclepius was held divine because physicians needed to have the 'ability of the diviner': 'this is a reference to medical prognosis'

26 See van Bladel, *The Arabic Hermes*, esp. ch. 4.

27 Orosius, could not have made the mistake Ibn Abī Uṣaybiʿah attributes to him most carelessly of thinking that Christianity preceded paganism. In fact the passage (Ch. 2.1.2) is taken from Ibn Juljul, who was not so casual himself.

28 Ḥunayn ibn Isḥāq, *Risālah*, 40/32.

29 Cf. Ch. 2.1.2.

(2.1.6.1). But when it came to Galen's 'discussion of pictorial representations of Asclepius' (2.1.6.2–3), Ḥunayn struggled partly because Galen's Commentary did the work itself but partly because the material was not easy to tie to science (cf. the prosaic comment at 2.1.6.2 that, 'the marshmallow plant has warming properties in some degree, etc.'). None of this matters to Ibn Abī Uṣaybiʿah for he is in the company of two favourites and he concludes with a section of the commentary that is a paean to good health (2.1.6.4).

His job is completed by the briefest note on Apollo (2.2). He makes no connection between Apollo and divinity. He might have done so given that the deformed name transmitted here (*A-y-l-q*) is identified earlier as a deity (2.1.2 'Temple of *A-y-l-q*, which was dedicated to the sun'). But he is now in a different register, following Ibn Juljul and a Judaeo-Christian chronology that placed Apollo alongside the Old Testament leader, Barak the Judge. His mind is on the legacy of Asclepius and the last of his line in Hippocrates, the man who made medicine a fully-fledged discipline.

4 The Greek Physicians Descended from Asclepius (Chapter Three)

In Ch. 1 Ibn Abī Uṣaybiʿah had concluded that medicine developed from divine inspiration and human experience and at the start of Ch. 2 he had highlighted the empiricist credentials of medicine's founder, Asclepius. In case readers have not been following, he begins Ch. 3 with a reminder that medicine depended on 'practical experience'. Asclepius' successors abided by 'the empirical system' (3.1, cf. 3.2), for the time being at any rate. For information about these successors Ibn Abī Uṣaybiʿah had to turn to a new source and one he regarded as unreliable, John the Grammarian (Yaḥyā al-Naḥwī). John's life itself is narrated in Ch. 6 on the late-antique Alexandrian school of medicine. Thirty five works of medical and philosophical commentary are attributed to him there (6.2).

There is no reason to doubt that a real late-antique commentator called John did exist. It is also quite certain that in the Arab tradition this medical John, who is unknown except in Arabic, was confused with the John the Grammarian who is more usually known as John Philoponus ('Lover of Toil'), i.e. the distinguished Christian commentator on Aristotle. The philosophical John died around 575 but a transmission error in Arabic of a date he gives in one of his commentaries catapulted him into the era of the Arab conquest of Egypt.[30] The confusion

30 See Ch. 6.1.2 with n. 16: the Arabic of his commentary on the *Physics*, as quoted by Ibn al-Nadīm, placed him in AD 627/28 rather than the 517/18 of the Greek text.

was welcome to anyone looking for a thread from Alexandria to the Arabs, as we shall see. In the list of John's books in Ch. 6 (quite a few being authentic works of John Philoponus), Ibn Abī Uṣaybiʿah makes no reference to John the Grammarian's chronology of physicians, the work he uses in Chs. 3–5 for dates and names. The omission might have been the result of familiarity or disapproval, but it is also clear that Ibn Abī Uṣaybiʿah knew this chronological account only through Isḥāq ibn Ḥunayn's *History of Physicians*, a work surviving as an epitome and partial text in the *Ṣiwān al-ḥikmah*.[31] In his introduction to the vizier Abū l-Ḥusayn al-Qāsim ibn ʿUbayd Allāh (d. 291/904),[32] Isḥāq follows Galen's commentary on the *Oath* for the origins of medicine. He does not acknowledge the source; but we may assume he expected readers to have known it. He had a problem, however: his addressee is known to have been uncompromising, and had instructed the philosopher to tell him who was the first and who was the last physician: 'Let me have it in a hurry and do not postpone' (trans. Rosenthal). Thus knowing that John the Lover of Toil did a good job and had been around when the Arabs came, Isḥāq looked to the chronology of physicians by John for the right answers. That this John had nothing to do with Philoponus was something he was presumably unaware of. In any case the credentials of Philoponus discouraged him from having doubts.

Ibn Abī Uṣaybiʿah makes good use of John's information in Chs. 3–4. But he was not happy with it. The discontent boils over in Ch. 5. As we shall see below, the life of Galen posed a problem that is true of all ancient physicians: a dearth of independent accounts. Hippocrates had lived long enough ago in the past for the biographical machine to produce fictitious letters and a decent quantity of semi-fictional biography, which are the typical productions of the Hellenistic or early Roman imperial eras.[33] Galen, who lived into the 3rd century AD, was a little late for this and anyway gave too many facts about himself in his often self-referential writings. At some point in late antiquity somebody saw the opportunity to pull off a master stroke by inventing a few additional

31 Rosenthal, 'Isḥāq b. Ḥunayn' (epitome, including Isḥāq's introduction); Dunlop, *Ṣiwān al-ḥikmah*, §§19–26 for a fuller text including numerous mainly unclear Greek names (cf. Ibn Abī Uṣaybiʿah Chs. 3–4) but lacking Isḥāq's preamble. Isḥāq's biography is at *ʿUyūn* Ch. 8.30.

32 *EI*[2] art. 'Wahb', no. 3 (C.E. Bosworth). Cf. Ch. 8.30.1, 5 for al-Qāsim ibn ʿUbayd Allāh's acquaintance with Isḥāq. One of the interlocutors in the discussion pictured in Isḥāq's forword is ʿAbd Allāh ibn Shamʿūn, to whom Isḥāq addressed a treatise on preserving one's memory (8.30.6 no. 11).

33 Including Soranus' Life of Hippocrates, which survives in a number of versions: Pinault, *Hippocratic Lives*; Smith, *Hippocrates*.

facts about him in order to appropriate him for Christianity. The potential relationship of the two greatest healers – Galen and Jesus – was a story too good to ignore, not to mention the dividend for true believers of a medical bent. It is possible that Galen's use of a book ascribed to the last of the Ptolemies, Queen Cleopatra, in his *Composition of Drugs by Places*[34] assisted the story, for if we wish to be vague about chronology, she is almost in the right time zone. It is highly likely that the chronological John had made Galen her pupil, and therefore that John is indeed the chronological culprit.[35] But the Christian Isḥāq took things a stage further. At the end of his *History* and speaking in his own person he pegs the dates of the physicians to the dates of Abraham, Moses, and Jesus. From this it is apparent that Galen in his 20s or 30s overlapped with Christ and was, as others developed the story, at the right age to be the uncle of Paul the Apostle (or Luke).[36]

The effect of this linkage was highly dangerous: medicine was sectarianized. We shall see below how Ibn Abī Uṣaybiʿah dealt with the case of Galen in particular. What is relevant now – since John is so important in the early chapters we are discussing – is Ibn Abī Uṣaybiʿah's view of John. In Ch. 5.1.3 he swiftly demolishes the chronographer's neat division of the lives of the physicians into years of study and years of practice by quoting Galen himself.

> The length of Galen's life was 'eighty-seven years. He was a youth and a student for seventeen years and a scholar and a teacher for seventy years'. So runs the account of John the Grammarian. In the same way I have based myself on John's reports for the divisions in the lives of the previously mentioned great master physicians down to my own time with respect to their studying and their teaching. This matter requires careful scrutiny for information on it cannot be restricted ... Indeed, logic dictates that some of it is inconceivable and this includes his information that Galen was a youth and a student for seventeen years and a scholar and a teacher for seventy years. The only way to check what he says is to use the reports of Galen himself.[37]

34 See Ch. 5.1.21.1 and note.

35 So in al-Mubashshir ibn Fātik, quoted Ch. 5.1.21.1 ('studied medicine with a woman called Cleopatra'). Cf. Rosenthal, 'Isḥāq b. Ḥunayn' 69/79 with the fuller version of the text at Dunlop, *Ṣiwān al-ḥikmah* ll. 364–365, where she is named.

36 See Ch. 5.1.4 with notes.

37 John's 87 for Galen is no more reliable than Alexander of Aphrodisias' remark that Galen had a Socratic revelation at 80 (Dunlop, *Ṣiwān al-ḥikmah*, l. 100).

Ibn Abī Uṣaybiʿah then begins a lengthy session refuting both John and Isḥāq for their assumptions and sloppy scholarship. The reader who has learnt so late that John is not a good source may want to know why Ibn Abī Uṣaybiʿah used him in the first place. Ibn Abī Uṣaybiʿah does raise the possibility at Ch. 3.5 that John is not accurate but lets it pass, and first readers would certainly miss any doubts. John was indispensable for supplying the names of Greek physicians and Ibn Abī Uṣaybiʿah had no reason to doubt these were real. He does not question in any way the sequence of the 'great master physicians' that John records since he had no evidence to the contrary.

Thus it is that he sets out in Ch. 3 the scheme of Ghūrus, Mēnas, Parmenides, Plato the Physician, Asclepius the Second, Hippocrates and Galen. The origins of this scheme are obscure. The presence of the real pre-Socratic philosopher Parmenides and the Plato doublet indicate philosophers were on the mind of its inventor, as they would be for anyone trained in the Galenic tradition. In which case, and given the level of deformation Greek names can suffer in the process of transmission, the philosophers Anaxagoras and Anaximenes may lie behind the forms we have somewhat arbitrarily vocalized as Ghūrus and Mēnas.[38] The result is fantastic. Its purpose was to establish the combination of experience and logic that typify the Hippocratic-Galenic physician. Menas concluded that the 'empirical system ... should be supplemented with analogy (*qiyās*)'. Parmenides (3.4) takes this too far claiming only 'analogy' (deductive reasoning) is useful. Our inventor knew the basics of the doxographical tradition and names as Parmenides' pupils physicians associated with promoting both the two fundamental approaches of Empiricism and Rationalism and, third, the Galenic bugbear of Methodism: Acron (an older contemporary of Hippocrates who is no more than a name to us), Diocles (a major figure of the 4th century BC), and Thessalus (a physician influential in 1st century AD Rome). Only in John's fantasy world could these three become Parmenides' students. It is Plato the Physician (3.5) who comes to the rescue: by reviewing earlier writings he realizes (as some ancient texts averred) that both experience and reason are necessary. At this point Ibn Abī Uṣaybiʿah pauses to note that, if there is truth (*ḥaqīqah*) in John's account, the view that Hippocrates was the first to write down medicine must be wrong. The problem thus raised is for the time being shelved, which is odd as Ibn Abī Uṣaybiʿah clearly did not believe John on the specific matter.[39] Asclepius the Second is then brought on (3.6) to transition us to Ch. 4 and the real medicine of Hippocrates.

38 Anaxagoras actually occurs in full in 3.5 (and is a likely error for the physician Praxagoras), and at Ch. 4.1.10.1 we again find *gh-w-r-s* and read (Prax)agoras.

39 Cf. Ch. 4.1.2 the report of Ibn Riḍwān that medicine was taught orally before Hippocrates

At this point in Ch. 3 ('At his death, Plato left six pupils …'), as in the rest of the chapter, John serves up a mishmash of names, some of which are obvious, real forms but are not necessarily identifiable. There is no sense to these: some are solidly Greek, some (as far as we can tell) Roman (Furius, Gallus, Marcus, Matthaeus, Maximinus, Sergius, Severus). A majority cannot be construed. If one assumes the names were all real and found in some lost doxographical account, it is clear that our author has mixed them up and defied any chronological reality, whether by will or by ignorance.

One gets the feeling that the names of Ch. 3 partially function as decoration for the lives of the eight great master physicians in John's scheme. The overall level of fiction or confusion and the Christianizing tendency may reveal in John the chronographer a non-Greek, Syriac author, who was untroubled by historical accuracy and mainly interested in claiming medicine for his religion. Isḥāq gives no information about the language he read John's work in.[40] But it would be strange for an Alexandrian Greek physician to have produced such nonsense. It is surely very doubtful that this unchronological John is the same John who wrote commentaries on Galen (Ch. 6). These are Greek works in origin and no Greek author would have come up with John's farrago. The truth of his identity is beyond recovery, but it is clear that those who look to Ibn Abī Uṣaybiʿah and John for a chronography of real lost Greek physicians will be disappointed. Quite a few names can be guessed at with varying degrees of plausibility. Not a few are genuine and identifiable, but all in all John's grasp of dates was at least weak.

5 Greek Physicians to Whom Hippocrates Transmitted the Art of Medicine (Chapter Four)

Ch. 4 is a complex achievement. Hippocrates takes up about one fifth of the chapter. We then have accounts of a number of physicians between him and Galen. This in turn is followed by a long section on the philosophers concentrating, Ibn Abī Uṣaybiʿah says (4.1.11.4), on Empedocles, Pythagoras, Socrates, Plato, and Aristotle, and including important figures such as Theophrastus and Alexander of Aphrodisias. The philosophical list is determined by Ṣāʿid al-

decided to write it down, 4.1.9.1 Ibn Abī Uṣaybiʿah on the fact that 'Hippocrates was the first to set down the art of medicine in writing'.

40 Note Ibn Abī Uṣaybiʿah's report that John was *al-Askalānī* (Ch. 6.1.1), i.e. from Ascalon in Palestine. The spelling of the town, against the normal *ʿAsqalān* in Arabic, perhaps reflects the Aramaic spelling which begins with the *ʾĀlep̄*.

Andalusī's *K. Ṭabaqāt al-umam*, and all its members were well known to Arabic authors. Aristotle, the embodiment of logic, fills not far short of one third of the whole.

The philosophers have nothing much to do with medicine or physicians, but their presence is easy to explain. Galen emphasized the importance of philosophy to the practice and theory of medicine, especially Aristotelian logic. The leading physicians of the Islamic Middle Ages often combined philosophical and medical learning. The tools of Aristotle were the key parts of Christian and Muslim learning. The philosophers Ibn Abī Uṣaybiʿah focusses on were all known but Ibn Abī Uṣaybiʿah's readers surely expected to learn more about them from him. He in turn enjoys telling them. For Pythagoras and Aristotle in particular he had excellent sources in the form of Porphyry's history of philosophy and Ptolemy's life of Aristotle (see below). He does not stint.

It is Hippocrates alone who is named in the title of the chapter. Ibn Abī Uṣaybiʿah offers biographical details that are similar to the fact and fiction preserved in ancient sources. But biography is not quite his goal. At 4.1.2 he turns to one of his most valued Arabic sources, the 11th-century Egyptian physician Ibn Riḍwān, to affirm that Hippocrates was the first to promulgate medicine in writing. The quotation, from an unknown work,[41] finishes with Hippocrates' concern for the morality of medical students and the behaviour of practitioners. The following generous citation by Ibn Abī Uṣaybiʿah from three deontological works (*Oath, Law, Testament*) is extremely important for the *ʿUyūn* because here for the first time Ibn Abī Uṣaybiʿah sets out his key expectations about the conduct of a medical doctor in any period. Examples of Hippocrates' 'practice and treatment' follow along with his establishment of the rules (*qawānīn*) of medicine, rejection of royal patronage, treatment of the poor, and continence ('embracing virtue in hardship and distress', 4.1.4–6). The section ends with the description, according with his own injunctions, of Hippocrates' appearance and comportment, quoted from Mubashshir ibn Fātik (4.1.7).

Two other new features of the life of Hippocrates recur in subsequent biographies: lists of sayings and books. His fifty-three 'wise sayings' are divided into three parts. First is a section that has some overlaps with material in the *Ṣiwān al-ḥikmah* but is of unclear provenance (4.1.8.1), followed by sayings preserved in Ḥunayn's *Nawādir al-falāsifah* (4.1.8.2),[42] which are enlivened with the longish discussion of 'passionate love' (*ʿishq*) drawn from there (4.1.8.2–3). The third section is taken from Mubashshir ibn Fātik's *Choicest Maxims*. The

41 Cf. the lost *M. fī madhhab Abuqrāṭ fī taʿlīm al-ṭibb*, Ch. 14.25.9 no. 21.

42 Note Ḥunayn's non-standard translation of the famous first maxim of the *Aphorisms*, Ch. 4.1.8.2 no. 33.

content of this sayings material, blending moralizing sententiae with more 'scientific' pronouncements, presents Hippocrates the man of virtue and expertise. It consolidates the picture that has emerged and prepares us for the fruits of his work, his books, which combine (4.1.9.1) 'enigmatic language' (*lughz*) – the often obscure material of works like *Aphorisms* and *Epidemics* – with conciseness and explanatory information.

The canon of Hippocrates' twelve most famous books is almost certainly derived from Ibn Riḍwān,[43] a source who is used extensively in Ch. 6 for details of the Alexandrian curriculum. The information about each of the twelve gives this first proper biographical list a 'strong' feel which is enforced by Ibn Abī Uṣaybiʿah's assertion that these books show 'the right principles and order'. The rest of the corpus, as Ibn Abī Uṣaybiʿah knew it, is registered for the most part by title alone. Ibn Abī Uṣaybiʿah notes that 'some of them are falsely attributed' (4.1.9.2) but presumably felt that most of these works were genuine for he includes among them those he quoted earlier (*Law*, *Testament*, *Oath*, the last two with important comments).

In our translation we have included a new heading at 4.1.10.1, 'The Followers of Hippocrates'. But for Ibn Abī Uṣaybiʿah the followers were Hippocrates' family and students, with names taken largely from John, and are part of a single Hippocrates section. Some of the names of immediate successors can be restored on the basis of the largely fictional Brussels Life of Hippocrates[44] but regrettably not the name of the daughter of Hippocrates.[45] Most of the names of those 'between Hippocrates and Galen' (4.1.10.2–5) can also be guessed at reasonably, and they show the now familiar jumbled chronology. The main gain of these sections for us is the list of fifty-seven titles written by Rufus of Ephesus, drawn largely from Ibn al-Nadīm (4.1.10.2). The other important figure is Dioscorides (4.1.11.1), whose *De materia medica* was extremely well known to Arabic medicine.[46] After him the list of physicians 'between Hippocrates and Galen' continues. Some of these (Cassius, Alexander of Tralles) lived long after Galen.

It is Isḥāq ibn Ḥunayn who says he has added philosophers to John's history and Ibn Abī Uṣaybiʿah names him for those listed at 4.1.11.3. But a more congenial source, Ṣāʿid al-Andalusī's *Ṭabaqāt al-umam*, is cited to prepare readers for what comes next. Ṣāʿid al-Andalusī is the immediate basis for Ibn Abī

43 See notes ad loc. Ch. 4.1.9.1. On the canon of twelve, cf. Ullmann, *Medizin*, 27.

44 Pinault, *Hippocratic Lives*, 24–28, 131–134.

45 Ch. 4.1.10.1 *M-ā-l-ā-n-ā Ā-r-s-ā*; cf. similar forms in the *Ṣiwān* and the *Fihrist* in the notes to the Arabic text.

46 We have again introduced a new heading for the English.

Uṣaybiʿah's brief report of Empedocles (4.2) and he is cited to introduce Pythagoras (4.3.1). Ibn Abī Uṣaybiʿah has a trick up his sleeve at this point: Porphyry's *K. Akhbār al-falāsifah wa-qiṣaṣihim wa-ārāʾihim*, the great history of philosophy which is lost in Greek bar the biography of Pythagoras.[47] Ibn Abī Uṣaybiʿah's Pythagoras refines the model established by his Hippocrates because the account of Porphyry, mediated for the most part through al-Mubashshir ibn Fātik, for the first time allows details of a life actually led. Thus the opening summary by Qāḍī Ṣāʿid quickly yields to Porphyry's account, which is soon credited directly to al-Mubashshir (4.3.4.1–3). Mubashshir was too good a source to let go and Ibn Abī Uṣaybiʿah draws on him next for Pythagoras' *ḥikam*. But he turns back to Porphyry for the books of Pythagoras (4.3.6.1–2). The two passages quoted, of unknown origin in Arabic, are important because they are not in our Greek text.

The Socrates who follows (4.4) also offers little of medical interest. Ibn Abī Uṣaybiʿah again begins with Ṣāʿid al-Andalusī, but thereafter is wholly dependent on Mubashshir. The account is dominated by the gripping narration of the last days of Socrates (4.4.2.3–4) and by sayings material, which is cut down from al-Mubashshir's extensive list but is still substantial. Ibn Abī Uṣaybiʿah's Plato (4.5) is broadly similar, except that the introduction is made for Ibn Abī Uṣaybiʿah by Ibn Juljul. Mubashshir provides what biographical details there are, and Ibn Abī Uṣaybiʿah then offers a selection from his also very extensive complement of sayings. But Ibn Abī Uṣaybiʿah, an author who was inordinately fond of his bibliographical lists, turned to his own unidentified source for the book catalogue, an ordering that is found only here in Arabic literature.

Next comes Aristotle. The Aristotle is the *pièce de résistance* of Ch. 4 and is the fourth longest life overall, running to some 6850 words in our Arabic text (cf. Ibn Sīnā at approximately 7200, Ḥunayn ibn Isḥāq 7700, Galen 15700). Although for sound chronological reasons Ibn Abī Uṣaybiʿah follows it with Theophrastus and Alexander of Aphrodisias, we should see it as a mirror of the Hippocrates, which (if we include every biography in Ch. 4.1 as Ibn Abī Uṣaybiʿah intended) amounts to some 6000 words. The Aristotle shows Ibn Abī Uṣaybiʿah at his most competent. Ibn Juljul's introduction holds up the main act for no more than a few sentences. Ptolemy the Stranger's *Epistle to Gallus on the Life of Aristotle* – a work surviving only in an epitomized form in Greek – is brought on immediately.[48] This provides Ibn Abī Uṣaybiʿah with a real life once again. He splices Ptolemy with al-Mubashshir ibn Fātik, then briefly imports

47 See A. Segonds' appendix on the Greek tradition, in Des Places, *Porphyre*.

48 For basic bibliography and discussion, see Toulouse, 'Ptolémée'.

Ḥunayn ibn Isḥāq before embarking on a discussion of Aristotelian philosophy courtesy of Ṣāʿid al-Andalusī and al-Fārābī, on whose *Iḥṣāʾ* he draws extensively (4.6.5.1). Job done, he tails the life with Aristotle's will (from Ptolemy, 4.6.6.1). But he was clearly determined to bulk out the biography of his non-medical subject and turns to some light reading in the form of fictional history from Ḥunayn's *Ādāb* (here: *Nawādir*) *al-falāsifah*. This work serves to remind us that even as knowledgeable a scholar as Ḥunayn was happy to include a good deal of pseudo-history to make his books of philosophers' wisdom more appealing. Ibn Abī Uṣaybiʿah makes no comment on the fiction of Plato's clash with Aristotle at the court of King Rufistanes, but since he knew better history perfectly well from Ptolemy we may assume he included the tale simply to entertain readers in the middle of a very long account.[49] As with Plato and Socrates, he extracts from al-Mubashshir a list of sayings (4.6.11) before turning back to Ptolemy for a bibliography to which he triumphantly adds 'many other books by Aristotle that were not listed by Ptolemy' (4.6.13.3). The bibliographer in him has to have the final word and he goes the extra mile by including bibliography for Theophrastus (4.7) and Alexander of Aphrodisias (4.8). This last figure is not only a prominent Aristotelian but also 'the contemporary of Galen, whom he met' and with whom he argued bitterly (4.8.2, titles nos. 20–22 for philosophical disputes).

6 Galen – Physicians from or after the Time of Galen (Chapter Five)

6.1 *The Limits of Fiction*

The biography of Galen of Pergamum (129 to ca. 216 AD[50]) is arguably the most important single life in the *ʿUyūn*. Its meaning to Ibn Abī Uṣaybiʿah is clear from the fact that he effectively allotted it a whole chapter by itself. The chapter actually consists of two parts: a section on Galen's life and doings, including a lengthy enquiry into the chronology of his lifetime, using arguments drawn from the chronographers and from Galen's own works to prove that he was not a contemporary of Christ; and a long section listing and commenting on Galen's writings, for which Ibn Abī Uṣaybiʿah relied principally on two works of Ḥunayn ibn Isḥāq. The chapter is closed by a short section on the physicians who came

49 Cf. below on fiction about Galen in the *Ṣiwān al-ḥikmah*.

50 The current orthodoxy among classicists that Galen lived to the age of 87 rests on John the Grammarian. John's division is useless (see above for Ibn Abī Uṣaybiʿah's rejection of it at 5.1.3), but it is clear from internal evidence in the corpus that Galen died around 216 or perhaps a little earlier.

after Galen, most of them from late antiquity, which leads neatly to the short Ch. 6 on the late Alexandrian curriculum, itself a kind of appendix to the 'Galen' inasmuch as the curriculum it discusses was based, of course, on his works.

The biography of Galen is present in all three versions of the *'Uyūn*, and like the other 'Greek' chapters in essence belongs to the first. It is interesting to note some signs of revision in Version 3. That Ibn Abī Uṣaybiʿah re-read his 'Galen' late in his life for the final edition of his book is unsurprising. And although several other chapters are longer (Chs. 4, 8, 10, 11, 13, 14, 15), there is no individual biography that is longer; indeed 'Galen' is twice the length of the second longest, Ḥunayn ibn Isḥāq.[51] This length is achieved by incorporating in the first half of the life long quotations from Galen that show the deep pleasure Ibn Abī Uṣaybiʿah drew from repeating the words of his hero relating to his conduct and ability. The list of Galen's works in the second half – taken from Ḥunayn's *Risālah* on the translations he made and from the short treatise he wrote on material 'not listed by Galen in the Catalogue of his books'[52] – is a comparable investment, which highlights the achievements of one hero through the comments of another. The commissioner of MS Gc (Version 2) heavily abbreviated this list, presumably (and not unreasonably) taking the view that the information was available elsewhere and that he could therefore omit much of it beyond the book titles themselves.[53]

With the exception of the abbreviated book-list in MS Gc, the manuscripts of the three versions of the Ibn Abī Uṣaybiʿah's history are highly consistent. MSS Ga (Version 2) and Gc have a number of readings in common, and R (Version 3) occasionally shows trivial errors and signs of haste in copying out this longest individual life. There are, however, some interesting examples of differences between the three versions. I shall consider several illustrative passages later and also say something about Ibn Abī Uṣaybiʿah's use of other authors, for, as we have already discussed,[54] he relies on many standard works and in Ch. 5 we are able to compare his text with these major authors: Ḥunayn ibn Isḥāq, Isḥāq ibn Ḥunayn, Ibn Juljul, al-Mubashshir ibn Fātik, Ibn al-Nadīm, and Ibn al-Qifṭī. Where there are differences (esp. for Ibn Juljul and al-Mubashshir ibn Fātik), it is difficult to know how far Ibn Abī Uṣaybiʿah represents innovation

---

51 For an approximate Arabic word count of Galen, Ḥunayn, Ibn Sīnā, Aristotle, Hippocrates, see above. Cf. e.g. al-Fārābī (2848), Ibn al-Muṭrān (3070), Ibn al-Haytham (3980), 'Abd al-Laṭīf al-Baghdādī (5590).

52 See Ch. 5.1.37 n. 228, 5.1.38 n. 428.

53 Though even he retains in full the romantic vignette of Ḥunayn's search for manuscripts of *On Demonstration* (5.1.37 no. 115).

54 Above, Ch. 5, on Ibn Abī Uṣaybiʿah's written sources.

as opposed to imperfect quotation from memory or a variant written tradition of the work in question. Ibn Abī Uṣaybiʿah also drew on sources now lost such as the physician ʿUbayd Allāh ibn Jibrīl ibn Bukhtīshūʿ[55] for the chronology of Galen and the courtier Yūsuf ibn al-Dāyah[56] for the long anecdote about the location and dimensions of Galen's house. Of the lengthy extracts from Galen's *On My Own Books*, *Good and Bad Juices*, and *Examination of the Best of Physicians*,[57] comparison may be made with the unedited Arabic of the first and the edition of the last.[58]

No-one had written a biography of Galen. But Ibn Abī Uṣaybiʿah was undaunted. As he put it towards the half-way point of his account (5.1.27),

> There is a vast amount of information about Galen. There are stories useful to people who examine them, and snippets and anecdotes scattered all over his books and throughout the reports passed down to us about him. There are many stories detailing his treatment of the sick and testifying to his ability and skill in the art of medicine.

He continues by saying that it is his intention to compose a separate monograph on all of this material 'if God the Exalted wills it'. Yet it is difficult to see what kind of treatment such a work could have been given. He might have assembled additional passages from Galen or drawn on the fairly imaginative stock of stories about Galen that circulated in salon literature and were presumably invented to address the absence of a comprehensive biography from ancient times and thus cater to the demand to know more about one of the giants of antiquity. Had Ibn Abī Uṣaybiʿah mined all the references Galen made about himself, he would surely have had trouble pulling them together into a satisfactory whole. Had he included fictional accounts in his own rather serious study of the 'seal' of physicians, he would have seemed incoherent.

As an example of the problems of harmonizing sources in the overall context of Ibn Abī Uṣaybiʿah's presentation, consider for a moment 5.1.17, where he includes an extract from Ibn Juljul misdating Galen to the reign of Nero, information that runs counter to his huge effort to find accurate dates that takes up the opening sixth of the life. Ibn Abī Uṣaybiʿah did not spot the problem,

55 See Ch. 5.1.7. He was the 11th-century member of a family of physicians active from the 760s. On his place within the family, see Ullmann, *Medizin*, 110 no. 7. His biography is at Ch. 8.6.

56 See Ch. 5.1.16.1. Client of Ibrāhīm ibn al-Mahdī, the half-brother of Hārūn al-Rashīd.

57 Ch. 5.1.37 nos. 1, 76, 112.

58 *Good and Bad Juices* is lost in Arabic.

perhaps because the name of the emperor was transmitted in a form so different from its appearance in ʿUbayd Allāh ibn Jibrīl's chronology, which he had quoted earlier. Again, whereas Ibn Abī Uṣaybiʿah's citation of al-Bayhaqī's statement (5.1.5) that St Paul was Galen's nephew and that Galen 'believed in Jesus' is a spur to turn to the evidence of Galen's own writings and begin the task of an historical dating, the repetition of the claim at 5.1.21.2 ('so he left Rome intent on Jerusalem') as a bridge to take readers from al-Mubashshir ibn Fātik on Galen's travels and death/burial to the story of the disease that killed him (5.1.22) shows a quite careless judgement, even if the story is signalled as dubious ('another source'[59]). Of course the tale of the great physician and the great healer of bodies and souls was so tempting that readers must have expected it to be mentioned. But having artfully introduced the fiction as a springboard for rejecting it in the first part of the life, it was not a good idea to let it stand unchallenged later. The following anecdote of Galen making ice in summertime in a futile attempt to cure himself is explicitly labelled 'a fiction' (5.1.22), and the mixture of fiction and scientific knowledge (the observation that saltpetre removes heat from water) relocates Ibn Abī Uṣaybiʿah momentarily from the library to the dinner party, where of course many of his anecdotes and quotations of poetry throughout the *ʿUyūn* naturally reside.

Ibn Abī Uṣaybiʿah certainly wanted to entertain, but he was not usually prepared to include outright fictions about Galen in a chapter designed to edify. It is easy to demonstrate this by comparison with the (deliberate) nonsense we find in a purely belletristic work like the *Cabinet of Wisdom*, which seems to have been composed around the middle of the 11th century[60] The difference from Ibn Abī Uṣaybiʿah is that, even though the fictional material on display in the *Cabinet* was intended as entertainment, to have included such stories in the *ʿUyūn* would have unbalanced the careful image of the physician hero. Consider the story in the *Cabinet*[61] of Galen's relations with two fictional kings of his own day, Nīfās and Bāz. Galen, we are told, lived in a town called 'Macedonia' in Egypt 'where he was the favourite doctor of good king Nīfās'. Nasty king Bāz lived in the Magrib. The good king was obliged to send his favourite physician to him after Bāz threatened war if Galen did not cure one of his girls with signs of leprosy on her face. Various stories follow. The tyrant cannot bear the thought that Galen will have to look at the girl, so Galen cleverly uses a mirror. Galen refuses to eat the tyrant's unhealthy food. He writes him a 'memo' (*dustūr*) to

59 In fact the source is Ibn Juljul.

60 The work was written in the circle of al-Sijistānī al-Manṭiqī (cf. Ch. 11.7): see Kraemer, *Philosophy* 119–123, al-Qāḍī, *'Kitāb Ṣiwān al-ḥikma'*.

61 Dunlop, *Ṣiwān al-ḥikmah*, ll. 2200–2340.

remind him that he had predicted he too would suffer from leprosy, and a 'treatise' (*maqālah*) on the prognosis of epidemic diseases. The tyrant fails to heed Galen's dietary advice and suffers for it: his hair thins, his eyelids lose shape, his nails shrink. Now Galen escapes disguised as a black man. Bāz, ill, follows him and pretends to enrol as a pupil but Galen recognizes him and turns him into a civilized Greek! Further stories focus on Bāz's son Glaucon, who becomes Galen's star pupil and Nīfās' heir. The second part of this narrative contains nineteen sayings of Galen (including an extract from *Character Traits*). It is followed by one further piece of information: 'Abū l-Nafīs said, "Galen tended to lisp".'[62]

The *Cabinet* throws in a few facts of the Galenic story to add to the entertainment. Glaucon is the philosopher to whom Galen dedicated the *Ad Glauconem, De methodo medendi*, which was part of the Alexandrian introductory syllabus.[63] The deposit of the 'memo' in the royal stores recalls Galen's account of storing his books in the royal depositories of the Temple of Peace at Rome, which was well known in his biographical tradition because of the great fire which consumed them and Galen's sanguine reaction to the loss.[64] The last words of the biography offer a tantalizing remark that perhaps alludes to Galen's interest in what he calls *to traulizein*.[65] It no doubt raised a smile to have this remark follow the sayings material, for how better to have fun at the expense of a wise sayer than to mention his speech impediment?

6.2 *Galen in Time and Space*

Ibn Abī Uṣaybiʿah was fully aware that there were entertaining fictions in circulation about his hero. He himself cheerfully admits one tale of this kind, concerning Galen's death, and carefully labels it, as we have noted. In general, though, amusement of this sort was not part of his plan any more than it was in other 'Greek' chapters.[66] The chronological linkage of Galen and Christ as found in Isḥāq ibn Ḥunayn's history of physicians on the basis of the chronology of John the Grammarian,[67] and the specific claim in other sources that

62 l. 2398 *kāna Jālīnūs althagh*. Cf. l. 3555 Abū l-Nafīs 'had a wonderful memory for the stories, best lines, and witty sayings of the philosophers'.

63 No. 6 in the list of Galen's works in Ch. 5.1.37.

64 See Ch. 5.1.26.

65 I.e. 'lisping'; see Wollock, *Noblest Animate Motion*, 191–192, 283–285; and cf. Ullmann, *WGAÜ* 681 s.v. τραυλός for the translation of the word by the root *l-th-gh*.

66 Cf. the tale of Plato, Aristotle, King Rufistanes and his son Nitaforas in Ch. 4.6.7.2. The story also circulated independently, see note ad loc.

67 See above, p. 152, Ch. 5.1.4 with notes.

Galen was the uncle of Paul (or Luke)[68] was an altogether different type of fiction and had to be included to be rebutted. It could not be ignored for, as has been noted, it involved a Christian appropriation of Galen.[69] The best way to meet this fictitious presentation of his hero was to use Christian chronographical material against it. Ibn Abī Uṣaybiʿah introduces the story in 5.1.5 from a no longer extant history of Persia by the Muslim littérateur al-Bayhaqī. In another of his works, the surviving *Tatimmat Ṣiwān al-ḥikmah*, al-Bayhaqī ascribes it to the Christian intellectual Ibn al-Ṭayyib (d. 1043), noting that Ibn al-Ṭayyib claimed descent from Paul.[70] The legend came to be known also in Byzantium. Our sources for it are late and that may suggest influence from Arabic literature.[71]

A man as prominent as Isḥāq ibn Ḥunayn could not be sidelined. In his text as preserved there is no explicit attempt to exploit the linkage but the sums make it clear. Luckily, then, help was at hand for Ibn Abī Uṣaybiʿah from ʿUbayd Allāh ibn Bukhtīshūʿ. ʿUbayd Allāh is known for his *Memorandum for the Sedentary and Provision for the Traveller* (from which he abstracted the definitions titled the *Medical Garden*) and a treatise on the maladies of the soul. His lost *Manāqib al-aṭibbāʾ* (*Merits of Physicians*) was an important source for Ibn Abī Uṣaybiʿah.[72] In an otherwise unknown epistle he addressed head-on the question 'whether Galen's lifetime coincided with Christ', and from this Ibn Abī Uṣaybiʿah quotes extensively at 5.1.7–13. With the help of other Christian authors ʿUbayd Allāh presents a fairly accurate summary of the regnal years of rulers from Alexander to Antoninus (= Antoninus Pius), the last however being confused with his successor Antoninus whom we call Marcus Aurelius. ʿUbayd Allāh alleviated the dullness of too much chronology by recounting the legend of Ptolemy Philadelphus and the creation of the Septuagint, which Ibn Abī

68 See 5.1.5 with note.

69 Cf. Bürgel, *Ärztliches Leben*, 401, on the 'Christianisierung der Antike'.

70 See note to Ch. 5.1.5.

71 In the 12th century Tzetzes devoted a section of his *Chiliades* (§ 397 ed. Leone) to proving that Galen was not a contemporary of Nero – people who said this 'talk out of their stomachs' – by noting that Galen had cited Nero's contemporary Andromachus the Elder; while Michael Glycas, *Annals* (p. 430 ed. Bekker), dismisses the 'popular' tale of Galen's meeting with Mary Magdalen and his deduction that Christ possessed medical earths from her report of his healing of the blind. The Byzantine encyclopaedia called the *Suda* (turn of the 11th century) s.v. γ 32 confirms the true dating. George the Monk followed by Tzetzes puts Galen under Caracalla (211–217); cf. Boudon-Millot, *Galien* pp. lxxxviii–xc.

72 For the works of ʿUbayd Allāh ibn Bukhtīshūʿ see Ullmann, *Medizin*, 110–111, 230; Ullmann, *Natur- und Geheimwissenschaften*, 28; Klein-Franke, *Über die Heilung*. Further, p. 88 above.

Uṣaybiʿah included for the same reason. He finished his discussion of Galen's dates with two passages from Galen's well-known *Character Traits*, which contain firm dates, and from his summary of the *Republic*, where he mentions the Christians' quasi-philosophical perfection.

So pleased was Ibn Abī Uṣaybiʿah with ʿUbayd Allāh as an antidote to Isḥāq that he unconsciously allowed him a paragraph of Christian apologetic propaganda on Christians' possession of a double bliss, the happiness of the Law and the happiness of intellectual ethical superiority (5.1.13). His techniques commended themselves and in 5.1.6 he himself took from Ibn al-Nadīm an extract from *Character Traits* as an immediate counter to al-Bayhaqī's report that Paul was Galen's nephew. ʿUbayd Allāh's citation of the summary of the *Republic* also provided inspiration, for in 5.1.14 Ibn Abī Uṣaybiʿah invokes an autograph manuscript of Ibn al-Muṭrān, in which he had collected various passages where Moses and Christianity are mentioned in Galen's works.[73] The several allusions to Christians and Jews in Galen are in fact extremely important,[74] for they are among the earliest observations of Christianity by pagan authors. In his *Ecclesiastical History* Eusebius reports from a treatise written by an orthodox apologetic writer of the early 2nd century, possibly Hippolytus of Rome, that Galen 'is perhaps even being offered obeisance by some' Christians, apparently wishing to polish up their logic, which Galen both specialized in and reproved Christians and Jews for ignoring in favour of blind faith.[75] Ibn Abī Uṣaybiʿah does not say if Ibn al-Muṭrān assembled the passages in order to prove that Galen lived at a time when Christianity was well established or whether he wished to indicate Galen's attitude towards two of the religions of the Book. In fact the three passages quoted are fairly critical. But as far as Ibn Abī Uṣaybiʿah in concerned, they backed up the arguments of ʿUbayd Allāh, and put distance between his hero and Jesus.

It is clear from the next section (5.1.16.1–5) that it was just as important for Ibn Abī Uṣaybiʿah to locate Galen in space. His information comes in the form of a long extract from the Baghdadi intellectual Yūsuf ibn al-Dāyah, who wished to set readers straight on the extent of the ancient Roman empire in the East and the size and appearance of Galen's house in Smyrna.[76] The discussion turns

73 See below on the late addition of this passage.

74 Walzer, *Jews and Christians*, is the basis of most modern discussion on this topic.

75 Ch. 5.28.14; Euclid, Aristotle, and Theophrastus were also mentioned. Cf. Barnes, 'Galen', 407–411, 417.

76 No record of a *manzil* or *dār* in Smyrna survives in Greek, but cf. 5.1.25 for Galen's period of study there. It was common for wealthy families to have widely distributed landholdings in his period, including multiple local citizenships. His home town of Pergamum is not far from Smyrna. On Ibn al-Dāyah, see p. 88.

humorously on the relative status of physician and emperor *versus* that of the caliph (i.e. Hārūn al-Rashīd) and his own physician, Jibrīl ibn al-Bukhtīshū', whom Ibn Abī Uṣaybi'ah knew to be great great grandfather of 'Ubayd Allāh ibn Jibrīl.[77] This forms a neat bridge to the next section (5.1.18), where Ibn Abī Uṣaybi'ah turns to Galen's character and ethics and starts by citing a long extract from *Examination of the Best Physician* (*De cognoscendo optimo medico*) on his avoidance of noblemen's levees at Rome and his preference for studying and furthering the art of medicine, as illustrated by his unmasking of a charlatan and his love of 'direct practice' rather than 'rote learning' (5.1.19–20). This allows the introduction of various biographical details and for this Ibn Abī Uṣaybi'ah turns to a favourite book, al-Mubashshir ibn Fātik's *Choicest Maxims*.[78]

We have considered how Ibn Abī Uṣaybi'ah uses al-Mubashshir ibn Fātik in general above.[79] His function in the first extract used by Ibn Abī Uṣaybi'ah here (5.1.21.1) is to provide basic information about Galen's travels and to introduce the topic and location of his death. Since he also mentions Galen's father's passion for education, when Ibn Abī Uṣaybi'ah returns to al-Mubashshir in 5.1.23 he uses him to emphasize the role of the father and to picture the young Galen as an anti-social swot. He backs up al-Mubashshir with a long quotation from Galen's *Good and Bad Juices* lauding his father's wisdom and his advice to steer clear of his 'young friends' excessive appetites and their demanding and irresponsible behaviour' (5.1.24.1), and then returns to al-Mubashshir for details of Galen's teachers and library.

It is curious, as we have seen, that Ibn Abī Uṣaybi'ah pauses at this point (5.1.27) to inform us, as if in conclusion, that he cannot provide all the useful and entertaining stories scattered throughout Galen's work but will one day pen a separate work on them. For in fact he starts up again, referring to Galen's wonderful skills in *On Prognosis* (*De praecognitione*; 5.1.28), and quoting at length once more from the *Examination of the Best Physician* to illustrate his amazing cures and predictions (5.1.29). He ends with two short quotations from a lost work on his generosity to pupils and patients alike (5.1.30). The first half of the chapter is topped with al-Mubashshir ibn Fātik on Galen's personal appearance, a disquisition on the form of his name in Greek with the help of Rhazes, and some verses from a lost poem of the great al-Ma'arrī in celebration of the

77 For Jibrīl, see Ch. 8.3.

78 *Mukhtār al-ḥikam wa-maḥāsin al-kalim* (Ch. 14.23.4 no. 2). See Rosenthal, 'Prolegomena to an Abortive Edition', 145–147, 156–157 for a list of Ibn Abī Uṣaybi'ah's extensive quotations.

79 Above, p. 86.

enduring legacy of Galen and Hippocrates. This, as MS A notes, brings to a close 'the first volume [*juz'*] according to the division in the exemplar'.[80]

6.3 *Sayings and Books*

A large part of the material produced by the biographical machinery of antiquity and the Middle Ages consisted of sayings designed to illustrate the sayer's wit and wisdom. In the Greek tradition, however, as we see it in the gnomologies of the Greco-Roman and Byzantine periods, Galen does not feature. The reason is probably that in comparison with the figures of classical and Hellenistic Greece he already lived too late to be included, especially as it seems that his position as the king of medicine was not assured till the 4th century.[81] But in the Arabic tradition Jālīnūs was accorded a small number of pronouncements. Ibn Abī Uṣaybiʿah makes use of those in the earliest collection, Ḥunayn's *Stories of the Philosophers and the Sages and Examples of the Manners and Culture of the Teachers of Old*,[82] and from al-Mubashshir ibn Fātik.

Since the preservation of utterances is a fundamental part of Islamic culture and enormous care was taken to demonstrate the veracity of such material, there is no reason to suspect that Ibn Abī Uṣaybiʿah doubted the authenticity of what he includes. But he is also clear that the sayings have an overriding function of providing evidence of Galen's intelligence and knowledge. The fourteen aphorisms from Ḥunayn centre on two topics much discussed in Arabic literature, worry (*hamm*) and sorrow (*ghamm*), and love as a disease of the soul.[83] For the physician the interest lies in the relationship of mind and body posed by such problems. Some of the sayings reflect actual commentary by Galen, and one of these sayings is a quotation from *Character Traits*. The sixteen sayings recorded from al-Mubashshir ibn Fātik are more pithy but there are also echoes of the Galenic corpus including two reworkings of passages in the *Adhortatio ad artes addiscendas*. To these Ibn Abī Uṣaybiʿah adds some from other, unspecified sources including finally a quotation from Galen's commentary on the *Oath*. The words are well chosen – 'we will not find that everybody is recept-

80 MS A, fols 58b–59a.

81 For example in Oribasius.

82 *Nawādir al-falāsifah wa-l-ḥukamā' wa-ādāb al-muʿallimīn al-qudamā'*.

83 They are part of a wider subgenre called *faraj baʿd al-shiddah* (relief/deliverance from distress); see e.g. the work of this title by the 10th-century qadi al-Tanūkhī, *Faraj* or al-Kindī's *R. fī l-ḥīlah li-dafʿ al-aḥzān* with Adamson & Pormann, *The Philosophical Works of al-Kindī*, 245–266. On the subliterature of love as a disease of the soul, see Klein-Franke, *Über die Heilung*, Biesterfeldt, Gutas, 'Malady', Rosenthal, 'As-Saraḫsî on Love'. In general, Biesterfeldt, *Galens Traktat*.

ive to the art of medicine but we need someone whose body and soul are well suited to the purpose'. So ends the presentation of his model's sentiments.

Ibn Abī Uṣaybiʿah loved bibliography. For the first serious list of his book he manages 61 works (Hippocrates). In Ch. 4 he then peaks with 142 titles for Aristotle. We may compare among others Avicenna (11.13): Ibn Abī Uṣaybiʿah lists 101 works; while for Ibn al-Haytham (14.23) he has 182 plus 30 summaries of Galenic works, there are 226 for Rhazes (11.5), and for al-Kindī (10.1) no fewer than 283. Galen has only 188; but in no other biography does Ibn Abī Uṣaybiʿah include as much information. For this he was able to draw on the descriptive catalogue of Ḥunayn, as he happily admits. He stays close to Ḥunayn's words and ordering, keeping to the record of the books and omitting extraneous comments. Ḥunayn was as much an enthusiast for Galen as Ibn Abī Uṣaybiʿah and several times revised the description of his life's work of translating Galen.[84] It is clear that Ibn Abī Uṣaybiʿah kept texts similar to our two surviving versions open on his desk: thus in title no. 110 he begins following version 'A' as usual but switches to 'B' to include further information found there, signalling the switch with the phrase, 'Ḥunayn says …'

The first eighteen titles following the two (bio-)bibliographical works with which Ḥunayn/Ibn Abī Uṣaybiʿah begins (*De libris propriis*, *De ordine librorum suorum*) are essentially the books that formed the Alexandrian canon, though details and order differ somewhat from the famous structure of seven stages quoted from Ibn Riḍwān's *Kitāb al-nāfiʿ* in Ch. 6.[85] Ibn Abī Uṣaybiʿah includes full information on these works and fairly full information for other key or interesting writings (*De anatomicis administrationibus*, *De usu partium*, *De simplicium medicamentorum temperamentis et facultatibus*, *De compositione medicamentorum*, *De demonstratione*), adding comments as necessary (e.g. *On My Own Books* used to amplify Ḥunayn on title no. 21). He was no doubt pleased to import into the list title no. 102a – it is not in the *Risālah* as transmitted to us – to illustrate Galen's reaction to criticism of title no. 102.

So important was Galen that Ibn Abī Uṣaybiʿah finishes the life with an extensive list of books attributed to him, drawing first (5.1.38) on Ḥunayn's supplementary bibliographical study,[86] and noting 34 additional works alluded to or translated, he says, by Ḥunayn (but not in his bibliographical works) or by others (titles 155–184). Of these he pauses to give full information on no. 167 *Secret Remedies* (*K.fī l-adwiyah al-maktūmah*) on the elements of Galen's prac-

84 See Ḥunayn ibn Isḥāq, *Galen Translations* (Lamoreaux), pp. xxi–xxvi for interpretation of the publication history, cf. Gutas, 'New "Edition"'.

85 See below.

86 Ḥunayn ibn Isḥāq, *Neue Materialen*, 84–98.

tice that he shared with those 'who possessed the right qualities and the soundest judgment'. He ends the study of his hero quietly:

> In conclusion I would note that Galen certainly wrote many other books that translators did not find or that were lost in the course of time … Should anyone wish to refer to their titles or to Galen's aim in each of them, he may consult (*On My Own Books*).
>
> 5.1.40

6.4 *Variations between Versions*

Ibn Abī Uṣaybiʿah's interest in his Galen is obvious from the evidence for revision between the different versions of the *ʿUyūn*. Here we must distinguish between scribal or patrons' preferences and authorial changes. It is clear that the scribe of MS Gc wished, or was instructed, to reduce the length of Ch. 5. His preference for cutting the information Ibn Abī Uṣaybiʿah gives on Galen's books in 5.1.37–39 has been mentioned: from title no. 15 onward this reduction is drastic and includes whole entries (title no. 19), though he does take care to preserve most of no. 79 *De compositione medicamentorum* (perhaps indicating an interest in pharmacology) and was intrigued by the lost *De demonstratione* (no. 115) both because of the story of Ḥunayn's search for its manuscripts and because of an apparent philosophical bent that is evident from the inclusion of all of title no. 124 (Galen's summaries of the Platonic dialogues). He also copied out most of no. 167 *Secret Remedies*, which illustrates Galen's esotericism. There are also significant omissions by this scribe in the first *juzʾ*. Half of the quotation from *On My Own Books* in 5.1.10[87] and most of 5.1.12 and all of 5.1.13 are left out,[88] thereby seriously undermining Ibn Abī Uṣaybiʿah's chronological investigation into Galen's dates. It looks as if he or his patron had had enough of this long disquisition. MS Gc also omits the whole of the story of Galen's residence quoted from Yūsuf ibn al-Dāyah (5.1.16). Whether this was a case of pruning is unclear, since 5.1.15 and 5.1.17 are short excerpts from Ibn Juljul and the omission may therefore have been caused by accidentally skipping over the intervening text.

MS B (our representative of Version 1) unsurprisingly lacks some items included later. It does not include Ibn al-Muṭrān's collection of passages in Galen pertaining to Christ and Moses (5.1.14). It also lacks most of Ibn Abī Uṣaybiʿah's

---

87 *On My Own Books* 1.11–16, 2.1 ed. Boudon-Millot = Ch. 5.1.10.2 'One day it happened … when I was thirty-seven years old'.

88 From "Ubayd Allāh notes that …'

discussion of the form of Galen's name in Greek and Arabic (5.1.32, information from Qāḍī Najm al-Dīn) and the poem of al-Maʿarrī that ends the first *juz'* (5.1.33). MS B omits titles 187–188 in 5.1.39. Perhaps Ibn Abī Uṣaybiʿah did not know of these pieces of information at the time of writing.[89] Some omissions and additions pertain to Christianity. B along with Gc and Ga omits the word 'Apostles' in 5.1.8.1. It looks as if the term was added by Ibn Abī Uṣaybiʿah in a later revision; but no change of attitude towards Christianity is evident and we have rather a difference between MSS BGaGc and AR. Thus BGaGc read 'crucified' in 5.1.8.1 against the revised 'raised up high' in AR, which was acceptable to Islam, and they add a note about Christ's Resurrection 'according to the witness of the Apostles', which was removed in Ibn Abī Uṣaybiʿah's final revision. Where the last version reads 'Christ was baptized in the Jordan by the hand of John the Baptist', the words 'in the Jordan ... Baptist' are omitted by BGaGc, pointing to an addition in AR. Here GaGc substitute in parenthesis 'he speaks as he believes', referring unflatteringly to Ibn Abī Uṣaybiʿah's Christian source, ʿUbayd Allāh ibn Jibrīl. A little later, with reference to Vespasian's destruction of the Jews, where Ibn Abī Uṣaybiʿah (ʿUbayd Allāh) notes God's promise 'about the coming of Christ (Peace Be Upon Him!)', for the words in brackets BGaGc substitute 'and there was no return for them' i.e. the Jews. Further evidence of a difference between BGaGc and AR is clear from 5.1.20, a passage added from Ibn Juljul. Although Ibn Juljul drew on a Galenic work of interest to Ibn Abī Uṣaybiʿah, *De compositione medicamentorum*,[90] Ibn Abī Uṣaybiʿah failed to identify the extract, which his source had mangled. But it seems, to judge by this addition, that he felt the need to include in his final version a further witness to Galen's preference for what the text calls 'direct practice'. Sometimes a revised reading shows up in R and A margin: so under Domitian 'the city of ʿAyn Zarbā was razed'; in ABGaGc we read 'the city of Ra's al-ʿAyn was plundered'.[91] A margin and R margin present the same comment, 'migration from place to place means leaving the first for the second, and this migration path is that of the Prophet, God bless him and his family and preserve them!, from Mecca to Medina' to 5.1.4 (mention of the Hijra) or 5.1.5 (mention of Galen's belief in Christ).

The most interesting instance of revision by Ibn Abī Uṣaybiʿah is a marginal addition in MS A from Ḥunayn that is correctly located by a caret mark[92] but

89 MS R omits Titles 33 and 104 – but the habits of the scribe suggest that this is due to carelessness.

90 See the full description and comments by Ibn Abī Uṣaybiʿah at title no. 79.

91 See note to Ch. 5.1.8.2.

92 The *bā' Hindiyyah* (٢), indicating a second source copy: Gacek, *Manuscripts*, 250.

appears in Gc and R in the text body in two different and incorrect locations. In A it comes under title no. 7 (*De ossibus ad tirones*) in the list of Galenic works and consists of a longish acknowledged quotation ('Ḥunayn says …'). R however adds it in the text of title no. 8 (*De musculorum dissectione*) while Gc places it in title no. 6 (*Ad Glauconem de methodo medendi*). There are slight differences between all three. This intervention indicates that Gc and R had sight of a copy where the additional quotation was placed in the margin but the *signe de renvoi* was obscure or absent. The scribe made his own choice about where the extra text should go.[93] In all cases scribes may of course have consulted sources for themselves and that may explain differences or more accurate readings, for example in no. 16 (*The Great Book of the Pulse*), where BGaR offer Ḥunayn's *ajnās* where others have *aṣnāf* and a little later *aṣnāf* where the others have *ajnās*.

One notable feature of the text is the way Ibn Abī Uṣaybiʿah handles his sources. In general he stays close to the traditions as we have them. The lion's share of quotation consists of Galen's own works, *On My Own Books*, *Good and Bad Juices*, and *Examination of the Best of Physicians*. In the case of the last, Ibn Abī Uṣaybiʿah is accorded manuscript authority by the editor, Iskandar.[94] The value of Ibn Abī Uṣaybiʿah's text for editions of the other works he uses is well known. Bergsträsser/Lamoreaux (Ḥunayn ibn Isḥāq), Rosenthal (Isḥāq ibn Ḥunayn/John the Grammarian), Sayyid (Ibn Juljul), Lippert (Ibn al-Qifṭī), Badawī (al-Mubashshir ibn Fātik) all benefit strongly from the text of *Best Accounts*. The editor of the *Cabinet of Wisdom*, Dunlop, cited Ibn Abī Uṣaybiʿah extensively for the parts of the work that draw on John's *History of Physicians*. But though Ibn Abī Uṣaybiʿah stays very close to these texts, there are changes, which are due to differing recensions, imperfect copies, quotation from memory, or the introduction of deliberate changes for reasons of style or to fit the context. The case of the lengthy quotation from ʿUbayd Allāh ibn Jibrīl (5.1.7–13) on the question of whether Galen lived at the same time as Christ is a particularly interesting case because ʿUbayd Allāh himself included citations of other authors in his answer. We have a very much shorter version of this *ris-*

93 Cf. Müller, 'Lesarten', 14 records the presence of the addition in R (his 'd') but places it at i, 91 l. 16, where it is in MS A. He records it as a marginal addition in his 'n' (a descendant of A) and in Paris, BnF Suppl. ar. 674, which appears to belong to Version 2 in general but here at least will represent an addition present in Ibn Abī Uṣaybiʿah's final text. In MS A the marginal addition is introduced by the word *ḥāshiyah* ('marginal note', 'gloss'), which is usually used for the scribe's own observations and comments. Note that in both A and R the addition is signalled after a similar phrase, *li-l-mutaʿallimīn* (A) / *ilā l-mutaʿallimīn* (R), which may account for R's confusion.

94 Iskandar, *Examinations*.

ālah in Ibn al-Qifṭī (126–128 Lippert), who begins by saying that ʿUbayd Allāh 'wrote an excessively long account of the matter' (126.13). It is obvious that Ibn Abī Uṣaybiʿah strongly disagreed about ʿUbayd Allāh's usefulness. It would be nice to know if he looked up any of the cited material for himself, but it is at any rate difficult to believe that ʿUbayd Allāh's long citation of Galen's *On My Own Books* (5.1.9–10) has not been amplified by Ibn Abī Uṣaybiʿah himself.

We know that Ibn Abī Uṣaybiʿah used different recensions of Ḥunayn's *Risālah* (cf. above) and where there are considerable differences between his quotation of an author and the text as transmitted to us, Ibn Abī Uṣaybiʿah may well be using a recension we do not know of. Take al-Mubashshir ibn Fātik, from whom he quotes extensively and accurately in 5.1.21.1, 23, 25–26, 31, 35 (sayings material). In 5.1.31 on Galen's appearance and attitude towards emperors half of the quotation is not found in the edited text of Badawī, who notes Ibn Abī Uṣaybiʿah has additional material but does not say whether he thinks it is genuine.[95] A more problematical case is Ibn Juljul, who is used in 5.1.15, 17, 20, 21.2. There is absolutely no doubt that Ibn Abī Uṣaybiʿah knew Ibn Juljul directly, but it is not easy to determine why there are differences. Quotation from memory may be indicated. The sayings material in 5.1.34–35 reveals a little more of Ibn Abī Uṣaybiʿah's choice. In 5.1.34 Ibn Abī Uṣaybiʿah uses material from Ḥunayn's *Nawādir al-falāsifah*. In 5.1.35 the material comes from al-Mubashshir ibn Fātik but Ibn Abī Uṣaybiʿah omits sayings also found in Ḥunayn and includes them in 5.1.34 using Ḥunayn's version. Finally a word about Ibn al-Nadīm. He is used verbatim in the important counter to al-Bayhaqī in 5.1.6 (see above) and then in 5.2.1 for the books of Oribasius and in 5.2.2 for Philagrius, though we see Ibn Abī Uṣaybiʿah intervening to reorder slightly and note the error in his text under Philagrius title no. 9 'structure of the womb', showing a faulty copy or sloppy note-taking and leading Ibn Abī Uṣaybiʿah unthinkingly to miss the very common (Galenic-)Hippocratic doctrine of 'suffocation of the womb'.[96] Faulty memory is again a possibility: when Ibn Abī Uṣaybiʿah records a comment by al-Masʿūdī 'in his book on routes and realms', he is in fact referring to the geographer al-Iṣṭakhrī (5.1.21.1).

6.5 *The Value of Galen*

Current scholarship interprets Galen as a prime witness to the scientific, intellectual, and literary cultures of his time.[97] He is seen as a key figure in the

95 Badawī, *Al-Mubashshir ibn Fātik*, 293 n. 1.

96 Cf. Galen, *Hippocratis aphorismi et Galeni in eos commentarii* 17b.824 Kühn 'not real suffocation but lack of breath' (οὐ πνίγα … ἀλλ' ἄπνοιαν).

97 See in general Dihle, *Greek and Latin Literature* (Galen index s.v.) and specifically on Galen

development of medical theory and practice (in particular in the areas of anatomy and physiology), as a hugely important codifier of now largely lost Hellenistic work on pharmacology and dietetics with his own original contributions to make, as a systematizer of Hippocratic notions of pathology who shaped the legacy of Hippocrates for subsequent generations (in particular the theory of the four humours and the three organic systems of brain, heart, and liver), and as a philosopher who demanded that physicians should be philosophically trained and presented his own independent thoughts on logic with a determination to modernize ancient authorities by bringing Plato, Aristotle, and Posidonius into harmony with his own deep knowledge of Hippocratic physiology.[98] He is viewed as a pivotal figure in the vibrant Greek culture of the early Roman empire (which is often referred to, following a remark by the 3rd-century author Philostratus, as the 'second sophistic'),[99] a man whose command of all fields of knowledge, including the study of correct language that was so valued by the elites of the age, propelled him into a spectacular career from his early days treating gladiators in Pergamum up to the court life of imperial Rome and long service as physician to the emperors. Like others of this age, Galen was obsessed with understanding 'the ancients', but like Ibn Abī Uṣaybiʿah in his day, was also sure of his own era's advances and excellence. His egotism and vicious criticism of others, which are excessive but not untypical of ancient Greek culture at any time, show an ambition to enjoy status and success in the deeply competitive and dangerous cultural-political circumstances where he contended for primacy.

Ibn Abī Uṣaybiʿah was in tune with Galen's circumstances, for most of his physicians operated in the environs of dynasts of one sort or another and had to fight for glory and repute, as many of the biographies recount. This is why he dwells on examples of Galen's feuds (5.1.10 Martialius). It is why he devotes so much space to the disquisition of Galen's status relative to the emperor and in comparison with the situation of the court physician of the Caliphate (5.1.16). It is why he records with obvious admiration his 'miracle' cures in the context of his repudiation of the luxury and laziness of contemporary physicians (5.1.18, 28–29). The great effort to pin down Galen's dates serves to insert him fully into the history of the emperors (5.1.8–9).

himself and the aspects mentioned here, with full bibliographies, *inter alios* Barnes & Jouanna, *Galien*; Boudon-Millot, *Galien de Pergame*; Gill, Whitmarsh, & Wilkins, *Galen*; Hankinson, *Galen*; Nutton, *Medicine*, chs. 15–16.

98 *De placitis Hippocratis et Platonis* (Ch. 5.1.37 no. 46).

99 Schmitz, *Bildung und Macht*; Swain, *Hellenism and Empire*; Whitmarsh, *Second Sophistic*.

As has been remarked, Ibn Abī Uṣaybiʿah mainly keeps his Galen free of popular stories or rejects them. As part of his preservation of Galen's scientific legitimacy he offers no criticism at all of his verbosity or boastfulness. He is a model for all physicians. Thus it is entirely wrong to couch Ibn Abī Uṣaybiʿah's biography in debates about the 'decline' of Greek culture in Galen's own day such as they were voiced in the 19th and earlier 20th centuries.[100] Classical scholarship in that period, especially in Germany, was quick to condemn Greek intellectuals of the empire and the condemnations too often coincided with racial slurs of impurity. These old debates should not be resurrected, since they imposed views of their own time to an excessive degree and represented a deeply inadequate understanding of the richness of imperial Greek society. Bürgel's regressive suggestion that 'guardians of "genuine" Humanism in the West today' show a well-known disparagement of Galen is simply wrong.[101] His appeal to the 'free Spirit' that signifies the West, as opposed to the love of dogmatic thought and absolute regard for authority that characterize near eastern and Islamic intellectual life and in particular the Islamic reception of Galen,[102] has but one valid point if one is charitable: though Galen experimented in theory and practice, though he questioned authority and sustained it by bringing it up to date, later generations were reluctant to question Galen himself and thus handed him an almost total dominance in the field of medicine. And so for Ibn Abī Uṣaybiʿah, who was not backward looking and who publicized the considerable merits of his contemporaries and their predecessors in the Islamic empire, Galen was his hero and his dominance was unchallengeable; but medicine, nonetheless, continued to produce representatives of the highest quality.

7 Alexandrian Physicians and Their Christian and Other Contemporaries (Chapter Six)

To reach Islam, Ibn Abī Uṣaybiʿah had to go via Alexandria, and this is the theme of Ch. 6. His readers knew that Greek medicine had been shaped in the late-

100 As in the recent treatment by Bürgel, *Ärztliches Leben*, 400, drawing on Ilberg, *Galens Praxis* (from 1905). Ilberg saw Galen's self-regard as proof that he was a *Graeculus*, a 'Greekling', rather than a true Hellene of old; there is nothing to this.

101 Bürgel, *Ärztliches Leben*, 400, on Galen who 'bekanntlich im Abendland heute von manchen Hütern des "echten" Humanismus gern etwas abschätzig beurteilt wird'.

102 Bürgel, *Ärztliches Leben*, 409: 'der "autonome Geist" ein Signum des Abendlandes … umgekehrt Dogmatisierung und Verabsolutierung von Autoritäten ein charakteristischer Zug des vorderorientalischen und damit auch islamischen Geisteslebens'.

antique world. But it is not until well into Ch. 5 that he mentions Alexandrian interventions (for example incorporating Galen's *On Muscles* in the collection known as *The Small Book of Anatomy*, Ch. 5.1.37 no. 8[103]). In Ch. 5.2 he adds a short appendix on the major physicians after Galen, beginning with the Alexandrians of late antiquity. These were 'among the commentators on Galen's books, and summarized and abridged them'.[104] The list, however, is slightly different from that at the start of Ch. 6, which includes John the Grammarian. John has been mentioned a handful of times in Chs. 3–4, and his presence is felt behind the lists of names of physicians. His importance as a source is signalled and questioned at 3.5.[105] But nothing prepares us for his ascendancy in Ch. 6.1–2. The explanation for it is simple: he embodies the transfer of medicine from Greek to Arabic because 'he was still living at the time of the advent of Islam'. The confusion here between John the Grammarian who was better known (and better known to us) as the philosopher John Philoponus with a John who had an interest in medicine and wrote a chronology of physicians taken up and adapted by Isḥāq ibn Ḥunayn has been discussed above where it was suggested that this latter John was perhaps a Syriac author with a sectarian agenda to align medicine and Christianity and very unlikely to be identical with the medical author also called John the Grammarian, who wrote several commentaries on Galen listed at Ch. 6.2. A date from one of the philosopher John Philoponus' works placed him just after the Hijra owing to a transmission error in the text.[106] This was just what the biographical machine needed. It somehow generated fictions about John recounting his religious differences with the Byzantine authorities. These tales are cited by Ibn Abī Uṣaybiʿah from the *Fihrist* and in particular and at length from ʿUbayd Allāh ibn Jibrīl's *Manāqib al-aṭibbāʾ*. As we have it, ʿUbayd Allāh ibn Jibrīl does not make John live at the time of the conquest of Alexandria. Indeed he places him in the midst of the Christological disputes of the mid-5th century, confusing him, wilfully it appears, with a Constantinopolitan archimandrite called Eutyches (Ch. 6.1.3). But this did not stop a good story.

---

103 Cf. title nos. 10, 14, 79.

104 We cannot tell if the three major focusses of Ch. 5.2, Oribasius ('physician to King Julian'), Paul of Aegina, Philagrius ('close to this time'), who were not Alexandrians, are associated by Ibn Abī Uṣaybiʿah with Alexandrian scholarship. It looks as if he is filing some known, important names as best he could; cf. Ch. 4.10.1–11.3 where key figures and their bibliographies are woven into the lists of names, and below on the quiet end to Ch. 6.

105 He appears as John of Alexandria at Ch. 4.1.11.3 in a list of philosophers, some of whom are identical with physicians mentioned at Chs. 5.2.1 or 6.1.1.

106 See p. 150.

Content that he had understood his man, Ibn Abī Uṣaybiʿah goes on to quote 35 book titles by John. First come real philosophical commentaries by the real John Philoponus, most of which survive in Greek. There follows a series of commentaries on Galen which are undoubtedly genuine and – why not? – by a medical John writing in Greek in late antiquity. Finally the list, taken perhaps from ʿUbayd Allāh ibn Jibrīl, offers additional philosophical titles, some of which are certainly by John Philoponus (e.g. against Aristotle). Curiously Ibn Abī Uṣaybiʿah does not include the chronology of physicians in the listing, perhaps because he knew it only through Isḥāq ibn Ḥunayn. The main achievement was a full identity for a crucial bridge between antiquity and what succeeded it.

After John, Ibn Abī Uṣaybiʿah turns to his ever trustworthy Ibn Riḍwān for an account of what the Alexandrians were teaching at the time of the Arab invasion, quoting from the *Kitāb al-nāfiʿ* (6.3.1–4). The account is a detailed exposition of seven 'levels' of instruction used by the 'Alexandrians'. It is Ibn Riḍwān's way of rationalizing the Sixteen Books represented by the Alexandrian Summaries. But it is surely no more than a teacher's fantasy of a planned syllabus read by perfect students, and when the teacher concludes by saying that, 'the approach adopted by the Alexandrians was well calculated to motivate learners, stimulate their interest in the art of medicine, and impel them to study all of Galen's works', we enter his dreams where the world of summaries, extracts, and reworkings of the canonical materials for convenience and speed of learning and use is ignored and his own collection of snippets (the *Fawāʾid*) conveniently forgotten.

Ibn Abī Uṣaybiʿah knew how medicine was taught for he tells us in later books that he studied this or that text. For now he quotes another trusted author, Ibn Hindū, from his *Miftāḥ al-ṭibb wa-minhāj al-ṭullāb*. Ibn Hindū's book was itself a summary of medicine with much useful information for learners and teachers. Ibn Abī Uṣaybiʿah's understatedness should not obscure the consciousness of its juxtaposition with the rigorous Ibn Riḍwān. He loved Ibn Hindū and quotes a good deal of poetry in his biography of him at Ch. 11.9. Ibn Hindū maintains he is quite happy with the summaries of the Sixteen and he points out to his own teacher, the physician Abū l-Khayr ibn al-Khammār,[107] who he records voicing an objection to the content of the Alexandrian curriculum, that students may progress to the hard stuff in due course. Diplomatically Ibn Hindū avers, with an attempt at a literary flourish (cf. 11.9.1), that his teacher was right as well, of course, by citing how Nature encourages learning.

107 Biography at Ch. 11.8.

Ibn Abī Uṣaybiʿah no doubt smiled quietly as he noted in conclusion the great range of the Alexandrian summaries and their focus on Galen and Hippocrates (6.5.2).

To end the chapter Ibn Abī Uṣaybiʿah, as usual doing the best he can, gathers a motley selection of physicians from Antyllus (2nd century) to Yūḥannā ibn Sarābiyūn (9th century), telling us that extracts from them can be found in 'the great compendium entitled *al-Ḥāwī*, by al-Rāzī'.

8 To Islam

The end of Ch. 6 marks a major break in Ibn Abī Uṣaybiʿah's book. Ch. 7 bears the title 'Arab and other physicians of the earliest Islamic period', and with the first of these biographies, al-Ḥārith ibn Kaladah, we are in a firmly Arab world of tribes and the Prophet and, above all, a lengthy provision of wisdom by the semi-legendary physician to the famous Sassanian ruler, Kisrā Anūshirwān (Chosroes Anushirvan), who had a significant career in Arabic literature but in reality was long dead by the time of al-Ḥārith. This section, completely fictional, circulated independently, which shows its popular nature (advice on food, wine, sex, as well as the four humours). And although the physician himself was historical, his medical interests are completely unknown and he is used instead to show the quality of 'the physician of the Arabs' at this early point in Islamic history.[108] The next three names are also Muslim, but the last of them, ʿAbd al-Malik ibn Abjar al-Kinānī, is a convert who was the successor in the Alexandrian school 'carrying on the tradition of the Alexandrians who have been mentioned earlier in this work'. In the year 99 (717) 'he moved the school of medicine to Antioch and Ḥarrān; since then, medical instruction has become widespread in various regions'. With a quiet nod to the narrative constructed in Abbasid times to explain how and why (i.e. as a result Christian hostility to learning) Greek philosophy came to be transferred to Islam, Ibn Abī Uṣaybiʿah notes the diffusion of Greek medicine into the different lands (*fī l-bilād*).[109] But the story from this point on is also the story of the entry of medicine – especially of Christian physicians – into the Islamic court.[110] Hippocrates

108 Ullmann, *Medizin*, 19–20.

109 Gutas, 'Alexandria', esp. 187–188 (it is not clear why Antioch and Ḥarrān are the named stages of the transmission to Baghdad); cf. above and note again Ibn Abī Uṣaybiʿah's own avoidance of negative stereotyping. Cf. Ch. 15.1.2 for al-Fārābī's famous version of the story, with Gutas, 155–156.

110 Biographies 5.5 to 5.9. The chapter ends, however, with the brief note on the very tradi-

had pointed out the danger of consorting with kings. The flagship biography of Ḥunayn ibn Isḥāq (8.29) represented the problem through Ḥunayn's (alleged) autobiography. Rivalry and danger at court are constant themes in the *ʿUyūn*. The perfect balancing act had been performed, of course, by Galen. He served kings but they allowed him to go when and where he wished. They honoured him but he was not beholden to them. If Galen was not replicable in full in later times, his example as *khātam al-aṭibbaʾ* remained universal, and Ibn Abī Uṣaybiʿah made a successful effort to keep it so by both magnifying and moderating the status of the physician wherever he lived in accordance with the model his Jālīnūs had established.

---

tional female practitioner, Zaynab 'the physician of the Banū Awd'. Once again we see Ibn Abī Uṣaybiʿah filing a name he felt was important at the very end of a chapter.

CHAPTER 8

The Practice of Medicine as Seen through the *ʿUyūn al-anbāʾ*

Emilie Savage-Smith

For over twenty years, Ibn Abī Uṣaybiʿah collected material for his *ʿUyūn al-anbāʾ fī ṭabaqāt al-aṭibbāʾ* concerned with the lives led by physicians, past and present, and their education, their social standing, and their achievements. He opens his world history of physicians with reflections on how humans have come to know of medical cures. While he is a great admirer of Greek learning (particularly, of course, that of the most influential of physicians, Galen) he does not allow any one group of people – including the Greeks – to claim to be the originators of medicine. He sees medicine as being practised, in one form or another, by all human societies.

To Ibn Abī Uṣaybiʿah, the practice of medicine is the noblest of occupations – one that was continually evolving and developing up to and including his own day. He was clearly proud of the role of physicians in society and proud of the Damascene physicians of his day, who stood at the culmination of two millennia of slowly developing medical knowledge.

Because the *ʿUyūn* is at heart a social and literary history of physicians, focusing upon the trials and tribulations, successes and failures, of medical practitioners throughout the ages, it was not intended as a systematic presentation of medical theories and therapies. Nonetheless, in amongst the hundreds of anecdotes presented, and the seemly endless lists of writings undertaken by physicians, we can find ample evidence of the various ailments that plagued the societies of Ibn Abī Uṣaybiʿah's day and the attempts made to cure them.

Of course, the physicians with which he associated, or about whom he could obtain any information, were (with a single exception) male and were urban physicians serving the elite. In other words, the *ʿUyūn* does not present the full spectrum of medical care. It is reasonable that there would be more information available regarding court physicians or physicians at major hospitals or otherwise prominent in a city than there would be about physicians in rural areas or those tending to the poor or to the women's quarters. The role of women as providers of medical care is essentially invisible and inaudible; invisible and inaudible because not one single document or treatise written by a Muslim female medic or midwife has been preserved from the Middle

 | DOI:10.1163/9789004410312_009

East before the nineteenth century (which contrasts markedly with medieval European sources where midwives do speak through preserved documents). Having said that, information does emerge from this history regarding concern for medical care in rural areas, and while only one semi-legendary female medical practitioner is named, women do figure prominently as patients and patrons.[1] Some conditions described in the *ʿUyūn*, however, are distinct to women, including retained menstruation or excessive loss of blood during menstruation. Being able to diagnose and predict if a wife or concubine was pregnant or not, seems to have been a skill highly prized amongst court physicians – an example being the physician to the caliph al-Mahdī who was able to predict through urine analysis the pregnancy and the sex of the births of both Hārūn al-Rashīd and his elder brother.[2]

Much is learned of hospital medical care, but, judging from the *ʿUyūn*, the medical scene itself was quite pluralistic. Some physicians engaged with medicine solely on a theoretical basis; some were known for their cures; some were known for skills in treating a particular ailment; some were known for their diagnostic skills or ability to detect apparent death, Moreover, it was customary for a physician to be a very learned man who composed treatises on a wide range of topics, many of which were not directly concerned with medicine – Ibn Sīnā being a prime example. Indeed, some included in this history appear better known for their poetry (or in one case being the brunt of a satirical poem) than for their medical successes.

Therefore, it must be constantly kept in mind that the *ʿUyūn* is a literary history of medicine whose main purpose was to promote knowledge of the importance of the medical profession and, along the way, to entertain the reader. It was not intended to provide a guide to medical care or to be a dull record of medical statistics.

Moreover, rivalry and disputes between physicians form an important and appealing literary topic exploited by Ibn Abī Usaybiʿah to the fullest. In fact, the disputes often display more about the characters of the disputants than they do about everyday medical care. Within this background of competition and disputation the reader can glimpse the religiously mixed medical communities of the early Islamic world, where Jews, Christians, Sabians, Zoroastrians and Muslims interacted closely with each other in a variety of settings, especially at the courts of rulers and in hospitals.[3]

---

1 For specific references to passages concerning topics mentioned in the following discussions, consult the Subject Index to Vols. 2 and 3 at the back of Volume 1.

2 Ch. 8.8.

3 For the general interaction of *dhimmīs* (non-Muslims) and Muslims in early Islamic society, see Sahner, *Christian Martyrs under Islam*.

1 Hospitals and Institutions

The *ʿUyūn* is one of the most important sources preserved today for the history of hospitals. The term used for hospital, *bīmāristān*, is a Persian word meaning literally 'place for the sick', and there was no Greek or Arabic word used for the institution at this time.[4] It is in Islamic lands that we first see the establishment of hospitals for teaching, training, and treatment, as well as serving as places of convalescence for those recovering from illness or accidents, or as charitable housing for the indigent and those without families to care for them.

Only the Abbasid foundations in Baghdad and the Ayyubid in Syria and Egypt are mentioned in the *ʿUyūn*. The famous Mamluk hospital in Cairo, the Manṣūrī, receives no attention because it was not founded until sixteen years after Ibn Abī Uṣaybiʿah's death, and by this time attitudes toward medical care and physicians appear to have changed from those of the Ayyubid period, of which Ibn Abī Uṣaybiʿah was so proud. The foundation document of the Manṣūrī hospital, dated 12 Ṣafar 685/9 April 1286 specified that only Muslim physicians were to be employed and only Muslims to be taken as patients.[5] This restriction comes at a point when the Muslim population had probably reached a majority for the first time, and it contrasts markedly with the attitudes of the Abbasid and Ayyubid rulers, where the directives were to extend medical care to non-Muslims (*dhimmīs*) as well as Muslims, and where Christian, Jewish, and Muslim physicians often worked together in hospitals and at the courts of rulers (though not always amicably, it has to be said).

In other words, the *ʿUyūn* was composed before the Islamization of medical practice. After the third quarter of the thirteenth century Muslims come to dominate the medical scene and non-Muslims appear marginalized.

In the course of composing the *ʿUyūn*, Ibn Abī Uṣaybiʿah provided especially important information on the early establishment of *bīmāristāns*. Of the Abbasid hospitals in Baghdad, Chapter Eight mentions[6] one possibly built on the Karkhāyah canal around the year 800 by Yaḥyā ibn Khālid ibn Barmak, vizier to Hārūn al-Rashīd (*reg.* 170–193/786–809) and refers to hospitals in Gondēshāpūr as forming models for the early Baghdadi hospital. Greater detail is provided in Chapter Ten: a *bīmāristān* founded by Badr al-Muʿtaḍidī, the

4 It was not until the modern period that the Arabic word *mustashfā*, 'place where healing is to be sought', became the normal word for 'hospital'.

5 Northrup, 'Al-Bīmāristān al-Manṣūrī'; Ragab, *Medieval Islamic Hospital*. 163–170; Gutas, 'Ibn al-Nafīs's Scientific Method', 142–144; Lewicka, 'Medicine for Muslims'.

6 See the Subject Index to Vols. 2 and 3 at the back of Volume 1 under 'hospital' for specific references to chapters and subsections.

steward (*ghulām*) of caliph al-Muʿtaḍid bi-Allāh (*reg.* 279–289/892–902) and three foundations between 914 and 920 established by the caliph al-Muqtadir bi-Allāh (*reg.* 295–320/908–932), by his mother Shaghab, and by his 'good vizier' ʿAlī ibn ʿĪsā, as well as a fifth hospital founded by Ibn al-Furāt (d. 312/924), several times vizier to al-Muqtadir, and yet another established by Abū l-Ḥasan Bajkam.[7] The most famous and important of the Baghdadi *bīmāristāns* was no doubt the ʿAḍudī hospital founded by ʿAḍud al-Dawlah in 371/981, and both Chapters Ten and Eleven contain biographies related to work in the ʿAḍudī *bīmāristān*.

The Ayyubid hospitals in Cairo are discussed in Chapter Fourteen, with particular emphasis on the Nāṣirī hospital founded in 566/1171 in Old Cairo by al-Malik al-Nāṣir Ṣalāḥ al-Dīn (Saladin), after whom it was named. The Nāṣirī hospital was created by modifying a part of the palace built by the Fatimid caliph al-ʿAzīz in 384/994 so that it could serve as a hospital.[8] In Chapter Thirteen there is a mention of an unspecified *bīmāristān* in Fustat.

The fullest accounts of hospitals and their administration and functioning are to be found in Chapter Fifteen, for both Ibn Abī Uṣaybiʿah and his family had direct experience with the *bīmāristān* established in Damascus by Nūr al-Dīn ibn Zangī around 569/1174. A small hospital, sometimes referred to as al-Bīmāristān al-Ṣaghīr, was situated to the west of the Great Mosque of the Umayyads and had been functioning about half a century before Nūr al-Dīn ibn Zangī founded what came to be called the Nūrī hospital to the east of the Great Mosque.[9] The accounts provided by Ibn Abī Uṣaybiʿah are the major evidence for what physicians may have been doing within the hospital walls. There are also scattered references in the *ʿUyūn* to other hospitals: in al-Raqqah, Rayy, Marrakesh, Mayyāfāriqīn, and Jerusalem.[10]

Recently, Ahmed Ragab has used Ibn Abī ʿUsaybiʿah as a major source for his thesis that the circle of Damascene physicians connected to Muhadhdhab al-Dīn al-Dakhwār (d. 628/1230), the often-mentioned teacher of Ibn Abī Uṣaybiʿah, developed a practice for hospitals that dominated in the Levant and Egypt in the thirteen and fourteenth centuries. That practice was based on a revival of interest in the 'practical' writings, such as the *Ḥāwī* of Abū Bakr Muḥammad ibn Zakariyyā al-Rāzī (d. 313/925), whose lengthy biography is given in Ch. 11.5, as well as the *Qānūn fī l-ṭibb* of Ibn Sīnā, whose biography is given by Ibn Abī Uṣaybiʿah in Ch. 11.13.[11]

---

7 See Pormann, 'Islamic Hospitals'.

8 See Rageb, *Medieval Islamic Hospital*, 84–89.

9 Pormann & Savage-Smith, 96–100; Ragab, *Medieval Islamic Hospital*, 52.

10 See the Subject Index to Vols. 2 and 3 at the back of Volume 1 under 'hospital'.

11 Ragab, *Medieval Islamic Hospital*, 139–175; see also review article by Nahyan Fancy in

We have no real data on any of these institutions – no statistics on patients, nor on what diseases were treated, or the success rates of treatments. The hospitals were for those travelling, or without families, or the poor. It is unlikely that any truly wealthy person would have gone to a hospital unless taken ill while travelling far from home, for in most circumstances the medical needs of the wealthy and powerful would have been administered at home. Until the hospitals developed as teaching centres, and probably long after that, physicians and medics of all sorts would conduct their practices either in the homes of the elite or in a shop (*dukkān*), where they also held teaching sessions (*majālis*).

2 'Prophetic Medicine', Magic & Astrology

It is after the *'Uyūn* was composed that interest really grew in what is called 'Prophetic Medicine' (*al-ṭibb al-nabawī*), a genre of medical writing in which clerics and legal scholars advocated the traditional medical practices of the Prophet's day rather than the medical ideas assimilated from the Greeks and extended by learned physicians into Islamic lands.[12] While some treatises on Prophetic Medicine were composed in the ninth century, the genre did not become widely popular until the thirteenth and fourteenth centuries, and these works continue to be so today. The treatises on the topic by the legal scholar Ibn Qayyim al-Jawziyyah (d. 751/1350), the historian al-Dhahabī (d. 749/1348), and the theologian Jalāl al-Dīn al-Suyūṭī (d. 911/1505) are still available in modern printings. It is significant that this alternative form of medical thinking is not even mentioned in the *'Uyūn* and is essentially invisible in the lists of compositions by the subjects of the biographies.[13]

Nor is any magical therapy described in the *'Uyūn* – no talismans, no amulets, no magic-medicinal bowls, though we know that at least one magjc-medicinal bowl was made in 565/1169 for Nūr al-Dīn ibn Zangī, the Ayyu-

---

Nazariyat 3/1 (2016), 137–146 and response by Ragab in *Nazariyat* 3/2 (2017), 125–134; and Savage-Smith, 'The Working Files of Rhazes' for a different interpretation of the Rāzī materials.

12 See Perho, *The Prophet's Medicine*; Pormann & Savage-Smith, *Medieval Islamic Medicine*, 71–75, 150–152.

13 Only twice does the topic come up in the lists of hundreds of books: a commentary on Prophetic traditions concerned with medical matters is listed amongst the writings of Amīn al-Dawlah ibn al-Tilmīdh (Ch. 10.64.20 no. 10) and amongst the books composed by Najm al-Dīn al-Minfākh (Ch. 15.56.2 no. 4). Neither are preserved today.

bid founder of the great hospital in Damascus that bore his name (the Nūrī Bīmāristān) and which plays a large role in the later chapters of the *ʿUyūn*.[14]

Astrology, on the other hand, and its usefulness in diagnosis and prognosis, was part of the Greek medical scene and continued to be so in later centuries. While medical magic may be absent from the *ʿUyūn*, there are examples for the use of illusions to cure psychosomatic ailments and one instance of the visual illusions created by a magician to cause amazement. In other words, the dominance of the learned professional system of medicine is reinforced by Ibn Abī Uṣaybiʿah throughout the *ʿUyūn*.

3 Brain, Mind, and Psychosomatic Disorders

Ibn Abī Uṣaybiʿah was particularly attracted (as probably were his readers also) to the dramatic reports of physicians able to detect if a person were still alive when others thought they had died. Avoiding pre-mature burial by being able to detect that a person had not actually yet died is a matter of concern found from antiquity through to modern times and formed a particular category of diagnosis and prognosis.[15] Amongst the works attributed to Galen is a *Treatise on Prohibiting Burial Within Twenty-Four Hours* (*M. fī taḥrīm al-dafn qabl arbaʿ wa-ʿishrīn sāʿah*).[16] This treatise later formed the centrepiece for a discourse on apparent death by ʿUbayd Allāh ibn Jibrīl ibn ʿUbayd Allāh ibn Bukhtīshūʿ (d. 450s/1058–1066), which has recently been edited and translated.[17] Curiously, however, this treatise on apparent death and avoiding pre-mature burial is not amongst the compositions attributed to him in his biography given in Ch. 8.6, though there is no reason to deny its attribution.

Apparent death (*saktah*) was defined by Ibn Hindū (d. 423/1032), whose biography is given in Ch. 11.9,[18] as occurring 'when there is loss of sensation and movement caused by an excess of blood or a cold thick humour filling the ventricles of the brain, thus preventing the passage of the psychic pneuma and causing the person to look as though he is asleep, although he is not'.[19] In

---

14 Now in the Khalili Collection of Islamic Art, Inv. no. MTW1443; see Maddison & Savage-Smith, *Science, Tools and Magic*, i:82–83; Savage-Smith, 'Medicine in Medieval Islam', 161 fig. 5.3.

15 See Bürgel, *Ärztliches Leben* 266–272.

16 See Ch. 5.1.39 no. 163.

17 ʿUbayd Allāh ibn Bukhtīshūʿ, *On Apparent Death*.

18 Ch. 11.9.

19 Ibn Hindū, *Miftāḥ* (Tibi), 71.

Chapter Ten[20] we read of the Sabian physician Thābit ibn Qurrah (d. 288/901) performing, on his way to the palace of the caliph, a 'Christ-like act' by reviving a butcher who people thought to be dead. In Chapter Twelve[21] an Indian physician revives a cousin of Hārūn al-Rashīd who was thought to have died, while in Chapter Fourteen[22] the Jewish physician Ibn Jumayʿ, sitting near his shop in the candlestick market in Fustat, stops a funeral procession to resuscitate a person about to be buried. In Chapter Fifteen we learn of the eleventh-century Jacobite Christian physician in Damascus, al-Yabrūdī, using induced vomiting to revive people who have lost consciousness because they overate some food (horse meat washed down by beer; too many carrots; or apricots eaten with hot bread).[23] *Saktah* (apparent death) also makes less spectacular appearances in the form of strokes, apoplexy, and fainting.

Physicians who could cure psychosomatic ailments with ingenious ruses formed another lively topic for Ibn Abī Uṣaybiʿah. In Chapter Eight,[24] a physician cures one of Rashīd al-Dīn's concubines of her inability to lower an arm by seizing the hem of her robe 'as though intending to lift it', at which point she recovered the function of her limbs. In Chapter Fourteen,[25] the Egyptian physician al-Tamīmī relates a story about his father who, when drunk, fell from a considerable height onto the floor of an inn and lost consciousness; upon awakening he rode off with no recollection of the event until the innkeeper later reminded him of it and told him he was lucky to be alive, at which point his father was wracked with pain and required many days for recovery. This anecdote then reminds Ibn Abī Uṣaybiʿah of another story[26] in which a group of travelling merchants stop to rest and one of them is bitten by a snake; another of the merchants tells the one bitten that nothing had happened except he hit his foot against a thorn while sleeping, and so the victim continues with no pain or ill effect until a few days later when it is revealed that he was bitten by a snake, at which point he is in excruciating pain and ultimately dies.

In Chapter Ten,[27] we have the extraordinary account of Abū l-Barakāt al-Baghdādī curing a patient who suffered from melancholia and was convinced that at all times he had an amphora on his head; the cure was achieved by having one servant put an amphora on the roof of the house, which he would

20 Ch. 10.3.9.
21 Ch. 12.6.
22 Ch. 14.32.3.
23 Ch. 15.3.1.2; 15.3.1.3; 15.3.1.5.
24 Ch. 8.3.3.
25 Ch. 14.14.4.
26 Ch. 14.14.5
27 Ch. 10.66.2.

break as a second servant swung a club above the head of the patient. Ibn Abī Uṣaybiʿah ends the account with the intriguing statement: 'This is a great type of treatment and a number of the ancient physicians such as Galen and others had similar cases of treatment by illusion, and I have mentioned a great deal of this in another book.' Regrettably, that book is not preserved today.

While hiccups are not considered a psychosomatic ailment, psychological cures were employed in curing them (as they often are today). There is, for example, the case[28] of a woman with a severe case of hiccups cured when servants startled her by simultaneously pushing over the edge of the roof and breaking a number of large jars filled with water. In another instance, obesity was cured by instilling a great fear in the patient so that he would not eat.[29]

4 Categories of Bodily Illness

As for the more mundane ailments and illnesses, they are what we might expect in a pre-modern society. It is the everyday complaints that dominate. When we try to interpret the medical concerns in the medieval period we should keep in mind the problems of obtaining good quality drinking water and the limited and monotonous diet. Malaria was endemic throughout the Mediterranean basin and slowly spread elsewhere, while in the pre-antibiotic era relatively minor infections could have very serious consequences. There were no x-rays or CT scans or any way of looking 'inside' the body, and no general anaesthetics and no reliable analgesics. Yet, even given these very limited conditions, the *physiological* longevity held at around 80 years of age – that is, the length of life usually attained when the principal causes of death other than aging are ruled out (childhood infections, childbearing, epidemic infectious diseases, war and famine). The biological limit of life expectancy in the adult male does not seem to have changed since proto-historical times, for it is more a *biological* than social parameter – that is, until very recent modern times in modern wealthy and well-fed societies.[30] Thus the average life expectancy of about 80 years for the adult male population is reflected in the biographies presented by Ibn Abī Uṣaybiʿah.

As for the afflictions to which the populations were subject, eight general categories seem to emerge from the *ʿUyūn*:[31]

28 Ch. 8.4.11.

29 Ch. 8.8.5.

30 Grmek, *Diseases in the Ancient Greek World*, 105–109.

31 See the Subject Index to Vols. 2 and 3 at the back of Volume 1 for specific references to

1. Gastro-intestinal complaints dominate: general abdominal pain, 'weakness of the stomach' (*ḍuʿf al-maʿidah*), flatulence (*rīḥ*), indigestion (*quṣūr fī l-haḍm*, and other terms), nausea (*ghathayān*), rumbling in the belly (*qarāqir*), pain in the belly (*wajaʿ al-baṭn*), and vomiting (*qayʾ*). Diarrhoea and dysentery (*is'hāl, dharab*), chronic as well as uncontrollable, were especially prevalent. Dropsy (*istisqāʾ*) and colic (*qawlanj*) are very frequently mentioned, but just what was meant by these terms is uncertain. Three types of dropsy were distinguished by Ibn Hindū (d. 423/1032): *ziqqī* (like a water-bag) when the fluid can be heard to rumble in the belly, *laḥmī* (like meat) with a hard swelling 'on the liver' as well as fluid in belly, and *ṭablī* (like a drum) with a large quantity of wind as well as fluid, producing drum-like sound when tapped,[32] Today, dropsy is defined as a pathological accumulation of diluted lymph in body tissues and cavities, but such precision is not applicable to medieval accounts, where it remains an imprecise term for any accumulation of fluid in the abdominal cavity.

The same vagueness and imprecision apply to the term 'colic', which covered a range of complaints of the bowels that were accompanied by pain in the belly and other symptoms. As William Webb stated in a medical dissertation on colic submitted in 1798, when discussing the history of colic:[33] 'The word colic might with propriety be blotted out of the annals of medicine; but as custom has established it as a word to express certain symptoms or stages of an intestinal disorder, it will perhaps be proper too retain it.' And so the term has been retained here.

Haemorrhoids (*bawāsīr*) and fissures caused by them appear to have been a very troubling problem to a large part of the population in classical antiquity and in medieval Europe as well as in the medieval Middle East. Yet, somewhat surprisingly, haemorrhoids are only mentioned twice in the course of the *ʿUyūn*, and one of those occurs in a book title.

Bladder and kidney stones (calculi, *ḥajar fī l-mathānah/ḥaṣāh*) are mentioned more times than a modern reader might expect. It has been noted that idiopathic bladder stone disease, especially in children, was and is endemic in India, Iran, Turkey and Egypt, and appears also to have

passages where various illnesses are discussed. It should be kept in mind that retrospective diagnosis of a condition is fraught with difficulties and exact identifications are often impossible; for a discussion of these issues, see Álvarez Millán, 'Disease in tenth-century Iran and Irak', 59–73.

32 Ibn Hindū, *Miftāḥ* (Tibi), 74.

33 Webb, 'An inaugural dissertation on the colic', 7.

been common in classical Greece, especially in boys.[34] Yet, in none of the instances mentioned in the *'Uyūn* is surgery undertaken, though in one instance[35] much is made of the fact that surgery was avoided. The procedure for surgically removing a calculus impacted in the urethra[36] was an ancient one, well described in antiquity and in Arabic surgical tracts,[37] yet not undertaken in the accounts given in the *'Uyūn*.

2. Respiratory ailments, including those of the nasal cavity and paranasal sinuses, form the next largest category of complaints. These include: asthma, bad breath, breathlessness, catarrh and head colds, persistent coughs, nosebleed, putrid or bad humours in the chest, and especially what has been translated as 'pleurisy', a translation of *shawṣah* or *birsām*. In the case of the latter term there has been much confusion in the medical literature with the term *sirsām* or *sarsām* meaning 'severe headache' and rendered by some as 'phrenitis'.[38] As a result, the condition being referred to often remains open to speculation.
3. Pains of all sorts form another category of ailments, including headaches, earaches, sore joints and the pain of gout. Curiously, toothache is not mentioned in the accounts in the *'Uyūn*, though there are several references to the use of a toothbrush (*siwāk*) and toothpick (*khilāl*), as well as one occurrence of a discharge from teeth. The care of the teeth and gums are topics of treatises mentioned in book-lists.
4. Skin conditions form the fourth category: carbuncles, erysipelas, scabies, severe itching, scrofula. Ulcers and ulcerations (*qarḥ* pl *qurūḥ*) of all sorts, including suppurating and inflammatory ulcers (*dubaylah*/*naghlah*) play a large role in the medical profile. A whitlow (an ulcerated inflammation of the area around the nail of a toe or finger, an uncommon complaint in modern urban society) is depicted as potentially causing death if the finger involved does not heal properly or had not been amputated in time.[39] The transmissible skin conditions of leprosy (*judhām*, which can also be elephantiasis), measles (*ḥaṣbah*), smallpox (*judarī*) and another form of pox called *ḥumayqā'* are mentioned several times.

---

34 Grmek, *Diseases in the Ancient Greek World*, 112.
35 Ch. 10.76.2
36 A calculus in the kidney or ureter would have been inaccessible and untreatable.
37 See Bitschai & Brodny, *Urology in Egypt*; Savage-Smith, 'The Practice of Surgery in Islamic Lands', 315.
38 See Dols, *Majnūn*, 57–58, Carpentieri, 'On the Meaning of *Birsām* and *Sirsām*'.
39 Ch. 10.66.3.

Cancer, on the other hand, gets only a single mention, where it is said that the early tenth-century physician Yūsuf al-Sāhir, who is the subject of the biography, 'suffered from a cancer (*saraṭān*) on his forehead which made it impossible for him to sleep, and it was for that reason that he was dubbed al-Sāhir ('the insomniac')'.[40] The disease is not mentioned in later entries. It has been suggested that the infrequency of cancerous disease in ancient and medieval populations as compared with our contemporary societies is due in part to the chemical pollutants of modern societies[41] and current methods of diagnosis.

5. The fifth category comprises the general conditions: fevers and swellings on various parts of the body, many of them inflamed. These were generally treated with cooling compresses. A swollen penis, however, gets treated by being hit with a stone so as to exude pus that has formed round a bit of grain acquired while sodomizing an animal.[42]

Paralysis, particularly partial paralysis (*fālij*, hemiplegia), figures very frequently in the accounts, as well as spasms, convulsions, and cramping. Epilepsy (*ṣarʿ*), a disorder well-known in antiquity, occurs a number of times in the *ʿUyūn* as well.[43]

Gout (*niqris*) is mentioned in chapters covering all time periods. A general 'wasting disease' (*sill*), sometimes rendered as 'consumption', occurs several times, but not in the later chapters. A more precise identification cannot be made at this distance, but the term could have included what today we call pulmonary tuberculosis, which has been known since the arrival of mankind and would have been contagious. Jaundice and a general paleness of the skin are also mentioned (for which the consumption of mutton was the cure).

Of the pestilential diseases, the vague terms *ṭāʿūn* and *wabāʾ* occur in various accounts. Identifying the precise nature of the pestilences and epidemics is difficult if not impossible without further evidence; interpretations vary from measles or smallpox, to influenza, pneumonia, tuberculosis, ergotism, and famine, as well as bubonic plague itself, depending upon time and place.[44]

---

40 Ch. 8.39. *Saraṭān* also occurs in some book titles, such as a treatise by Philagrius (Ch. 5.2.2 no. 11).

41 Grmek, *Diseases in the Ancient Greek World*, 72.

42 Ch. 13.21.2.1.

43 For a study of epilepsy in medieval Islam, see Jolin, 'Epilepsy in Medieval Islamic History.'

44 See Conrad, '*Ṭāʿūn* and *Wabāʾ*'; Conrad, 'Epidemic disease in formal and popular thought'.

6. Various mental and emotional disorders form another group of ailments: melancholia, insomnia (*sahar*), anxiety and worry (*hamm*), agitation (*qalaq*), distress (*karb*), sadness and sorrow (*ḥuzn/ghamm*), corruption or disorder of the mind (*fasād al-ʿaql / ikhṭilāt*), and hypochondria. These tended to be treated with various regimens.
7. The treatment of abscesses (*khurūj*) and small growths, as well as snakebites and the bites of wild animals form yet another category.
8. Eye ailments make up the final group of disorders, including ophthalmia (*ramad*), sore eyes, pain in the eye, blindness in general, and cataracts. More will be said below about the latter. At this point, we should pause to note what is *absent*; pannus (*sabal*) gets not a single mention, while pterygium, trachoma, prolapsis, chalazion, hypopyon, fistulas, and trichiasis[45] occur *only* as direct quotations from Galen's treatise *On Examinations by Which the Best Physicians are Recognized*[46] – that is, the conditions are mentioned only in the context of Greek medicine. This is most surprising, for Ibn Abī Uṣaybiʿah himself was trained as an oculist, and he came from a family of oculists. The omission of these conditions for which Syrian oculists in particular made considerable advances in their treatment is puzzling. We can only assume that the author's interest in recording the social lives and achievements of his family overrode any interest in the details of practice. There is also the possibility that the oculists of his day did not themselves perform the therapies described in the technical literature for these eye ailments, but others actually carried out the procedures. A third possibility is that the intricate therapeutic techniques described in the technical literature were more in the category of 'thought experiments' than actual procedures performed routinely.

5 Surgery and Amputation

Amputation of gangrenous limbs gets only a single mention, and that in the course of students disagreeing with a teacher, Abū l-Barakāt al-Baghdādī, regarding the proper treatment of whitlow.[47] In this account the students learn

45 For these eye ailments and their treatments in Arabic medical literature contemporaneous with IAU, see Savage-Smith, 'Ibn al-Nafīs's *Perfected Book on Ophthalmology*'; Savage-Smith, 'The Practice of Surgery', 316–319.

46 See Iskandar, *Examinations*, 116–117, and Ch. 15.51.3.

47 See above, under skin conditions, no. 4 in Categories of Bodily Illness.

that their teacher was correct and that if the finger is not amputated in time, then the patient may lose the entire arm or even die.[48] Other accounts in the *ʿUyūn* speak of treating a limb following an amputation carried out as punishment.

Missing entirely from the *ʿUyūn* is the setting of broken bones, except for a statement that 'Plato the Physician' taught one Phanias, 'the skills of setting broken bones and reducing luxations'.[49] This omission probably reflects the fact that in some localities bone setting was performed by a specialist medic. The Syrian physician Muwaffaq al-Dīn ʿAbd al-ʿAzīz al-Sulamī (d. 604/1208),[50] for example, clearly distinguished bone-setters, oculists and surgeons from general physicians in his *The Experts' Examination for All Physicians*., and this division of labour is borne out by the accounts given elsewhere in the *ʿUyūn*.

There is also no mention of the removal of splinters and arrow heads[51] – perhaps again explained by the separation of functions amongst the medical practitioners, or by the fact that most of the patients were not working-class manual labourers. The treatment of an arrow wound, however, does get a mention.[52] The relatively little attention given these topics when compared to medical compendia may also reflect the fact that the physicians forming the focus of the *ʿUyūn* were, for the most part, court physicians or attending a ruler (and not the soldiers) while the ruler was on military campaign.

While psychological cures and occasionally music would improve the condition of a patient, for most ailments and disorders at this time the first order of therapy was diet, followed by drugs, with surgery resorted to only when absolutely necessary.[53] Even cautery (*kayy*)[54] – a type of minor surgery consisting of the application of a heated metal instrument or caustic substances to a given part of the body – is essentially never mentioned. Cupping (*ḥijāmah*) is mentioned once as a means of staunching bleeding.

Surgery, in fact, plays a very minor role in the *ʿUyūn*. Even a tonsillectomy, which is essentially a procedure for removing a growth, is not mentioned, though it was described in antiquity and continued, according to the Arabic

48 Ch. 10.66.3. Amputation of gangrenous limbs was a topic in nearly every surgical manual; see Savage-Smith, 'The Practice of Surgery', 316 and sources cited there.

49 Ch. 3.5. There are however, some book titles that refer to setting broken bones (Chs. 4.1.9.1 no. 12; 4.1.9.2 no. 29; 4.1.10.2 no. 39; 11.5.25 no. 69 and no. 182).

50 *Imtiḥān al-alibbāʾ li-kāffat al-aṭibbāʾ*. His biography is given in Ch. 15.34, although this treatise is not named. For the treatise, see Leiser & al-Khaledy, *Questions and Answers*.

51 A book on arrowhead extraction, however, is attributed to Hippocrates (Ch. 4.1.9.2 no. 59).

52 Ch. 10.10.2.

53 See, for example, Ch. 15.51.3.

54 See *EI Three* art. 'cautery' (C. Álvarez Millán).

technical literature, to be practised (with better instrumentation) and was a procedure quite within the capabilities of the day.[55]

The fact that a given physician was skilled with the knife is recorded a couple of times,[56] but there is, in fact, in the *ʿUyūn* only one mention of invasive surgery actually taking place, and that is for dropsy (or ascites).[57] In the earlier Greek literature dropsy was said to be treated by inserting a metal tube, or cannula, through an incision in the abdominal wall and drawing off the liquid (paracentesis). The procedure was repeated in Arabic medical manuals, though most writers, such as Ibn Sīnā, warned against it except under special circumstances. In the major treatise by the Andalusian physician Khalaf ibn ʿAbbās al-Zahrāwī (fl. c. 1000), which Ibn Abī Uṣaybiʿah praised as 'a perfect book as regards its content' in his biography given in Ch. 13.56, al-Zahrāwī said the following:[58]

> When you have been trying medical treatment for ... dropsy and your treatment is ineffective, then consider: and if the patient be in a low state or suffer from another disease besides dropsy, such as cough or diarrhoea or some such, then beware of attempting operative treatment since it will be of uncertain outcome. But if you see that the patient has plenty of strength and suffers from no other complaint beside the dropsy, and is of neither tender nor advanced age, then the operative procedure is [as follows] ...

Ibn Abī Uṣaybiʿah's teacher, al-Dakhwār, who had been a student of the physician Ibn al-Muṭrān (d. 578/1191)[59] told Ibn Abī Uṣaybiʿah[60] that he had accompanied Ibn al-Muṭrān to the great Nūrī hospital in Damascus where he worked, He was present when a surgeon (*jarāʾiḥī*) by the name of Ibn Ḥamdān treated a patient with dropsy (*istisqāʾ ziqqī*) by puncturing the swollen abdomen and draining off a yellow liquid, while Ibn al-Muṭrān watched the patient's pulse during the operation. The patient weakened and so the surgeon stopped the draining and applied a dressing. The patient's wife was with him, and Ibn Muṭrān instructed her not to remove the dressing or change it in any way until he

55 Savage-Smith, 'The Practice of Surgery in Islamic Lands', 315–316.

56 Chs. 14.17; 14.31.1

57 Or ascites. See Ch. 15.23.4.2.

58 See Savage-Smith, 'The Practice of Surgery in Islamic Lands', p. 311 and n. 15 where specific references are given to this passage and to similar statements in other Greek and Arabic medical writers.

59 His biography is given in Ch. 15.23.

60 Ch. 15.23.4.2.

could examine the patient the next day. The patient, however, felt so much better during the night that he insisted his wife remove the dressing, which she did, and he died.

This case recorded in the *ʿUyūn* is one of the very few testimonies to paracentesis actually being performed, though in this case the patient dies – not, it should be noted, because the physician did anything wrong, but because the patient's wife failed to follow instructions. This case history also reflects a division of labour amongst the medics, for the person performing the paracentesis is designated a surgeon (*jarāʾiḥī*) while the physician whose biography is being presented at this point (Ibn al-Muṭrān) only observes the operation and gives directions for the after care. Particularly worth noting is that Ibn al-Muṭrān monitors the pulse during the operation, and this function of a physician has not been previously noted in the literature.

Dropsy, with its associated distended abdomen, was either a relatively common complaint or one that caught people's imagination, for the illustration on the cover of the present publication is a manuscript painting made in Baghdad in 1224 inserted into a copy of the Arabic translation of Dioscorides' treatise on medicine substances.[61] It shows a physician with two patients, one with an enormous swollen stomach, possibly awaiting paracentesis.[62]

6 Drugs and Medicaments

As for the medicinal drugs used by the physicians, a large proportion of the plants described by Dioscorides and Galen would not have been known in various regions of the Middle East. The differing climatic conditions of the desert, marsh, mountain and coastal communities meant that the species of medicinal plants, as well as animal species and mineral resources, varied greatly from one region to another. Sometimes there were related local species and varieties that could be identified as similar to those described by Dioscorides

61 Washington D.C., Arthur M. Sackler Gallery, Smithsonian Institution, S1986.97a. See also Pormann & Savage-Smith, *Medieval Islamic Medicine*, 123–124 and fig. 4.1.

62 In James Morier's entertaining *The Adventures of Hajji Baba of Ispahan*, first published in 1824, the picaresque hero becomes a physician's assistant; in Ch. xix his master mocks a Frankish doctor at the Persian court who ignores Galen and Avicenna, and 'stabs the belly with a sharp instrument for wind in the stomach'. To this, Morier made a note saying: 'This alludes to tapping in cases of dropsy; an operation unknown among the Persians, until our surgeons taught it them', apparently unaware of the antiquity of paracentesis and convinced that it must have been invented by Europeans. See Morier, *The Adventures of Hajji Baba*, 97–98.

or Galen, but in some instances the substances described in the Greek sources meant little to an Arab practitioner. Conversely, the broader and different geographic horizons of Islamic writers brought them into contact with new drugs. These 'new' drugs included camphor, musk, the myrobalans (a type of plum) and the lemon (which has a mild antiseptic property), as well as commodities previously unknown to Europe, such as cotton.[63] A wide range of medicinal substances can be seen in the *ʿUyūn*, many of which were also foods and aromatic substances.

Laxatives, purgatives, emetics, and 'digestives' or other stomachics (*jawārish*; *jawārishn*) were usually simple drugs, though syrups made of honey mixed with cooked apples or other fruit were popular. Oxymel (*sakanjubīn* or *sakanjabīn*), a syrup made of honey and vinegar, often combined with other ingredients, such as quince, was a popular remedy throughout the ancient and medieval literature for fevers and indigestion.[64] A compound remedy employing very many ingredients was often referred to as a 'theriac' (*tiryāq*) and could be used as an antidote for poisoning or as a panacea for virtually any ailment. Theriacs frequently had special names and particular physicians became famous for compounding them. The formulation call *tiryāq al-fārūq*, for example, became particularly popular in Egypt, and the Jewish physician Ibn Jumayʿ compounded it for Saladin.[65]

When it comes to pain management, there is surprisingly little use of opium in the *ʿUyūn*, though in the general medical manuals opium was considered the most powerful of pain killers,[66] along with (in descending order of effectiveness) mandrake, other varieties of poppy, hemlock, the soporific black night shade, and lettuce seeds. Of these alternative analgesics, only mandrake (*luffāḥ*) occurs in the *ʿUyūn*, where in one instance it is included in a mixture of oxymel and other ingredients compounded to aid sleep, and in the second instance it is used as an aromatic incense.[67]

As for opium (*afyūn/abyūn*) itself, it is said that Galen 'journeyed to Egypt and stayed there for a while examining Egyptian medicaments, particularly opium in the territory of Asyūṭ, which is one of the provinces of Upper

63 See King, 'The New *materia medica* of the Islamicate Tradition'; Amar & Lev, *Arabian Drugs*; Paavilainen, *Medieval Pharmacotherapy*.

64 See Chipman, *Pharmacists in Mamluk Cairo*, 185–186, 193–196; Tibi, *Medicinal Use of Opium*, 93, 152; Lev & Amar, *Materia Medica*, 60; Marin & Waines, 'Ibn Sīnā on *Sakanjabīn*'.

65 Ch. 14.32.1; see also 14.54 and 14.14.6.

66 Tibi, *Medicinal Use of Opium*.

67 Ch. 10.13.3 and Ch. 8.4.7. Hemlock (*qūniyūn*) is mentioned in Ch. 4.4.3.2 in connection with the death of Socrates.

Egypt',[68] and in another instance it is said that the eighth-ninth century military commander, poet and musician Abū Dulaf suffered from a chronic stomach complaint (*mabṭūn*) for which opium was one of several options recommended by his friends and physicians.[69] The only clinical application of it in the *ʿUyūn* is when al-Dakhwār, the teacher of Ibn Abī Uṣaybiʿah, directed that opium should be added to barley water and given to a patient in the hospital suffering from mania (*māniyā*), and it is said the patient improved at once.[70]

In Chapter Seven there is a mention of edible earths or clays,[71] but otherwise medicinal earths are noticeably lacking in the *ʿUyūn*. In the Greek medical literature several types of medicinal earths, usually called *terra sigillata*, were included amongst the *materia medica*, and in Chapter Five it is said that Galen travelled to Chios to see the production of *terra sigillata*.[72] While in the Arabic translations of Greek treatises the medicinal earths were discussed in fairly comprehensible terms, only in Persia and the eastern provinces did Islamic medieval physicians display any direct experience with medicinal earths, and even then they did not apparently employ the ceremonies and make the intricate distinctions between different types of medical earths that are given in the Greek literature. In most of the medieval Arabic literature from central and western Islamic lands the medicinal earths were simply a dead literary tradition. However, after the Ottomans took direct control of the north Aegean island of Lemnos in 1479, it seems that Mehmed the Conqueror took a direct interest in reviving the custom of the annual harvest (6 August) of the Lemnian medicinal earth, possibly inspired by a reading of Galen's book on *On Simple Drugs*. And the production of special pottery containers for the earth developed in the seventeenth century. But this remarkable activity, which caused Europe to be interested in the medicinal earths as well, was not an outgrowth of the medieval Arabic medical tradition.[73]

7 Regimen and Treatment

Regimen and diet were fundamental to the maintenance of health and recovery from illness. Much is said about the topic throughout the *ʿUyūn*. Regimen,

68 Ch. 5.1.21.1. In Ch. 1.8, opium is used not as a medicament or sedative but as a lethal drug.

69 Ch. 8.20.11.

70 Ch. 15.50.4.

71 Ch. 7.9.2.

72 Ch. 5.1.21.1

73 See Raby, 'Terra Lemnia'.

on occasion, included bloodletting at established regular intervals.[74] Cupping (*ḥijāmah*)[75] employed in general regimen is mentioned a number of times in the chapters covering the earlier periods, but there is little mention of it in the later chapters.

Diet was, as in Greek medical theory, of great importance in maintaining health. Opinions differed, however, as to when and what you should eat and how much. Numerous aphorisms throughout the *'Uyūn* provide guidance on eating habits, regimen, bathing, and sexual intercourse. For example: al-Ḥārith ibn Kaladah al-Thaqafī, a legendary Arab physician and contemporary of the prophet Muḥammad, is alleged to have said:[76]

> Shovelling in more food on top of food already taken will be the ruin of any creature; it is enough to kill wild beasts in the heart of the steppe.

and also:

> Never enter the bath when you have just eaten, and moreover, never sleep with your women while intoxicated, or rise from your bed at night while naked, or sit down to eat while angry. Avoid stress, for you will be more relaxed; eat sparingly, and you will sleep better.

Two hundred years later we find in Baghdad the Christian physician Yūḥannā ibn Māsawayh (d. 243/857) saying:[77]

> You should have food that is new and wine that is old.

A teacher of Ibn Sīnā, the Christian physician Abū Sahl al-Masīḥī (d. 410/1010), is credited with saying:[78]

> Taking a nap during the day after eating is better than swallowing a beneficial syrup.

Amongst the many aphorisms on diet, regimen and general medical etiquette attributed to Abū Bakr Muḥammad ibn Zakariyyā al-Rāzī (d. 313/925) are:[79]

---

74 For example, Ch. 8.3.22; 8.20.
75 See *EI Three* art. 'Bloodletting and Cupping' (C. Álvarez Millán).
76 Ch. 7.1.3.
77 Ch. 8.26.19 no. 4.
78 Ch. 11.12.
79 Ch. 11.5.23 nos. 3, 7, 18, and 1.

> Life is too short to understand the effect of every plant growing on earth, so use the most well-known for which there is a consensus and avoid the unusual. Confine yourself to what you have tried and tested.
>
> When convalescents crave a certain food which is harmful to them the physician should contrive to manage the situation in such a way as to replace that foodstuff with something whose qualities are appropriate, and not simply refuse the patients what they crave.
>
> If the physician is able to treat a patient using foodstuffs rather than medicines, he has been truly fortunate.
>
> Certainty in medicine is an unattainable goal, and the treatment of patients according to what is written in books, without the skilful physician using his own judgement, is fraught with danger.

Physicians often disagreed regarding diet and regimen. For example, Ibn Abī Uṣaybiʿah himself was of the opinion that it was not desirable to eat unless you had a genuine appetite. In the biography of the Syrian physician Raḍī al-Dīn al-Raḥbī (d. 631/1233), Ibn Abī Uṣaybiʿah says:[80]

> Once, I was with Raḍī al-Dīn, studying some of al-Rāzī's remarks on the order of eating food. Al-Rāzī states that a person should eat twice a day and then once on the following day. 'Pay no attention to that advice,' Raḍī al-Dīn said to me. 'On the contrary, remember to eat whenever you have a real appetite, at all times, no matter whether it is once or twice during the day, day or night; for it is eating with a real appetite that benefits the body, whereas the opposite is harmful'. And he was right.

Another Syrian physician, Sadīd al-Dīn ibn Raqīqah (d. 636/1238), put his 'good advice' into poetry:[81]

> Beware of eating your fill, shun it!
> Digest one kind of food before eating another.[82]

80 Ch. 15.36.1.2.

81 Ch. 15.46.3.6.

82 Literally, 'inserting food upon food' (*idkhāl al-ṭaʿām ʿalā l-ṭaʿām*)a practice condemned earlier by al-Ḥārith ibn Kaladah (see n. 76 above).

Do not have sex often, for by doing it
continually one invites illness.
Don't drink water straight after eating
and you will be safe from great harm,
Nor on an empty stomach and being hungry,
unless you have a light snack with it.
Take a little of it: that is useful
when you have an aching, burning thirst.
Make sure your digestion is sound, that is the basic principle.
Purge yourself with laxatives once a year.
…
But do not rest continually, for this makes that
every humour in you will be made unhealthy.
Drink as little water as possible after exercise
and abstain from drinking wine.
Balance the mixing of your wine with water, for this preserves
the innate heat that always burns in you.
But do not become inebriated, shun it forever,[83]
for drunkenness is something for common people.
Keep your soul well away from its cravings,
and you will attain eternity in the Abode of Wellbeing.[84]

A topic that was common in Greek treatises on regimen and hygiene but conspicuously absent from most Arabic materials, including the *ʿUyūn*, is organized physical exercise. including wrestling. Exercises continued to be described in great detail in the Arabic versions of Greek treatises, even though there were neither gymnasiums nor palaestras in Islamic towns. They represent repetitions of a practice that had little relevance to life in medieval Islam, particularly in the western provinces. It is true that wrestling was extensively discussed in Persian literature, but even in the eastern provinces the concept of a gymnasium was not equivalent to that in earlier Greek societies.

83 *Maliyyan* means 'for a long time', which would sound odd.

84 *Dār al-Salām*, i.e. Paradise (see Q Yūnus 10:25); *salām* is often, but not quite accurately, translated as 'peace' (which here would also fail to convey the physical association).

8 Breaks with Tradition

While medical theory remained anchored in the teachings of earlier Greek figures, particularly Galen,[85] over the passing centuries medical practices were modified and adapted to changing circumstances and available medicaments.

For example, in the early eleventh century a major change of regimen for certain illnesses was introduced. In the biography[86] of Ṣāʿid ibn Bishr ibn ʿAbdūs, a Baghdadi hospital physician, we read that he was the first person in Baghdad to realize that certain conditions – in particular, facial paralysis, lassitude, and partial paralysis – that had previously been treated with hot medicines responded better to a cold regimen, even if that went against the precepts of the Ancients.

It is in the area of ophthalmology that we can see in the accounts given by Ibn Abī Uṣaybiʿah the greatest innovations that had taken place since the time of the great Galen. Cataracts were and are a common problem throughout the Middle East. Though today we know that cataracts are due to an opaque lens, in medieval literature it was said that an opaque fluid was interposed between the lens (their 'crystalline humour') and the pupil of the eye. When treating cataracts, the technique commonly used was an ancient one, known to classical antiquity and possibly originating in India. This ancient technique of 'couching' pushed the lens to one side using a thin needle, rather than removing it.[87] Curiously, couching (*qadḥ al-ʿayn*) is mentioned only once in the *ʿUyūn*, where it is said that the tenth-century oculist Ibn Waṣīf al-Ṣābiʾ refused to perform couching on a patient because the potential patient had been dishonest with him about how much money he had with him to pay for the treatment.[88] This passage does suggest the procedure was commonly done, since in the anecdote the patient refused treatment was one of seven being couched that day.

In the fifteenth chapter of the *ʿUyūn*, however, there is an account[89] of the Syrian physician Sadīd al-Dīn ibn Raqīqah, mentioned above for giving 'good advice' in poetic form, also removing cataracts (*al-māʾ al-nāzil*):

---

85 The concept of disease was based upon the Greek notion of disease arising from a disturbance in the mixture of the four fundamental humours; see Grmek, 'The Concept of Disease'.

86 Ch. 10.13.2; see also Ch. 10.23.2; and 10.38.6 no. 7.

87 For a general discussion of the couching of cataracts in the medieval period, see Pormann & Savage-Smith, 127–128 and sources cited in the notes.

88 Ch. 10.9.2.

89 Ch. 15.46.1.

> from the eyes of many persons, who, thanks to his skill, were able to see again. The instrument that he used for that purpose was hollow and curved, so that during the operation, the fluid could be more efficiently extracted, with the result that the treatment was more effective.

This passage is important evidence that at least some physicians attempted to remove a soft cataract by suction through a hollow needle[90] rather than push it to one side with a couching needle. The removal of a cataract through a hollow needle was also discussed by the Egyptian oculist ʿAmmār ibn ʿAlī al-Mawṣilī, who worked during the reign of al-Ḥākim (996–1021); Ibn Abī Uṣaybiʿah provides a biography of him and mentions his ophthalmological manual, in which ʿAmmār claimed much success with the technique, but Ibn Abī Uṣaybiʿah does not mention the hollow needle technique in association with ʿAmmār or any other oculist other than Sadīd al-Dīn ibn Raqīqah.

If such a procedure was in fact actually successfully carried out, it could only have worked on a soft juvenile cataract. Caution is needed, however, when interpreting the evidence for such procedures.[91]

9 Conclusion

The ailments and disorders in the *ʿUyūn* are not terribly dissimilar to those found in the collection of nearly 900 case histories by Abū Bakr Muḥammad ibn Zakariyyā al-Rāzī (d. ca. 313/925) reflecting the medical concerns in Iraq and Iran.[92] It is of course never possible to establish a precise disease profile of a country or region on the basis of written treatises, whether they are medical treatises or literary compositions, such as the *ʿUyūn*.[93] Moreover, the sources used by Ibn Abī Uṣaybiʿah span hundreds of years and extend across the Mediterranean basin and beyond, thus making comparisons very fragile. In

90 See Pormann & Savage-Smith, *Medieval Islamic Medicine*, 126 fig. 4.2 for an illustration of a couching needle and also a 'hollow needle', illustrating an Arabic ophthalmological treatise written in Syria between 1256 and 1275 by Khalīfah ibn Abī al-Maḥāsin al-Ḥalabī, who is not given a biography in the *ʿUyūn*.

91 For texts suggesting that the use of a 'hollow' instrument was experimental at best, see Pormann & Savage-Smith, *Medieval Islamic Medicine*, 131–135; Savage-Smith, 'The Practice of Surgery', 318–319; Savage-Smith, 'Medicine in Medieval Islam' 164–165.

92 Álvarez Millán, 'Disease in tenth-century Iran and Irak', 73–88.

93 See Biraben, 'Diseases in Europe', 345–348 for a discussion of this issue in the context of our knowledge of medicine as practised in the monasteries. Even palaeopathological evidence is not a representative sample.

addition, we cannot use the high incidence of psychosomatic illnesses or detections of apparent death portrayed in the *ʿUyūn* as evidence for their frequency within the society, or even as evidence that these physicians were especially good at treating such conditions, for these illnesses provided entertaining narrative frameworks – much as amnesia becomes a modern convenient literary narrative tool.

Unfortunately we lack hospital records and similar documents that would allow us to better estimate the frequency of illnesses and the success rate of therapies described in the *ʿUyūn*. Nonetheless, the limited evidence does suggest that a rather practical medical care was being dispensed. To a large extent the syrups of honey and other ingredients, the salves of camphor and herbs, are much like the patent medicine of not so long ago, while many of the treatments – the hot and cold compresses for headaches and swellings, the induced vomiting with quince when the stomach required emptying, the use of laxatives to open the bowels – are not unlike the domestic medicine dispensed in homes today.

The medicine on view in the *ʿUyūn* is not modern urban medicine, but nonetheless practical and basic medicine, particularly when you think about the lack of the modern technologies and chemicals that wealthy and affluent parts of the world today take so for granted. The Abbasid and Ayyubid physicians portrayed in the *ʿUyūn* were for the most part practising a rational, reasonable, and sensible medicine, and one that in many cases did help the patient.

List of Biographies by Chapter in Ibn Abī Uṣaybiʿah's *ʿUyūn al-anbāʾ*

Note: Numbers have been assigned to the biographies by the editors; they are not part of the original Arabic text. Other physicians are discussed as well, but in contexts that did not merit assigning them their own numbered biography.

CHAPTER ELEVEN: *Physicians in the Lands of the Persians* (*Bilād al-ʿajam*)

CHAPTER FOURTEEN: *Famous Physicians in Egypt*

14.1 Politianus (Balīṭīyān)
14.2 Ibrāhīm ibn ʿĪsā
14.3 al-Ḥasan ibn Zīrak
14.4 Saʿīd ibn Tawfīl
14.5 Khalaf al-Ṭūlūnī
14.6 Nasṭās ibn Jurayj
14.7 Isḥāq ibn Ibrāhīm ibn Nasṭās
14.8 al-Bālisī
14.9 Mūsā ibn al-ʿĀzār al-Isrāʾīlī
14.10 Yūsuf al-Naṣrānī
14.11 Saʿīd ibn al-Biṭrīq (Eutychius)
14.12 ʿĪsā ibn al-Biṭrīq
14.13 Aʿyan ibn Aʿyan
14.14 al-Tamīmī
14.15 Sahlān
14.16 Abū l-Fatḥ Manṣūr ibn Sahlān ibn Muqashshir
14.17 ʿAmmār ibn ʿAlī al-Mawṣilī
14.18 al-Ḥaqīr al-Nāfiʿ
14.19 Abū Bishr, 'the Physician to Rulers'
14.20 Ibn Maʿshar the Physician
14.21 ʿAlī ibn Sulaymān
14.22 Ibn al-Haytham
14.23 al-Mubashshir ibn Fātik
14.24 Isḥāq ibn Yūnus
14.25 ʿAlī ibn Riḍwān
14.26 Ifrāʾīm ibn al-Zaffān
14.27 Salāmah ibn Raḥmūn
14.28 Mubārak ibn Salāmah ibn Raḥmūn
14.29 Ibn al-ʿAynzarbī
14.30 Bulmuẓaffar ibn Muʿarrif
14.31 al-Shaykh al-Sadīd Raʾīs al-Ṭibb (Abū l-Manṣūr ʿAbd Allāh)
14.32 Ibn Jumayʿ
14.33 Abū l-Bayān ibn al-Mudawwar
14.34 Abū l-Faḍāʾil ibn al-Nāqid
14.35 al-Raʾīs Hibat Allāh
14.36 al-Muwaffaq ibn Shūʿah
14.37 Abū l-Barakāt ibn al-Quḍāʿī
14.38 Abū l-Maʿālī ibn Tammām
14.39 al-Raʾīs Mūsā ibn Maymūn (Maimonides)

Bibliography

For the alphabetic order all definite and indefinite articles, in any language and in any form, are ignored.

Abattouy, *Greek mechanics in Arabic context* = Mohamed Abattouy, *Greek mechanics zin Arabic context: Thabit ibn Qurra, al-Isfizari and the Arabic tradition of Aristotelian and Euclidean mechanics*. Berlin: Max-Planck Institute for the History of Science, 2001.

ʿAbbās, *Shadharāt* = Iḥsān ʿAbbās, *Shadharāt min kutub mafqūdah fī l-taʾrīkh*. Beirut: Dār al-Gharb al-Islāmī, 1408/1988.

al-ʿAbbāsī, *Maʿāhid al-tanṣīṣ* = ʿAbd al-Raḥīm ibn ʿAbd al-Raḥmān ibn Aḥmad al-ʿAbbāsī, *Maʿāhid al-tanṣīṣ: Sharḥ shawāhid al-Talkhīṣ*. 2 vols. Cairo: Maṭbaʿah al-Bahiyyah, AH 1316.

Abbott, 'Two Būyid Coins' = Nadia Abbott, 'Two Būyid Coins in the Oriental Institute', *The American Journal of Semitic Languages and Literatures*, 56/4 (1939), 350–364.

Abbott, *Two Queens of Baghdād* = Nadia Abbott, *Two Queens of Baghdād: Mother and Wife of Hārūn al-Rashīd*. Chicago: University of Chicago Press, 1946.

ʿAbd al-Jabbār, *Tathbīt* = ʿAbd al-Jabbār, *Tathbīt dalāʾil al-nubuwwah*, ed. ʿAbd al-Karīm ʿUthmān. Beirut: Dār al-ʿArabiyyah li-l-ṭibāʿah wa-l-nashr, 1966.

Abḥāth al-Muʾtamar = *Abḥāth al-Muʾtamar al-Sanawī al-Thāmin: al-munʿaqid fī Jāmiʿat Ḥalab, 25–26 Nīsān (Abrīl) 1984*, ed. Muḥammad ʿIzzat ʿUmar and Khālid Māghūṭ. Aleppo: Jāmiʿat Ḥalab, Maʿhad al-Turāth al-ʿIlmī al-ʿArabī, 1984.

al-Ābī, *Nathr al-durr* = Abū Saʿd Manṣur ibn al-Ḥusayn al-Ābī, *Nathr al-durr*, ed. Muḥammad ʿAlī Quranah et al. 7 vols. Cairo: al-Hayʾah al-Miṣriyyah al-ʿĀmmah li-l-Kitāb, 1980–1990.

al-Abīwardī, *Dīwān* = Abū l-Muẓaffar Muḥammad ibn Aḥmad al-Abīwardī, *Dīwān*. Beirut: al-Maṭbaʿah al-ʿUthmāniyyah, AH 1317 [/1899].

Abou Aly, 'A few notes on Ḥunayn's translation' = Amal Abou Aly, 'A few notes on Ḥunayn's translation and Ibn al-Nafīs's commentary on the first book of the *Aphorisms*', *Arabic Sciences and Philosophy*, 10 (2000) 13–22.

Abou-Aly, 'The Medical Writings of Rufus of Ephesus' = Abdullah Abou-Aly, 'The Medical Writings of Rufus of Ephesus'. Unpublished PhD thesis, University of London, 1992.

Abou-Elgheit, *Gedanken aus Ibn Abi Useibia* = Ahmed A. Abou-Elgheit, 'Gedanken aus Ibn Abi Useibia über die arabischen Ärzte und ihre Kunst'. Inaug.-Diss., Ludwig-Maximilians-Universität zu München, 1940. Reprinted in Sezgin, *Studies* II, 281–308.

Abū Aḥmad al-ʿAskarī, *Maṣūn* = Abū Aḥmad al-ʿAskarī, *al-Maṣūn fī l-adab*, ed. ʿAbd al-Sallām Muḥammad Hārūn. Kuwayt: Wizārat al-Iʿlān, 1984.

Abu-Asab, Amri, & Micozzi, *Avicenna's Medicine* = Mones Abu-Asab, Hakima Amri, Marc S. Micozzi, *Avicenna's Medicine: A New Translation of the 11th-Century Canon with Practical Applications for Integrative Health Care*. Rochester, Vermont; Toronto: Healing Arts Press, 2013.

Abū l-Aswad al-Du'alī, *Dīwān* = Abū l-Aswad al-Du'alī, *Dīwān*, ed. ʿAbd al-Karīm al-Dujaylī. Baghdad: Sharikat al-Nashr wa-l-Ṭibāʿah al-ʿIrāqiyyah, 1954.

Abū l-ʿAtāhiyah, *Dīwān* = Abū l-ʿAtāhiyah, *Ashʿāruhū wa-akhbāruhū*, ed. Shukrī Fayṣal. Damascus: Dār al-Mallāḥ, n.d.

Abū l-Barakāt al-Baghdādī, *Muʿtabar* = Abū l-Barakāt al-Baghdādī, *al-Muʿtabar fī l-ḥikmah*, ed. S. Yaltkaya. 3 vols. Hyderabad: Dā'irat al-Maʿārif al-ʿUthmāniyyah, 1357/1939.

Abū Firās, *Dīwān* = Abū Firās al-Ḥamdānī, *Dīwān*, ed. ʿAlī al-ʿUsaylī. Beirut: Mu'assasat al-Aʿlamī, 1418/1997.

Abū Ḥātim al-Rāzī, *Proofs of Prophecy* = Abū Ḥātim al-Rāzī, *Abū Ḥātim al-Rāzī: The Proofs of Prophecy. A Parallel Arabic-English Text*. Ed. and tr. Tarif Khalidi. Chicago: Chicago University Press, 2012.

Abū l-Khayr al-Ishbīlī, *Kitāb ʿumdat al-ṭabīb* = Abū l-Khayr al-Ishbīlī, *Kitāb ʿumdat al-ṭabīb fī maʿrifat al-nabāt li-kull labīb* (*Libro base del médico para el conocimiento de la botánica por todo experto*), ed. and trad. Joaquín Bustamante, Federico Corriente, and Mahomed Tilmatine. 2 vols. Madrid: CSIC, 2004–2007.

Abū Maʿshar, *Great Introduction* = *The Great Introduction to Astrology* by Abū Maʿshar. Edited and translated by Keijo Yamamoto and Charles Burnett, with an edition of the Greek version by David Pingree. [Islamic Philosophy, Theology and Science, 106/1–2] 2 vols. Leiden/Boston: Brill, 2019.

Abū Maʿshar, *Kitāb al-madkhal al-kabīr* (Lemay) = Abū Maʿshar, *Kitāb al-madkhal al-kabīr ilá ʿilm aḥkām al-nujūm, Liber introductorii maioris ad scientiam judiciorum astrorum*, ed. and tr. R. Lemay. 9 vols. Naples: Istituto Universitario Orientale, 1995.

Abū Maʿshar, *On Historical Astrology* = *Abū Maʿshar on Historical Astrology: The Book of Religions and Dynasties* (*On the Great Conjunctions*), ed. and tr. Keiji Yamamoto and Charles Burnett (Islamic Philosophy, Theology and Science: Texts and Studies, 33–34). 2 vols. Leiden: E.J. Brill, 2000.

Abū Nuwās, *Dīwān* = Abū Nuwās al-Ḥasan ibn Hāni' al-Ḥakamī, *Dīwān*, ed. Ewald Wagner and Gregor Schoeler, 5 vols. Wiesbaden-Cairo: Franz Steiner and Berlin-Beirut: Klaus Schwarz, 1958–2003 + Index, 2 vols. Beirut-Berlin: Klaus Schwarz, 2006.

Abū Rashīd al-Nīsābūrī, *al-Masā'il* = Abū Rashīd al-Muʿtazilī Saʿīd ibn Muḥammad ibn Saʿīd, *al-Masā'il fī l-khilāf bayna al-Baṣriyyīn wa-l-Baghdādiyyīn*, ed. Maʿn Ziyādah & Riḍwān al-Sayyid. Beirut: Maʿhad al-Inmā' al-ʿArabī, 1979.

Abū l-Ṣalt Umayyah, *al-Risālah al-Miṣriyyah* = Abū l-Ṣalt Umayyah ibn ʿAbd al-ʿAzīz, *al-Risālah al-Miṣriyyah*, in *Nawādir al-makhṭūṭāt*, ed. ʿAbd al-Salām Hārūn. 2 vols. Cairo: Maktabat Muṣṭafā l-Bābī l-Ḥalabī, 1972.

Abū l-Ṣalt Umayyah, *Dīwān Umayyah Ibn Abī l-Ṣalt* = Umayyah ibn ʿAbd al-ʿAzīz ibn Abī l-Ṣalt, *Dīwān*, ed. ʿAbd Allāh Muḥammad al-Hūnī. Doha: Dār al-Awzāʿī, 1990.

Abū Shāmah, *Dhayl* = Abū Shāmah, *al-Dhayl ʿalā l-Rawḍatayn*, ed. Ibrāhīm Shams al-Dīn. Beirut: Dār al-Kutub al-ʿIlmiyyah, 2002.

Abū Shāmah, *Dhayl* [Kawtharī] = Abū Shāmah, *al-Dhayl ʿalā l-Rawḍatayn*, ed. Muḥammad Zāhid al-Kawtharī, as *Tarājim rijāl al-qarnayn al-sādis wa-l-sābiʿ*. Cairo: Dār al-Kutub al-Mālikiyyah, 1366/1947.

Abū Tammām, *Dīwān* = Abū Tammām, *Dīwān bi-sharḥ al-Khaṭīb al-Tibrīzī*, ed. Muḥammad ʿAbduh ʿAzzām. 4 vols. Cairo: Dār al-Maʿārif, 1976.

Abū Tammām, *Ḥamāsah see* al-Marzūqī, *Sharḥ*.

Abū ʿUbayd al-Bakrī, *Faṣl al-maqāl fī sharḥ Kitāb al-amthāl* = Abū ʿUbayd al-Bakrī, *Faṣl al-maqāl fī sharḥ Kitāb al-amthāl* (*li-Abū ʿUbayd al-Qāsim ibn Sallam*), ed. Iḥsān ʿAbbās & ʿAbd al-Majīd ʿĀbidīn. Beirut: Dār al-Amānah, 1983.

Abū ʿUbayd al-Bakrī, *Muʿjam mā istaʿjam* = Abū ʿUbayd ʿAbd Allāh ibn ʿAbd al-ʿAzīz al-Bakrī, *Muʿjam mā istaʿjam min asmāʾ al-bilād wa-l-mawāḍiʿ*, ed. Muṣṭafā al-Saqqā, 4 vols. Cairo: 1945–1951, repr. Beirut: ʿĀlam al-Kutub, n.d.

Abū ʿUbayd al-Bakrī, *Simṭ al-laʾālī* = Abū ʿUbayd al-Bakrī, *Simṭ al-laʾālī fī sharḥ Amālī al-Qālī*, ed. ʿAbd al-ʿAzīz al-Maymanī, Beirut: Dār al-Kutub al-ʿIlmiyyah, 1984 (reprod. of ed. Cairo, 1936).

Abū ʿUbayd al-Qāsim b. Sallām, *K. al-amthāl* = Abū ʿUbayd al-Qāsim b. Sallām, *K. al-amthāl*, ed. ʿAbd al-Majīd Qaṭāmish. Beirut: Dār al-Maʾmūn, 1980.

Abū l-Walīd ibn Rushd, *Talkhīṣ K. al-nafs* = Aḥmad Fuʾād al-Ahwānī, *Talkhīṣ K. al-nafs li-Abī l-Walīd ibn Rushd wa-arbaʿ rasāʾil*. Cairo: Maktabat al-Nahḍah al-Miṣriyyah, 1980.

Adamson, *Health* = Peter Adamson (ed.), *Health: A History*. Oxford: Oxford University Press USA, 2019.

Adamson, *Al-Kindī* = Peter Adamson, *Al-Kindī*. New York – Oxford: Oxford University Press, 2007.

Adamson, 'Miskawayh on Pleasure' = Peter Adamson, 'Miskawayh on Pleasure', *Arabic Sciences and Philosophy*, 25 (2015), 199–223.

Adamson, *Philosophy* = Peter Adamson, *Philosophy in the Islamic World: A History of Philosophy without any gaps*, vol. 3. Oxford: Oxford University Press, 2016.

Adamson, *Studies on Plotinus and al-Kindī* = Peter Adamson, *Studies on Plotinus and al-Kindī*. Farnham: Ashgate, 2014.

Adamson & Key, 'Philosophy of language in the medieval Arabic tradition' = Peter Adamson and Alexander Key, 'Philosophy of language in the medieval Arabic tradition', in Margaret Cameron and Robert J. Stainton (eds.), *Linguistic Content: New Essays on the History of Philosophy of Language*. Oxford: Oxford University Press, 2015, 74–99.

Adamson & Pormann, *Philosophy and Medicine* = Peter Adamson & Peter E. Pormann (eds.), *Philosophy and Medicine in the Formative Period of Islam*. London: Warburg Institute, 2017.

Adamson & Pormann, *The Philosophical Works of al-Kindī* = Peter Adamson and Peter E. Pormann, *The Philosophical Works of Al-Kindī*. Karachi: Oxford University Press, 2012.

Afnan, *Philosophical Lexicon* = Soheil M. Afnan, *A Philosophical Lexicon in Persian and Arabic*. Beirut: Dār al-Mashriq, 1968.

Agrimi & Crisciani, *Consilia Médicaux* = Jole Agrimi et Chiara Crisciani, *Les consilia médicaux* (Typologie des sources du Moyen Âge occidental, fasc. 69). Turnhout: Brepols, 1994.

Ahlwardt, *The Divans* = Wilhelm Ahlwardt (ed.), *The Divans of the Six Ancient Arabic Poets*. London: Trübner, 1870.

Aḥmad ʿĪsā, *Muʿjam al-aṭibbāʾ* = Aḥmad ʿĪsā, *Muʿjam al-aṭibbāʾ: min sanat 650 H ilā yawminā hādhā*. [Cairo:] Maṭbaʿat Fatḥ Allāh Ilyās Nūrī wa-Awlādihi, 1942.

Ahmed, *Avicenna's* Deliverance: Logic = Asad Q. Ahmed, *Avicenna's* Deliverance: Logic, translated and with a commentary and glossary by Asad Q. Ahmed. Oxford: Oxford University Press, 2010.

Ahsan, *Social Life under the Abbasids* = Muhammad Manazir Ahsan, *Social Life under the Abbasids: 170–289 AH, 786–902 AD*. London: Longman, 1979.

Ajami, *The Neckveins of Winter* = Mansour Ajami, *The Neckveins of Winter: The Controversy over Natural and Artificial Poetry in Medieval Arabic Literary Criticism*. Leiden: Brill, 1984.

Akasoy, 'Ibn Sabʿīn's *Sicilian Questions*' = Anna Akasoy, 'Ibn Sabʿīn's *Sicilian Questions*: The Text, its sources, and their historical context', *Al-Qanṭara*, 29 (2008), 115–146.

Akasoy & Fidora, *Arabic Version* = Anna Akasoy and Alexander Fidora, *The Arabic Version of the "Nicomachean Ethics"*. Introduction by Douglas M. Dunlop. Leiden: Brill, 2005.

Āl Khalīfah, *Umarāʾ al-Kūfah* = Muḥammad ʿAlī Āl Khalīfah, *Umarāʾ al-Kūfah wa-ḥukkāmuhā*, ed. Yāsīn Ṣalawātī. Tehran: Muʾassasat al-Ṣādiq, 1425/2004.

Alavi, 'Some Aspects'= Mohd. Badruddin Alavi, 'Some Aspects of the Literary and Poetical Activities of Avicenna', in *Avicenna Commemoration Volume*. Calcutta: Iran Society, 1956, 65–72.

ʿAlāwī, *Maqālāt* = Jamāl al-Dīn al-ʿAlāwī, *Maqālāt fī l-manṭiq wa-l-ʿilm al-ṭabīʿī li-Abī l-Walīd Ibn Rushd*. Casablanca: Dār al-Nashr al-Maghribiyyah, 1983.

ʿAlawī, *Rasāʾil* = Jamāl al-Dīn al-ʿAlawī, *Rasāʾil falsafiyyah li-Abī Bakr ibn Bājjah. Nuṣūṣ falsafiyyah ghayr manshūrah*. Beirut: Dār al-Thaqāfah, 1983.

Albertini, 'Quadrature du cercle' = Tamara Albertini, 'La Quadrature du cercle d'Ibn al-Haytham: Solution mathématique ou philosophique?', *Journal for the History of Arabic Science*, 9 (1991), 5–21.

Alf laylah wa-laylah = *Alf laylah wa-laylah*, 4 vols. Cairo: Maktabat Muḥammad ʿAlī Ṣubayḥ, n.d.

Alfonso, *Islamic Culture* = Esperanza Alfonso, *Islamic Culture through Jewish Eyes: al-Andalus from the Tenth to the Twelfth Century*. London and New York: Routledge, 2008.

ʿAlī ibn Abī Ṭālib, *Dīwān* = ʿAlī ibn Abī Ṭālib, *Dīwān*, ed. Naʿīm Zarzūr. Beirut, 1985.

ʿAlī ibn ʿĪsā al-Kaḥḥāl, *Memorandum Book* = ʿAlī ibn ʿĪsā al-Kaḥḥāl, *Memorandum Book of a Tenth-Century Oculist for the Use of Modern Ophthalmologists: A translation of the Tadhkirat of Ali ibn Isa of Baghdad*, translated by Casey A. Wood. Chicago: Northwestern University, 1936, repr. Birmingham: Classics of Medicine Library, 1985

ʿAlī ibn Sahl Rabban al-Ṭabarī, *Firdaws al-ḥikmah* (ed. al-Ṣiddīqī) = ʿAlī ibn Sahl Rabban al-Ṭabarī, *Kitāb Firdaws al-ḥikmah*, ed. Muḥammad Zubayr al-Ṣiddīqī (Berlin: Sonne Press, 1928)

ʿAlī ibn Sahl Rabban al-Ṭabarī, *Firdaws al-ḥikmah* (ed. al-Jundī) = ʿAlī ibn Sahl Rabban al-Ṭabarī, *Kitāb Firdaws al-ḥikmah*, ed. ʿAbd al-Karim Sāmī al-Jundī. Beirut: Dār al-Kutub al-ʿIlmiyyah, 2002.

Allemann, *Mudjādalat al-ḥakīmayn* = Franz Allemann, "Abdallaṭīf al-Baġdādī: *Ris. fī Mudjādalat al-ḥakīmayn al-kīmiyāʾī wan-naẓarī* ("Das Streitgespräch zwischen dem Alchemisten und dem theoretischen Philosophen"). Eine textkritische Bearbeitung der Handschrift: Bursa, Hüseyin Çelebi 823, fol. 100–123 mit Übersetzung und Kommentar'. Ph.D. dissertation, University of Bern, 1988.

Alon, *Socrates Arabus* = Ilai Alon, *Socrates Arabus: Life and Teachings*. Jerusalem: The Hebrew University of Jerusalem, 1995.

Alon, *Socrates in Mediaeval Arabic Literature* = Ilai Alon, *Socrates in Mediaeval Arabic Literature*. Leiden: Brill, 1991.

Altmann & Stern, *Isaac Israeli* = Alexander Altmann and Samuel M. Stern, *Isaac Israeli: A Neoplatonic Philosopher of the Early Tenth Century*. Chicago: University of Chicago Press, 2009 (original edition 1958).

Álvarez Millán, 'The Case History' = Cristina Álvarez Millán, 'The Case History in Medieval Islamic Medical Literature: *Tajārib* and *Mujarrabāt* as Source', *Medical History*, 54/2 (2010), 195–214.

Álvarez Millán, 'Disease in tenth-century Iran and Irak' = Cristina Álvarez-Millán, 'Disease in tenth-century Iran and Irak according to al-Rāzī's Casebook', *Suhayl*, 14 (2015), 49–88.

Alwishah & Hayes, *Aristotle and the Arabic Tradition* = Ahmed Alwishah and Josh Hayes (eds.), *Aristotle and the Arabic Tradition*. Cambridge: Cambridge University Press, 2015.

Amar & Lev, *Arabian Drugs* = Zohar Amar and Efraim Lev, *Arabian Drugs in Early Medieval Mediterranean Medicine*. Edinburgh University Press, 2017.

Amar & Serri, *Ibn al-Suri* = Zohar Amar and Yaron Serri, 'Ibn al-Suri, Physician and Botanist of al-Sham', in *Palestine Exploration Quarterly*, 135/2 (2003), 124–130.

Ambros, 'Beobachtungen' = Arne A. Ambros, 'Beobachtungen zu Aufbau und Funktionen der gereimten klassisch-arabischen Buchtitel', *Wiener Zeitschrift für die Kunde des Morgenlandes*, 80 (1990), 13–57.

Amedroz, 'On a dirham of Khusru Shah' = H.F. Amedroz, 'On a dirham of Khusru Shah of 361 A.H., Etc.', *Journal of the Royal Asiatic Society* 37/3 (1905): 471–484

al-Āmidī, *al-Mu'talif wa-l-mukhtalif* = Abū l-Qāsim al-Ḥasan ibn Bishr al-Āmidī, *al-Mu'talif wa-l-mukhtalif fī asmā' al-shu'arā' wa-kunāhum wa-alqābihim wa-ansābihim wa-ba'ḍ shi'rihim*, ed. F. Krenkow. Repr. Beirut: Dār al-Kutub al-'Ilmiyyah, 1402/1982.

al-'Āmilī, *Kashkūl* = Muḥammad Bahā' al-Dīn al-'Āmilī, *al-Kashkūl*. Beirut: Dār al-Kitāb al-Lubnānī, 1403/1983.

Ammann, *Vorbild und Vernunft* = Ludwig Ammann, *Vorbild und Vernunft: Die Regelung von Lachen und Scherzen im mittelalterlichen Islam*. Hildesheim: Georg Olms, 1993.

'Ammār ibn 'Alī al-Mawṣilī, *Las operaciones de catarata* = Max Meyerhof, *Las operaciones de catarata de 'Ammâr ibn 'Alî al-Mauṣilî*. Barcelona, 1937, reprinted in Fuat Sezgin (ed.), *Augenheilkunde im Islam. Texte, Studien und Übersetzungen*. Frankfurt am Main: Institut für Geschichte der Arabisch-Islamischen Wissenschaften, 1986, iii:590–711.

Anastassiou & Irmer, *Testimonien* = Anargyros Anastassiou and Dieter Irmer, *Testimonien zum Corpus Hippocraticum III. Nachleben der hippokratischen Schriften in der Zeit vom 4. bis zum 10. Jahrhundert n. Chr.* Göttingen: Vandenhoeck and Ruprecht, 2012.

Anawati, 'Arabic alchemy' = Georges C. Anawati, 'Arabic alchemy' in Roshdi Rashed (ed.), *Encyclopedia of the History of Arabic Science*. 3 vols. London / New York: Routledge 1996. Vol. 3, pp. 853–885.

Anawati, *Kitāb al-jamāhir* = Georges C. Anawati, 'The Kitāb al-jamāhir fī ma'rifat al-jawāhir of al-Bīrūnī', in *Al-Bīrūnī Commemorative Volume*, ed. Hakim Mohammad Said. Karachi: Hamdard National Foundation, 1979, 437–453.

Anawati & Zāyid, *Rasā'il Ibn Rushd al-ṭibbiyyah* = Georges C. Anawati (Jūrj Qanawātī) and Sa'īd Zāyid (Zayed), *Rasā'il Ibn Rushd al-ṭibbiyyah* (*Les traités médicaux d'Averroes*). Cairo: al-Hay'ah al-Miṣriyyah al-'Āmmah l-il-Kitāb, 1987.

al-Anbārī, *Nuzhat* = al-Anbārī, *Nuzhat al-alibbā' fī ṭabaqāt al-udabā'*, ed. 'Aṭiyyah 'Āmir. Stockholm-Göteborg-Uppsala: Almqvist & Wiksell, 1963.

Angellus de Ravenne, *Lectures galéniques* = Angellus de Ravenne, *Lectures galéniques: le "De pulsibus ad tirones"*, ed. and tr. Ivan Garofalo & Nicoletta Palmieri. Saint-Étienne: Publications de l'Université de Saint-Étienne, 2005.

Anon., *Book of Gifts and Rarities* = Anon., *Book of Gifts and Rarities* (Kitāb al-Hadāyā wa-l-Tuḥaf)*: Selections Compiled in the Fifteenth Century from an Eleventh-Century Manuscript on Gifts and Treasures*, tr. with introd., annot., glossary, appendices and indices by Ghāda al-Ḥijjāwī al-Qaddūmī. Cambridge, Massachusetts: Harvard University Press, 1996.

al-Anṭākī, *Histoire continuateur Saʿid Ibn Bitriq*, ed. & tr. Kratchkovskii and Vasiliev *see* Yaḥyā ibn Saʿīd al-Anṭākī, *Histoire* (Kratchkovsky & Vasiliev)

al-Anṭākī, *Histoire continuateur Saʿid Ibn Bitriq*, tr. Micheau & Troupeau *see* Yaḥyā ibn Saʿīd al-Anṭākī, *Histoire* (Micheau & Tropeau)

Antoon, *Poetics of the Obscene* = Sinan Antoon, *The Poetics of the Obscene in Premodern Arabic Poetry: Ibn al-Ḥajjāj and Sukhf*. New York: Palgrave Macmillan, 2014.

Antrim, *Routes* = Zayde Antrim, *Routes and Realms: The Power of Place in the Early Islamic World*. New York: Oxford University Press, 2012.

ʿĀnūtī, *Ibn Abī Usaybiʿah* = Usāmah ʿĀnūtī, *Ibn Abī Usaybiʿah: taʿrīf wa-taqwīm*. Beirut: Dār al-Nafāʾis, 1975.

Aouad, 'La *Rhétorique*' = M. Aouad, 'Aristote de Stagire – La *Rhétorique*. Tradition Syriaque et Arabe', in Goulet, *Dictionnaire*, i:455–472, and Suppl. 220–223.

Aouad, 'La *Théologie d'Aristote*' = M. Aouad, 'Aristote de Stagire – La *Théologie d'Aristote* et autres textes du Plotinus Arabus', in Goulet, *Dictionnaire*, i:541–590.

Aouad & Woerther, 'Le commentaire par Averroès' = Aouad, M. and Woerther, F., 'Le commentaire par Averroès du chapitre 9 du livre x de *l'Éthique à Nicomaque*: pédagogie de la contrainte, habitudes et lois', in *Mélanges de l'Université Saint-Joseph*, 62 (2009). 303–330.

Arberry, *Avicenna on Theology* = A.J. Arberry, *Avicenna on Theology*. London: John Murray, 1951.

Arberry, *Poems of al-Mutanabbī* = A.J. Arberry, *Poems of al-Mutanabbī: A Selection with Introduction, Translations and Notes*. Cambridge: University Press, 1967.

Arberry, *Spiritual Physick* = Arthur J. Arberry, *The Spiritual Physick of Rhazes*: Translated from the Arabic by Arthur J. Arberry. London: John Murray, 1950.

Arjona Castro, 'Iglesia de San Acisclo' = Antonio Arjona Castro, 'Sobre la localización de la iglesia de San Acisclo y del *fahs al-suradiq* (campamento militar de la Córdoba islāmica)', *Boletín de la Real Academia de Córdoba de Ciencias, Bellas Letras y Nobles Artes*, 144 (2003), 161–172.

Arnaldus de Villanova, *opera medica omnia, v. 17* = *Arnaldi de Villanova opera medica omnia*, 17: *Translatio libri Albuzale de medicinis simplicibus*, ed. J. Martínez Gázquez and M.R. McVaugh, *Kitāb al-adwiya al-mufrada* [of] *Abū-l-Salt Umayya*, ed. A. Labarta, and *Libre de'Albumesar de simples medecines*, ed. L. Cifuentes. Barcelona: Universitat de Barcelona, 2004.

Arndt, *Judah ha-Cohen and the Emperor's Philosopher* = Sabine Arndt, 'Judah ha-Cohen and the Emperor's Philosopher: Dynamics of Transmission at Cultural Crossroads'. D. Phil. thesis, University of Oxford, 2016.

Arnzen, '*De Anima*' = Rüdiger Arnzen, 'Aristote de Stagire – *De Anima*. Paraphrase arabe anonyme', in Goulet, *Dictionnaire*, Suppl., 359–365.

Arnzen, 'Plato's Timaeus in the Arabic Tradition' = Rüdiger Arnzen, 'Plato's Timaeus in the Arabic Tradition. Legends–Testimonies–Fragments', in F. Celia and A. Ulacco, *Il Timeo. Esegesi greche, arabe, latine*. Pisa: Pisa University Press, 2012, 181–267.

al-Aʿshā Maymūn, *Dīwān* = al-Aʿshā [Maymūn ibn Qays], *Dīwān*. Beirut: Dār Ṣādir, 1966, repr. 1994.

Asín Palacios, 'Avempace botánico' = Miguel Asín Palacios, 'Avempace botánico', *al-Andalus*, 5 (1940), 255–299.

Asín Palacios, 'Carta del adios' = Miguel Asín Palacios, 'Carta del adios', *al-Andalus*, 8 (1943), 1–87.

Asín Palacios, 'Un texto de Avempace' = Miguel Asín Palacios, 'Un texto de Avempace sobre la unión del intelecto con el hombre', *al-Andalus*, 7 (1942), 1–47.

al-ʿAskarī, Abū Aḥmad *see* Abū Aḥmad al-ʿAskarī.

al-ʿAskarī, *Dīwān al-maʿānī* = Abū Hilāl al-ʿAskarī, *Dīwān al-maʿānī*, 2 vols. Cairo: Maktabat al-Qudsī, n.d.

al-ʿAskarī, *Jamharah* = Abū Hilāl al-Ḥasan ibn ʿAbd Allāh ibn Sahl al-ʿAskarī, *Jamharat al-amthāl*, ed. Aḥmad ʿAbd al-Salām and Abū Hājar Muḥammad Saʿīd ibn Basyūnī Zaghlūl, 2 vols. Beirut: Dār al-Kutub al-ʿIlmiyyah, 1408/1988.

al-ʿAskarī, *Ṣināʿatayn* = Abū Hilāl al-Ḥasan ibn ʿAbd Allāh ibn Sahl al-ʿAskarī, *Kitāb al-ṣināʿatayn al-kitābah wa-l-shiʿr*, ed. ʿAlī Muḥammad al-Bajāwī. Cairo: ʿĪsā al-Bābī al-Ḥalabī, 1971.

Assemani, *Bibliotheca Orientalis Clementino-Vaticana* = Assemani, Giuseppe Simone, *Bibliotheca Orientalis Clementino-Vaticana: In Qua Manuscriptos Codices Syriacos, Arabicos, Persicos, Turcicos, Hebraicos, Samaritanos, Armenicos, Aethiopicos, Graecos, Aegyptiacos, Ibericos, & Malabaricos, Jussu Et Munificentia Clementis XI. Pontificis Maximi Ex Oriente conquisitos, comparatos, avectos, & Bibliothecae Vaticanae addictos / Recensuit, digessit & genuina scripta à spuriis secrevit, Addita Singulorum Auctorum Vita Joseph Simonius Assemanus Syrus Maronita Sacrae Theologia Doctor, atque in eadem Bibliotheca Vaticana linguarum Syriacae & Arabicae Scriptor*. Roma: Typis Sacrae Congregationis de Propaganda Fidei, 1719–1728.

al-Astarābādī, *Sharḥ al-Kāfiyah* = Raḍī al-Dīn al-Astarābādī, *Sharḥ al-Kāfiyah li-Ibn al-Ḥājib*, ed. Yaḥyā Bashīr Miṣrī, 2 vols. n. pl.: Jāmiʿat al-Imām Muḥammad ibn Suʿūd al-Islāmiyyah, 1996.

al-Asṭurlābī, *Durrat al-tāj* = Hibat Allāh Badīʿal-Zamān, al-Asṭurlābī, *Durrat al-tāj min shiʿr Ibn al-Ḥajjāj*, ed. ʿAlī Jawād al-Ṭāhir. Cologne: Manshūrāt al-Jamal, 2009.

Atallah, *Ibn Abi Useibià* = Sadik Atallah, 'Ibn Abi Useibià – Schrift über die arabischen Ärzte und ihre Kunst. Abschnitt über die Anfänge'. Inaug.-diss., Ludwig-Maximilians-Universität zu München, 1939. Repr. in Sezgin, *Studies* II, 261–280.

Atiyeh, *Al-Kindi* = George N. Atiyeh, *Al-Kindi: The Philosopher of the Arabs*. New Delhi: Kitab Bhavan Publishers, 1994.

Avenary, 'The Hebrew Version' = H. Avenary, 'The Hebrew Version of Abū l-Ṣalt's Treatise on Music', *Yuval*, 3 (1974), 7–82.

Avicenna, *Poème de la médicine* = Avicenna, *Poème de la médicine* (*al-Urjūzah fī l-ṭibb*), texte arabe, trad. française, trad. latine du XIIIe siècle avec introd., notes et index par H. Jahier & A. Noureddine. Paris: Les Belles-Lettres, 1956.

Avicenna *see also* Ibn Sīnā.

Aydin, *Sergius of Reshaina* = Sami Aydin, *Sergius of Reshaina: Introduction to Aristotle and his Categories, Addressed to Philotheos*. Leiden: Brill, 2016.

Azar, *Ibn Zuhr (Avenzoar)* = Henry Amin Azar, 'Ibn Zuhr (Avenzoar), "Supreme in the science of medicine since Galen": The translation of his work into Latin and his image in medieval Europe', Ph.D. diss., Univ. of North Carolina, 1998.

Azar, *The Sage of Seville* = Henry A. Azar, *The Sage of Seville: Ibn Zuhr, His Time and his Medical Legacy*. Cairo: AUC Press, 2008.

al-Azdī, *Ḥikāyat Abī l-Qāsim al-Baghdādī* = Abū l-Muṭahhar al-Azdī, *Ḥikāyat Abī l-Qāsim al-Baghdādī* / *Abulḳâsim: ein bagdâder Sittenbild*, ed. Adam Mez. Heidelberg: Carl Winter's Universitätsbuchhandlung, 1902.

al-Azdī, *Ḥikāyat Abī l-Qāsim al-Baghdādī*, ed. al-Shāljī = Abū l-Muṭahhar al-Azdī, *Ḥikāyat Abī l-Qāsim al-Baghdādī* (as *al-Risālah al-Baghdādiyyah* and attrib. to Abū Ḥayyān al-Tawḥīdī), ed. ʿAbbūd al-Shāljī. Köln: Manshūrāt al-Jamal, 1997.

Baarda, 'The Author of the Arabic Diatessaron' = Tjitze Baarda, 'The Author of the Arabic Diatessaron', in Tjitze Baarda *et alii*, *Miscellanea Neotestamentica*, Vol. 1, Leiden: Brill, 1978, 61–103; reprint in: Tjitze Baarda, *Early Transmission of Words of Jesus. Thomas, Tatian and the Text of the New Testament*, selected and edited by Jan Helderman and Sybolt Jan Noorda. Amsterdam: VU Boekhandel/Uitgeverij, 1983, 207–249.

Bābānī, *Hadiyyat al-ʿārifīn* = Ismāʿīl Bāshā Bābānī, *Hadiyyat al-ʿārifīn: asmāʾ al-muʾallifīn wa-athār al-muṣannifīn*. Istānbūl: 2 vols. Wikālat all-Maʿārif, 1951–1955.

Bacher, *Vier Abhandlungen über Abulwalîd ibn Ganâh* = Wilhelm Bacher, *Vier Abhandlungen über Abulwalîd ibn Ganâh, ca. 990–1050*. Amsterdam: Philo Press, 1970 (first published in *Jahresbericht der Landesrabbinerschule in Budapest*, 1884/85–1888/89).

Bachmann, *Galens Abhandlung* = Peter Bachmann, *Galens Abhandlung darüber, daß der vorzügliche Artz Philosoph sein muß*. Göttingen: Vandenhoeck & Ruprecht, 1965.

Badawī, *Ādāb* = ʿAbd al-Raḥman Badawī, *Ḥunayn ibn Isḥāq, Ādāb al-falāsifah* (*Sentences des philosophes*). Kuwait: Maʿhad al-Makhṭūṭāt al-ʿArabiyyah, 1985.

Badawī, 'Akhlāq' = Galen, 'Mukhtaṣar min Kitāb al-akhlāq li-Jālīnūs', in Badawī, *Dirāsāt wa-nuṣūṣ fī l-falsafah wa-l-ʿulūm ʿinda l-ʿArab*. Beirut: al-Muʾassasah al-ʿArabiyyah li-l-Dirāsāt wa-l-Nashr, 1981, 190–211.

Badawī, *Arisṭū ʿinda al-ʿArab* = ʿAbd al-Raḥman al-Badawī, *Arisṭū ʿinda al-ʿArab. Dirāsah wa-nuṣūṣ ghayr manshūrah*. Cairo: Maktabat al-Nahḍah al-Miṣriyyah, 1947.

Badawī, *Commentaires* = *Commentaires sur Aristote perdus en grec et autres épîtres* (Shurūḥ ʿalā Arisṭū mafqūdah fī al-yūnāniyyah wa-rasāʾil ukhrā). Beirut: Dar el-Machreq, 1971.

Badawī, *Manṭiq Arisṭū* = *Manṭiq Arisṭū*, ed. ʿAbd al-Raḥmān Badawī. 3 vols. Cairo: Dār al-Kutub al-Miṣriyyah, 1948–1952.

Badawī, *Al-Mubashshir ibn Fātik* = ʿAbd al-Raḥman Badawī, *Abū l-Wafāʾ al-Mubashshir ibn Fātik, Mukhtār al-ḥikam wa-maḥāsin al-kalim / Los Bocados de Oro* (*Mujtār al-Ḥikam*). Madrid: Maṭbaʿat al-Maʿhad al-Miṣrī li-l-Dirāsāt al-Islāmiyyah, 1958.

Badawī, *Orosius* = ʿAbd al-Raḥman Badawī, *Ūrūsiyūs, Taʾrīkh al-ʿĀlam / Orosii Historia contra paganos*. Beirut: al-Muʾassasah al-ʿArabīyah li-l-Dirāsāt wa-l-Nashr, 1982.

Badawī, *Platon* (*Aflāṭūn fī l-Islām*) = ʿAbd al-Raḥman Badawī, *Platon en pays d'Islam. Textes publiés et annotés*. (*Aflātūn fī l-Islām, nuṣūṣ*). Tehran: Institute of Islamic Studies, McGill, Tehran Branch, 1974.

Badawī, *Rasāʾil falsafiyyah* = ʿAbd al-Raḥman Badawī, *Rasāʾil falsafiyyah li-l-Kindī wa-l-Fārābī wa-Ibn Bājjah wa-Ibn ʿAdī*. Beirut: Dār al-Andalus, 1980².

Badawī, *Sirr al-asrār* = Pseudo-Aristotle, *Kitāb al-siyāsah fī tadbīr al-riʾāsah al-maʿrūf bi-sirr al-asrār*, in ʿAbd al-Raḥmān Badawī (ed.), *al-Uṣūl al-yunāniyyah li-l-naẓariyyāt al-siyāsiyyah fī-l-islam*. Cairo: Maktabat al-Kutub al-Miṣriyyah, 1954, 65–171.

Badr al-Dīn ibn Qāḍī Baʿlabakk, *Mufarriḥ al-nafs* (*Ḥanūn*) = Badr al-Dīn ibn Qāḍī Baʿlabakk, *Mufarriḥ al-nafs*, ed. ʿAbd al-Fattāḥ ʿAbd al-Razzāq Ḥanūn. Beirut: Dār al-Kutub al-ʿIlmiyyah, 2006.

Baffioni, 'Il "*Liber introductorius*"' = Carmela Baffioni, 'Il "Liber introductorius in artem logicae demonstrationis": problemi storici e filologici', *Studi filosofici*, 17 (1994), 69–90.

al-Baghdādī, *The Eastern Key* = ʿAbd al-Laṭīf ibn Yūsuf al-Baghdādī, *Kitāb al-Ifādah wa-l-iʿtibār fī l-umūr al-mushāhadah wa l-ḥawādith al-muʿāyanah bi-arḍ Miṣr*. Facsimile edition of the autograph manuscript at the Bodleian Library, Oxford, and English translation by Kamal Hafuth Zand and John A. and Ivy E. Videan, as *The Eastern Key*. London: George Allen and Unwin, 1965.

al-Baghdādī, *Khizānah* = ʿAbd al-Qādir al-Baghdādī, *Khizānat al-adab wa-lubāb lisān al-ʿArab*, ed. ʿAbd al-Salām Muḥammad Hārūn. 13 vols. Cairo: Dār al-Kātib al-ʿArabī / al-Hayʾah al-Miṣriyyah al-ʿĀmmah li-l-Kitāb, 1967–1986.

Bahāʾ al-Dīn Zuhayr, *Dīwān* = Behà-ed-Dìn Zoheir of Egypt, *The Poetic Works, I: Arabic tekst, II: Metrical English translation, notes and introduction*, by Edward Henry Palmer. Cambridge, 1876–1877, repr. Amsterdam: Oriental Press, 1971.

Bahmanyār, *K. al-Taḥṣīl* = Bahmanyār ibn al-Marzubān, *Kitāb al-Taḥṣīl*, ed. Shahīd Murtaḍā Muṭahharī. Tehran: Dānishkada-i Ilāhiyyāt u-Maʿārif-i Islāmiyya, 1375/1996.

al-Bākharzī, *Dumyah* = Abū l-Ḥasan ʿAlī ibn al-Ḥasan ibn ʿAlī al-Bākharzī, *Dumyat al-qaṣr wa-ʿuṣrat ahl al-ʿaṣr*, ed. ʿAbd al-Fattāḥ al-Ḥulw. 2 vols. Cairo: Dār al-Fikr al-ʿArabī, 1968, 1971.

Ball, 'Saint Sabas' = Jennifer Ball, 'Saint Sabas and the Monks of the Holy Land', in Barbara Drake Boehm and Melanie Holcomb (eds.), *Jerusalem, 1000–1400: Every People under Heaven*. New York: The Metropolitan Museum of Art / New Haven: Yale University Press, 2016, 76–78.

al-Balādhurī, *Ansāb* = Aḥmad ibn Yaḥyā al-Balādhurī, *Ansāb al-ashrāf*, ed. Maḥmūd al-Firdaws al-ʿAẓm, 25 vols. Damascus: Dār al-Yaqaẓah, 1996–2004.

al-Balādhurī, *Ansāb* (ed. Zakkār & Ziriklī) = Aḥmad ibn Yaḥyā al-Balādhurī, *Jumal min ansāb al-ashrāf*, ed. Suhayl Zakkār and Riyāḍ Ziriklī, 13 vols. Beirut: Dār al-Fikr, 1996.

al-Balādhurī, *Ansāb* (ed. al-Dūrī et al.) = Aḥmad ibn Yaḥyā al-Balādhurī, *Ansāb al-ashrāf*, ed. ʿAbd al-ʿAzīz al-Dūrī et al. Wiesbaden: Franz Steiner, 1978–, 7 vols (?). (vols. 1–2: Beirut – Berlin: Klaus Schwarz)

al-Balawī, *Sīrat Aḥmad ibn Ṭūlūn* = Abū Muḥammad ʿAbd Allāh ibn Muḥammad al-Madīnī al-Balawī, *Sīrat Aḥmad ibn Ṭūlūn*, ed. Muḥammad Kurd ʿAlī. Cairo: Maktabat al-Thaqāfah al-Dīniyyah, n. d. [reprinted several times, including Damascus: al-Maktabah al-ʿArabiyyah fī Dimashq, 1939].

Balty-Guesdon, 'Les *Ṭabaqāt al-Aṭibbā*'' = Marie Geneviève Balty-Guesdon, 'Les *Ṭabaqāt al-Aṭibbāʾ wa-l-Ḥukamāʾ* d'Ibn Ǧulǧul: une condamnation du régime ʿāmiride'. In: Denis Aigle et al. (Eds.), *Cahiers d'Onomastique Arabe 1988–1992*. Paris: Ed. du CNRS, 1993, 49–59.

Bancel, 'Le traité sur la théorie du levier' = Faïza Bancel, 'Le traité sur la théorie du levier d'al-Muẓaffar al-Isfizārī: une réécriture du *Kitāb fī al-qarasṭūn* de *Thābit* ibn *Qurra*?' In *De Zénon d'Élée à Poincaré: recueil d'études en hommages à Roshdi Rashed*. Éd. Régis Morelin et Ahmad Hasnawi. Louvain – Paris: Peeters, 2004, 321–353.

Bang, *Islamic Sufi Networks* = Anne K. Bang, *Islamic Sufi Networks in the Western Indian Ocean (c. 1880–1940): Ripples of Reform* (Islam in Africa, 16). Leiden / Boston: Brill, 2014.

Bar Bahlūl *see* Duval, Rubens.

Barhebraeus, *Chronicon Ecclesiasticum* = Bar Hebraeus, *Chronicon Ecclesiasticum*, ed. J.B. Abbeloos and Th.J. Lamy. Leuven – Paris: Peeters, 1877.

Barhebraus, *Chronicum Syriacum* = Barhebraeus, *Gregorii Barhebræi Chronicon Syriacum, emendatum adnotationibusque locupletatum*, ed. Paul Bedjan. Paris: Maisonneuve 1890.

Barhebraeus, *Chronography* = *The chronography of Gregory Abūʾl Faraj, the son of Aaron, the Hebrew physician, commonly known as Bar Hebraeus*, ed. Ernest Alfred Thompson Wallis Budge, 2 vols. Oxford / London: Oxford University Press, 1932.

Barhebraeus, *Mukhtaṣar* (Beirut edn.) = Ibn al-ʿIbrī (Bar Hebraeus), *Mukhtaṣar tārīkh al-duwal*, ed. Anṭūn Ṣālḥānī. Beirut: al-Maṭbaʿah al-Kāthūlīkiyyah li-l-Ābāʾ al-Yasūʿiyyīn, 1890; 2nd ed. al-Ḥāzimiyyah: Dār al-Rāʾid al-Lubnānī, 1994.

Barnes, 'Galen' = Jonathan Barnes, 'Galen, Christians, logic', in T.P. Wiseman (ed.), *Classics in Progress. Essays on Ancient Greece and Rome*. Oxford: Oxford University Press for the British Academy, 2002, 399–418.

Barnes, *Terms and Sentences* = Jonathan Barnes, *Terms and sentences: Theophrastus on Hypothetical Syllogisms*. London: Oxford University Press, 1984.

Barnes & Jouanna, *Galien* = Jonathan Barnes and Jacques Jouanna, *Galien et la philosophie* (Entretiens Hardt, XLIX). Vandoeuvres / Geneva: Fondation Hardt, 2003.

Baumstark, *Aristoteles bei den Syrern* = Anton Baumstark, *Aristoteles bei den Syrern* vol. 1: *Syrisch-arabische Biographieen des Aristoteles, Syrische Commentare zur Εἰσαγωγή des Porphyrios*. Leipzig: B.G. Teubner, 1900.

Baumstark, *Geschichte* = Anton Baumstark, *Geschichte der syrischen Literatur, mit Ausschluss der christlich-palästinensischen Texte*. Bonn: Marcus und Weber, 1922.

Bautz, *Biographisch-bibliographisches Kirchenlexikon* = Friedrich W. Bautz and Traugott Bautz (eds.), *Biographisch-bibliographisches Kirchenlexikon*. Nordhausen: Bautz 1970–.

al-Bayḍāwī, *Niẓām al-tawārīkh* = ʿAbd Allāh ibn ʿUmar al-Bayḍāwī, *Niẓām al-tawārīkh*, ed. Hāshim Muḥaddith. Tehran: Bunyād-i Mawqūfāt-i Duktur Maḥmūd Afshār, 1382/2003.

al-Bayhaqī *see* Shafīʿ.

al-Bayhaqī, *al-Jāmiʿ li-shuʿab al-īmān* = Aḥmad ibn al-Ḥusayn al-Bayhaqī, *al-Jāmiʿ li-shuʿab al-īmān*. Riyad/Cairo: Maktabat al-Rushd, Nāshirūn, 2003. 14 vols.

al-Bayhaqī, *Sunan* = Abū Bakr Aḥmad ibn al-Ḥusayn ibn ʿAlī al-Bayhaqī, *al-Sunan al-kubrā*. Ed. Muḥammad ʿAbd al-Qādir ʿAṭā. 11 vols. Beirut: Dār al-Kutub al-ʿIlmiyyah, 1994.

al-Bayhaqī, *Tatimmah see* Shafīʿ, *Tatimmat Ṣiwān*.

Beck, *Dioscorides* = L.Y. Beck, *Pedanius Dioscorides of Anazarbus, De materia medica*. 2nd ed. Hildesheim: Olms-Weidmann.

Beeston, *Arabic Nomenclature* = A.F.L. Beeston, *Arabic Nomenclature: A summary guide for beginners*. Oxford: Oxford University Press, 1971.

Behrens-Abouseif, *The Book in Mamluk Egypt and Syria* = Doris Behrens-Abouseif, *The Book in Mamluk Egypt and Syria (1250–1517): Scribes, Libraries and Market*. Leiden and Boston: Brill, 2018.

Bekker, *Aristotelis opera* = Immanuel Bekker, *Aristotelis opera*. Vols. 1, 2, 4, 5. Berlin: Apud Georgium Reimerum, 1831–1870.

Bellosta, 'Ibrāhīm ibn Sinān' = Hélène Bellosta, 'Ibrāhīm ibn Sinān: On analysis and synthesis', *Arabic Sciences and Philosophy*, 1 (1991), 211–232.

Ben Ahmed, *Compendium on Logic* = Fouad Ben Ahmed, *Ibn Ṭumlūs (Alhagiag Bin Thalmus, d. 620/1223) Compendium on Logic (Al-Mukhtaṣar fī al-manṭiq), edited with introductions, notes, and indexes*. (Islamic Philosophy, Theology and Science, 110). Leiden: Brill, 2019.

Benevich, *Essentialität und Notwendigkeit* = Fedor Benevich, *Essentialität und Notwendigkeit: Avicenna und die aristotelische Tradition*. Leiden: Brill, 2018.

Berggren, 'Numbers at work' – Len Berggren, 'Numbers at work in medieval Islam', *Journal for the History of Arabic Science*, 11 (1995–1997), 45–51.

Bergsträsser, Gotthelf *see* Ḥunayn ibn Isḥāq, *Risālah* and Ḥunayn ibn Isḥāq, *Neue Materialen*.

Bernheimer, *The 'Alids* = Teresa Bernheimer, *The 'Alids: the first family of Islam, 750–1200*. Edinburgh: Edinburgh University Press, 2013.

Berman, 'Lost Arabic Original of Ibn Rushd's Middle Commentary' = V. Berman, 'Excerpts of the Lost Arabic Original of Ibn Rushd's Middle Commentary on the Nicomachean Ethics', *Oriens*, 20 (1967), 31–59.

Berti, *L'au-delà de l'âme* = Berti, V., *L'au-delà de l'âme et l'au-deça du corps: approches d'anthropologie chrétiennes de la mort dans l'église syro-orientale*. Fribourg: Academic Press Fribourg, 2015.

Bertier, '*Opuscules*' = J. Bertier, 'Aristote de Stagire – *Opuscules*', in Goulet, *Dictionnaire*, Suppl., 487–498.

Bertolacci, 'From al-Kindī to al-Fārābī' = Amos Bertolacci, 'From al-Kindī to al-Fārābī: Avicenna's Progressive Knowledge of Aristotle's *Metaphysics* according to his Autobiography', *Arabic Sciences and Philosophy* 11 (2001) 257–291.

Biblioteca de al-Andalus = Jorge Lirola Delgado and José Miguel Puerta Vílchez (eds), *Biblioteca de al-Andalus*. 9 vols. Almería: Enciclopedia De La Cultura Andalusí, 2004.

Bibliotheca Orientalis = G.S. Assemani, *Bibliotheca Orientalis Clementina-Vaticana*. 3 parts in 4, Rome: Sacrae Congregationis de Propaganda Fide, 1719–1728.

Biesterfeldt, *Galens Traktat* = Hans Hinrich Biesterfeldt, *Galens Traktat "Dass die Kräfte der Seele den Mischungen des Körpers folgen" in arabischer Übersetzung*. Wiesbaden: Franz Steiner, 1973.

Biesterfeldt et al., 'The Beginnings of Islamic Philosophy in the Tradition of al-Kindī' = Hans Hinrich Biesterfeldt, Elvira Wakelnig, Gerhard Endress, and Cleophea Ferrari, 'The Beginnings of Islamic Philosophy in the Tradition of al-Kindī' in (ch. 5) Ulrich Rudolph, Rotraud E. Hansberger, and Peter Adamson (eds.), *Philosophy in the Islamic World*. Leiden: Brill, 2017.

Biesterfeldt & Gutas, 'Malady' = Hans Hinrich Biesterfeldt and Dimitri Gutas, 'The Malady of Love', *Journal of the American Oriental Society*, 104/1 (1984), 21–55.

Bin Murād, *Buḥūth fī ta'rīkh al-ṭibb* = Ibrāhīm Bin Murād, *Buḥūth fī ta'rīkh al-ṭibb wa-l-ṣaydalah 'inda al-'Arab*. Beirut: Dār al-Gharb al-Islāmī, 1991.

Biographical Encyclopedia of Astronomers = *The Biographical Encyclopedia of Astronomers*, ed. Thomas Hockey et al. New York: Springer, 2007, available on-line at: http://islamsci.mcgill.ca/RASI/BEA/

Biraben, 'Diseases in Europe' = Jean-Noël Biraben, 'Diseases in Europe: Equilibrium and Breakdown of the Pathocenosis', in *Western Medical Thought from Antiquity to the Middle Ages*, ed. Mirko D. Grmek and Bernardino Fantini. Translated by Antony Shugaar. Cambridge, Mass.: Harvard University Press, 1998, 319–353.

al-Bīrūnī, *Alberuni's India* = al-Bīrūnī, [*Kitāb Ta'rīkh al-Hind*] *Alberuni's India: An Account of the Religion, Philosophy, Literature, Geographhy, Chronology, Astronomy, Customs, Laws and Astrology of India about A.D. 1030*, ed. Eduard Sachau. 2 vols. London: Trübner, 1888; repr. Delhi: S. Chand, 1964.

al-Bīrūnī, *al-As'ilah wa-l-ajwibah* = *al-Bīrūnī, Al-As'ilah wa-l-ajwibah (Questions and Answers), including the further answers of al-Bīrūnī and al-Ma'ṣūmī's defence of Ibn Sīnā*, ed. with English and Persian introductions by Seyyed Hossein Nasr and Mehdi Mohaghegh. Kuala Lumpur: International Institute of Islamic Thought and Civilization, 1995.

al-Bīrūnī, *al-Āthār al-bāqiyah* = al-Bīrūnī, *Chronologie orientalischer Völker von Albêrûnî*, ed. C. Eduard Sachau. Leipzig: F.A. Brockhaus, 1878; repr. Wiesbaden: Harrassowitz, 1923.

al-Bīrūnī, *The Chronology of Ancient Nations* = al-Bīrūnī, *The Chronology of Ancient Nations: An English Version of the Arabic Text of the 'Athār-ul-bākiya' of Albīrūnī, or 'Vestiges of the Past'. Collected and Reduced to Writing by the Author in A.H. 390–391, A.D. 1000*, ed. and tr. Eduard Sachau. London: W.H. Allen, 1879; repr. Frankfurt am Main: Minerva, 1969.

al-Bīrūnī, *Fi taḥqīq mā baqiya li-l-Hind* = al-Bīrūnī, *Fi taḥqīq mā baqiya li-l-Hind*. Hyderabad: Dā'irat al-Ma'ārif al-'Uthmāniyyah, 1958.

al-Bīrūnī, *Ifrād al-maqāl fī amr al-ẓilāl* (Kennedy) = *The Exhaustive Treatise on shadows by Abū Rayḥān Muḥammad b. Aḥmad al-Bīrūnī*, translated with commentary by Edward Stewart Kennedy. 2 vols. Aleppo: Institute for the History of Arabic Science, 1976.

al-Bīrūnī, *Kitāb al-jamāhir* (Hyderabad) = al-Bīrūnī, *Kitāb al-jamāhir fī ma'rifat al-jawāhir*, ed. F. Krenkow. Hyderabad: Dā'irat al-Ma'ārif al-'Uthmāniyyah, 1936.

al-Bīrūnī, *Kitāb al-jamāhir* (Cairo) = al-Bīrūnī, *Kitāb al-jamāhir fī ma'rifat al-jawāhir*. Cairo: Maktabat al-Mutanabbī, n.d.

al-Bīrūnī, *Kitāb al-jamāhir* (trns. Said) = al-Bīrūnī, *The book most comprehensive in knowledge on precious stones: al-Beruni's book on mineralogy*, translated by Hakim Mohammad Said. Islamabad: Pakistan Hijra Council, 1989.

al-Bīrūnī, *Maqālīd 'ilm al-hay'ah* = al-Bīrūnī, *K. Maqālīd 'ilm al-hay'ah*, ed. and tr. Marie-Thérèse Debarnot. Damascus: Institut français de Damas, 1985.

al-Bīrūnī, *al-Qānūn al-Mas'ūdī* (Hyderabad) = al-Bīrūnī, *Kitāb al-Qānūn al-Mas'ūdī*. 3 vols. Hyderabad: Dā'irat al-Ma'ārif al-'Uthmāniyyah, 1954–1956.

al-Bīrūnī, *al-Qānūn al-Mas'ūdī* (al-Jundī) = al-Bīrūnī, *Kitāb al-Qānūn al-Mas'ūdī*, ed. 'Abd al-Karīm Sāmī al-Jundī. Beirut: Dār al-Kutub al-'Ilmīyah, 2002.

al-Bīrūnī, *al-Qānūn al-Mas'ūdī* (tr. Bulgakov & Rozhanskaya) = al-Bīrūnī, *al-Qānūn al-Mas'ūdī*, Russian trans. by P.G. Bulgakov and M.M. Rozhanskaya, in Абу Рейхан Бируни избранные произведения [selected works of Abu Rayhan Biruni], Vol. 5 *Canon Masudicus*, parts 1–2. Tashkent: Fan, 1973, 1976.

al-Bīrūnī, *Risālah* = Paul Kraus, *Risālah li-l-Bīrūnī fī fihrist kutub Muḥammad ibn Zakariyyā l-Rāzī (Épître de Bērūnī contenant le repertoire des ouvrages de Muḥammad b. Zakariyyā ar-Rāzī)*. Paris: Imprimerie Orientaliste au Calame, 1936.

al-Bīrūnī, *al-Ṣaydalah* = *K. al-ṣaydalah fī l-ṭibb. Al-Biruni's Book on Pharmacy and Mater-*

ica Medica. ed. and tr. Hakim Muḥammad Said. Karachi: Hamdard National Foundation, 1973.

al-Bīrūnī, *K. al-tafhīm li-awāʾil ṣināʿat al-tanjīm* (Ramsey Wright) = al-Bīrūnī, *The Book of Instruction in the Elements of the Art of Astrology*, translated with reproduction of BM MS. Or. 8349, by R. Ramsey Wright. London: Luzac & Co., 1934.

Bischoff, *Ergebnis des Nachdenkens* = Hans Dieter Bischoff, *Das Ergebnis des Nachdenkens über die Behandlung der Augenkrankheiten von Fatḥ ad-Dīn al-Qaisī* (Europäische Hochschulschriften, Asiatische und Afrikanische Studien, 27). Frankfurt am Main / Bern: Peter Lang, 1988.

Bitschai & Brodny, *Urology in Egypt* = J. Bitschai and M. Leopold Brodny, *A History of Urology in Egypt*. New York: Riverside Press, 1956.

Block Friedman, *Monstrous* = John Block Friedman, *The Monstrous Races in Medieval Art and Thought*. Syracuse, NY: Syracuse University Press, 2000.

Blodi & Wafai, *Arabian ophthalmologists* = Frederick C. Blodi & M. Zafer Wafai, *The Arabian Ophthalmologists*. Riyadh, 1993.

de Blois, *Burzōy's Voyage* = François de Blois, *Burzōy's Voyage to India and the origin of the book of Kalīlah wa Dimnah*. London: Royal Asiatic Society, 1990.

Bloom, *Paper before Print* = Jonathan M. Bloom, *Paper before Print: The History and Impact of Paper in the Islamic World*. New Haven, Conn.: Yale University Press, 2001.

Boilot, 'L'Œuvre d'al-Beruni' = D.J. Boilot, 'l'Œuvre d'al-Beruni. Essai bibliographie', *Mideo*, 2 (1955), 161–256.

Bonebakker, 'Adab and the Concept of belles-lettres' = S.A. Bonebakker, 'Adab and the Concept of belles-lettres', in Julia Ashtiany et al. (eds), *ʿAbbasid Belles-Lettres* (The Cambridge History of Arabic Literature). Cambridge: Cambridge University Press, 1990, 16–30.

Bora, 'A Mamluk Historian's Holograph' = Fozia Bora, 'A Mamluk Historian's Holograph. Messages from a *Musawwada* of *Taʾrīkh*', *Journal of Islamic Manuscripts*, 3/2 (2012), 119–153.

Bos, 'Maimonides' Medical Works' = Gerrit Bos, 'Maimonides' Medical Works and their Contribution to his Medical Biography', *Maimonidean Studies*, 5 (2008), 243–266.

Bos, 'Reception of Galen in Maimonides' = Gerrit Bos, 'The Reception of Galen in Maimonides' *Medical Aphorisms*' in *The Unknown Galen*, ed. Vivian Nutton (Bulletin of the Institute of Classical Studies, Suppl. 77). London: University of London, 2002, 139–152.

Bos & Käs, 'Arabic Pharmacognostic Literature' = Gerrit Bos and Fabian Käs, 'Arabic Pharmacognostic Literature and its Jewish Antecedents', *Aleph*, 16/1 (2016), 145–229.

Bosch, Carswell & Petherbridge, *Islamic Bookmaking* = Gulnar Bosch, John Carswell, & Guy Petherbridge, *Islamic Bindings & Bookmaking. A Catalogue of an Exhibition, The Oriental Institute, The University of Chicago, May 18 – August 18, 1981*. Chicago: The Oriental Institute Museum, 1981.

Bosworth, *Dynasties* = Clifford E. Bosworth, *The New Islamic Dynasties: A Chronological and Genealogical Manual*. Edinburgh: Edinburgh University Press, 2004.

Boudon, *Galien, Exhortation* = V. Boudon, *Galien, tome II. Exhortation à l'Étude de la médicine, Art médical*. Paris: Les Belles Lettres, 2002.

Boudon-Millot, *Galien* = V. Boudon-Millot, *Galien, tome I. Introduction générale, Sur l'ordre de ses propres livres, Sur ses propres livres, etc*. Paris: Les Belles Lettres, 2007.

Boudon-Millot, *Galien de Pergame* = V. Boudon-Millot, *Galien de Pergame. Un médecin grec à Rome*. Paris: Les Belles Lettres, 2012.

Boudon-Millot, Jouanna, Pietrobelli, *Ne pas se chagriner* = V. Boudon-Millot, J. Jouanna, A. Pietrobelli, *Galien, tome IV. Ne pas se chagriner*. Paris: Les Belles Lettres, 2010.

Bouras-Vallianatos & Xenophontos, *Greek Medical Literature* = Petros Bouras-Vallianatos and Sophia Xenophontos (eds.), *Greek Medical Literature and its Readers: From Hippocrates to Islam and Byzantium*. Abingdon / New York: Routledge, 2018.

Bouras-Vallianatos, 'Galen's Reception' = Petros Bouras-Vallianatos, 'Galen's Reception in Byzantium: Symeon Seth and his Refutation of Galenic Theories on Human Physiology', *Greek, Roman, and Byzantine Studies*, 55 (2015), 431–469.

Bowersock, *Hellenism* = Glen W. Bowersock, *Hellenism in Late Antiquity*. Cambridge: Cambridge University Press, 1990.

Brabant, 'La *urǧūza*' = Rosa Kuhne Brabant, 'La *urǧūza fī l-ṭibb* de Saʿīd Ibn ʿAbd Rabbihi', *al-Qanṭara*, 1 (1980), 279–338.

Brafman, 'The Arabic "De Mundo"' = David Brafman, 'The Arabic *De mundo*: An Edition with Translation and Commentary', Ph.D thesis, Duke University, 1985.

Brain, *Bloodletting* = P. Brain, *Galen on Bloodletting. A Study of the Origins, Development and Validity of his Opinions, with a Translation of the Three Works*. Cambridge: Cambridge University Press, 1986.

Bravo-Villasante, *Utilidades de la animales* = Carmen Ruiz Bravo-Villasante (tr.), *Libro de la utilidades de los animales*. Madrid: Fundación Universitaria Española, 1980.

Brentjes, 'Ayyubid princes' = Sonja Brentjes, 'Ayyubid princes and their scholarly clients from the ancient sciences', in Albrecht Fuess and Jan-Peter Hartung (eds.), *Court Cultures in the Muslim World: seventh to nineteenth centuries*. London / New York: Routledge, 2011, 326–356.

Brentjes, 'Karābīsī' = Sonja Brentjes, 'Aḥmad al-Karābīsī's Commentary on Euclid's "Elements"', in Menso Folkerts and Richard Lorch (eds.), *Sic itur ad astra: Studien zur Geschichte der Mathematik und Naturwissenschaften. Festschrift für den Arabisten Paul Kunitzsch zum 70. Geburtstag*. Wiesbaden: Harrassowitz, 2000, 31–75.

Brentjes, 'Narratives' = Sonja Brentjes, 'Narratives of Knowledge in Islamic Societies: What do They tell Us about Scholars and their Contexts?', *Almagest*, 4 (2013). 75–95.

Brentjes, 'Notes on Thābit ibn Qurrah and his rule for amicable numbers' = Sonja Brentjes, 'Notes on Thābit ibn Qurrah and his rule for amicable numbers', *Historia Mathematica*, 16 (1989), 373–378.

Brentjes, 'Prison' = Sonja Brentjes, 'The Prison of Categories – Decline and its Company', in Felicita Opwis and David Reisman (eds.), *Islamic Philosophy, Science, Religion and Culture: Studies in Honor of Dimitri Gutas*. Leiden: Brill, 2012, 131–156.

Brentjes, 'Teaching the Sciences in Ninth-century Baghdad' = Sonja Brentjes, 'Teaching the sciences in ninth-century Baghdad as a question in the history of the book: the case of Abū Yūsuf Yaʿqūb b. Isḥāq al-Kindī (d. after 256/870),' *Intellectual History of the Islamicate World*, 5 (1–2), 2017, 1–27.

Brett, *The Rise of the Fatimids* = Michael Brett, *The Rise of the Fatimids. The World of the Mediterranean & the Middle East in the Fourth Century of the Hijra, Tenth Century* CE. Leiden: Brill, 2001.

Breydy, *Études sur Saʿīd ibn Baṭrīq see* Saʿīd ibn al-Biṭrīq, *Etudes* (Breydy).

Brill's New Pauly, ed. Hubert Cancik & Helmith Schneider (Antiquity) & Manfred Landfester (Classical Tradition), Eng. tr. ed. by Christine F. Salazar and Francis G. Gentry (Engl. edition of *Der Neue Pauly*, Stuttgart: J.B. Metzler, 1996–), Leiden: Brill, online at https://referenceworks.brillonline.com/browse/brill-s-new-pauly

Broadhurst, *Ayyūbid Sultans* = Ronald J.C. Broadhurst, *A History of the Ayyūbid Sultans of Egypt*. Translated from the Arabic of al-Maqrīzī. Boston: Twayne, 1980.

Brock, 'Syriac into Greek at Mar Saba' = Sebastian Brock, 'Syriac into Greek at Mar Saba', in J. Patrick (ed.), *The Sabaite Heritage in the Orthodox Church*. Louvain: Peeters, 2001, 201–208.

Browne, *A Literary History of Persia* = E.G. Browne, *A Literary History of Persia*, 4 vols. Cambridge: University Press, 1951–1953.

Brugman & Drossaart Lulofs, *Aristotle. Generation of Animals* = Jan Brugman and Hendrik Joan Drossaart Lulofs (eds.), *Aristotle. Generation of Animals: the Arabic Translation Commonly Ascribed to Yaḥyā ibn al-Biṭrīq*. Leiden: Brill, 1971.

Brunschwig, 'Métiers' = R. Brunschwig, 'Métiers vils dans l'Islam', *Studia Islamica*, 16 (1967), 41–60.

al-Buḥturī, *Dīwān* = al-Buḥturī, *Dīwān*, ed. Ḥasan Kāmil al-Ṣayrafī, 5 vols. Cairo: Dār al-Maʿārif, 1972–1978.

Bukhārī, *Ṣaḥīḥ* = Muḥammad ibn Ismāʿīl al-Bukhārī, *al-Jāmiʿ al-musnad al-ṣaḥīḥ*, 9 vols. Ed. Muḥammad Zuhayr ibn Nāsir. Beirut: Dār Ṭawq al-Najāh, 1422/2001–2002.

Bukhārī, *al-Tārīkh al-kabīr* = Muḥammad ibn Ismāʿīl al-Bukhārī, *Kitāb al-Tārīkh al-Kabīr*. Hydarabad: Maṭbaʿat Jamʿiyyat Dāʾirat al-Maʿārif Al-ʿUthmāniyyah, 1941.

Bulliet, *Patricians* = Richard W. Bulliet, *The Patricians of Nishapur: A Study in Medieval Islamic Social History*. Cambridge, Mass.: Harvard University Press, 1972.

Bürgel, *Allmacht und Mächtigkeit* = Johann C. Bürgel, *Allmacht und Mächtigkeit. Religion und Welt im Islam*. Munich: Beck, 1991.

Bürgel, *Ärtzliches Leben* = Johann C. Bürgel, *Ärtzliches Leben und Denken im arabischen Mittelalter*, ed. F. Käs. Leiden: Brill, 2016.

Bürgel, *Averroes, 'Contra Galenum'* = Johann C. Bürgel, *Averroes, 'Contra Galenum':*

das Kapitel von der Atmung im Colliget *des Averroes als ein Zeugnis mittelalterlich-islamischer Kritik an Galen*. Nachrichten von der Akademie der Wissenschaften in Göttingen, Philol.-hist. Kl., Jg. 1967, Nr. 9. Göttingen, 1968.

Burgess, *Studies* = R. Burgess, *Studies in Eusebian and Post-Eusebian Chronography*. Stuttgart: Franz Steiner, 1999.

Burnett, 'Arabic, Greek and Latin Works' = Charles Burnett, 'Arabic, Greek and Latin Works on Astrological Magic Attributed to Aristotle', in Kraye et al. (eds.), *Pseudo-Aristotle in the Middle Ages*, 84–96.

Burnett, 'Hermann of Carinthia and the *Kitāb al-Isṭamāṭis*. Further Evidence for the Transmission of Hermetic Magic', in *Magic and Divination in the Middle Ages. Texts and Techniques in the Islamic and Christian Worlds*. London: Routledge, 1996, repr. of *Journal of the Warburg and Courtauld Institutes*, 44 (1981), 167–169.

Burnett, 'Albumasar in Sadan in the Twelfth Century' = Charles Burnett, '"Albumasar in Sadan" in the Twelfth Century', in G. Marchetti, O. Rignani, V. Sorge (eds.), *Ratio et superstitio: Essays in Honor of Graziella Federici Vescovini*. Turnhout: Brepols Publishers, 2003, 59–67.

Burnett, *Ibn Baklarish's Book of Simples* = Charles Burnett (ed.), *Ibn Baklarish's Book of Simples: Medical Remedies between three Faiths in Twelfth-Century Spain*. Oxford: Oxford University Press, 2008.

Burnett & Jacquart, *Constantine the African and ʿAlī ibn ʿAbbās al-Mağūsī* = Charles Burnett and Danielle Jacquart (eds.), *Constantine the African and ʿAlī ibn ʿAbbās al-Mağūsī: The Pantegni and related texts*. Leiden: Brill, 1994.

Burton, *Personal Narrative of a Pilgrimage* = Richard Burton, *A Personal Narrative of a Pilgrimage to El Medina and Meccah*. 2nd ed. London: Longman, Brown, Green and Longmans, 1857.

Bustānī, *Dāʾirat al-maʿārif* = *Dāʾirat al-maʿārif: qāmūs ʿāmm li-kull fann wa-maṭlab*, ed. Fuʾād Afrām Bustānī. Beirut: s.n., 1956–.

Cacouros, '*De virtutibus et vitiis*' = M. Cacouros, 'Aristote de Stagire – Le traité Pseudo-Aristotélicien *De virtutibus et vitiis*', in Goulet, *Dictionnaire*, Suppl. 506–546 (511–513 for the Arabic tradition).

Cadden, *Sex Difference* = Joan Cadden, *Meanings of Sex Difference in the Middle Ages: Medicine, Science, and Culture*. Cambridge: Cambridge University Press, 1993.

Cahen, *formation of Turkey* = Claude Cahen, *The Formation of Turkey. The Seljukid Sultanate of Rum: Eleventh to Fourteenth Century*. Harlow, Essex: Routledge, 2001.

Canby et al., *Court and Cosmos* = Sheila Canby, C.S. Peacock, Deniz Beyazit & Martina Rugiadi, *Court and Cosmos: The Great Age of the Seljuqs*. New Haven / London: Yale University Press, 2016.

Carra de Vaux *see* al-Masʿūdī, *L'avertissement*.

Carpentieri, 'On the Meaning of *Birsām* and *Sirsām*' = Nicola Carpentieri, 'On the Meaning of *Birsām* and *Sirsām*: A Survey of the Arabic Commentaries on the Hippocratic *Aphorisms*', *Mélanges de l'Institut dominicain d'études orientales*, 21 (2017), 81–92.

Carpentieri & Mimura, 'Phrenitis' = Nicola Carpentieri and Taro Mimura, 'Arabic Commentaries on the Hippocratic *Aphorisms*, vi. 11: A Medieval Medical Debate on Phrenitis', *Oriens*, 45 (2017), 176–202.

Carra de Vaux, 'La Ḳaçîda' = B. Carra de Vaux, 'La Ḳaçîda d'Avicenne sur l'âme', *Journal Asiatique*, n. s., 14 (1899), 157–173.

Carey, 'The Gold and Silver Lining' = Moya Carey, 'The Gold and Silver Lining: Shams al-Dīn Muḥammad b. Mu'ayyad al-'Urḍi's inlaid celestial globe (c. AD 1288) from the Ilkhamid observatory at Marāgha', *Iran*, 47, (2009), 97–108.

Casiri, Bibliotheca Arabico-Hispana Escurialensis = Miguel Casiri, Bibliotheca Arabico-Hispana Escurialensis sive Librorum omnium mss. quos Arabicè ab auctoribus magnam partem Arabo-Hispanis ompositos Bibliotheca Cœnobii Escurialensis complectitur, recensio et explanatio. 2 vols. Madrid: Antonius Perez de Soto imprimebat, 1760–1770.

Caskel, *Ǧamharat an-nasab* see Ibn al-Kalbī, *Ǧamharat an-nasab*.

Castells & Samsó, 'Seven Chapters' = M. Castells and J. Samsó, 'Seven Chapters of Ibn al-Ṣaffār's lost *zīj*', *Archives Internationales d'Histoire des Sciences*, 45 (1995), 229–262.

Celentano, *Due scritti medici di al-Kindī* = Guiseppe Celentano, *Due scritti medici di al-Kindī*. Napoli: Istituto orientale di Napoli, 1979.

Celentano, 'Le petit traité … sur la prophylaxie et la thérapie de dents' = Guiseppe Celentano, 'Le petit traité de Ḥunayn ibn Isḥāq sur la prophylaxie et la thérapie de dents', *Arabica*, 21/3 (1975), 245–251.

Chadwick, *Church* = Henry Chadwick, *The Church in Ancient Society from Galilee to Gregory the Great*. Oxford: Oxford University Press, 2003.

Chamberlain, *Knowledge and Social Practice* =Michael Chamberlain, *Knowledge and Social Practice in Medieval Damascus, 1190–1350*. Cambridge: Cambridge University Press, 1994.

Chase, 'Der Gottheit lebendiges Kleid' = Michael Chase, 'Der Gottheit lebendiges Kleid Orpheus Arabicus, or myths of weaving in Greco-Arabic philosophy'. Unpublished paper available on-line at: http://www.academia.edu/968206/Der_Gottheit_lebendiges_Kleid_Orpheus_Arabicus_or_myths_of_weaving_in_Greco-Arabic_philosophy

Cheikho, *Traités inédits* = Louis Cheikho et al., *Traités inédits d'anciens philosophes arabes, musulmans et chrétiens*. 2nd ed. Beirut: Imprimerie catholique, 1911.

Chew Barry, *Syriac Medicine* = Samuel Chew Barry, *Syriac Medicine and Ḥunayn ibn Isḥāq's Arabic Translation of the Hippocratic Aphorisms* (Journal of Semitic Studies Supplement 39). Oxford: Oxford University Press, 2018.

Chipman, 'Jewish Presence in Arabic Writings on Medicine' = Leigh Chipman, 'Jewish

Presence in Arabic Writings on Medicine and Pharmacology during the Medieval Period', *Religion Compass*, 7/9 (2013), 394–401.

Chipman, *Pharmacists in Mamluk Cairo* = Leigh Chipman, *The World of Pharmacy and Pharmacists in Mamluk Cairo* (Sir Henry Wellcome Asian Series, 8). Leiden: Brill, 2010.

Christys, *Christians* = Ann R. Christys, *Christians in al-Andalus (711–1000)*. London: Routledge, 2002.

Chroust, *Aristotle* = Anton-Hermann Chroust, *Aristotle: New Light on his Life and On Some of his Lost Works*. vol. 1. Notre Dame: University of Notre Dame Press, 1973.

Cilliers & Retief, 'Nervous System' = Louise Cilliers & François Retief, 'The nervous system in antiquity', *SAMJ Forum* 98, 10 (2008), 768–772.

Cloarec, *Bīmāristāns* = F. Cloarec, *Bīmāristāns, lieux de folie et de sagesse. La folie et ses traitements dans les hôpitaux médiévaux au Moyen-Orient*. Paris: L'Harmattan, 1998.

CMG = *Corpus Medicorum Graecorum* (Berlin); the majority of texts cited are available online at http://cmg.bbaw.de/epubl/online/editionen.html.

Compton, *Andalusian Lyrical Poetry* = Linda Fish Compton, *Andalusian Lyrical Poetry and Old Spanish Love Songs: The Muwashshaḥ and its Kharja*. New York: New York University Press, 1976.

Conrad, 'Epidemic disease in formal and popular thought' = Lawrence I. Conrad, 'Epidemic disease in formal and popular thought in early Islamic Society' in *Epidemics and ideas: Essays on the historical perception of pestilence*, ed. Terence Ranger and Paul Slack. Cambridge: Cambridge University Press, 1992, 77–99.

Conrad, 'Ibn Buṭlān in *Bilād al-Shām*' = Lawrence Conrad, 'Ibn Buṭlān in *Bilād al-Shām*: The Career of a Travelling Christian Physician', in David Thomas (ed.), *Syrian Christians under Islam. The First Thousand Years*. Leiden: Brill, 2001, 131–158.

Conrad, 'Scholarship and social context' = Lawrence I. Conrad, 'Scholarship and social context: a medical case from the eleventh-century Near East' in Don Bates (ed.), *Knowledge and the scholarly medical traditions*. Cambridge: Cambridge University Press, 1995, 84–100.

Conrad, '*Ṭāʿūn* and *Wabā*'' = Lawrence I. Conrad, '*Ṭāʿūn* and *Wabā*': Conceptions of Plague and Pestilence in Early Islam', *Journal of the Economic and Social History of the Orient*, 25 (1982), 268–307.

Contadini, 'The Ibn Bakhtīshuʿ Bestiary Tradition' = Anna Contadini, 'The Ibn Bakhtīshuʿ Bestiary Tradition: The Text and its Sources', *Medicina nei Secoli: Arte e Scienze*, 6 (1994), 349–364.

Contadini, *World of Beasts* = Anna Contadini, *A World of Beasts: A Thirteenth-Century Illustrated Arabic Book on Animals (the* Kitāb Naʿt al-Ḥayawān*) in the Ibn Bakhtīshūʿ Tradition*. Leiden: Brill, 2012.

Cook, 'A Survey of Muslim Material on Comets and Meteors' = David Cook, 'A Survey of Muslim Material on Comets and Meteors', *Journal for the History of Astronomy*, 30/2 (1999), 131–160.

Cooper, 'Rational and Empirical Medicine' = Glen M. Cooper, 'Rational and Empirical Medicine in Ninth-Century Baghdad: Qusṭā ibn Lūqā's Questions on the Critical Days in Acute Illnesses', *Arabic Sciences and Philosophy* 24, (2014), 69–102.

Cooperson, 'Purported Autobiography' = Michael Cooperson, 'The Purported Autobiography of Ḥunayn ibn Isḥāq', *Edebiyât.* 7/2 (special issue: 'Arabic Autobiography', ed. D.F. Reynolds) (1996), 235–249.

Cooperson & Toorawa, *Arabic Literary Culture* = Michael Cooperson and Shawkat M. Toorawa (eds.), *The Dictionary of Literary Biography: Arabic Literary Culture, 500–925* (Dictionary of Literary Biography, 311). Detroit: Thomson Gale, 2005.

Cooperson, *Al-Ma'mūn* = Michael Cooperson, *Al-Ma'mūn* (Makers of the Muslim World). Oxford: Oneworld, 2005.

Cooperson, *Classical Arabic Biography* = Michael Cooperson, *Classical Arabic Biography: The Heirs of the Prophet in the Age of al-Ma'mun.* Cambridge: Cambridge Univdersity Press, 2000.

The Coptic Encycl. = *The Coptic Encyclopedia*, ed. Aziz S. Atiya. 8 vols. New York: MacMillan and Toronto: Collier Macmillan Canada, c. 1991; available on-line as the Claremont Coptic Encyclopedia at http://ccdl.libraries.claremont.edu/cdm/landingpage/collection/cce

Corriente, *A Dictionary of Andalusi Arabic* = F. Corriente, *A Dictionary of Andalusi Arabic.* Leiden: Brill, 1997.

Cottrell, 'Abū Sulaymān' = Emily J. Cottrell, 'Abū Sulaymān al-Sijistānī al-Manṭiqī' in *Encyclopedia of Medieval Philosophy: Philosophy between 500 and 1500.* Berlin: Springer, 2011, 17–20.

Cottrell, 'Pythagoras' = Emily Cottrell, 'Pythagoras, the Wandering Ascetic: A Reconstruction of the Life of Pytrhagoras According to al-Mubashshir Ibn Fātik and Ibn Abī Uṣaybiʿa', in Almut-Barbara Renger and Alessandro Stavro (eds.), *Forms and Transformations of Pythagorean Knowledge.* Wiesbaden: Harrassowitz, 2016, 489–512.

Craik, *Hippocratic Corpus* = Elizabeth M. Craik, *The 'Hippocratic' Corpus. Content and Context.* London / New York: Routledge, 2015.

Crone, "Abbāsid Abnā'" = Patricia Crone, 'The 'Abbāsid Abnā' and Sāsānid Cavalrymen', *Journal of the Royal Asiatic Society*, 8/1 (1998), 1–19.

Crone, *Slaves on Horses* = Crone, Patricia, *Slaves on Horses: The Evolution of the Islamic Polity.* Cambridge: Cambridge University Press, 1980.

Cruz Hernández, *Abū l-Walīd Ibn Rushd* = Miguel Cruz Hernández, *Abū l-Walīd Ibn Rushd (Averroes): vida, obra, pensamiento, influencia.* Córdoba: Monte de Piedad y Caja de Ahorros de Córdoba, 1986.

Cureton, *Catalogus* = William Cureton, *Catalogus codicum manuscriptorum orientalium qui in Museo Britannico asservantur.* Pars secunda, codices Arabicos amplectens. 2 vols. Londini: Impensis curatorum Musei Britannici, 1846.

Cureton, 'Extract' = William Cureton, 'Extract from the work entitled *'Uyūn al-anbā' fī ṭabaqāt al-aṭibbā'*, or, *Fountains of information respecting classes of physicians* by ... Ibn Abī Uṣaybiâh; with remarks by Horace Hayman Wilson', *Journal of the Royal Asiatic Society*, 6 (1841), 105–119. Repr. in Sezgin, *Studies* III, 1–15.

Curtis, *Garum* = Robert I. Curtis, *Garum and Salsamenta: Production and Commerce in Materia Medica*. Leiden: Brill, 1991.

al-Ḍabbī, *Bughyat al-multamis* = Abū Ja'far Aḥmad ibn Yaḥyā al-Ḍabbī, *Bughyat al-multamis fī ta'rīkh rijāl ahl al-Andalus*. Cairo: Dār al-Kātib al-'Arabī, 1967.

Daftary, *The Ismāʿīlīs: Their History and Doctrines* = Farhad Daftary, *The Ismāʿīlīs: Their History and Doctrines*. 2nd ed. Cambridge: Cambridge University Press, 2007.

Dagher & Troupeau, *Réponses aux questions posées par Ibn Butlan dans le Banquet des médecins* = Joseph Dagher and Gérard Troupeau (eds.), *Réponses aux questions posées par Ibn Butlan dans le Banquet des médecins*. Paris: Geuthner, 2011.

Daiber, 'Abū Bakr al-Rāzī' = Hans Daiber, 'Abū Bakr al-Rāzī' in Rudolph, Hansberger & Adamson, *Philosophy in the Islamic World*, pp. 381–420.

Daiber, *Aetius Arabus* = Hans Daiber, *Aetius Arabus. Die Vorsokratiker in arabischer Überlieferung*. Wiesbaden: Franz Steiner, 1980.

Daiber, *Bibliography* = Hans Daiber, *Bibliography of Islamic Philosophy*. 3 vols with a Supplement. Leiden: Brill, 1999–2006.

Daiber, *Collection* = *Daiber Collection Database. The Arabic Manuscripts in the Daiber Collection, Institute of Oriental Culture, University of Tokyo*. Online at http://ricasdb.ioc.u-tokyo.ac.jp/daiber/db_index_eng.html

Daiber, *Islamic Thought* = Hans Daiber, *Islamic Thought in the Dialogue of Cultures: A Historical and Bibliographical Survey* (Themes in Islamic Studies). Leiden / Boston: Brill, 2012.

Daiber, *Ein Kompendium der aristotelischen Meteorologie* = Hans Daiber, *Ein Kompendium der aristotelischen Meteorologie in der Fassung des Hunain ibn Isḥâq*. Amsterdam / Oxford: Noord-Hollandsche Uitgevers Maatschappij, 1975.

Daiber, '*Meteorology*' = Hans Daiber, 'The *Meteorology* of Theophrastus in Syriac and Arabic Translation', in W.W. Fortenbaugh and D. Gutas (eds), *Theophrastus, His Psychological, Doxographical, and Scientific Writing*. New Brunswick / London: Transaction Publishers, 1992, 166–293.

Daiber, 'Survey of Theophrastean material' = Hans Daiber, 'A survey of Theophrastean material indirectly transmitted in Arabic', in Fortenbaugh, Huby & Long, *Theophrastus*, 103–114.

al-Dakhwār, *Sharḥ taqdimat al-maʿrifah* = al-Dakhwār, *Kitāb sharḥ taqdimat al-maʿrifah*, ed. Māhir 'Abd al-Qādir Muḥammad, 'Alī 'Abd al-Muʿṭī Muḥammad, and Muḥammad 'Abduh Maḥjūb. Alexandria: Dār al-Maʿrifah al-Jāmiʿiyyah, 2000.

al-Dalajī, *al-Falākah* = Aḥmad ibn 'Alī al-Dalajī, *al-Falākah wa-l-mufallakūn*, Cairo: Maṭbaʿat al-Shaʿb, AH 1322.

al-Damīrī, *Ḥayāt al-ḥayawān* = Kamāl al-Dīn al-Damīrī, *Ḥayāt al-ḥayawān al-kubrā*, 2 vols. Cairo: al-Maktabah al-Tijāriyyah al-Kubrā, repr. Beirut: Dār al-Fikr, n.d.

D'Ancona & Taylor, *'Liber de Causis'* = Cristina D'Ancona and Richard C. Taylor *'Liber de Causis'*, in Goulet, *Dictionnaire*, Suppl., 599–654.

Dāneshpazūh, *Manṭiq* = Muḥammad Taqī Dāneshpazūh (ed.), *al-Manṭiq li-Ibn al-Muqaffaʿ wa-Ḥudūd al-manṭiq li-Ibn Bihrīz*. Tehran: Anjuman-i Shāhanshāhī-i Falsafa-i Īrān, 1978.

Daremberg & Ruelle, *Rufus* = Charles Daremberg and Emile Ruelle, *Oeuvres de Rufus d'Ephèse*. Paris: L'Imprimerie Nationale, 1879.

Das & Koetschet, 'Two Pharmacological Texts on Whey' = Aileen Dan and Pauline Koetschet, 'Two Pharmacological Texts on Whey by Abū Bakr Muḥammad ibn Zakariyyā' al-Rāzī (d. 925)', *Journal of the American Oriental Society* 137 (2017), 25–42.

Davies 'Traits' = Davies, D. 'Character Traits', in *Galen: Psychological Writings* ed. P.N. Singer. Cambridge: Cambridge University Press, 2013, 107–201.

Dāwūd al-Anṭākī, *Tadhkirah* = Dāwūd ibn 'Umar al-Anṭākī, *Tadhkirat Dāwūd al-Anṭākī al-musammā Tadhkirat ulī l-albāb wa-l-jāmiʿ li-l-ʿajab al-ʿujāb*, ed. Aḥmad Shams al-Dīn. Beirut: Dār al-Kutub al-ʿIlmiyyah, 2000.

Dāwūd al-Anṭākī, *Tazyīn* = Dāwūd al-Anṭākī, *Tazyīn al-aswāq bi-tafṣīl ashwāq al-ʿushshāq*, ed. Muḥammad Altūnjī. 2 vols. Beirut: 'Ālam al-Kutub, 1413/1993.

De Callataÿ, 'Again on Maslama Ibn Qāsim al-Qurṭūbī' = De Callataÿ, 'Again on Maslama Ibn Qāsim al-Qurṭūbī, the Ikhwān al-Ṣafā' and Ibn Khaldūn: New Evidence from Two Manuscripts of *Rutbat al-ḥakīm*', in *al-Qanṭara*, 37/2 (2016), 329–372.

De Cillis, *Free will and Predestination* = Maria de Cillis, *Free will and predestination in Islamic thought: theoretical compromises in the works of Avicenna, Ghazālī and Ibn 'Arabī*. Abingdon / New York: Routledge, 2014.

de Koning, *Traité sur le calcul* = Pieter de Koning, *Traité sur le calcul dans les reins et dans la vessie par Abū Bekr Muḥammed ibn Zakarīyā al-Rāzī*. Leiden: E.J. Brill, 1896.

de Koning, *Trois traités* = Pieter de Koning, *Trois traités d'anatomie arabes*. Leiden: E.J. Brill, 1903.

De Lacy, *Galeni de placitis* = Philip De Lacy, *Galeni de placitis Hippocratis et Platonis. On the Doctrines of Hippocrates and Plato*, 3 vols. CMG V 4,1,2. Berlin: Corpus Medicorum Graecorum, 1978–2005.

de Prémare, 'Un Andalou en Égypte' = Alfred Luis de Prémare, 'Un Andalou en Égypte à la fin du XIe siècle: Abu l-Ṣalt Omayya de Denia et son Êpître Égyptienne', in *Mélanges de l'Institut Dominicain d'Études Orientales du Caire*, 3 (1964–1966), 179–208.

De Smet, *Empedocles* = Daniel de Smet, *Empedocles Arabus. Une lecture néoplatonicienne tardive*. Brussels: Koninklijke Akademie voor Wetenschappen, Letteren en Schone Kunsten van België, 1998.

De Vos, 'The "Prince of Medicine"' = Paula De Vos, 'The "Prince of Medicine": Yūḥannā ibn Māsawayh and the Foundations of the Western Pharmaceutical Tradition', *Isis* 104 (2013), 667–712.

De Young, 'Isḥāq ibn Ḥunayn' = G. De Young, 'Isḥāq ibn Ḥunayn, Ḥunayn ibn Isḥāq, and the Third Arabic Translation of Euclid's Elements', in *Historia Mathematica* 19 (1992): 188–199.

Delaporte, *Chonographie* = J.-L. Delaporte, *La Chronographie d'Élie Bar-Šinaya Métropolitain de Nisibe*, Bibliothèque de l'Ecole des Hautes Etudes, Sciences historiques et philologiques, 181e fasc. Paris: Librairie Honoré Champion, 1910.

Denooz, *Transmission de l'art médical de la Grèce à l'Islam* = L. Denooz, *Transmission de L'art médical de la Grèce à l'Islam: Étude du traité Sur Les Foetus de Sept Mois de Galien, de sa traduction par Ḥunayn b. Isḥâq et son commentaire par Ṯâbit b. Qurra.* Liège: Université de Liège, Section Orientale, 1999.

des Places, *Porphyre* = Édouard des Places, *Porphyre. Vie de Pythagore. Lettre à Marcella.* Paris: Les Belles Lettres, 1982.

al-Dhahabī, *Mīzān al-iʿtidāl* = Muḥammad ibn Aḥmad al-Dhahabī, *Mīzān al-iʿtidāl fī naqd al-rijāl*, ed. ʿAlī Muḥammad al-Bijāwī. 4 vols. [repr.] Beirut: Dār al-Maʿrifah, n.d.

al-Dhahabī, *Siyar* = Muḥammad ibn Aḥmad al-Dhahabī, *Siyar aʿlām al-nubalāʾ*, ed. Shuʿayb al-Arnaʾūṭ et al., 25 vols. Beirut: Muʾassasat al-Risālah, 1981–1988.

al-Dhahabī, *Tadhkirah* = Muḥammad ibn Aḥmad al-Dhahabī, *Tadhkirat al-ḥuffāẓ*, ed. Zakariyyā ʿUmayrāt, 5 vols. Bayrūt: Dār al-Kutub al-ʿIlmiyyah, 1998.

al-Dhahabī, *Tārīkh al-Islām* = Muḥammad ibn Aḥmad al-Dhahabī, *Tārīkh al-Islām wa-wafayāt al-mashāhīr wa-l-aʿlām*, ed. ʿUmar al-Tadmurī, 52 vols. Beirut: Dār al-Kitāb al-ʿArabī, 2001.

al-Dhahabī, *Tārīkh al-Islām* = Muḥammad ibn Aḥmad al-Dhahabī, *Tārīkh al-Islām wa-wafayāt al-mashāhīr wa-l-aʿlām*, ed. Bashshār ʿAwwād Maʿrūf. Beirut: Dār al-Gharb al-Islāmī, 2003.

al-Dhahabī, *Tārīkh al-Islām, Ḥawādith wa-wafayāt 421–430 H, 431–440 H* = Muḥammad ibn Aḥmad al-Dhahabī, *Tārīkh al-Islām, Ḥawādith wa-wafayāt 421–430 H, 431–440 H*, ed. ʿUmar Tadmurī, Beirut: Dār al-Kitāb al-ʿArabī, 1998.

al-Dhahabī, *Tārīkh al-Islām, Ḥawādith wa-wafayāt 621–630 H* = Muḥammad ibn Aḥmad al-Dhahabī, *Tārīkh al-Islām, Ḥawādith wa-wafayāt 621–630 H*, ed. ʿUmar Tadmurī. Beirut: Dār al-Kitāb al-ʿArabī, 1998.

Di Martino, '*Parva Naturalia*' = Carla Di Martino, 'Aristote de Stagire – *Parva Naturalia*. Tradition arabe', in Goulet, *Dictionnaire* Suppl., 375–378.

Diels, *Handschriften* = Hermann Diels, *Die Handschriften der antiken Ärtze. 1. Teil, Hippocrates und Galenos*. Berlin: Königliche Akademie der Wissenschaften, 1905.

Dietrich, *Medicinalia Arabica* = Albert Dietrich, *Medicinalia Arabica: Studien über arabische medizinische Handschriften in türkischen und syrischen Bibliotheken* (Abhandlungen der Akademie der Wissenschaften in Göttingen, phil.-hist. Klasse, Dritte Folge, 66). Göttingen: Vandenhoeck & Ruprecht, 1966.

Dihle, *Greek and Latin Literature* = Albrecht Dihle, *Greek and Latin Literature of the Roman Empire from Augustus to Justinian*. London: Routledge, 1994.

al-Dīnawarī, *Nabāt* = Abū Ḥanīfah Aḥmad ibn Dāwūd al-Dīnawarī, *Kitāb al-nabāt: al-juzʾ al-thālith, wa-l-niṣf al-awwal min al-juzʾ al-khāmis / The Book of Plants: Part of the Monograph Section*, ed. Bernhard Lewin, Wiesbaden: Franz Steiner, 1394/1974.

Dioscorides *see* Beck, Wellmann.

Dīwān Ibn Sīnā = Ibn Sīnā, *Dīwān*, ed. Ḥusayn ʿAlī Maḥfūẓ. Tehran: Maṭbaʿat al-Ḥaydarī, 1957.

Djebbar, 'al-Risālah fī l-taksīr' = Ahmed Djebbar, 'al-Risālah fī l-taksīr li-Ibn ʿAbdūn, shāhid ʿalā l-mumārasāt al-sābiqah li-l-taqlīd al-jabrī l-ʿarabī', *Suhayl*, 5 (2005), 7–68 [Arabic section] and *Suhayl*, 6 (2006), 81–86 [Arabic section].

Djebbar, 'La géometrie' = Ahmed Djebbar, 'La géometrie du mesurage et du découpage dans les mathématiques d' Al-Andalus (X^e^–XIII^e^s.)', in P. Radelet de Grave (ed.), *Liber Amicorum Jean Dhombres*, Turnhout: Brepols, 2008, 113–147.

Dodds, *Gorgias* = Eric R. Dodds, *Plato, Gorgias*. Oxford: Clarendon Press, 1959.

Dols, 'The Leper' = Michael W. Dols, 'The Leper in medieval Islamic society', *Speculum*, 58 (1983), 891–916.

Dols, 'Leprosy' = Michael W. Dols, 'Leprosy in Medieval Arabic Medicine', *Journal of the History of Medicine and Allied Sciences*, 34 (1979), 314–333.

Dols, *Majnūn* = Michael W. Dols, *Majnūn: The Madman in Medieval Islamic Society*, ed. Diana E. Immisch. Oxford: Clarendon Press, 1992.

Dols, *Medieval Islamic Medicine* = Michael W. Dols, *Medieval Islamic Medicine*: *Ibn Riḍwān's treatise, "On the Prevention of Bodily Ills in Egypt"*, translated, with an introduction by Michael W. Dols, Arabic text edited by Adil S. Gamal. Berkeley: University of California Press, 1984.

Donohue, *The Buwayhid Dynasty in Iraq* = John J. Donohue, *The Buwayhid Dynasty in Iraq 334h., 945 to 403h., 1012: Shaping Institutions for the Future*. Leiden / Boston: Brill, 2003.

Dooley, *Astrology* = B. Dooley (ed.), *A Companion to Astrology in the Renaissance*. Leiden: Brill, 2014.

Dozy, *Dictionnaire détaillé* = R.P.A. Dozy, *Dictionnaire détaillé des noms des vêtements chez les Arabes*. Amsterdam: Jean Müller, 1845.

Dozy, *Supplément* = R. Dozy, *Supplément aux dictionnaires arabes*. 2 vols. Leiden: E.J. Brill, 1881; 2me éd. 1927.

Druart, 'La fin humaine selon Ibn Bajjah' = Thérèse-Anne Druart, 'La fin humaine selon Ibn Bajjah (Avenpace)', *Bulletin de Philosophie Mediévale*, 23 (1981), 59–64.

DSB = *Dictionary of Scientific Biography*, ed. C.C. Gillispie. 18 vols. New York: Charles Scribner's Sons, 1970–1980; see also *New* DSB = *New Dictionary of Scientific Biography*. Detroit: Charles Scribner's Sons, 2008. Available online through Gale Virtual Reference Library.

Ducène, 'Isidore' = Jean-Charles Ducène, 'Al-Bakrī et les *Étymologies* d'Isidore de Séville', *Journal Asiatique*, 297 (2009), 379–397.

Dunlop, 'Arabic Medicine in England' = Douglas M. Dunlop, 'Arabic Medicine in England', *Journal of the History of Medicine and Allied Sciences*, 11 (1956), 166–182.

Dunlop, "Arabian Tailor" = Douglas M. Dunlop, 'The "Arabian Tailor", Henry Wild', *Bulletin of the School of Oriental and African Studies*, 19 (1957), 579–581.

Dunlop, *Dialogues on Astrology* = Douglas M. Dunlop, 'The *Mudhākarāt fī 'Ilm an-Nujūm* (Dialogues on Astrology) attributed to Abū Ma'shar al-Balkhī (Albumasar)' in C.E. Bosworth (ed.), *Iran and Islam. In memory of the late Vladimir Minorsky*. Edinburgh: Edinburgh University Press, 1971, 229–246.

Dunlop, 'Philosophical Predecessors' = Douglas M. Dunlop, 'Philosophical Predecessors and Contemporaries of Ibn Bājjah', *The Islamic Quarterly*, 2 (1955), 100–116.

Dunlop, *Ṣiwān al-ḥikmah* = Douglas M. Dunlop, *The Muntakhab Ṣiwān al-ḥikmah of Abū Sulaimān as-Sijistānī*. The Hague / New York: Mouton Publishers, 1979.

Dunlop, 'The Translations of al-Biṭrīq' = Douglas M. Dunlop, 'The Translations of al-Biṭrīq and Yaḥyā (Yuḥannā) b. al-Biṭrīq', in *Journal of the Royal Asiatic Society*, 3–4 (1959), 140–150.

Düring, *Aristotle* = Ingemar Düring, *Aristotle in the Ancient Biographical Tradition* (Studia Graeca et Latina Gothoburgensia, 5). Göteborg: Elanders, 1957.

Duval, *Lexicon syriacum* = Rubens Duval, *Lexicon syriacum auctore Hassano bar Bahlule voces syriacas græcasque cum glossis syriacis et arabicis complectens*. 3 vols. Parisiis: e Reipublicæ typographæo, 1888–1901.

EAL = *Encyclopedia of Arabic Literature*, ed. Julie Scott Meisami and Paul Starkey. London: Routledge, 1998.

Eastmond, *Tamta* = Antony Eastmond, *Tamta's World: The Life and Encounters of a Medieval Noblewoman from the Middle East to Mongolia*. Cambridge: Cambridge University Press, 2017.

Ebied & Thomas, *The polemical works of 'Alī al-Ṭabarī* = *The polemical works of 'Alī al-Ṭabarī*, ed. Rifaat Ebied and David Thomas (History of Christian-Muslim Relations, 27). Leiden / Boston: Brill, [2016].

Eddé, *La principauté Ayyoubide* = Anne-Marie Eddé, *La principauté Ayyoubide d'Alep (579/1183–658/1260)*. Stuttgart: Franz Steiner Verlag, 1999.

Eddé, *Saladin* = Anne-Marie Eddé, *Saladin*. Cambridge, Mass. / London: Belknap, 2011 (tr. from the French by Jane Marie Todd).

Edelstein, *Asclepius* = E.J. and L. Edelstein, *Asclepius. Collection and Interpretation of the Testimonies*. 2 vols. Baltimore: Johns Hopkins University Press, 1945.

Edelstein, 'Greek Medicine' = Ludwig Edelstein, 'Greek Medicine in its Relation to Religion and Magic', in Ludwig Edelstein *Ancient Medicine: Select Papers of Ludwig Edelstein*. Baltimore: Johns Hopkins University Press, 1967, 205–246.

Edelstein, 'The Methodists' = Ludwig Edelstein, 'The Methodists', in Ludwig Edelstein

Ancient Medicine: Select Papers of Ludwig Edelstein. Baltimore: Johns Hopkins University Press, 1967, 173–191.

Edzard, 'A new look at the Greek, Syriac, and Arabic versions of Aristotle's Poetics' = Lutz Edzard, 'A new look at the Greek, Syriac, and Arabic versions of Aristotle's Poetics', in Lutz Edzard and Janet Watson (eds.), *Grammar as a window onto Arabic humanism. A collection of articles in honour of Michael G. Carter*. Wiesbaden: Harrassowitz, 2006, 222–264.

Ehrenkreutz, 'Knowledge of the Fiscal Administration of Egypt' = A.S. Ehrenkreutz, 'Contributions to the Knowledge of the Fiscal Administration of Egypt in the Middle Ages', *Bulletin of the School of Oriental and African Studies* 16 (1954), 502–514

EI^2 = *The Encyclopaedia of Islam*, New [= Second] Edition. 13 vols. Leiden: Brill, 1960–2009. Available on-line at http://referenceworks.brillonline.com/encyclopaedia-of-islam-2

EI Three = *The Encyclopaedia of Islam*, Third Edition. Leiden: Brill, 2007–.

Elamrani-Jamal, '*Catégories*' = A. Elamrani-Jamal, 'Aristote de Stagire – *Catégories. Tradition arabe*', in Goulet, *Dictionnaire*, i:510–513.

Elamrani-Jamal, '*De Anima*' = A. Elamrani-Jamal, 'Aristote de Stagire – *De Anima*. Tradition arabe', in Goulet, *Dictionnaire*, Suppl., 346–358.

Elamrani-Jamal, 'Les *Seconds Analytiques*' = A. Elamrani-Jamal, 'Aristote de Stagire – *Les Seconds Analytiques*', in Goulet, *Dictionnaire*, i:520–520.

Elamrani-Jamal, '*Les Topiques*' = A. Elamrani-Jamal, 'Aristote de Stagire – *Les Topiques*', in Goulet, *Dictionnaire*, i:524–526.

El-Bizri, 'Defence of the Sovereignty of Philosophy'= Nader El-Bizri, 'In defence of the sovereignty of philosophy: al-Bagdādī's Critique of Ibn al-Haytham's geometrisation of place', *Arabic Sciences and Philosophy*, 17 (2007), 57–80.

El-Bizri, *Epistles* = Nader El-Bizri (ed.), *Epistles of the Brethren of Purity. The Ikhwān al-Ṣafā' and their Rasā'il. An Introduction*. Oxford: Oxford University Press, 2008.

El-Gammal, 'Preparation' = S.Y. El-Gammal, 'Preparation of ethereal oils (Al-Duhoun) by Ibn al-Quff (13th century A.D.)', *Bulletin of the Indian Institute of History of Medicine (Hyderabad)*, 26/1–2 (1996), 59–64.

Elgood, *Medical History of Persia* = Cyril Elgood, *A Medical History of Persia and the Eastern Caliphate*. Cambridge: Cambridge University Press, 1951.

Elias of Nisibis, *Opus Chronologicum* = Elias of Nisibis, *Eliae metropolitae Nisibeni opus chronologicum*. Vol. 1, ed. Ernest Walter Brooks. Paris: E Typographeo Reipublicae / Leipzig: Harrassowitz, 1909–1910.

Elisséeff, *Nūr ad-Dīn* = Nikita Elisséeff, *Nūr ad-Dīn: un grand prince musulman de Syrie au temps des croisades (511–569 H./1118–1174)*. 3 vols. Damascus: Institut francais, 1967.

Elkhadem, *Tacuini sanitatis* = Hosam Elkhadem, *Le Taqwīm al-ṣiḥḥa* (*Tacuini sanitatis*) *d'Ibn Buṭlān: un traité médical du XIe siècle: histoire du texte, édition critique, traduction, commentaire*. Leuven: Peeters, 1990.

Ellenblum, *The Collapse of the Eastern Mediterranean* = Ronnie Ellenblum, *The Collapse of the Eastern Mediterranean: Climate change and the decline of the East, 950–1072*. Cambridge: Cambridge University Press, 2012.

El-Rouayheb & Schmidtke, *The Oxford Handbook of Islamic philosophy* = Khaled El-Rouayheb and Sabine Schmidtke (eds.), *The Oxford Handbook of Islamic philosophy*. Oxford: Oxford University Press, 2017.

El-Tawil, Music of Avicenna = Mahfouz el-Tawil, *Music of Avicenna: Ibn Sīnā (Avicenna) and medieval music (370–428 A.H./980–1037 A.D.). A comprehensive study of Ibn Sīnā's life and his works on music; including: introduction, surveys, parallels, translation and annotated glossary of the Arab musical terminology*. Torquay: White Windmill, 2015.

Emmerick, 'Rapigupta's *Siddhasāra* in Arabic' = R. Emmerick, 'Rapigupta's *Siddhasāra* in Arabic', in H.R. Roemer and A. Noth (Eds.), *Studien zur Geschichte und Kultur des vorderen Oriens* (Leiden: Brill, 1981): 28–32.

Encycl. Hist Arabic Science = *The Encyclopedia of the History of Arabic Science*, ed. Roshdi Rashed. 3 vols. London / New York: Routledge, 1996.

Encycl. Iranica = *Encyclopaedia Iranica*, ed. Ehsan Yarshater. 15 vols+. London: Routledge & Kegan Paul and Costa Mesa, CA: Mazda Publishers, 1985–; available on-line at http://www.iranicaonline.org/

Encycl. Islamica = *Encyclopaedia Islamica*, editors-in-chief, Wilferd Madelung, Farhad Daftary. Leiden: Brill, in association with The Institute of Ismaili Studies, London, 2008–; available on-line at http://referenceworks.brillonline.com/browse/encyclo paedia-islamica

Encycl. Judaica = *Encyclopaedia Judaica*, ed. Michael Berenbaum and Fred Skolnik. 2nd ed., 22 vols. Detroit: Macmillan Reference, 2007; available on-line through Gale Virtual Reference Library.

Encycl. Qur'ān = *Encyclopaedia of the Qur'ān*, ed. Jane Dammen McAuliffe. Leiden and Boston: Brill, 2001–2006.

Endress, *Abū 'Alī Miskawayh* = Gerhard Endress, 'Ancient Ethical Traditions for Islamic Society: Abū 'Alī Miskawayh', in Rudolph, Hansberger & Adamson, *Philosophy in the Islamic World*, 304–344.

Endress, 'Averrois Opera' = Gerhard Endress, 'A Bibliography of Editions and Contributions to the Text', in G. Endress and J.A. Aertsen (eds.), *Averroes and the Aristotelian Tradition: Sources, Constitution, and Reception of the Philosophy of Ibn Rushd (1126–1198): Proceedings of the Fourth Symposium Averroicum, Cologne, 1996*. Leiden: Brill, 1999, 339–381.

Endress, 'Die Bagdader Aristoteliker' = Gerhard Endress, 'Die Bagdader Aristoteliker', in Ulrich Rudolph (ed.), *Philosophie in der islamischen Welt*. Band 1, 8.–10. Jahrhundert. Basel: Schwabe, 2012, 290–362.

Endress, 'Building the Library' = Gerhard Endress, 'Building the Library of Arabic Philosophy. Platonism and Aristotelianism in the Sources of al-Kindī', in C. D'Ancona, *The Libraries of the Neoplatonists*. Leiden 2007, 319–350.

Endress, 'The Circle of al-Kindī' = Gerhard Endress, 'The Circle of al-Kindī: Early Arabic Translations and the Rise of Islamic Philosophy', in Gerhard Endress and Remke Kruk (eds.), *The Ancient Tradition in Christian and Islamic Hellenism: Studies in the Transmission of Greek Philosophy and Sciences*. Leiden: Research School CNWS, 1997, 43–76.

Endress, 'Die wissenschaftliche Literatur' = Gerhard Endress, 'Die wissenschaftliche Literatur', in H. Gätje, *Grundriß der arabischen Philologie. Band II: Literaturwissenschaft der arabischen Philologie*. Wiesbaden: Dr. Ludwig Reichert, 1987, 400–506; and 'Die wissenschaftliche Literatur', in Wolfdietrich Fischer, *Grundriß der arabischen Philologie. Band III: Supplement*. Wiesbaden: Dr. Ludwig Reichert, 1992, 3–152.

Endress & Ferrari, 'The Baghdad Aristotelians' = Gerhard Endress and Cleophea Ferrari, 'The Baghdad Aristotelians' in Rudolph, Hansberger & Adamson, *Philosophy in the Islamic World*, pp. 421–525.

Ephrat, *A Learned Society* = Daphna Ephrat, *A learned Society in a Period of Transition: The Sunni 'Ulama' of Eleventh-Century Baghdad* (Medieval Middle East History). Albany N.Y: State University of New York Press, 2000.

Eudemus (ed. Bodnár & Fortenbaugh) = *Eudemus of Rhodes*, ed. István Bodnár and William M. Fortenbaugh. New Brunswick, NJ: Transaction Publishers, 2002.

Eutychius, *Alexandrini Annales*, ed. Breydy *see* Saʿīd ibn al-Biṭrīq, *Annalenwerk* (Breydy)

Evetts & Butler, *The Churches & Monasteries of Egypt* = B.T.A. Evetts and Alfred J. Butler, *The Churches & Monasteries of Egypt and Some Neighboring Countries: Attributed to Abū Ṣāliḥ the Armenian*. Oxford: Oxford University Press, 1969.

Ezzaher, *Three Arabic treatises on Aristotle's Rhetoric* = Lahcen E. Ezzaher, *Three Arabic treatises on Aristotle's Rhetoric: the commentaries of al-Fārābī, Avicenna, and Averroes. Translated, with introduction and notes by Lahcen Elyazghi Ezzaher*. Carbondale: Southern Illinois University Press, 2015.

Fabricius, *Galens Excerpte* = Caius Fabricius, *Galens Excerpte aus älteren Pharmakologen*. Berlin: De Gruyter, 1972.

Fakhry, *Ethical Theories* = Majid Fakhry, *Ethical Theories in Islam*. Leiden: Brill, 1994.

Fakhry, *History* = Majid Fakhry, *A History of Islamic Philosophy*, 3rd ed. New York: Columbia University Press, 2004.

Fakhry, *Paraphrase* = Majid Fakhry & Ibn Bājjah, *Paraphrase of Aristotle's Physics*. Beirut: Dār al-Nahār, 1973.

Fakhry, *Rasā'il* = Majid Fakhry, *Opera Metaphysica* (*Rasā'il Ibn Bājjah al-ilāhiyyah*). Beirut: Dār al-Nahār, 1968.

Fancy, 'Medical commentaries' = Nahyan Fancy, 'Medical commentaries: a preliminary examination of Ibn al-Nafīs's *shurūḥ*, the *Mūjaz* and commentaries on the *Mūjaz*', *Oriens*, 41 (2013), 525–524.

Fancy, *Science and Religion* = Nahyan Fancy, *Science and Religion in Mamluk Egypt. Ibn al-Nafīs, Pulmonary Transit and Bodily Resurrection.* London / New York: Routledge, 2013.

Fansa, *Aleppo* = Mamoun Fansa, *Aleppo. Ein Krieg zerstört Weltkulturerbe: Geschichte-Gegenwart-Perspektiven.* Mainz: Nünnerich-Asmus, 2013.

al-Fārābī, *Iḥṣāʾ* = al-Fārābī, *Iḥṣāʾ al-ʿulūm*, ed. ʿUthmān Amīn. Cairo: Dār al-Fikr al-ʿArabī, 1949.

Faracovi, 'Return to Ptolemy' = Ornella Faracovi, 'The return to Ptolemy', in B. Dooley (ed.), *A Companion to Astrology in the Renaissance*, Leiden: Brill, 2014, Ch. 3.

Farmer, 'The Music of Islam' = Henry George Farmer, 'The Music of Islam' in *The New Oxford History of Music. Volume 1: Ancient and Oriental Music.* Oxford – London: Oxford University Press, 1957, pp. 421–477.

Farmer, *Liber introductorius* = Henry George Farmer, 'Who was the Author of the "*Liber introductorius in artem logicae demonstrationis*"?' *Journal of the Royal Asiatic Society*, 1934, 553–556.

Faultless, 'Ibn al-Tayyib', = Julian Faultless, 'Ibn al-Tayyib', in David Thomas, Alexander Mallett et al. (eds.), *Christian-Muslim Relations. A Bibliographic History*. Vol. 2 [900–1050]. Leiden: Brill, 2010, 667–697.

Faultless, 'Ibn al-Ṭayyib's Commentary on the Gospels' = Julian Faultless, 'The Two Recensions of the Prologue to John in Ibn al-Ṭayyib's Commentary on the Gospels', in David Thomas (ed.), *Christians at the Heart of Islamic Rule. Church Life and Scholarship in ʿAbbasis Iraq*. Leiden: Brill, 2003, 177–198.

Fazzo, 'Alexandros d'Aphrodisias' = S. Fazzo 'Alexandros d' Aphrodisias', in Goulet, *Dictionnaire*, Suppl., 61–70.

Fazzo & Zonta, *La provvidenza* = Alessandro di Afrodisia, *La provvidenza; Questioni sulla provvidenza*, ed. Silvia Fazzo & Mauro Zonta. Milan: Biblioteca Universale Rizzoli, 1998.

Fear, *Seven Books of History* = Andrew Fear, *Seven Books of History against the Pagans* (Translated Texts for Historians, 54). Liverpool: Liverpool University Press, 2010.

Ferrari, *Die Kategorienkommentar von Abū l-Farağ ʿAbdallāh ibn aṭ-Ṭayyib* = Cleophea Ferrari, *Die Kategorienkommentar von Abū l-Farağ ʿAbdallāh ibn aṭ-Ṭayyib. Text und Untersuchungen*. Leiden: Brill, 2006.

Fichtner, *Corpus Galenicum* = Gerhard Fichtner, *Corpus Galenicum. Bibliographie der galenischen und pseudogalenischen Werke*. Berlin: Berlin-Brandenburgische Akademie der Wissenschaften, 2015; available online only, at http://cmg.bbaw.de/online-publications/hippokrates-und-galenbibliographie-fichtner

Fichtner, *Corpus Hippocraticum* = Gerhard Fichtner, *Corpus Hippocraticum. Bibliographie der hippokratischen und pseudohippokratischen Werke*. Berlin: Berlin-Brandenburgische Akademie der Wissenschaften, 2015; available online only, at http://cmg.bbaw.de/online-publications/hippokrates-und-galenbibliographie-fichtner

Fiey, *Chrétiens syriaques* = Jean M. Fiey, *Chrétiens syriaques sous les Abbassides, surtout à Bagdad (749–1258)*. Louvain: Peeters, Secrétariat du Corpus SCO, 1980.

Fierro, 'Bāṭinism in Al-Andalus' = Maribel Fierro, 'Bāṭinism in Al-Andalus. Maslama b. Qāsim al-Qurṭubī (d. 353/964), Author of the "Rutbat al-Ḥakīm" and the "Ghāyat al-Ḥakīm (Picatrix)"', in *Studia Islamica*, 84 (1996), 87–112.

al-Fīrūzābādī, *al-Bulghah* = Majd al-Dīn Muḥammad ibn Ya'qūb al-Fīrūzābādī, *al-Bulghah fī tarājim a'immat al-naḥw wa-l-lughah*, ed. Muḥammad al-Miṣrī. Kuwait: Markaz al-Makhṭūṭāt wa-l-Turāth, 1987.

Fischer & Weisser, 'Vorwort' = Karl Dietrich Fischer & Ursula Weisser, 'Das Vorwort zur lateinischen Übersetzung von *Rhazes*' Liber continens (1282): Text, Übersetzung und Erläuterungen', *Medizinhistorisches Journal*, 21 (1986), 211–241.

Fonahn, *Anatomical terminology* = Adolf Mauritz Fonahn, *Arabic and Latin anatomical terminology chiefly from the Middle Ages*. Kristiania: Broggers, 1922.

Forcada, 'Ibn Bājja on medicine' = Miquel Forcada, 'Ibn Bājja on medicine and medical experience', *Arabic Sciences and Philosophy*, 21 (2011), 111–148.

Ford, *Aristotle as Poet* = Andrew L. Ford, *Aristotle as Poet. The Song for Hermias and its Contexts*. Oxford: Oxford University Press, 2011.

Fortenbaugh, Huby & Long, *Theophrastus, Life* = William W. Fortenbaugh, Pamela M. Huby, and Anthony A. Long (eds.), *Theophrastus of Eresus. On his Life and Work*. New Brunswick, NJ: Transaction Books, 1985.

Fortenbaugh, Huby, Sharples, & Gutas, *Theophrastus, Sources* = W. Fortenbaugh, P. Huby, R. Sharples, D. Gutas, *Theophrastus of Eresus: Sources for His Life, Writings, Thought and Influence*. Leiden: Brill 1992–.

Fraenkel, *Die aramäische Fremdwörter* = Siegmund Fraenkel, *Die aramäische Fremdwörter im Arabischen*, Leiden: Brill, 1886.

Frank, 'Bodies and Atoms' = Richard M. Frank, 'Bodies and Atoms: the Ash'arite Analysis', in Michael E. Marmura (ed.), *Islamic Philosophy and Theology: Studies in Honor of George F. Hourani*. Albany: State University of New York, 1984, 39–52.

Freind, *History of Physic* = J. Freind, *The History of Physic, from the time of Galen to the beginning of the sixteenth century: chiefly with regard to practice*. 3rd ed. London: J. Walthoe, 1726.

Frend, *Monophysite Movement* = W.H.C. Frend, *The Rise of the Monophysite Movement. Chapters in the History of the Church in the Fifth and Sixth Centuries*. Cambridge: Cambridge University Press, 1972.

Freudenthal & Zonta, 'Avicenna Among Medieval Jews' = Gad Freudenthal and Mauro Zonta, 'Avicenna Among Medieval Jews: The Reception of Avicenna's Philosophcal, Scientific and Medical Writings in Jewish Cultures, East and West', *Arabic Sciences and Philosophy*, 22 (2012), 217–287.

Gacek, *Manuscripts* = Adam Gacek, *Arabic Manuscripts. A Vademecum for Readers*. Leiden: Brill, 2009.

GAL = Carl Brockelmann, *Geschichte der arabischen Litteratur*. First Edition, 2 vols. Weimar / Berlin / Leiden, 1898–1936. Second Edition, 2 vols. & 3 Supplement vols. Leiden: Brill, 1943–1949.

Galadza, 'Eastern Catholic Christianity' = Peter Galadza, 'Eastern Catholic Christianity' in Ken Parry (ed.), *The Blackwell Companion to Eastern Christianity*. Oxford / Chichester: Wiley-Blackwell, 2010, 291–318.

Galen, Kühn ed. = Galen, *Claudii Galeni Opera Omnia*, ed. Carl Gottlob Kühn. 22 vols. Leipzig: Karl Knobloch, 1821–1833.

Galen, *Properties* = *Galen, On the Properties of Foodstuffs* (*De alimentorum facultatibus*). Introduction, Translation and Commentary by Owen Powell with a foreword by John Wilkins. Cambridge: Cambridge University Press, 2003.

Garbutt, 'Ibn Jazlah: The Forgotten 'Abbāsid Gastronome' = Nina Garbutt, 'Ibn Jazlah: The Forgotten 'Abbāsid Gastronome', *Journal of the Economic and Social History of the Orient*, 39/1 (1996), 42–44.

Garnsey, *Social Status* = Peter Garnsey, *Social Status and Legal Privilege in the Roman Empire*. Oxford: Oxford University Press, 1970.

Garofalo, *Anat. admin.* = Ivan Garofalo, *Anatomicarum administrationum libri qui supersunt novem earundem interpretatio arabica Hunaini Isaaci filio ascripta*, 2 vols. Naples: Istituto universiatario orientale, 1986–2000.

Garofalo, 'Il *sunto*' = Ivan Garofalo, 'Il *sunto* di Ioannes 'Grammatikos' delle opere del canone di Galeno', in Daniela Manetti (ed.), *Studi su Galeno: scienza, filosofia, retorica e filologia. Atti del seminario, Firenze, 13 novembre 1998*. Florence: Università degli studi di Firenze, Dipartimento di scienze dell'antichità 'Giorgio Pasquali', 2000, 135–151.

Garofalo, 'I sommari' = Ivan Garofalo, 'I Sommari degli alessandrini', in Ivan Garofalo and Amneris Roselli (eds.), *Galenismo e medicina tardoantica: fonti greche, latine e arabe*. Naples: Istituto Universitario Orientale, 2003, 203–227.

Gätje, *Die arabische Übersetzung der Schrift des Alexander von Aphrodisias über die Farbe* = Helmut Gätje, *Die arabische Übersetzung der Schrift des Alexander von Aphrodisias über die Farbe*. Göttingen: Vandenhoeck und Ruprecht, 1968.

Gätje, *Studien zur Überlieferung der aristotelischen Psychologie* = Helmut Gätje, *Studien zur Überlieferung der aristotelischen Psychologie im Islam*. Heidelberg: C. Winter, 1971.

Gaube & Wirth, *Aleppo* = Heinz Gaube and Eugen Wirth, *Aleppo. Historische und geographische Beiträge zur baulichen Gestaltung, zur socialen Organisation und zur wirtschaftlichen Dynamik einer vorderasiatischen Fernhandelsmetropole* (TAVO Beihefte, Reihe B, No. 58). Wiesbaden: Dr Ludwig Reichert, 1984.

Gayangos, *History* = Pascual de Gayangos, *The History of the Mohammedan Dynasties in Spain*. 2 vols. London: The Oriental Translation Fund, 1840.

Genequand, *On the Cosmos* = Charles Genequand, *Alexander of Aphrodisias, On the Cosmos*. Leiden: Brill, 2001.

Geoffroy, *Averroès, l'Islam et la raison* = Marc Geoffroy, *Averroès, l'Islam et la raison. Anthologie de textes juridiques, théologiques et polemiques*. Paris: Flammarion, 2000.

Gero, 'Galen on the Christians' = Stephen Gero, 'Galen on the Christians. A Reappraisal of the Arabic Evidence', *Orientalia Christina Periodica*, 56 (1990), 371–411.

Ghaleb, *Dictionnaire des sciences de la nature* = Edouard Ghaleb (Idwār Ghālib), *Dictionnaire des sciences de la nature / al-Mawsūʿah fī ʿulūm al-ṭabīʿah*, 3 vols. Beirut: al-Maṭbaʿah al-Kāthūlīkiyyah, 1965–1966.

Ghalioungui & Abdou, *Maqālah fī l-ḥawāss* = ʿAbd al-Laṭīf al-Baghdādī, *Maqālah fī l-ḥawāss wa-masāʾil ṭabīʿiyyah, Risālah li-l-Iskandar fī l-faṣl, Risālah fī l-maraḍ al-musammā diyābīṭis*, ed. Paul Ghalioungui and Said Abdou (Saʿīd ʿAbduh). Kuwait: Maṭbaʿat Ḥukūmat al-Kuwayt, 1972.

Ghalioungui, *Questions on Medicine* = Ghalioungui, Paul, *Questions on Medicine for Scholars* (…) *from a critical edition by Galal M. Moussa*. Cairo: A.R.E. Al-Ahram Center for Scientific Translations, 1980.

Ghaly, *Islam and Disability* = Mohammed Ghaly, *Islam and Disability: pespectives in theology and jurisprudence*. Abingdon: Routledge, 2010.

al-Ghazālī, *Iḥyā'* = Abū Ḥāmid Muḥammad ibn Muḥammad al-Ghazālī, *Iḥyāʾ ʿulūm al-dīn*, 5 vols. Cairo: Maktabat al-Mashhad al-Ḥusaynī, n.d.

al-Ghazālī, *al-Tibr* = *al-Tibr al-masbūk fī naṣīḥat al-mulūk*, ed. Aḥmad Shams al-Dīn. Beirut: Dār al-Kutub al-ʿIlmiyyah, 1988.

Ghersetti, *Il Kitāb Arisṭāṭalīs* = Antonella Ghersetti, *Il Kitāb Arisṭāṭalīs al-faylasūf fī l-firāsa nella traduzione di Ḥunayn ibn Isḥāq*. Venice: Herder Editrice, 1999.

Ghersetti, 'Istanbul Polemon' = Antonella Ghersetti, 'The Istanbul Polemon (TK Recension): Edition and Translation of the Introduction', in Swain, *Seeing the Face*, 465–485.

Ghersetti, 'Medici sagaci' = Antonella Ghersetti, 'Medici sagaci. Aneddoti dal Kitāb Aḫbār al-Aḏkiyāʾ di Ibn al-Ǧawzī', *Quaderni di Studi Arabi*, N.S. 11 (2016): 111–126.

Ghersetti, 'Physiognomy and Medicine in Islamic Culture' = Antonella Ghersetti, 'The Semiotic Paradigm: Physiognomy and Medicine in Islamic Culture' in Swain, *Seeing the Face*, 281–308.

Ghouchani & Adle, 'Sphero-conical vessel as fuqqāʿa' = A. Ghouchani and C. Adle, 'A Sphero-Conical Vessel as *fuqqāʿa* or a Gourd for "Beer"', *Muqarnas*, 9 (1992), 72–92.

al-Ghuzūlī, *Maṭāliʿ* = ʿAlāʾ al-Dīn ʿAlī ibn ʿAbd Allāh al-Ghuzūlī, *Maṭāliʿ al-budūr fī manāzil al-surūr*, 2 vols. Cairo: Maṭbaʿat al-Waṭan, AH 1299–1300.

Gibb, 'Argument from Design' = Hamilton A.R. Gibb, 'The Argument from Design', in S. Löwinger (ed.), *Ignaz Goldziher Memorial Volume*, vol. 1, Budapest: Globus Nyomdai Müintézet, 1948, i:149–162.

Gigandet, *La Risāla al-Hārūniyya* = Suzanne Gigandet, *La Risāla al-Hārūniyya de Masīḥ b. Ḥakam al-Dimašqī, médecin*. Damascus: Institut Français de Damas, 2001.

Gigon, *Aristotelis opera* = Olof Gigon, *Aristotelis opera. Volumen tertium: Librorum deperditorum fragmenta*. Berlin / New York: De Gruyter, 1987.

Gil, *Jews in Islamic Countries* = Moshe Gil, *Jews in Islamic Countries in the Middle Ages*, Leiden: Brill, 2004.

Giladi, *Muslim Midwives* = Avner Giladi, *Muslim Midwives: The Craft of Birthing in the Premodern Middle East*. Cambridge: Cambridge University Press, 2015.

Gill, Whitmarsh, & Wilkins, *Galen* = Christopher Gill, Tim Whitmarsh, John Wilkins (eds.), *Galen and the World of Knowledge*. Cambridge: Cambridge University Press, 2009.

Gimaret, 'Sur un passage' = Daniel Gimaret, 'Sur un passage énigmatique du Tabyīn d'Ibn 'Asākir', *Studia Islamica* 47 (1978), 143–163.

Gohlmann, *The Life of Ibn Sina* = William E. Gohlmann, *The Life of Ibn Sina. A Critical Edition and Annotated Translation*. Albany, New York: SUNY Press, 1974.

Goldstein, 'Evidence for a supernova' = B. Goldstein, 'Evidence for a supernova in A.D.1006', *The Astronomical Journal*, 70 (1965), 105–114.

González Palencia, *Rectificación* = Ángel González Palencia, *Rectificación de la mente, tratado de lógica por Abussalt de Denia* (*Kitāb taqwīm al-dhihn*). Madrid: Centro de Estudios Históricos, 1915.

Gordon, *Robes* = Stewart Gordon, *Robes and Honor: The Medieval World of Investiture*. New York: Palgrave, 2001.

Gottschalk, *al-Malik al-Kāmil* = Hans L. Gottschalk, *Al-Malik al-Kāmil von Egypten und seine Zeit*. Wiesbaden: Otto Harrassowitz, 1958.

Goulet, 'Artémon' = Richard Goulet, 'Artémon', in Goulet, *Dictionnaire*, i:615–616.

Goulet & Aouad, 'Alexandros' = Richard Goulet and Maroun Aouad, 'Alexandros d'Aphrodisias', in Goulet, *Dictionnaire*, i:125–139.

Goulet, *Dictionnaire*, vol. 1 = Richard Goulet (ed.), *Dictionnaire des Philosophes Antiques*, vol. 1. Paris: CNRS éditions, 1989.

Goulet, *Dictionnaire* Suppl. = Richard Goulet (ed.), *Dictionnaire des philosophes antiques*. Suppl. Paris: CNRS éditions, 2003.

Goulet, *Dictionnaire* vol 5a = Richard Goulet (ed.), *Dictionnaire des Philosophes Antiques*, vol. 5a. Paris: CNRS éditions, 2012.

Gourevitch, *Triangle hippocratique* = Danielle Gourevitch, *Le triangle hippocratique dans le monde gréco-romain: le malade, sa maladie et son médecin*. Rome: École française de Rome, 1984.

Graciani García, 'Consideraciones ... tapia' = Amparo Graciani García, 'Consideraciones iniciales y reflexiones sobre la tapia', in S. Huerta and F. López Ulloa (eds.), *Actas del Octavo Congreso Nacional de Historia de la Construcción*, Madrid, 2013, 439–446.

Graf, *Geschichte* = G. Graf, *Geschichte der christlichen arabischen Literatur* (Studi e testi, 118, 133, 146, 147, 172). 5 vols. Vatican: Biblioteca apostolica Vaticana, 1944–1953.

Graille, *Le Livre des simples* = Barbara Graille, *Le Livre des simples de Umayya b. 'Abd al-*

ʿAzīz b. Abī al-Ṣalt al-Dānī al-Išbīlī, *Bulletin d'Études Orientales*, Suppl. 55, Damascus: Institut Français du Proche-Orient, 2003.

Graziani, *Arabic medicine in the eleventh century* = Joseph Salvatore Graziani, *Arabic medicine in the eleventh century as represented in the works of Ibn Jazlah*. [Karachi]: Hamdard Academy, 1980.

Graziani, *Ibn Jazlah's eleventh-century tabulated medical compendium* = Joseph Salvatore Graziani, 'Ibn Jazlah's eleventh-century tabulated medical compendium, *Taqwīm al-abdān*'. D.Phil Thesis, University of California, Los Angeles, 1973.

Greenhill, MS GREEW = William A. Greenhill, London, Royal College of Physicians, MS GREEW/264/149, 150, 152, and 153. [unpublished translations and notes made about 1810 of the lives of ʿAbd Laṭīf al-Baghdādī, Ibn Juljul, and Ibn al-Bayṭār].

Griffel, 'Fakhr al-Dīn al-Rāzī's Life' = Frank Griffel, 'On Fakhr al-Dīn al-Rāzī's Life and the Patronage He Received', in *Journal of Islamic Studies*, 18/3 (2007), 313–344.

Griffel, 'Fakhr al-Dīn al-Rāzī' = Frank Griffel, 'Fakhr al-Dīn al-Rāzī' in: Henrik Lagerlund (ed.), *Encyclopedia of Medieval Philosophy: Philosophy between 500 and 1500*. Berlin: Springer, 2011, 341–345.

Griffel & Hachmeier, 'Prophets as Physicians of the Soul' = Frank Griffel and Klaus Hachmeier, 'Prophets as Physicians of the Soul: A Dispute About the Relationship Between Reason and Revelation Reported by al-Tawḥīdī in his *Book of Delightful and Intimate Conversations* (*Kitāb al-Imtāʿ wa-l-muʾānasa*)', *Mélanges de l'Université Saint-Joseph*, 223–258.

Griffith, 'Sergios of Rešʿaina' = Sidney H. Griffith, 'Sergios of Rešʿaina', in A.P. Kazhdan (ed.), *The Oxford Dictionary of Byzantium*. Oxford: Oxford University Press, 2011, iii:1879–1888.

Grmek, 'The Concept of Disease' = Mirko D. Grmek, 'The Concept of Disease', in *Western Medical Thought from Antiquity to the Middle Ages*, ed. Mirko D. Grmek and Bernardino Fantini. Translated by Antony Shugaar. Cambridge, Mass.: Harvard University Press, 1998, 241–258.

Grmek, *Diseases in the Ancient Greek World* = Mirko D. Grmek, *Diseases in the Ancient Greek World*, translated by Mireille Muellner and Leonard Muellner. Baltimore/London: The Johns Hopkins Universisty Press, 1989.

Gruendler, '*Tawqīʿ* (Apostille)' = Beatrice Gruendler, '*Tawqīʿ* (Apostille): Royal Brevity in the Pre-modern Islamic Appeal Court', in Lale Behzadi and Vahid Behmardi (eds.), *The Weaving of Words: Approaches to Classical Arabic Prose*, Beirut – Würzburg: Ergon, 2009, 101–129.

Guigues, 'La guérison' = Pierre Guigues, 'La guérison en une heure de Razès', *Janus*, 8 (1903), 363–370, 411–418.

Gutas, 'Alexandria' = Dimitri Gutas, 'The "Alexandria to Baghdad" Complex of Narratives. A Contribution to the Study of Philosophical and Medical Historiography among the Arabs', *Documenti e studi sulla tradizione filosofica medievale*, 10 (1999), 155–193.

Gutas, *Avicenna and the Aristotelian Tradition* = Dimitri Gutas, *Avicenna and the Aristotelian tradition: introduction to reading Avicenna's philosophical works*. Leiden: Brill, 2014.

Gutas, 'Galen's Synopsis of Plato's Laws' = Dimitri Gutas, 'Galen's Synopsis of Plato's Laws and Fārābī's *Talḫīṣ*', in Remke Kruk and Gerhard Endress (eds.), *The Ancient Tradition in Christian and Islamic Hellenism: Studies on the Transmission of Greek Philosophy and Sciences, Dedicated to H.J. Drossaart Lulofs on His Ninetieth Birthday*. Leiden: Brill, 1997:101–119.

Gutas, *Greek Thought, Arabic Culture* = Dimitri Gutas, *Greek Thought, Arabic Culture. The Graeco-Arabic Translation Movement in Baghdad and Early 'Abbāsid Society (2nd–4th/8th–10th centuries)*. London: Routledge, 1998.

Gutas, 'Ibn al-Nafīs's Scientific Method' = Dimitri Gutas, 'Ibn al-Nafīs's Scientific Method', *Nazafiyat*, 4 (2017), 130–145.

Gutas, 'New "Edition"' = Dimitri Gutas, 'A New "Edition" of Ḥunayn's *Risāla*', *Arabic Sciences and Philosophy*, 28 (2018), 279–284.

Gutas, 'The Life, Works and Sayings of Theophrastus' = Dimitri Gutas, 'The Life, Works and Sayings of Theophrastus in the Arabic Tradition', in Fortenbaugh, Huby & Long, *Theophrastus, Life*, 63–102.

Gutas, *'Maḏhab'* = Dimitri Gutas, 'Avicenna's "Maḏhab" with an Appendix on the Question of his Date of Birth', *Quaderni Di Studi Arabi*, 5/6 (1987), 323–336.

Gutas, 'Notes & Texts for Cairo Manuscripts' = Dimitri Gutas, 'Notes & Texts for Cairo Manuscripts II: Texts from Avicenna's library in a copy by 'Abd-ar-Razzāq aṣ-Ṣiġnāḫī', *Manuscripts of the Middle East*, 2 (1987), 8–17.

Gutas, 'Plato's *Symposion*', = Dimitri Gutas, 'Plato's Symposion in the Arabic Tradition', *Oriens* 31 (1988) 36–60; repr. in Dimitri Gutas, *Greek Philosophers in the Arabic Tradition* (Aldershot 2000) no. IV.

Gutas, 'Scholars' = Dimitri Gutas, 'Scholars as Transmitters of Philosophical Thought', in Rudolph, Hansberger & Adamson, *Philosophy in the Islamic World*.

Gutas, 'Ṣiwān al-Ḥikmah' = Dimitri Gutas, 'The Siwān al-Ḥikmah Cycle of Texts', *Journal of the American Oriental Society*, 102 /4 (1982): 645–650.

Gutas, 'The Spurious and the Authentic' = Dimitri Gutas, 'Aristotle and the Early Peripatos: The Spurious and the Authentic in the Arabic Lives of Aristotle', in Kraye et al. (eds.), *Pseudo-Aristotle in the Middle Ages*, 15–36; repr. in Dimitri Gutas, *Greek Philosophers in the Arabic Tradition*. Aldershot: Ashgate, 2000, no. VI.

Gutas, *Theophrastus on First Principles* = Dimitri Gutas, *Theophrastus on First Principles (known as his Metaphysics)*. Leiden: 2010.

Gutas, 'Tradition arabe' = Dimitri Gutas, 'Platon – Tradition Arabe', in Goulet, *Dictionnaire*, vol. 5a: 845–863.

Gyekye, Ibn al-Tayyib's Commentary on Porphyry's Eisagoge = Kwame Gyekye, *Ibn al-Tayyib's Commentary on Porphyry's Eisagoge. Arabic Text edited with introduction and a glossary of Greek-Arabic Logical Terms*. Beirut: Dār al-Mashriq, 1975.

Gyekye, *Arabic Logic* = Kwame Gyekye, *Arabic Logic. Ibn al-Tayyib's Commentary on Porphyry's Eisagoge*. Albany: State University of New York, 1979.

Haddawy, *The Arabian Nights* = *The Arabian Nights*, tr. Husain Haddawy, based on the text ed. by Muhsin Mahdi, New York: W.W. Norton, 1990.

Hafsi, 'Recherches' = I. Hafsi, 'Recherches sur le genre "*Ṭabaqāt*" dans la littérature arabe', *Arabica*, 23 (1976), 228–265; 24 (1977), 1–41, 150–186.

Ḥājjī Khalīfah, *Kashf al-ẓunūn* (Flügel) = Ḥājjī Khalīfah (Kâtip Çelebi), *Kashf al-ẓunūn: Lexicon bibliographicum et encyclopædicum*, ed. G. Flügel. 7 vols. Leipzig: Typis Frider. Chr. Guil. Vogeli / London: Richard Bentley for the Oriental Translation Fund of Great Britain and Ireland, 1835–1858.

al-Ḥākim, *Mustadrak* = Abū ʿAbd Allāh al-Ḥākim al-Nīsābūrī, *al-Mustadrak ʿalā l-Ṣaḥīḥayn*. Cairo: Dār al-Ḥaramayn, 1997.

Halfmann, *Die Senatoren* = H. Halfmann, *Die Senatoren aus dem östlichen Teil des Imperium Romanum bis zum Ende des 2. Jahrhunderts n. Chr.* Göttingen: Vandenhoeck & Ruprecht, 1979.

Halm, *Empire of the Mahdi* = Heinz Halm, *The Empire of the Mahdi: The Rise of the Fatimids*, tr. Michael Bonner. Leiden: Brill, 1996.

Halm, *Shīʿism* = Heinz Halm, *Shiʿism*, tr. Janet C.E. Watson and Marian Hill. 2nd ed. Edinburgh: Edinburgh University Press, 2004.

al-Hamadhānī, *Dīwān* = Badīʿ al-Zamān al-Hamadhānī, *Dīwān*, ed. Yusrī ʿAbd al-Ghanī ʿAbd Allāh. Beirut: Dār al-Kutub al-ʿIlmiyyah, 2003.

al-Hamadhānī, *Maqāmāt*, = Badīʿ al-Zamān al-Hamadhānī, *al-Maqāmāt*, ed. Muḥammad ʿAbduh, repr. Beirut: Dār al-Mashriq, 1973 [1st publ. 1889].

al-Hamadhānī, *Maqāmāt* (2006) = Badīʿ al-Zamān al-Hamadhānī, *al-Maqāmāt*, ed. Muḥammad ʿAbduh, repr. Beirut: Dār al-Kutub al-ʿImiyyah, 2006 [1st publ. 1889].

Hamarneh, 'Ibn al-ʿAyn Zarbi and his Definitions of Diseases' = Sami K. Hamarneh, 'Ibn al-ʿAyn Zarbi and his Definitions of Diseases and their Diagnoses', in *Proceedings for the First International Symposium for the History of Arabic Science, April 5–12, 1976. Volume II: Papers in European Languages*, ed. Ahmad Y. al-Hassan, Ghada Karmi, Nizar Namnum. Aleppo: Institute for the History of Arabic Science, University of Aleppo, 1976, 301–323. Repr. in *Health Sciences in Early Islam: Collected Papers of Sami K. Hamarneh*, ed. Munawar A. Anees. 2 vols. Blanco, TX: Noor Health Foundation/Zahra Publications, 1983, 315–333.

Hamarneh, *Ibn al-Quff* = Sami K. Hamarneh, *The Physician, Therapist, and Surgeon, Ibn al-Quff (1233–1286). An Introductory Survey of his Time, Life, and Works*. Cairo: Atlas Press, 1974.

Hamarneh & Ḥimṣī, *Fihris* = Sami K. Hamarneh and Asmāʾ al-Ḥumṣī, *Fihris makhṭūṭāt Dār al-Kutub al-Ẓāhiriyyah: al-ṭibb wa-l-ṣaydalah*. Damascus: Majmaʿ al-Lughah al-ʿArabiyyah bi-Dimashq, 1968.

Hamarneh, 'National Library of Medicine' = Sami K. Hamarneh, 'Arabic Manuscripts of the National Library of Medicine, Washington, D.C.', *Journal for the History of Arabic Science*, 1 (1977), 72–103.

Hamarneh & Sonnedecker, *A Pharmaceutical View* = Sami K. Hamarneh and Glenn A. Sonnedecker, *A Pharmaceutical View of Abulcasis Al-Zahrāwī in Moorish Spain: With Special Reference to the "Adhān"*. Amsterdam: *Janus* Suppl. vol. 5, 1963.

Hamilton, *The Leper King* = Bernard Hamilton, *The Leper King and his Heirs, Baldwin IV and the Crusader Kingdom of Jerusalem*. Cambridge University Press, 2000.

Ḥamzah al-Rāḍī, *Kitāb al-Mujarrad* = Fāṭimah Ḥamzah al-Rāḍī (ed.), *Kitāb al-Mujarrad li-lughat al-ḥadīth (li-ʿAbd al-Laṭīf al-Baghdādī)*. Baghdad: Maṭbaʿat al-Shaʿb, 1977.

Ḥamzah al-Rāḍī, *Min kitāb al-Mujarrad* = Fāṭimah Ḥamzah al-Rāḍī, 'Min Kitāb al-Mujarrad li-lughat al-ḥadīth', *al-Mawrid*, 8/2 (1979), 121–136.

Hankinson, 'Galen's Anatomy of the Soul' = R.J. Hankinson, 'Galen's Anatomy of the Soul', *Phronesis*, 36 (1991), 197–233.

Hankinson, *Galen* = R.J. Hankinson, *The Cambridge Companion to Galen*. Cambridge: Cambridge University Press, 2008.

Hansberger, 'The Arabic Aristotle and his Medical Readers' = Rotraud Hansberger, 'Length and Shortness of Life Between Philosophy and Medicine: The Arabic Aristotle and his Medical Readers', in Adamson & Pormann, *Philosophy and Medicine*, 48–74.

Hansberger, *'Kitāb al-Ḥiss wa-l-maḥsūs'* = Rotraud Hansberger, *'Kitāb al-Ḥiss wa-l-maḥsūs*. Aristotle's *Parva Naturalia* in Arabic guise', in Christophe Grellard and Pierre-Marie Morel (eds.), *Les Parva Naturalia d'Aristote. Fortune antique et médiévale*. Paris: Publications de la Sorbonne, 2010, 146–162.

al-Ḥarīrī, *Durrat al-ghawwāṣ* = al-Ḥarīrī, *Durrat al-ghawwāṣ fī awhām al-khawāṣṣ*, ed. Heinrich Thorbecke. Leipzig: F.C.W. Vögel, 1871.

Hassan & Hill, *Islamic Technology* = Al-Ahmad Y. Hassan and Donald R. Hill, *Islamic Technology: An Illustrated History*. Cambridge: Cambridge University Press, 1994.

Hasse, *King Avicenna* = Dag Nikolaus Hasse, 'King Avicenna: The Iconographic Consequences of a Mistranslation', *Journal of the Warburg and Courtauld Institutes*, 60 (1997), 230–243.

Hasse, 'Mosul and Frederick II Hohenstaufen' = Dag Nikolaus Hasse, 'Mosul and Frederick II Hohenstaufen: Notes on Atiraddin al-Abhari and Siragiddin al-Urmawi', in *Occident et Proche-Orient: contacts scientifiques au temps des Croisades*, ed. Isabelle, Draelants, Anne Tihon, Baudouin van den Abeele, and Charles Burnett. Turnhout: Brepols, 2000, 145–163.

Hasse, 'Plato arabicus-latinus' = Dag Nikolaus Hasse, 'Plato arabicus-latinus: Philosophy – Wisdom Literature – Occult Sciences', in Stephen Gersh and Maarten J.F.M. Hoenen (eds.), *The Platonic Tradition in the Middle Ages: A Doxographic Approach*. Berlin / New York: De Gruyter, 2002, 31–66.

Hasse, *Success and Suppression* = Dag Nikolaus Hasse, *Success and Suppression: Arabic*

Sciences and Philosophy in the Renaissance. Cambridge, Massachusetts: Harvard University Press, 2016.

Hau, 'Razis Gutachten über Rosenschnupfen' = Friedrun R. Hau, 'Razis Gutachten über Rosenschnupfen', *Medizinhistorisches Journal*, 10 (1975), 94–102.

Hava, *al-Farā'id* = J.G. Hava, *al-Farā'id al-durriyyah: Al-Faraid Arabic-English Dictionary*. Beirut: Catholic Press, 1964.

Hawting, 'Biography of al-Ḥārith ibn Kalada' = Gerald R. Hawting, 'The Development of the Biography of al-Harith Ibn Kalada and the Relationship Between Medicine and Islam', in C.E. Bosworth (ed.), *The Islamic World from Classical to Modern Times. Essays in honor of Bernard Lewis*. Princeton, NJ: Darwin Press, 1989, 127–140.

Heath, *Euclid's Elements* = T.L. Heath, *The Thirteen Books of Euclid's Elements*. Translated from the text of Heiberg, with introduction and commentary by T.L. Heath. 3 vols. Cambridge: Cambridge University Press, 1908.

Heck, *Construction of Knowledge* = Paul L. Heck, *The Construction of Knowledge in Islamic Civilization: Qudāma b. Ja'far & his* Kitāb al-kharāj wa-ṣinā'at al-kitāba. Leiden: Brill, 2002.

Heers, 'Thalāthah mujalladāt' = N. Heers, 'Thalāthah mujalladāt min Kitāb al-Shāmil l-Ibn al-Nafīs', *Majallat Ma'had al-Makhṭūṭāt al-'Arabiyyah*, 6 (1960), 203–210.

Heffening, 'Ephraem-Paraenesis' = W. Heffening, 'Die griechische Ephraem-Paraenesis gegen das Lachen in arabischer Übersetzung. Ein Beitrag zum Problem der arabischen Ephraemübersetzungen und ihrer Bedeutung für eine kritische Ausgabe des griechischen Ephraem', *Oriens Christianus*, 24 (1927), 94–119.

Heimgartner, 'Timothy I' = Martin Heimgartner 'Timothy I', in David Thomas and Barbara Roggema (eds.), *Christian-Muslim Relations: A Bibliographical History. Vol. 1 (600–900)*. Leiden: Brill, 2009, 515–517.

Heinen, 'Ibn al-Haiṯams Autobiographie' = Anton Heinen, 'Ibn al-Haiṯams Autobiographie in einer Handschrift aus dem Jahr 556 H./1161 A.D.', in *Die Islamische Welt zwischen Mittelalter und Neuzeit. Festschrift für Hans Robert Roemer zum 65. Geburtstag*, ed. Ulrich Haarmann and Peter Bachmann (Beiruter Texte und Studien, 22). Wiesbaden: Steiner Verlag, 1979, 254–277.

Hershkovits & Hadromi-Allouche, 'Divine Doctors' = Keren A. Hershkovits and Zohar Hadromi-Allouche, 'Divine Doctors: The Construction of the Image of Three Greek Physicians in Islamic Biographical Dictionaries of Physicians', *Al-Qanṭara*, 34 (2013), 35–63.

Herzig & Stewart, *The Age of the Seljuqs* = Edmund Herzig and Sarah Stewart, *The Age of the Seljuqs*. London / New York: I.B. Tauris, 2015.

Heym, *al-Rāzī and Alchemy* = Gerard Heym, 'Al-Rāzī and alchemy', in *Muḥammad ibn Zakarīyā' ar-Rāzī: texts and studies, I*. Collected and reprinted by Fuat Sezgin, in collaboration with Carl Ehrig-Eggert, Eckhard Neubauer, Farid Benfeghoul. Frankfurt am Main: Institut für Geschichte der Arabisch-Islamischen Wissenschaften, 2002, 176–183.

HIA = *An Historical Atlas of Islam*. ed. Hugh Kennedy. Leiden: Brill. 2002; available online at http://referenceworks.brillonline.com/browse/historical-atlas-of-islam

Hibri, *Reinterpreting Islamic Historiography* = Tayeb El-Hibri, *Reinterpreting Islamic Historiography: Harun al-Rashid and the Narrative of the 'Abbasid Caliphate*. Cambridge: Cambridge University Press, 1999.

Hilgenberg & Kirfel, *Vāgbhaṭa's Aṣṭāṅgahṛdayasaṃhitā* = Luise Hilgenberg and Wilibald Kirfel, *Vāgbhaṭa's Aṣṭāṅgahṛdayasaṃhitā, ein altindisches Lehrbuch der Heilkunde, aus dem Sanskrit ins Deutsche übertragen mit Einleitung, Anmerkungen und Indices*. Leiden: Brill, 1941.

Hill, *Arabic Water-Clocks* = Donald R. Hill, *Arabic Water-Clocks* (Sources & Studies in the History of Arabic-Islamic Science, History of Technology Series, 4). Aleppo: University of Aleppo, Institute for the History of Arabic Science, 1981.

Hill, *Ingenious Devices* = Donald R. Hill (translation and annotation), *The Book of Ingenious Devices (Kitāb al-Ḥiyal) by the Banū (Sons of) Mūsā bin Shākir*. Dordrecht / Boston / Leiden: D. Reidel Publishing Company, 1979.

Hilloowala, 'An Analysis of Ibn Abi Usaybiʿah' = Franak Hilloowala, 'An analysis of Ibn Abi Usaybiʿah's *'Uyūn al-anbā' fī ṭabaqāt al-aṭibbā'* (Ahmad ibn al-Qasim Ibn Abi Usaybiʿah, Syria, Egypt)'. Ph.D. diss., University of Arizona, 2000.

Hilloowala, *History of the Conquest of Egypt* = Yasmin Hilloowala, 'The History of the Conquest of Egypt, being a partial translation of Ibn 'Abd al-Hakam's *Futuh Misr* and an analysis of this translation', Ph.D. dissertation, University of Arizona, 1998.

Hinz, *Islamische Masse* = Walther Hinz, *Islamische Masse und Gewichte umgerechnet ins metrische System*. Handbuch der Orientalistik, I, 1. Leiden: E.J. Brill, 1955; tr. by Muhammad Ismail Marcinkowski as *Measures and Weights in the Islamic World*. Kuala Lumpur: ISTAC, 2003.

Hinz, *Islamische Währungen* = Walther Hinz, *Islamische Währungen des 11. bis 19. Jahrhunderts, umgerechnet in Gold: ein Beitrag zur islamischen Wirtschaftsgeschichte*. Wiesbaden: Harrassowitz, 1991.

Hippocrates, *Fuṣūl* = Hippocrates, *al-Fuṣūl al-Abuqrāṭiyyah fī l-uṣūl al-ṭibbiyyah.*, ed. John Tytler. Calcutta: Committee of Public Instruction, 1832.

Hippocrates, *Nature of Man* = J.N. Mattock and M.C. Lyons (eds. and trns.) *Kitāb Buqrāṭ fī ṭabīʿat al-insān (Hippocrates: On the Nature of Man)*. Cambridge: Heffer & Sons, 1968.

Hippocrates, 'Prognostikon' = Oliver Overwien and Uwe Vagelpohl, 'Hippokrates' Prognostikon', in *Hippocrate, Pronostic*, texte établi, trad. et annoté par J. Jouanna avec la collaboration de Anargyros Anastassiou et Caroline Magdelaine. Paris: Les Belles Lettres, 2013, 273–298.

Hirschberg, *Geschichte* = Hirschberg, Julius, *Geschichte der Augenheilkunde. Zweites Buch: Geschichte der Augenheilkunde im Mittelalter. 1. Geschichte der Augenheilkunde bei den Arabern* (Graefe-Saemisch, *Handbuch der gesamten Augenheilkunde*, ii, 13.2). Leipzig: Wilhelm Engelmann, 1908; repr. in Fuat Sezgin (ed.), *Augenheilkunde im Islam. Texte, Studien und Übersetzungen. Dritter Band: Beiträge von Hirschberg, Meyerhof und Prüfer*. Frankfurt am Main: Institute für Geschichte der Arabisch-Islamischen Wissenschaften, 1986, 1–251.

Hirschberg, Lippert & Mittwoch, *Die arabischen Augenärzte* = Julius Hirschberg, Julius Lippert, & Eugen Mittwoch, *Die arabischen Augenärzte nach den Quellen bearbeitet*. Leipzig: Veit & Co., 1905.

Hirschler, *Medieval Damascus* = Konrad Hirschler, *Medieval Damascus: Plurality and Diversity in an Arabic Library. The Ashrafiya Library Catalogue*. Edinburgh: Edinburgh University Press, 2016.

Hirschler, *Written Word* = Konrad Hirschler, *The Written Word in the Medieval Arabic Lands. A Social and Cultural History*. Edinburgh: Edinburgh University Press, 2012.

Hofer, 'Chief Sufi' = Nathan Hofer, 'The Origins and Development of the Office of the "Chief Sufi" in Egypt, 1173–1325', *Journal of Sufi Studies*, 3 (2014), 1–37.

Hogendijk, 'Amicable numbers' = Jan P. Hogendijk, 'Thābit ibn Qurra and the pair of amicable numbers 17296, 18416', *Historia mathematica*, 12/3 (1985), 269–273.

Hogendijk, 'Arabic traces of lost works of Apollonius' = Jan P. Hogendijk, 'Arabic Traces of Lost Works of Apollonius,' *Archive for History of Exact Sciences* 35.3 (1986), 187–253.

Hogendijk, 'Bankipore' = Jan P. Hogendijk, 'Rearranging the Arabic Mathematical and Astronomical Manuscript Bankipore 2468', *Journal for the History of Arabic Science*, 6 (1982), 133–159.

Hogendijk, 'Two Editions' = Jan P. Hogendijk, 'Two Editions of Ibn al-Haytham's *Completion of the Conics*', *Historia Mathmatica*, 29 (2002), 247–265.

Holes, *Dialect* = Clive Holes, *Dialect, Culture, and Society in Eastern Arabia*. 3 vols. Leiden: Brill, 2001–2016.

Holmes, 'Sympathy' = Brooke Holmes, 'Galen's Sympathy', in E. Schliesser (ed.), *Sympathy. A History*. Oxford: Oxford University Press, 2015, 61–69.

Horden, *Hospitals and Healing* = Peregrine Horden, *Hospitals and Healing from Antiquity to the Later Middle Ages*. Aldershot / Burlington: Ashgate Variorum, 2008.

Horst, 'Die Entstehung der adab-Literatur' = Heribert Horst, 'Die Entstehung der adab-Literatur und ihre Arten', in Helmut Gätje (ed.), *Grundriß der Arabischen Philologie, Band II: Literaturwissenschaft*. Wiesbaden: Ludwig Reichert, 1987, 208–220.

Houzel, 'New Astronomy of Ibn al-Haytham' = Christian Houzel, 'The New Astronomy of Ibn al-Haytham', *Arabic Sciences and Philosophy*, 19 (2009), 1–41.

Hoyland, 'Islamic Background to Polemon's Treatise' = Robert Hoyland, 'The Islamic Background to Polemon's Treatise', in Swain, *Seeing the Face*, 227–280.

Hoyland, 'Theomnestus of Nicopolis, Ḥunayn ibn Isḥāq, and the beginnings of Islamic veterinary science' = Robert G. Hoyland, 'Theomnestus of Nicopolis, Ḥunayn ibn Isḥāq, and the beginnings of Islamic veterinary science', in *Islamic Reflections, Arabic Musings: Studies in Honour of Professor Alan Jones*, ed. Robert G. Hoyland and Philip F. Kennedy. London: Gibb Memorial Trust, 2004, 150–169.

Huffman, *Archytas* = Carl A. Huffman, *Archytas of Tarentum: Pythagorean, Philosopher, and Mathematician King*. Cambridge: Cambridge University Press, 2005.

Huffman, *Pythagoreanism* = Carl A. Huffmann (ed.), *A History of Pythagoreanism*. Cambridge: Cambridge University Press, 2017.

Hugonnard-Roche, '*De Caelo*' = Henri Hugonnard-Roche, 'Aristote de Stagire – *De Caelo*. Tradition syriaque et arabe', in Goulet, *Dictionnaire*, Suppl., 283–294.

Hugonnard-Roche, '*De Interpretatione*' = Henri Hugonnard-Roche, 'Aristote de Stagire – *De Interpretatione*', in Goulet, *Dictionnaire*, i:514–516.

Hugonnard-Roche, 'Aristote de Stagire – Pseudo-Aristote, *De Plantis*' = Henri Hugonnard-Roche, 'Aristote de Stagire – Pseudo-Aristote, *De Plantis*', in Goulet, *Dictionnaire*, Suppl., 499–505.

Hugonnard-Roche, 'La *Poétique*' = Henri Hugonnard-Roche, 'Aristote de Stagire – La *Poétique*. Tradition syriaque et arabe', in Goulet, *Dictionnaire*, Suppl., 208–218.

Hugonnard-Roche, 'Les *Premieres Analytiques*' = Henri Hugonnard-Roche, 'Aristote de Stagire – *Les Premieres Analytiques*', in Goulet, *Dictionnaire*, i:516–520.

Hugonnard-Roche, 'Aristote de Stagire – Les *Réfutations sophistiques*' = Henri Hugonnard-Roche, 'Aristote de Stagire – *Les Premieres Analytiques*', in Goulet, *Dictionnaire*, i:526–528.

Humphreys, *From Saladin to the Mongols* = R. Stephen Humphreys, *From Saladin to the Mongols: the Ayyubids of Damascus, 1193–1260*. Albany: State University of New York Press, 1977.

Ḥunayn ibn Isḥāq, *Ādāb* = Ḥunayn ibn Isḥāq, *Ādāb al-falāsifah* (*Sentences des philosophes*), ed. ʿAbd al-Raḥmān Badawī. Kuwait: Maʿhad al-Makhṭūṭāt al-ʿArabiyyah, 1985.

Ḥunayn ibn Isḥāq, *Fī ḥifẓ al-asnān* (ed. al-Dhākirī) = Ḥunayn ibn Isḥāq, *Fī ḥifẓ al-asnān wa-l-lithah wa-istiṣlāḥihā*, ed. Muḥammad Fu'ād al-Dhākirī. Aleppo: Dār al-Qalam al-ʿArabī bi-Ḥalab, 1997.

Ḥunayn ibn Isḥāq, *Galen Translations* = John C. Lamoreaux (ed.), *Ḥunayn ibn Iṣḥāq on his Galen Translations*. Provo, Utah: Brigham Young University, 2016.

Ḥunayn ibn Isḥāq, *Neue Materialen* = Gotthelf Bergsträsser (ed.), *Neue Materialen zu Ḥunain ibn Iṣḥāq's Galen-Übersetzungen*. Leipzig: Brockhaus, 1932.

Ḥunayn ibn Isḥāq, *Risālah* = Gotthelf Bergsträsser (ed. and tr.), *Ḥunain ibn Isḥāq über die syrischen und arabischen Galen-Übersetzungen*. Leipzig: Brockhaus, 1925.

Ḥunayn ibn Isḥāq, *Ten Treatises* = *The book of the Ten Treatises on the Eye ascribed to Ḥunain ibn Is-hāq (809–877 A.D.). The earliest existing Systematic Text-book of Ophthalmology*, ed. and tr. by Max Meyerhof. Cairo: Government Press, 1928.

al-Ḥuṣrī, *Jamʿ al-jawāhir* = al-Ḥuṣrī, *Jamʿ al-jawāhir fī l-mulaḥ wa-l-nawādir*, ed. ʿAlī Muḥammad al-Bijāwī. Beirut: Dār al-Jīl, 1987.

al-Ḥuṣrī, *Zahr al-ādāb* = Abū Isḥāq Ibrāhīm ibn ʿAlī al-Ḥuṣrī al-Qayrawānī, *Zahr al-ādāb wa-thamar al-albāb*, ed. Zakī Mubārak and Muḥammad Muḥyī l-Dīn ʿAbd al-Ḥamīd, 4 vols. Repr. Beirut: Dār al-Jīl, 1972.

IAU, *Anecdotes & Antidotes = Ibn Abī Uṣaybiʿah, Anecdotes and Antidotes : A Medieval Arabic History of Physicians. A new translation.* Translated by Emilie Savage-Smith, Simon Swain et al., selected and edited by Henrietta Sharp Cockrell, with Introduction by Geert Jan van Gelder. [Oxford World's Classics] Oxford: Oxford University Press, 2020.

IAU Müller = Ibn Abī Uṣaybiʿah, *ʿUyūn al-anbāʾ fī ṭabaqāt al-aṭibbāʾ*, ed. August Müller. 2 vols. Cairo: al-Maṭbaʿah al-Wahbiyyah / Königsberg: Selbstverlag, 1882–1884.

IAU Najjār = Ibn Abī Uṣaybiʿah, *ʿUyūn al-anbāʾ fī ṭabaqāt al-aṭibbāʾ*, ed. ʿĀmir al-Najjār. 6 vols., vol. 1 Cairo: Dār al-Maʿārif, 1996; vols. 2–6. Cairo: al-Hayʾah al-Miṣriyyah al-ʿĀmmah li-l-Kitāb, 2001–2004.

IAU Riḍā = Ibn Abī Uṣaybiʿah, *ʿUyūn al-anbāʾ fī ṭabaqāt al-aṭibbāʾ*, ed. Nizār Riḍā. Beirut: Dār Maktabat al-Ḥayāt, 1965.

IAU Sūd = Ibn Abī Uṣaybiʿah, *ʿUyūn al-anbāʾ fī ṭabaqāt al-aṭibbāʾ*, ed. Muḥammad Bāsil ʿUyūn al-Sūd. Beirut: Dār al-Kutub al-ʿIlmiyyah, 1998.

IAU Wahb = Ibn Abī Uṣaybiʿah, *ʿUyūn al-anbāʾ fī ṭabaqāt al-aṭibbāʾ*, ed. Qāsim Muḥammad Wahb. 4 vols. Damascus: Wizārat al-Thaqāfah, 1997.

Ibn al-Abbār, *al-Takmilah li-kitāb al-ṣilah* = Ibn al-Abbār, *al-Takmilah li-kitāb al-ṣilah*, ed. ʿAbd al-Salām al-Harrās. 4 vols. Beirut: Dār al-Fikr, 1995.

Ibn ʿAbd al-Barr, *al-Tamhīd* = Ibn ʿAbd al-Barr, *al-Tamhīd li-mā fī l-Muwaṭṭaʾ min al-maʿānī wa-l-asānīd*, ed. Muṣṭafā ibn Aḥmad al-ʿAlawī and Muḥammad ibn ʿAbd al-Kabīr al-Bakrī. 24 vols. Rabat: Wizārat al-Awqāf wa-l-Shuʾūn al-Islāmiyyah, 1967–.

Ibn ʿAbd al-Ḥakam, *Futūḥ Miṣr* = *The History of the Conquest of Egypt, North Africa and Spain, known as the Futūḥ Miṣr of Ibn ʿAbd al-Ḥakam*, ed. Charles C. Torrey (Yale Oriental Series – Research, 3). New Haven: Yale University Press, 1922.

Ibn ʿAbd Rabbih, *ʿIqd* = Abū ʿUmar Aḥmad ibn Muḥammad Ibn ʿAbd Rabbih, *al-ʿIqd al-farīd*, ed. Aḥmad Amīn, Aḥmad al-Zayn, Ibrāhīm al-Ibyārī, 7 vols. Cairo, 1948–1953, repr. Beirut: Dār al-Kitāb al-ʿArabī, 1983.

Ibn Abī ʿAwn, *Tashbīhāt* = Ibn Abī ʿAwn, *al-Tashbīhāt*, ed. Muḥammad ʿAbd al-Muʿīd Khān, London: Luzac, 1950.

Ibn Abī l-Bayan, *Al-Dustūr al-bīmāristānī*, Sbath & Avierinos = Sahlān ibn Kaysān [false attribution], *Mukhtaṣar fī l-adwiyah al-murakkabah fī akthar al-amrāḍ*, ed. with French translation by Paul Sbath and Christos D. Avierinos, in *Sahlān Ibn Kaysān et Rašīd al-Dīn Abū Ḥulayqa: deux traités medicaux* (Publications de l'Institut français d'Archéologie orientale du Caire, Textes arabes et études islamiques, 10). Cairo, 1953, 7–74.

Ibn Abī l-Bayan, *El formulario de los hospitales* = José Luis Valverde and Carmen Peña Muñoz, *El formulario de los hospitales de Ibn Abī al-Bayān* (Cuadernos de Historia de la Farmacia, 8). Granada: Universidad de Granada, 1981.

Ibn Abī l-Bayān, 'A Hospital Handbook' = Efraim Lev, Leigh Chipman, and Friedrich Niessen, 'A Hospital Handbook for the Community: Evidence for the Extensive Use of Ibn Abī 'l-Bayān's *al-Dustūr al-bīmāristanī* by the Jewish Practitioners of Medieval Cairo', *Journal of Semitic Studies*, 53/1 (2008), 103–118.

Ibn Abī l-Ḥadīd, *Sharḥ Nahj al-balāghah* = 'Izz al-Dīn Abū Ḥāmid 'Abd al-Ḥamīd ibn Hibat Allāh Ibn Abī l-Ḥadīd al-Madā'inī, *Sharḥ Nahj al-balāghah, al-jāmi' li-khuṭab wa-ḥikam wa-rasā'il al-Imām Amīr al-Mu'minīn 'Alī ibn Abī Ṭālib*, ed. Ḥusayn al-A'lamī. 5 vols. Beirut: Mu'assasat al-A'lamī, 1415/1995.

Ibn Abī Ḥajalah, *Dīwān al-ṣabābah* = Shihāb al-Dīn Aḥmad Ibn Abī Ḥajalah al-Maghribī, *Dīwān al-ṣabābah*. Cairo: Dār al-Hilāl, 1999.

Ibn Abī l-Iṣba', *Taḥrīr al-taḥbīr* = Ibn Abī l-Iṣba' al-Miṣrī, *Taḥrīr al-taḥbīr fī ṣinā'at al-shi'r wa-l-nathr wa-bayān i'jāz al-Qur'ān*, ed. Ḥifnī Muḥammad Sharaf. Cairo: Lajnat Iḥyā' al-Turāth al-Islāmī, AH 1383.

Ibn Abī Ṭāhir Tayfūr, *K. Baghdād* = Ibn Abī Ṭāhir Ṭayfūr, *Das Kitāb Baghdād*, ed. H. Keller. Beirut: Dār al-Jinān s.d. (reprint of 1908 ed.).

Ibn Abī Uṣaybi'h *see* IAU

Ibn al-'Adīm, *Bughyah* = Kamāl al-Dīn 'Umar ibn Aḥmad Ibn al-'Adīm, *Bughyat al-ṭalab fī ta'rīkh Ḥalab*, ed. Suhayl Zakkar. 10 vols. Damascus: Dār al-Fikr, 1988.

Ibn al-'Adīm, *Bughyah* (facs.) = Kamāl al-Dīn 'Umar ibn Aḥmad Ibn al-'Adīm, *Bughyat al-ṭalab fī ta'rīkh Ḥalab*. Facsimile edition ed. by Fuat Sezgin et al. 11 vols. Frankfurt am Main: Institut für Geschichte der Arabisch-Islamischen Wissenschaften, 1986–1990.

Ibn al-'Adīm, *Zubdah* (ed. al-Manṣūr) = Kamāl al-Dīn 'Umar ibn Aḥmad Ibn al-'Adīm, *Zubdat al-ḥalab min tārīkh Ḥalab*, ed. Khalīl al-Manṣūr. Beirut: Dār al-Kutub al-'Ilmiyyah, 1996.

Ibn 'Asākir, *Tārīkh* = Ibn 'Asākir, *Tārīkh madīnat Dimashq*, ed. 'Umar al-'Amrāwī and 'Alī Shīrī. 80 vols. Beirut: Dār al-Fikr, 1995–2001.

Ibn al-Athīr, *Kāmil* = 'Izz al-Dīn Ibn al-Athīr, *al-Kāmil*, ed. C.J. Tornberg. 15 vols. Beirut: Dār Ṣādir, 1965–1967.

Ibn al-Azraq, *Tārīkh al-Fāriqī* = Aḥmad ibn Yūsuf ibn 'Alī ibn al-Azraq al-Fāriqī, *Tārīkh Mayyāfāriqīn wa-Āmid*, ed. 'Abd al-Laṭīf Badawī. Cairo: al-Hay'ah al-'Āmmah li-Shu'ūn al-Maṭābi' al-Amīriyyah, 1959.

Ibn al-Athīr, *al-Mathal al-sā'ir* = Ḍiyā' al-Dīn Ibn al-Athīr, *al-Mathal al-sā'ir fī adab al-kātib wa-l-shā'ir*, ed. Aḥmad al-Ḥūfī and Badawī Ṭabānah. 4 vols. Cairo: Dār Nahḍat Miṣr, 1973.

Ibn Bājjah, *Burhān* (Fakhrī) = Ibn Bājjah, *al-Burhān*, ed. Mājid Fakhrī, *al-Abḥāth*, 27 (1979), 23–42.

Ibn Bājjah, *K. al-nafs* (ed. Maʿṣūmī) = Ibn Bājjah, *Kitāb al-nafs, ṣannafahu Abū Bakr Ibn Bājjah al-Andalusī*, ed. Muḥammad Ṣaghīr Ḥasan Maʿṣūmī. Beirut: Dār Ṣādir, 1992.

Ibn Bājjah, *K. al-nafs* (tr. Maʿṣūmi) = Ibn Bājjah, *ʿIlm al-Nafs. English Translation and Notes*, tr. Muḥammad Ṣaghīr Ḥasan Maʿṣūmi. Karachi: Pakistan Historical Society, 1960.

Ibn Bājjah, *K. tadbīr al-mutawaḥḥid* (Asín Palacios) = Ibn Bājjah, *El regimen del solitario*, ed. M. Asín Palacios. Madrid: Escuelas de Estudios Árabes de Madrid y Granada, 1946.

Ibn Bājjah, *K. tadbīr al-mutawaḥḥid* (Ziyādah) = Ibn Bājjah, *Kitāb tadbīr al-mutawaḥḥid*, ed. Maʿn Ziyādah. Beirut: Dār al-Fikr al-Islāmī, 1978.

Ibn Bājjah, *K. tadbīr al-mutawaḥḥid* (tr. Lomba Fuentes) = Ibn Bājjah, *Avempace. El régimen del solitario* (*Tadbīr al-mutawaḥḥid*), tr. J. Lomba Fuentes. Madrid: Editorial Trotta, 1997.

Ibn Bājjah, *K. tadbīr al-mutawaḥḥid* (tr. Companini & Iluminati) = Ibn Bājjah, *Il regime del solitario*, ed. and tr. M. Campanini and A. Iluminati. Milan: Biblioteca Universale Rizzoli, 2002.

Ibn al-Bayṭār, *Colección de Medicamentos y Alimentos* = Ana María Cabo González, *Kitāb al-Ŷamiʿ li-mufradāt al-adwiya wa-l-agḏiya / Colección de Medicamentos y Alimento Ibn al-Baytar*. 2 vols. Seville: Mergablum Edición y Comunicación, 2002, 2005.

Ibn al-Bayṭār, *Die Dioskurides-Erklärung* = Albert Dietrich, *Die Dioskurides-Erklärung des Ibn al-Baiṭār: ein Beitrag zur arabischen Pflanzensynonymik des Mittelalters*. Göttingen: Vandenhoeck & Ruprecht, 1991.

Ibn al-Bayṭār, *Jāmiʿ li-mufradāt* (Būlāq) = Ibn al-Bayṭār, *Kitāb al-Jāmiʿ li-mufradāt al-adwiyah wa-l-aghdhiyah*. 4 vols. Cairo (Būlāq), 1291/1874

Ibn al-Bayṭār, *al-Jāmiʿ li-mufradāt* = Ibn al-Bayṭār, *al-Jāmiʿ li-mufradāt al-adwiyah wa-l-aghdhiyah*. 4 vols. Beirut: Dār al-Kutub al-ʿIlmiyyah, 1992.

Ibn al-Bayṭār, 'Traité des simples' = Lucien Leclerc, *Traité des simples par Ibn El-Beithar* [Notices et extraits des manuscrits de la Bibliothèque nationale et autres bibliothèques. xxiii, xxv and xxvi]. Paris: Imprimerie. nationale, 1877–1883; repr. Frankfurt am Main: Institut für Geschichte der Arabisch-Islamischen Wissenschaften, 1996.

Ibn Bishrī, *ʿUddat al-jalīs* = ʿAlī Ibn Bishrī, *ʿUddat al-jalīs / An Anthology of Andalusian Arabic* Muwashshaḥāt, ed. Alan Jones. Cambridge: E.J.W. Gibb Memorial Trust, 1992.

Ibn Buṭlān, *Daʿwat al-aṭibbāʾ* (tr. Klein-Franke) = Ibn Buṭlān, *Das Ärztebankett*, tr. Felix Klein-Franke. Stuttgart: Hippokrates-Verlag, 1984.

Ibn Buṭlān, *Daʿwat al-aṭibbāʾ* (ed. Klein-Franke) = Ibn Buṭlān, *Daʿwat al-aṭibbāʾ / The Physicians' Dinner Party*, ed. Felix Klein-Franke. Wiesbaden: Harrassowitz, 1985.

Ibn Buṭlān, *Daʿwat al-aṭibbāʾ* (trns. Dagher & Troupeau) = J. Dagher, J & Troupeau, G., trad., *Le banquet des Médecins*. Paris: Geuthner, 2007

Ibn Buṭlān, *Daʿwat al-aṭibbāʾ* (trns. Sedky) – M. Sedky, trad., abrégé, *Un banquet des médecins au temps de l'émir Naṣr al-Dawla*. Le Caire: Imprimerie Misr, 1928.

Ibn Buṭlān, *Le banquet des prêtres* = Ibn Buṭlān, *Le banquet des prêtres : une maqama chrétienne du XIe siècle*, tr. J. Dagher & G. Troupeau. Paris: Geuthner, 2004.

Ibn Dāwūd, *Zahrah* = Abū Bakr Muḥammad Ibn Dāwūd al-Iṣbahānī, *al-Zahrah*, ed. Ibrāhīm al-Sāmarrā'ī. 2 vols. al-Zarqā' (Jordan): Maktabat al-Manār, 1985.

Ibn al-Dāyah, *Mukāfa'ah* = Aḥmad ibn Yūsuf Ibn al-Dāyah al-Kātib, *Kitāb al-Mukāfa'ah wa-ḥusn al-ʿuqbā*, ed. Maḥmūd Muḥammad Shākir. Beirut: Dār al-Kutub al-ʿIlmiyyah, 1984.

Ibn Diḥyah, *Muṭrib* = Dhū l-Nasabayn Abū l-Khaṭṭāb ʿUmar ibn Ḥasan Ibn Diḥyah, *al-Muṭrib min ashʿār ahl al-Maghrib*, ed. Ibrāhīm al-Ibyārī, Ḥāmid ʿAbd al-Majīd, Aḥmad Aḥmad Badawī, and Ṭāhā Ḥusayn. Cairo: Dār al-ʿIlm li-l-Jamīʿ, 1955.

Ibn al-Dubaythī, *Ta'rīkh Baghdād* = Abū ʿAbd Allāh Muḥammad ibn Saʿīd Ibn al-Dubaythī, *Dhayl Tarīkh Madīnat al-Salām Baghdād*. 17 vols. Baghdād: Wizārat al-Iʿlām, 2001.

Ibn Durayd, *Jamharah* = Abū Bakr Muḥammad ibn al-Ḥasan Ibn Durayd, *Jamharat al-lughah*, ed. Ramzī Munīr Baʿlabakkī. 3 vols. Beirut: Dār al-ʿIlm li-l-Malāyīn, 1987.

Ibn Faḍl Allāh, *Masālik* = Shihāb al-Dīn Aḥmad ibn Yaḥyā Ibn Faḍl Allāh al-ʿUmarī, *Masālik al-abṣār fī mamālik al-amṣār*, ed. Kāmil Sulaymān al-Jubūrī et al., 27 vols. Beirut: Dār al-Kutub al-ʿIlmiyyah, 2010.

Ibn al-Faraḍī, *Ta'rīkh ʿulamā' al-Andalus* = ʿAbd Allāh Muḥammad ibn Yūsuf Ibn al-Faraḍī, *Ta'rīkh ʿulamā' al-Andalus*, ed. ʿIzzat al-ʿAṭṭār al-Ḥusaynī. Cairo: Maktabat al-Khānjī, 1988.

Ibn al-Faradī, *Ta'rīkh ʿulamā' al-Andalus* (ʿAwwād Maʿrūf) = ʿAbd Allāh Muḥammad ibn Yūsuf Ibn al-Faradī, *Ta'rīkh ʿulamā' al-Andalus*, ed. Bashshār ʿAwwād Maʿrūf. Tunis: Dār al-Gharb al-Islāmī: 2008.

Ibn Habal *see* Ibn Hubal.

Ibn Ḥabīb, *Asmā' al-mughtālīn* = Muḥammad ibn Ḥabīb, *Asmā' al-mughtālīn min al-ashrāf fī l-jāhiliyyah wa-l-Islām wa-asmā' man qutila min al-shuʿarā'*, in *Nawādir al-makhṭūṭāt*, ed. ʿAbd al-Salām Hārūn. 2 vols. Cairo: Maktabat Muṣṭafā al-Bābī al-Ḥalabī, 1972–1973, ii:105–278.

Ibn Ḥamdūn, *Tadhkirah* = Muḥammad ibn al-Ḥasan ibn Muḥammad ibn ʿAlī Ibn Ḥamdūn, *al-Tadhkirah al-Ḥamdūniyyah*, ed. Iḥsān ʿAbbās and Bakr ʿAbbās. 10 vols. Beirut: Dār Ṣādir, 1996.

Ibn Hānī, *Dīwān Ibn Hānī* (ed. al-Bustānī) = Ibn Hānī al-Andalusī, *Dīwān*, ed. Karam al-Bustānī. Beirut: Dār Bayrūt, 1980.

Ibn al-Haytham, 'Abhandlung über das Licht' = J. Baarmann, 'Abhandlung über das Licht von Ibn al-Haiṯam', *Zeitschrift der Deutschen Morgenländischen Gesellschaft*, 36 (1882), 195–237.

Ibn al-Haytham, 'Die Abhandlungen Thâbit b. Ḳurras' = H. Suter, 'Die Abhandlungen Thâbit b. Ḳurras und Abû Sahl al-Kûhîs über die Ausmessung der Paraboloide', *Sitzungsberichte der Physikalisch-medizinischen Sozietät in Erlangen*, 48–49 (1916–1917), 186–227.

Ibn al-Haytham, 'Die Ausmessung des Paraboloides' = H. Suter, 'Die Abhandlung über die Ausmessung des Paraboloides von el-Ḥasan b. el-Ḥasan b. el-Haitham', *Bibliotheca mathematica*, 3rd ser., 12 (1912), 289–332.

Ibn al-Haytham, *Alhacen on Image-Formation* = A. Mark Smith, *Alhacen on Image-Formation and Distortion in Mirrors: A Critical Edition, with English Translation and Commentary, of Book 6 of Alhacen's De Aspectibus*. Philadelphia: American Philosophical Society, 2008.

Ibn al-Haytham, *Alhacen on the Principles of Reflection* = A. Mark Smith, *Alhacen on the Principles of Reflection: A Critical Edition with English Translation and Commentary of Books 4 and 5 of Alhacen's De Aspectibus*. Philadelphia: American Philosophical Society, 2006.

Ibn al-Haytham, *Alhacen's Theory of Visual Perception*. A. Mark Smith, *Alhacen's theory of visual perception: a critical edition, with English translation and commentary, of the first three books of Alhacen's De aspectibus*. Philadelphia: American Philosophical Society, 2001.

Ibn al-Haytham, 'L'analyse et la synthèse' = Roshdi Rashed, 'L'analyse et la synthèse', *Mélanges de l'Institut Dominicain d'Études Orientales du Caire*, 20 (1991), 31–231; also in Rashed, *Les mathématiques infinitésimales*, iv:230–391.

Ibn al-Haytham, 'Behandlung einiger geometrischen Fragenpunkte' = Carl Schoy, 'Behandlung einiger geometrischen Fragenpunkte durch muslimische Mathematiker', *Isis*, 8 (1926), 254–263.

Ibn al-Haytham, 'Besondere Art des Gesellschaftsrechnens' = Eilhard Wiedemann, 'Über eine besondere Art des Gesellschaftsrechnens nach Ibn al Haiṯam', *Sitzungsberichte der Physikalisch-medizinischen Sozietät in Erlangen*, 58–59 (1926–1927), 191–196.

Ibn al-Haytham, *Completion of the Conics* = J.P. Hogendijk, *Ibn al-Haytham's Completion of the Conics*. New York: Springer, 1985.

Ibn al-Haytham, 'Les connus' = Roshdi Rashed, 'La philosophie mathématique d'Ibn al-Haytham, II: Les connus', *Mélanges de l'Institut Dominicain d'Études Orientales du Caire* 21 (1993), 87–275; also in Rashed, *Les mathématiques infinitésimales*, iv:444–583.

Ibn al-Haytham, 'Construction de l'heptagone régulier' = Roshdi Rashed, 'La construction de l'heptagone régulier par Ibn al-Haytham', *Journal for the History of Arabic Science*, 3 (1979), 309–387.

Ibn al-Haytham 'Criticism of Ptolemy's *Optics*' = A.I. Sabra, 'Ibn al-Haytham's Criticism of Ptolemy's *Optics*', *Journal of the History of Philosophy*, 4 (1966), 145–149.

Ibn al-Haytham, *de l-Optique d'Ibn al-Haytham* = Paul Pietquin, *Le Septième Livre du Traité De aspectibus d'Alhazen, Traduction Latine Médiévale de l'Optique d'Ibn al-Haytham*. Bruxelles: Académie Royale de Belgique, 2010.

Ibn al-Haytham, 'Le Discours de la lumière' = Roshdi Rashed, 'Le Discours de la lumière d'Ibn al-Haytham (Alhazen): Traduction Française Critique', *Revue d'histoire des Sciences et de Leurs Applications*, 21 (1968), 197–224.

Ibn al-Haytham, 'Discourse on the Concave Spherical Mirror' = H.J.J. Winter and W. Arafat, 'A Discourse on the Concave Spherical Mirror by Ibn Al-Haitham', *Journal of the Royal Asiatic Society of Bengal*, 3rd sers., 16 (1950), 1–16.

Ibn al-Haytham, *Doubts Concerning Ptolemy* = Don L. Voss, 'Ibn al-Haytham's Doubts Concerning Ptolemy: A Translation and Commentary'. Ph.D. dissertation, University of Chicago, 1985.

Ibn al-Haytham, 'Einige geometrische Aufgaben' = Heinrich Suter, 'Einige geometrische Aufgaben bei arabischen Mathematikern', repr. from *Bibliotheca mathematica* 3rd ser. 8 (1907), pp. 27–30.

Ibn al-Haytham, 'Finding the Direction of the Qibla' = Ahmad Dallal, 'Ibn al-Haytham's Universal Solution for Finding the Direction of the Qibla by Calculation', *Arabic Sciences and Philosophy* 5 (1995), 145–193.

Ibn al-Haytham 'Geometrische Instrumente' = E. Wiedemann, 'Über geometrische Instrumente bei den muslimischen Völkern', repr. from *Zeitschrift für Vermessungswesen*, 39 (1910), 585–592, 617–625.

Ibn al-Haytham, 'Kleinere Arbeiten von Ibn al Haiṯam' = Eilhard Wiedemann 'Kleinere Arbeiten von Ibn al Haiṯam' *Sitzungsberichte der physikalisch-medizinischen Sozietät Erlangen*, 41 (1909), 16–24.

Ibn al-Haytham, 'Kniga kommentariev k vvedeniyam knigi Evklida' = B.A. Rozenfeld 'Kniga kommentariev k vvedeniyam knigi Evklida "Nachala"', *Istoriko-matematicheskie issledovaniya*, 11 (1958), 743–762.

Ibn al-Haytham, 'Die Kreisquadratur' = Heinrich Suter, 'Die Kreisquadratur des Ibn el-Haiṯam', *Zeitschrift für Mathematik und Physik. Historisch-litterarische Abteilung*, 44 (1899), 33–47.

Ibn al-Haytham, 'The Light of the Stars' = W. 'Arafat, H.J.J. Winter, 'The Light of the Stars: A Short Discourse by Ibn Al-Haytham', *The British Journal for the History of Science*, 5 (1971), 282–288.

Ibn al-Haytham, *Les Mathématiques Infinitésimales* = Roshdi Rashed, *Les Mathématiques Infinitésimales du IXe au XIe Siècle*. London: Al-Furqan Islamic Heritage Foundation, 1993.

Ibn al-Haytham, *Muʿāmalāt* = Ulrich Rebstock, 'Der *Muʿāmalāt* des Ibn al-Haitams', *Zeitschrift für Geschichte der Arabisch-Islamischen Wissenschaften*, 10 (1995–1996), 61–121.

Ibn al-Haytham, *On the Configuration of the World* = Y. Tzvi Langermann, *Ibn al-Haytham's On the Configuration of the World*. New York: Garland, 1990.

Ibn al-Haytham, 'On the Paraboloidal Focusing Mirror' = H.J.J. Winter, 'Ibn al-Haitham on the Paraboloidal Focusing Mirror', *Journal of the Royal Asiatic Society of Bengal*, 3rd ser. [science], 15 (1949), 25–40.

Ibn al-Haytham, 'On the Premises of Euclid's Elements' = Barbara Hooper Sude, 'Ibn al-Haytham's commentary on the premises of Euclid's Elements: (*Sharḥ muṣādarāt kitāb Uqlīdis fī al-uṣūl*), Books I–VI'. Ph.D. diss., Princeton, 1974.

Ibn al-Haytham, 'On Seeing the Stars' = A.I. Sabra and Anton Heinen, 'On Seeing the Stars', *Zeitschrift für Geschichte der Arabisch-Islamischen Wissenschaften*, 7 (1991–1992), 31–72.

Ibn al-Haytham, 'On Seeing the Stars, II' = A.I. Sabra, '*On Seeing the Stars, II*. Ibn al-Haytham's 'Answers' to the 'Doubts' Raised by [Abū l-Qāsim] Ibn Maʿdān', *Zeitschrift für Geschichte der Arabisch-Islamischen Wissenschaften*, 10 (1995–1996), 1–59.

Ibn al-Haytham, *On the Shape of the Eclipse* = Dominique Raynaud, *A Critical Edition of Ibn al-Haytham's On the Shape of the Eclipse. The First Experimental Study of the Camera Obscura* (Sources and Studies in the History of Mathematics and Physical Sciences). Cham, Switzerland: Springer, 2018.

Ibn al-Haytham, 'Die Polhöhe mit grösster Genauigkeit zu bestimmen' = C. Schoy, 'Abhadndlung des Ḥasan ben al-Ḥasan ben al-Haiṯam über eine Methode, die Polhöhe mit grösster Genauigkeit zu bestimmen', *De Zee*, 10 (1920), 586–601.

Ibn al-Haytham, 'Die sphärischen Hohlspiegel' = E. Wiedemann, 'Ibn al Haiṯams Schrift über die sphärischen Hohlspiegel', *Bibliotheca mathematica*, 3rd ser., 10 (1909–1910), 293–307.

Ibn al-Haytham, 'Theorie des Regenbogens' = E. Wiedemann 'Theorie des Regenbogens von Ibn al Haiṯam', *Sitzungsberichte der physikalisch-medizinischen Sozietät Erlangen*, 46 (1914).

Ibn al-Haytham, *Die trigonometrischen Lehren* = Carl Schoy, *Die trigonometrischen Lehren des persischen Astronomen Abu'l-Raihân Muh. Ibn Ahmad al-Bîrûnî*. Hannover: Orient-Buchhandlung Heinz Lafaire K.-G., 1927.

Ibn al-Haytham, 'Über den Aufbau der Welt' = K. Kohl, 'Über den Aufbau der Welt nach Ibn al-Haiṯam', *Sitzungsberichte der Physikalisch-medizinischen Sozietät in Erlangen*, 54–55 (1992–1993), 140–179.

Ibn al-Haytham, 'Über die Beschaffenheit der Schatten' = Eilhard Wiedemann, 'Über eine Schrift von Ibn al Haiṯam "Über die Beschaffenheit der Schatten"', *Sitzungsberichte der Physikalisch-medizinischen Sozietät in Erlangen*, 39 (1907), 226–244.

Ibn al-Haytham, 'Über die Bestimmung der Richtung der Qibla' = Carl Schoy, 'Abhandlung des al-Ḥasan ibn al-Ḥasan ibn al-Haiṯam (Alhazen) über die Bestimmung der Richtung der Qibla', *Zeitschrift der Deutschen Morgenländischen Gesellschaft*, 75 (1921), 242–253.

Ibn al-Haytham, 'Über die Brechung des Lichtes in Kugeln' = E. Wiedemann, 'Über die Brechung des Lichtes in Kugeln nach Ibn al-Haiṯam und Kamāl al-Dīn al-Fārisī', *Sitzungsberichte der Physikalisch-medizinischen Sozietät in Erlangen*, 42 (1910), 15–58.

Ibn al-Haytham, 'Über die *Camera obscura*' = E. Wiedemann, 'Über die *Camera obscura* bei Ibn al Haiṯam', *Sitzungsberichte der Physikalisch-medizinischen Sozietät in Erlangen*, 46 (1914), 155–169.

Ibn al-Haytham, 'Über die Lage der Milchstrasse' = Eilhard Wiedemann, 'Über die Lage der Milchstrasse nach Ibn al Haitham', *Sirius*, 39 (1906), 113–115.

Ibn al-Haytham, 'Über das Licht des Mondes' = Karl Kohl, 'Über das Licht des Mondes. Eine Untersuchung von Ibn al-Haitham', *Sitzungsberichte der Physikalisch-medizinischen Sozietät in Erlangen*, 56–57 (1924–1925), 305–398.

Ibn al-Haytham, 'Über das Licht der Sterne' = Eilhard Wiedemann, 'Über das Licht der Sterne nach Ibn al Haitham', *Wochenschrift für Astronomie, Meteorologie und Geographie*, n.s., 33 (1890), 129–133.

Ibn al-Haytham, *Über die Natur der Spuren* [*Flecken*] = Carl Schoy, *Abhandlung des Schaichs Ibn ʿAlî al-Hasan ibn al-Hasan ibn al-Haitham: Über die Natur der Spuren* (*Flecken*), *die man auf der Oberfläche des Mondes sieht*. Hannover: Lafaire, 1925.

Ibn al-Haytham, 'Über parabolische Hohlspiegel' = J L Heiberg and E. Wiedemann, 'Ibn al Haiṯams Schrift über parabolische Hohlspiegel', *Bibliotheca mathematica*, 3rd ser., 10 (1909–1910), 201–237.

Ibn Ḥibbān, *K. al-Thiqāt* = Ibn Ḥibbān al-Bustī, *K. al-Thiqāt*, ed. Muḥammad ʿAbd al-Muʿīd Khān. 9 vols. Hyderabad: Dā'irat al-Maʿārif al-ʿUthmāniyyah, 1973–1983.

Ibn Ḥijjah, *Thamarāt* = Taqī al-Dīn Abū Bakr ibn ʿAlī Ibn Ḥijjah al-Ḥamawī, *Thamarāt al-awrāq*, ed. Muḥammad Abū l-Faḍl Ibrāhīm. Cairo: Maktabat al-Khānjī, 1971.

Ibn Hindū, *Miftāḥ* (Manṣūrī) = Abū l-Faraj ʿAlī ibn al-Ḥusayn ibn Hindū, *Miftāḥ al-ṭibb wa-minhāj al-ṭullāb*, ed. A. Manṣūrī. Beirut: Mu'assasat al-Balāgh, 2002.

Ibn Hindū, *Miftāḥ* (Mohaghegh) = Abū l-Faraj ʿAlī ibn al-Ḥusayn ibn Hindū, *Miftāḥ al-ṭibb wa-minhāj al-ṭullāb*, ed. M. Mohaghegh and M.T. Daneshpajuh. Tehran: Mu'assasah-i Muṭālaʿāt-i Islāmī-i Dānishgāh-i Mak Gīl bā hamkārī-i Dānishgāh-i Tihrān, 1989.

Ibn Hindū, *Miftāḥ* (Tibi) = Abū l-Faraj ʿAlī ibn al-Ḥusayn ibn Hindū, *The Key to Medicine and a Guide for Students* (*Miftāḥ al-ṭibb wa-minhāj al-ṭullāb*), tr. by Aida Tibi, reviewed by Emilie Savage-Smith. (The Great Books of Islamic Civilization). Reading: Garnet Publishing, 2010.

Ibn Hishām, *Sīrah* = Ibn Hishām, *al-Sīrah al-nabawiyyah*, ed. Muṣṭafā al-Saqqā, Ibrāhīm al-Ibyārī, and ʿAbd al-Ḥafīẓ Shalabī. 2 vols. Cairo: Muṣṭafā al-Bābī al-Ḥalabī, 1955.

Ibn Hishām, *Sīrah*, tr. Guillaume = *The Life of Muhammad, A Translation of Isḥāq's* [*sic*] Sirāt Rasūl Allāh, with introd. and notes by A. Guillaume. Karachi: Oxford University Press, 1978.

Ibn Hubal, *Kitāb al-Mukhtārāt* = Ibn Hubal, *Kitāb al-Mukhtārāt fī l-ṭibb li-ibn Hubal al-Baghdādī*. 4 vols. Hyderabad: Dā'irat al-Maʿārif al-ʿUthmāniyyah, 1943–1944.

Ibn al-ʿIbrī *see* Bar Hebraeus.

Ibn ʿIdhārī, *al-Bayān al-mughrib* = Ibn ʿIdhārī, *al-Bayān al-mughrib fī akhbār al-Andalus wa-l-Maghrib*, ed. G.S. Colin and E. Lévi-Provençal. Beirut: Dār al-Thaqāfah, 1983.

Ibn ʿIdhārī, *al-Bayān al-mughrib* (*Muwaḥḥidūn*) = Ibn ʿIdhārī, *al-Bayān al-mughrib fī akhbār al-Andalus wa-l-Maghrib: qism al-muwaḥḥidīn*, ed. Muḥammad Ibrāhīm al-Kattānī et al. Beirut: Dār al-Gharb al-Islāmī, 1985.

Ibn ʿIdhārī, *al-Bayān al-mughrib* (Tunis) = Ibn ʿIdhārī, *al-Bayān al-mughrib fī akhbār al-Andalus wa-l-Maghrib*, ed. Bashshār ʿAwwād Maʿrūf and Maḥmūd Bashshār ʿAwwād. Tunis: Dār al-Gharb al-Islāmī, 2013.

Ibn al-ʿImād, *Shadharāt* = Shihāb al-Dīn ʿAbd al-Ḥayy ibn Aḥmad Ibn al-ʿImād, *Shadharāt al-dhahab fī akhbār man dhahab*, ed. ʿAbd al-Qādir al-Arnāʾūṭ and Maḥmūd al-Arnāʾūṭ. 10 vols. Damascus: Dār Ibn Kathīr, 1986–1993.

Ibn al-Jarrāḥ, *Waraqah* = Muḥammad ibn Dāwūd Ibn al-Jarrāḥ, *al-Waraqah*, ed. ʿAbd al-Wahhāb ʿAzzām and ʿAbd al-Sattār Aḥmad Farrāj. Cairo: Dār al-Maʿārif, n.d.

Ibn al-Jawzī, *Adhkiyāʾ* = Abū l-Faraj Jamāl al-Dīn ʿAbd al-Raḥmān ibn ʿAlī Ibn al-Jawzī, *al-Adhkiyāʾ*, ed. Muḥammad ʿAbd al-Raḥmān ʿAwaḍ. Beirut: Dār al-Kitāb al-ʿArabī, 1406/1986.

Ibn al-Jawzī, *Mud'hish* = Ibn al-Jawzī, *al-Mud'hish*, ed. Marwān Qabbānī. Beirut: Dār al-Kutub al-ʿIlmiyyah, 1981.

Ibn al-Jawzī, *Muntaẓam* = Abū l-Faraj ʿAbd al-Raḥmān ibn ʿAlī ibn Muḥammad Ibn al-Jawzī, *al-Muntaẓam fī tārīkh al-mulūk wa-l-umam*, ed. Muḥammad ʿAbd al-Qādir ʿAṭā and Muṣṭafā ʿAbd al-Qādir ʿAṭā. 19 vols. Beirut: Dār al-Kutub al-ʿIlmiyya, 1992–1993.

Ibn al-Jawzī, *Ṣayd al-khāṭir* = Ibn al-Jawzī, *Ṣayd al-khāṭir*, ed. Nājī al-Ṭanṭāwī and ʿAlī al-Ṭanṭāwī. Beirut: Dār al-Fikr, 1399/1979.

Ibn al-Jazzār, *K. fī l-maʿidah* = Ibn al-Jazzār, *K. fī l-maʿidah wa-amrāḍihā wa-mudāwātihā*, ed. Salmān Qaṭābah. Baghdad: Wizārat al-Thaqāfah wa-l-Iʿlām/Dār al-Rashīd, 1980.

Ibn al-Jazzār, *al-Furūq* = Ibn al-Jazzār, *al-Furūq bayna al-ishtibāhāt fī l-ʿilal*, ed. Ramziyyah Aṭraqchī. Baghdad: Wizārat al-Taʿlīm al-ʿĀlī wa-l-Baḥth al-ʿIlmī/Jāmiʿat Baghdād, Bayt al-Ḥikmah, 1989.

Ibn al-Jazzār, *al-Iʿtimād fī l-adwiyah al-mufradah* (Qashsh) = Ibn al-Jazzār, *al-Iʿtimād fī l-adwiyah al-mufradah: al-ʿilāj bi-l-adwiyah al-ʿarabiyyah*, ed. Idwār Qashsh. Beirut: Sharikat al-Maṭbūʿāt li-l-Tawzīʿ wa-l-Nashr, 1998.

Ibn al-Jazzār, *Mujarrabāt al-ʿaṭṭārīn* = Ibn al-Jazzār, *Mujarrabāt al-ʿaṭṭārīn al-maʿrūf bi-Ṭibb al-fuqarāʾ*, ed. Muḥsin ʿAqīl. Beirut: Dār al-Rasūl al-Akram, 2002.

Ibn al-Jazzār, *Die Risāla fī l-Ḫawāṣṣ* = Ibn al-Jazzār, *Die Risāla fī l-Ḫawāṣṣ des Ibn al-Ǧazzār: die arabische Vorlage des Albertus Magnus zugeschriebenen Traktats De mirabilius mundi*, ed. and tr. Fabian Käs. Wiesbaden: Harrassowitz, 2012.

Ibn al-Jazzār, *Ṭibb al-fuqarāʾ* = Ibn al-Jazzār, *Ṭibb al-fuqarāʾ wa-l-masākīn*, ed. al-Rāḍī al-Jāzī and Fārūq ʿUmar al-ʿAsalī. Tunis: al-Majmaʿ al-Tūnisī li-l-ʿUlūm wa-l-Ādāb wa-l-Funūn, 2009.

Ibn al-Jazzār & Bos, *Fevers* = Ibn al-Jazzār, *Ibn al-Jazzār on Fevers: Provisions for the Trav-*

eller and Nourishment for the Sedentary, Book 7, Chapters 1–6, ed. and trans. Gerrit Bos (Sir Henry Wellcome Asian Series). London / New York: Kegan Paul International, 2000.

Ibn al-Jazzār & Bos, *Sexual Diseases* = Ibn al-Jazzār, *Ibn al-Jazzār on Sexual Diseases and their Treatment: Provisions for the Traveller and Nourishment for the Sedentary, Book 6*, ed. and tr. by Gerrit Bos (Sir Henry Wellcome Asian Series). London / New York: Kegan Paul International, 1997.

Ibn Jubayr, *Riḥlah* = *The travels of Ibn Jubayr ed. from a ms. in the university library of Leyden*, ed. William Wright, 2nd revised ed. M.J. de Goeje. Leiden: Brill, 1907.

Ibn Juljul, *Die Ergänzung Ibn Ǧulǧuls zur Materia medica* = Ibn Ǧulǧul, *Die Ergänzung Ibn Ǧulǧuls zur Materia medica des Dioskurides. Arabischer Text, kommentierter deutscher Übersetzung herausgegeben von Albert Dietrich*. Göttingen; Vandenhoeck und Ruprecht, 1993.

Ibn Juljul, *Libro de la explicación de los nombres* = Ibn Juljul, *Libro de la explicación de los nombres de los medicamentos simples tomados del libro de Dioscórides*, ed. and tr. Ildefonso Garijo Galán. Cordova: Area de Estudios Árabes e Islámicos, Cátedra de Lengua y Literatura Árabes, Universidad de Córdoba, 1992.

Ibn Juljul, *Ṭabaqāt al-aṭibbāʾ* = Sulaymān ibn Ḥassān Ibn Juljul, *Ṭabaqāt al-aṭibbāʾ wa-l-ḥukamāʾ*, ed. Fuʾād Sayyid. Beirut: Muʾassasat al-Risālah, 1985. Repr. of *Les générations de médecins et des sages*, ed. Fuʾād Sayyid (Publications de l'Institut d'Auteurs Orientaux, 10). Cairo: Imprimerie de l'Institut Français d'Archéologie Orientale, 1955.

Ibn Juljul, *Tratado Octavo*, = *Ibn Ŷulŷul, Tratado Octavo, Mencionamos en el lo que Dioscorides no cita en su libro*, ed. Ildefonso Garijo Galán. Cordova: Area de Estudios Árabes e Islámicos, Cátedra de Lengua y Literatura Árabes, Universidad de Córdoba, 1992.

Ibn Jumayʿ, 'Ibn Jamīʿ on the Skeleton' = C. Rabin, 'Ibn Jamīʿ on the Skeleton', in *Science, Medicine, and History: Essays on the Evolution of Scientific Thought and Medical Practice Written in Honour of Charles Singer*. London: Oxford University Press, 1953.

Ibn Jumayʿ, 'Kitab al-irshad' = Umberto de Martini, *'Kitāb al-irshād fī maṣāliḥ al-anfus wa-l-ajsād', Il trattato di medicina di Bin Giumai, medico arabo del XII secolo*. Rome: Istituto di storia della medicina dell'Università di Roma, 1960.

Ibn Jumayʿ, 'Sultan Saladin's Physician' = Max Meyerhof, 'Sultan Saladin's Physician on the Transmission of Greek Medicine to the Arabs', *Bulletin of the History of Medicine*, 18 (1945), 169–178.

Ibn Jumayʿ, *Treatise to Ṣalāḥ ad-Dīn* = Ibn Jumayʿ, *Treatise to Ṣalāḥ ad-Dīn on the Revival of the Art of Medicine*, ed. and tr. by Hartmut Fahndrich. [*Abhandlungen für die Kunde des Morgenlandes*] Wiesbaden: Kommissionsverlag F. Steiner, 1983.

Ibn al-Kalbī, *Jamharah* = Werner Caskel, *Ǧamharat an-nasab: Das genealogische Werk des Hišām ibn Muḥammad al-Kalbī*. 2 vols. (I: Tafeln, II: Register). Leiden: Brill, 1966.

Ibn al-Kalbī, *Nasab* (Ḥasan) = Abū l-Mundhir Hishām ibn Muḥammad ibn al-Sāʾib al-Kalbī, *Nasab Maʿadd wa-l-Yaman al-kabīr*. ed. Nājī Ḥasan. 2 vols. Beirut: ʿĀlam al-Kutub; Maktabat al-Nahḍah al-ʿArabiyyah, 1988.

Ibn Kathīr, *al-Bidāyah* = Ibn Kathīr, *al-Bidāyah wa-l-nihāyah*, ed. ʿAlī Shīrī. 14 vols. Beirut: Dār al-Kutub al-ʿIlmiyyah, 1994.

Ibn Kathīr, *al-Bidāyah wa al-Nihāyah* = Ibn Kathīr, *al-Bidāyah wa al-Nihāyah*, ed. ʿAlī Shīrī. Beirut: Dār Iḥyāʾ al-Turāth al-ʿArabī, 1988.

Ibn Khaldūn, *Muqaddimah*= ʿAbd al-Raḥmān ibn Muḥammad Ibn Khaldūn, *al-Muqaddimah*, ed. ʿAlī ʿAbd al-Raḥmān Wāfī. 4 vols. Cairo: Lajnat al-Bayān al-ʿArabī, 1957–1962.

Ibn Khaldūn, *Muqaddimah*, tr. Rosenthal = Ibn Khaldūn, *The Muqaddimah*, tr. Franz Rosenthal. 2nd ed., 3 vols. Princeton: Princeton University Press, 1967.

Ibn Khaldūn, *Taʾrīkh* = ʿAbd al-Raḥmān ibn Muḥammad Ibn Khaldūn, *Taʾrīkh Ibn Khaldūn al-musammā Dīwān al-mubtadaʾ wa-l-khabar fī taʾrīkh al-ʿArab wa-l-Barbar wa-man ʿāṣarahum min dhawī al-shaʾn al-akbar*, eds. Khalīl Shihādah and Suhayl Zakkār. 8 vols. Revised ed. Beirut: Dār al-Fikr, 2001.

Ibn Khallikān, *Wafayāt* = Abū l-ʿAbbās Shams al-Dīn Aḥmad ibn Muḥammad ibn Abī Bakr Ibn Khallikān, *Wafayāt al-aʿyān wa-anbāʾ abnāʾ al-zamān*, ed. Iḥsān ʿAbbās. 8 vols. Beirut: Dār al-Thaqāfah, 1968–1972.

Ibn Khallikān, *Biographical Dictionary* = William MacGuckin de Slane, *Ibn Khallikan's Biographical Dictionary*. 4 vols. Paris: printed for the Oriental Translation Fund of Great Britain and Ireland, 1842–1871.

Ibn Khamīs, *Manāqib al-abrār* = al-Ḥusayn ibn Naṣr ibn Muḥammad *al-maʿrūf bi*-Ibn Khamīs al-Mawṣilī, *Manāqib al-abrār wa-maḥāsin al-akhyār fī ṭabaqāt al-Ṣūfiyyah*, ed. Saʿīd ʿAbd al-Fattāḥ. Beirut: Dār al-Kutub al-ʿIlmiyyah, 2006.

Ibn al-Khaṭīb, *Jaysh al-tawshīḥ* = Lisān al-Dīn Ibn al-Khaṭīb, *Jaysh al-tawshīḥ / An Anthology of Andalusian Arabic* Muwashshaḥāt, ed. Alan Jones. Cambridge: E.J.W. Gibb Memorial Trust, 1997.

Ibn Manẓūr, *Lisān al-ʿArab* = Jamāl al-Dīn Muḥammad ibn Mukarram Ibn Manẓūr al-Anṣārī, *Lisān al-ʿArab*. 20 vols. Būlāq, AH 1300–1308, repr. Cairo: al-Dār al-Miṣriyyah li-l-Taʾlīf wa-l-Tarjamah, n.d.

Ibn Manẓūr, *Surūr* = Muḥammad ibn Mukarram Ibn Manẓūr, *Surūr al-nafs bi-madārik al-ḥawāss al-khams*, ed. Iḥsān ʿAbbās. Beirut: al-Muʾassasah al-ʿArabiyyah, 1980.

Ibn Maṭrūḥ, *Dīwān* = Ibn Maṭrūḥ, *Dīwān*, ed. Ḥusayn Naṣṣār. Cairo: Dār al-Kutub wa-l-Wathāʾiq al-Qawmiyyah, 2004.

Ibn Maymūn, *Muntahā l-ṭalab* = Abū Ghālib Muḥammad ibn al-Mubārak Ibn Maymūn, *Muntahā l-ṭalab min ashʿār al-ʿarab* (*The Utmost in the Search for Arab Poetry*), facs. ed. 3 vols. Frankfurt am Main: Institut für Geschichte der Arabisch-Islamischen Wissenschaften, 1986–1993.

Ibn al-Muqaffaʿ, *Adab* = Ibn al-Muqaffaʿ, *al-Adab al-ṣaghīr wa-l-adab al-kabīr*. Beirut: Dār Ṣādir, 1977.

Ibn al-Muqaffaʿ, *Kalīlah wa-Dimnah* = Baydabā al-Faylasūf al-Hindī, *Kitāb Kalīlah wa-Dimnah, tarjamahu ilā l-ʿarabiyyah fī ṣadr al-Dawlah al-ʿAbbāsiyyah ʿAbd Allāh ibn al-Muqaffaʿ*. Beirut: Maktabat Lubnān, 1991.

Ibn al-Muʿtazz, *Dīwān* = *Dīwān ashʿār al-amīr Abī l-ʿAbbās ʿAbd Allāh ibn Muḥammad al-Muʿtazz bi-Allāh al-khalīfāh al-ʿAbbāsī*, ed. Muḥammad Badīʿ Sharīf. 2 vols. Cairo: Dār al-Maʿārif, 1989.

Ibn al-Muʿtazz, *Ṭabaqāt* = Ibn al-Muʿtazz, *Ṭabaqāt al-shuʿarāʾ al-muḥdathīn*, ed. ʿAbd al-Sattār Aḥmad Farrāj. Cairo: Dār al-Maʿārif, 1968.

Ibn al-Muṭrān, *Bustān* = Asʿad ibn Ilyās ibn al-Muṭrān, *Bustān al-aṭibbāʾ wa-rawḍat al-alibbāʾ*, ed. ʿAbd al-Karīm Abū Shuwayrib. Ṭarābulus [Tripoli, Libya]: Jamʿiyyat al-Daʿwah al-Islāmiyyah al-ʿĀlamiyyah, 1993.

Ibn al-Muṭrān, *Bustān* (facs.) = Asʿad ibn Ilyās ibn al-Muṭrān, *Bustān al-aṭibbāʾ wa-rawḍat al-alibbāʾ* (*The Gardens of Physicians and Meadows of the Intelligent*), Facsimile edition of the manuscript held at Malik Public Library with an introduction by M. Mohaghegh (Encyclopaedia Islamica Foundation: Center for the publication of manuscripts, 5). Tehran, 1989.

Ibn Muṭrān, *Bustān al-aṭibbāʾ* = Asʿad ibn Ilyās ibn al-Muṭrān, *Bustān al-aṭibbāʾ wa-rawḍat al-alibbāʾ*. Tehran: Markaz-i Intishār-i Nusakh-i Khaṭṭī, 1988.

Ibn al-Nadīm, *Fihrist* (Ṭawīl) = Abū l-Faraj Muḥammad ibn Abī Yaʿqūb Isḥāq [Ibn] al-Nadīm, *al-Fihrist*, ed. Yūsuf ʿAlī Ṭawīl, indexes by Aḥmad Shams al-Dīn. Beirut: Dār al-Kutub al-ʿIlmiyyah, 2002.

Ibn al-Nadīm, *Fihrist* (Flügel) = Ibn al-Nadīm, *al-Fihrist*, ed. Gustav Flügel, Johannes Roediger, and August Müller. 2 vols. in 1. Leipzig: F.C.W. Vogel, 1871–1872; repr. Beirut: Maktabat Khayyāṭ, n.d.

Ibn al-Nadīm, *Fihrist* (Tajaddud) = Abū l-Faraj Muḥammad ibn Abī Yaʿqūb Isḥāq Ibn al-Nadīm al-Warrāq al-Baghdādī, *al-Fihrist*, ed. Riḍā Tajaddud. Tehran: Maṭbaʿat Dānishgāh, 1971.

Ibn al-Nadīm, *Fihrist* (Sayyid) = Abū l-Faraj Muḥammad ibn Abī Yaʿqūb Isḥāq Ibn al-Nadīm al-Warrāq al-Baghdādī, *al-Fihrist*, ed. Ayman Fuʾād Sayyid. 2nd ed., 4 vols. Cairo: Al-Furqān Islamic Heritage Foundation, 2014.

Ibn al-Nadīm, *The Fihrist*, tr. Bayard Dodge. 2 vols. New York: Columbia University Press, 1970.

Ibn al-Nafīs, *Sharḥ fuṣūl Abuqrāṭ* = Ibn al-Nafīs, *Sharḥ fuṣūl Abuqrāṭ*, ed. Yūsuf Zaydān. Cairo: Dār al-Miṣriyyah al-Lubnāniyyah, 1991.

Ibn Nubātah, *Sarḥ al-ʿuyūn* = Ibn Nubāta, *Sarḥ al-ʿuyūn fī sharḥ Risālat Ibn Zaydūn*, ed. Muḥammad Abū l-Faḍl Ibrāhīm. Cairo: Dār al-Fikr al-ʿArabī, 1964.

Ibn Qayyim al-Jawziyyah, *Rawḍat al-muḥibbīn* = Abū ʿAbd Allāh Muḥammad ibn Abī Bakr ibn Ayyūb Ibn Qayyim al-Jawziyyah, *Rawḍat al-muḥibbīn wa-nuzhat al-mushtāqīn*, ed. Muḥammad ʿUzayr Shams. Jedda: Majmaʿ al-Fiqh al-Islāmī, AH 1431.

Ibn Qayyim al-Jawziyyah, *al-Ṭibb al-nabawī* = Shams al-Dīn Muḥammad ibn Abī Bakr

Ibn Qayyim al-Jawziyyah, *al-Ṭibb al-nabawī*, ed. Shuʿayb al-Arnaʾūṭ and ʿAbd al-Qādir al-Arnaʾūṭ. Beirut: Muʾassasat al-Risālah, 1405/1985. See also Shams al-Dīn Muḥammad ibn Abī Bakr Ibn Qayyim al-Jawziyyah, *Zād al-maʿād fī hady khayr al-ʿibād, al-juzʾ al-rābiʿ: al-Ṭibb al-nabawī*, ed. Shuʿayb al-Arnaʾūṭ and ʿAbd al-Qādir al-Arnaʾūṭ. Beirut: Muʾassasat al-Risālah, 1418/1998.

Ibn al-Qifṭī, *Taʾrīkh al-ḥukamāʾ* = Ibn al-Qifṭī, *Taʾrīkh al-ḥukamāʾ*, ed. Julius Lippert. Leipzig: Dieterich'sche Verlagsbuchhandlung, 1903.

Ibn al-Qifṭī, *Inbāh* = Ibn al-Qifṭī, *Inbāh al-ruwāh ʿalā anbāh al-nuḥāh*, ed. Muḥammad Abū l-Faḍl Ibrāhīm. 4 vols. Cairo: Dār al-Kutub al-Miṣriyyah, 1950–1955.

Ibn al-Quff, *K. al-ʿUmdah* (Hamarneh) = Ibn al-Quff, *K. al-ʿUmdah fī ṣināʿat al-jirāḥah*, ed. Sāmī Khalaf al-Ḥamārinah. Ammān: al-Jāmiʿah al-Urdunniyyah, ʿImādat al-Baḥth al-ʿIlmī, 1994.

Ibn Qutaybah, *Faḍl al-ʿArab* = Abū Muḥammad ʿAbd Allāh ibn Muslim Ibn Qutaybah al-Dīnawarī, *Faḍl al-ʿArab wa-l-tanbīh ʿalā ʿulūmihā*, ed. Walīd Maḥmūd Khāliṣ. Abu Dhabi: Manshūrāt al-Majmaʿ al-Thaqāfī, 1998.

Ibn Qutaybah, *Shiʿr* = Abū Muḥammad ʿAbd Allāh ibn Muslim Ibn Qutaybah al-Dīnawarī, *al-Shiʿr wa-l-shuʿarāʾ*, ed. Aḥmad Muḥammad Shākir. 2 vols. Cairo: Dār al-Maʿārif, 1966–1967.

Ibn Qutaybah, *ʿUyūn* = Ibn Qutaybah, *ʿUyūn al-akhbār*. 4 vols. Cairo: Dār al-Kutub, 1925–1930, repr. al-Hayʾah al-Miṣriyyah al-ʿĀmmah li-l-Kitāb, 1972.

Ibn Rashīq, *ʿUmdah* = Abū ʿAlī al-Ḥasan Ibn Rashīq al-Qayrawānī al-Azdī, *al-ʿUmdah fī maḥāsin al-shiʿr wa-ādābihi wa-naqdih*, ed. Muḥammad Muḥyī l-Dīn ʿAbd al-Ḥamīd. 2 vols. Cairo: al-Maktabah al-Tijāriyyah al-Kubrā, 1955, repr. Beirut: Dār al-Jīl, 1972.

Ibn Riḍwān, *Kitāb al-Nāfiʿ* (Sāmarrāʾī) = Ibn Riḍwān, *K. al-Nāfiʿ fī kayfiyyat taʿlīm ṣināʿat al-ṭibb*, ed. Kamāl al-Sāmarrāʾī. Baghdad: Jāmiʿat Baghdād, Markaz Iḥyāʾ al-Turāth al-ʿIlmī al-ʿArabī, 1986.

Ibn Riḍwān, 'The Kitāb al-Nāfiʿ' = M.C. Lyons, 'The Kitāb al-Nāfiʿ of ʿAlī ibn Riḍwān', *Islamic Quarterly*, 6 (1961), 65–71.

Ibn Riḍwān, 'On the Nature of Man' = Malcolm C. Lyons, '"On the Nature of Man" in ʿAlī ibn Riḍwān's Epitome', *Al-Andalus*, 30 (1965), 181–188.

Ibn Riḍwān, *Weg zur Glückseligkeit* = Albert Dietrich, *Über den Weg zur Glückseligkeit durch den ärztlichen Beruf: arabischer Text nebst kommentierter deutscher Übersetzung*. Göttingen: Vandenhoeck & Ruprecht, 1982.

Ibn Rushd, *Averrois … in Aristotelis de Anima* (Crawford) = Ibn Rushd, *Averrois Cordubensis Commentarium Magnum in Aristotelis de Anima*, ed. F.S. Crawford. Cambridge, Mass.: Mediaeval Academy of America, 1953.

Ibn Rushd, *Averroes, Avicennae cantica. Texto árabe, versión latina y traducción española*, eds. and trs. J. Coullaut, E. Fernández Vallina, and C. Álvarez de Benito. Salamanca: Ediciones Universidad de Salamanca, 2010.

Ibn Rushd, *Bidāyat al-mujtahid* (1914) = Ibn Rushd, *Bidāyat al-mujtahid wa-nihāyat al-muqtaṣid*. Istanbul: Maṭbaʿat Aḥmad Kāmil, 1914.

Ibn Rushd, *Bidāyat al-mujtahid* (ʿAbd al-Ḥalīm & Maḥmūd) = Ibn Rushd, *Bidāyat al-mujtahid wa-nihāyat al-mujtahid*, ed. ʿAbd al-Ḥalīm Muḥammad ʿAbd al-Ḥalīm and ʿAbd al-Raḥmān Ḥasan Maḥmūd. Beirut: Dār Iḥyāʾ al-Turāth al-ʿArabī, 1992.

Ibn Rushd, *Commentaria Averrois in Galenum* (Vázquez de Benito) = Ibn Rushd, *Commentaria Averrois in Galenum*, ed. C. Vázquez de Benito. Madrid: CSIC, 1984.

Ibn Rushd, *Commentary on Aristotle's Book on the Heaven* (facs.) = Ibn Rushd, *Commentary on Aristotle's Book on the Heaven and the Universe*: Sharḥ Kitāb al-Samāʾ wa-l-ʿĀlam, facsimile ed. Fuat Sezgin with an introduction by G. Endress. Frankfurt am Main: Institut für Geschichte der Arabisch-Islamischen Wissenschaften, 1994.

Ibn Rushd, *The Distinguished Jurists' Primer* = Ibn Rushd, *The Distinguished Jurists' Primer*, tr. Imran Ahsan Khan Nyazee and Mohammad Abdul Raouf. 2 vols. Reading: Garnet, 1994–1996]

Ibn Rushd, *Faṣl al-maqāl* (tr. Alonso) = Ibn Rushd, *Teología de Averroes* (*textos y estudios*), trans. Manuel Alonso. Madrid: CSIC, 1947.

Ibn Rushd, *Faṣl al-maqāl* (tr. Gauthier) = Ibn Rushd, *Accord de la religion et de la philosophie, traité d'Ibn Rochd*, trans. L. Gauthier. Algiers: Carbonel, 1905.

Ibn Rushd, *Faṣl al-maqāl* (ed. Hourani) = Ibn Rushd, *Ibn Rushd* (*Averroes*) *Kitāb Faṣl al-Maqāl*, ed. George. F. Hourani. Leiden: Brill, 1959.

Ibn Rushd, *Faṣl al-maqāl* (tr. Hourani) = Ibn Rushd, *On the Harmony of Religions and Philosophy. A Translation with Introd. and Notes of Ibn Rushd's* Kitāb Faṣl al-Maqāl, tr. George F. Hourani. London: Luzac, 1961.

Ibn Rushd, *Faṣl al-maqāl* (tr. Lucchetta) = Ibn Rushd, *L'accordo della Legge divina con la filosofia*, ed. and tr. F. Lucchetta. Genoa: Marietti, 1994.

Ibn Rushd, *Faṣl al-maqāl* (Müller) = Ibn Rushd, *Philosophie and Theologie von Averroës*, tr. Marcus Müller. Munich: G. Franz, 1859.

Ibn Rushd, *Jawāmiʿ li-kutub Arisṭūṭālīs* = Ibn Rushd, *Jawāmiʿ li-kutub Arisṭūṭālīs fī l-jadal wa-l-khitābah wa-l-shiʿr. Averroes' Three Short Commentaries on Aristotle's "Topics", "Rhetoric", and "Poetics"*, ed. Charles Butterworth. Albany: State University of New York, 1977.

Ibn Rushd, *Kashf ʿan manāhij* (al-Jābirī) = Ibn Rushd, *al-Kashf ʿan manāhij al-adillah fī ʿaqāʾid al-millahed*, ed. Muḥammad ʿAbīd al-Jābirī. Beirut: Markaz Dirāsāt al-Waḥdah al-ʿArabiyyah, 1998.

Ibn Rushd, *Kashf ʿan manāhij* (tr. Najjar) = Ibn Rushd, *Faith and Reason in Islam. Averroes' Exposition of Religious Arguments*, tr. Ibrahim Najjar. Oxford: Oneworld, 2001.

Ibn Rushd, *Kashf ʿan manāhij* (Qāsim) = Ibn Rushd, *Manāhij al-adillah fī ʿaqāʾid al-millah*, ed. Maḥmūd Qāsim. Cairo: Maktabat al-Anglo al-Miṣriyyah, 1955.

Ibn Rushd, *Kitāb al-muqaddimāt fī l-falsafah, aw, al-Masāʾil fī l-manṭiq wa-l-ʿilm al-ṭabīʿī wa-l-ṭibb*, ed. As'ad Jum'ah. Tunis: Kulliyyat al-Ādāb wa-l-'Ulūm al-Insāniyyah bi-l-Qayrawān, 2008.

Ibn Rushd, *K. al-Kulliyyāt* (Fórneas Besteiro & Álvarez de Morales) = Ibn Rushd, *Kitāb al-Kulliyyāt fī l-ṭibb*, ed. J.M. Fórneas Besteiro and C. Álvarez de Morales. Madrid: CSIC, 1987.

Ibn Rushd, *K. al-Kulliyyāt* (al-Jābirī) = Ibn Rushd, *Kitāb al-Kulliyyāt fī l-ṭibb*, ed. Muḥammad 'Abīd al-Jābirī. Beirut: Markaz Dirāsāt al-Waḥdah al-'Arabiyya, 1999.

Ibn Rushd, *El libro de las generalidades* = Ibn Rushd, *El libro de las generalidades de la medicina*, tr. C. Vázquez de Benito and C. Álvarez de Morales. Madrid: Editorial Trotta, 2003.

Ibn Rushd, *Rasā'il* = Ibn Rushd, *Rasā'il*, ed. Jīrār Jihāmī and Rafīq al-'Ajam. Beirut: Dār al-Fikr al-Lubnānī, 1994.

Ibn Rushd, *Sharḥ Kitāb al-Samā' wa-l-'Ālam* (Jum'ah) = Ibn Rushd, *Sharḥ Kitāb al-samā' wa-l-'ālam*, ed. As'ad Jum'ah. Tunis: Markaz al-Nashr al-Jāmi'ī, 2003.

Ibn Rushd, *Sharḥ wa-talkhīṣ Kitāb al-Burhān* = Ibn Rushd, *Sharḥ wa-talkhīṣ Kitāb al-Burhān*, ed. 'Abd al-Raḥmān Badawī. Kuwait, 1984.

Ibn Rushd, *Tahāfut al-tahāfut* (Bouyges) = Ibn Rushd, *Tahafot al-tahafot/Tahāfut al-tahāfut*, ed. Maurice Bouyges. Beirut: Imprimerie Catholique, 1930, 2nd ed. 1987.

Ibn Rushd, *Tahāfut al-tahāfut* (tr. Campanini) = Ibn Rushd, *L'incoerenza della incoerenza dei filosofi*, tr. M. Campanini. Turin: Unione Tipografico-Editrice Torinese, 1997.

Ibn Rushd, *Tahāfut al-tahāfut* (Dunyā) = Ibn Rushd, *Tahāfut al-tahāfut*, ed. Sulaymān Dunyā. 2 vols. Cairo: Dār al-Ma'ārif, 1964, reedited in 1980 and 1981.

Ibn Rushd, *Tahāfut al-tahāfut* (tr. Horten) = Ibn Rushd, *Die Hauptlehren des Averroes nach seiner Schrift Die Wiederlegung des Gazali*, tr. Max Horten. Bonn: A. Marcus und E. Webers Verlag, 1913.

Ibn Rushd, *Tahāfut al-tahāfut* (al-Jābirī) = Ibn Rushd, *Tahāfut al-tahāfut*, ed. Muḥammad 'Abīd al-Jābirī. Beirut: Markaz Dirāsāt al-Waḥdah al-'Arabiyyah, 1998.

Ibn Rushd, *Tahāfut al-tahāfut* (tr. Puig) = Ibn Rushd, *L'ensorrament de l'Ensorrament*, tr. J. Puig. Barcelona: Servei de Publicacions de la Universitat Autònomade Barcelona, 2005.

Ibn Rushd, *Tahāfut al-tahāfut* (tr. 1954) = Ibn Rushd, *Averroes' Tahafut al-Tahafut* (*The Incoherence of the Incoherence*). London: Luzac, 1954.

Ibn Rushd, *Talkhīṣ* (Amīn) = Ibn Rushd, *Talkhīṣ mā ba'd al-ṭabī'ah*, ed. 'Uthmān Amīn. Cairo: Muṣṭafā al-Bābī al-Ḥalabī, 1958.

Ibn Rushd, *Talkhīṣ* (tr. Arnzen) = Ibn Rushd, *Averroes on Aristotle's Metaphysics. An Annotated Translation of the So-called Epitome*, tr. Rüdiger Arnzen. Berlin: De Gruyter, 2010.

Ibn Rushd, *Talkhīṣ* (tr. Horten) = Ibn Rushd, *Die Metaphysik des Averroes (1198)*, tr. Max Horten. Halle: M. Niemeyer, 1912.

Ibn Rushd, *Talkhīṣ* (Qabbānī) = Ibn Rushd, *Kitāb mā baʿd al-ṭabīʿah wa-huwa al-qism al-rābiʿ min talkhīṣ maqālāt Arisṭū*, ed. Muṣṭafā Qabbānī. Cairo: 190?

Ibn Rushd, *Talkhīṣ* (ed. tr. Quirós Rodríguez) = Ibn Rushd, *Compendio de metafísica*, ed. & tr. Carlos Quirós Rodríguez. Madrid, 1919.

Ibn Rushd, *Talkhīṣ* (tr. van den Bergh) = Ibn Rushd, *Die Epitome der Metaphysik des Averroes*, tr. Simon van den Bergh. Leiden: Brill, 1924.

Ibn Rushd, *Talkhīṣ K. al-samāʿ al-ṭabīʿī* (Beirut) = Ibn Rushd, *Talkhīṣ K. al-samāʿ al-ṭabīʿī*, ed. Rafīq al-ʿAjam and Jirār Jihāmī. Beirut: Dār al-Fikr al-Lubnānī, 1994.

Ibn Rushd, *Talkhīṣ K. al-samāʿ al-ṭabīʿī* (Hyderabad) = Ibn Rushd, *Talkhīṣ K. al-samāʿ al-ṭabīʿī*. Hyderabad: Dāʾirat al-Maʿārif al-ʿUthmāniyyah, 1946.

Ibn Rushd, *Talkhīṣ K. al-samāʿ al-ṭabīʿī* (Madrid) = Ibn Rushd, *Epitome in Physicorum libros*, ed. J. Puig. Madrid: CSIC, 1983.

Ibn Rushd, *Talkhīṣ K. al-samāʿ al-ṭabīʿī* (ed. tr. Puig) = Ibn Rushd, *Epítome de física* (*filosofía de la naturaleza*), ed. and tr. J. Puig. Madrid, CSIC, 1987.

Ibn Rushd, *Two Andalusian Philosophers. The Story of Hayy Ibn Yaqzan by Abu Bakr Muhammad Ibn Tufayl and The Definitive Statement by Abu'l-Walid Muhammad Ibn Rushd*, tr. J. Colville. London: Kegan Paul, 1999.

Ibn Rushd, *Talkhīṣ K. al-Qiyās* (Butterworth et al.) = Ibn Rushd, *Talkhīṣ K. al-Qiyās*, ed. Charles Butterworth et al. Cairo: al-Hayʾah al-Miṣriyyah al-ʿĀmmah l-il-Kitāb, 1983.

Ibn Rushd, *Talkhīṣ K. al-Qiyās* (Badawī) = Ibn Rushd, *Talkhīṣ K. al-Qiyās li-Arisṭūṭālis*, ed. ʿAbd al-Raḥmān Badawī. Kuwait: al-Majlis al-Waṭanī lil-Thaqāfah wa-l-Funūn wa-l-Ādāb, 1988.

Ibn Rushd, *Talkhīṣ K. al-Qiyās* (Jihāmī) = Ibn Rushd, *Talkhīṣ manṭiq Arisṭū*, ed. Jīrār Jihāmī. Beirut: Dāt al-Fikr al-Lubnānī, 1992.

Ibn al-Sāʿātī, *Kitāb ʿilm al-sāʿāt* = Riḍwān Muḥammad al-Sāʿātī, *Kitāb ʿilm al-sāʿāt wa l-ʿamal bihā* (*On the knowledge and use of clocks*), ed. Muḥammad Aḥmad Duhmān. Damascus: Dār al-Fikr al-Muʿāṣir, 1981.

Ibn Saʿd, *Ṭabaqāt* = Ibn Saad, *Biographien Muhammeds, seiner Gefährten und der späteren Träger des Islams: bis zum Jahre 230 der Flucht*, ed. Eduard Sachau. 9 vols. Leiden, E.J. Brill, 1904–1940.

Ibn Sahl, *Dīwān al-aḥkām al-kubrā* = Ibn Sahl, *Dīwān al-aḥkām al-kubrā*, ed. Yaḥyā Murād. Beirut: Dār al-Ḥadīth, 2007.

Ibn Sahlān al-Sāwī, *al-Baṣāʾir al-Naṣīriyyah* = Ibn Sahlān al-Sāwī, *Kitāb al-Baṣāʾir al-Naṣīriyyah fī ʿilm al-manṭiq*, or *The Naṣīrī Observations, concerning the Science of Logic*. Beirut: Dār al-Fikr al-Lubnānī, 1993.

Ibn al-Sāʿī, *Nisāʾ al-khulafāʾ* = Ibn al-Sāʿī, *Nisāʾ al-khulafāʾ/Consorts of the Caliphs*, ed. and tr. Shawkat Toorawa et al. (Library of Arabic Literature). New York: New York University Press, 2015.

Ibn Saʿīd, *Mughrib* = Ibn Saʿīd al-Andalusī, *al-Mughrib fī ḥulā l-maghrib*, ed. Shawqī Ḍayf, 2 vols. Cairo: Dār al-Maʿārif, 1980.

Ibn Saʿīd al-Andalusī, *al-Murqiṣāt wa-l-muṭribāt*. Cairo, AH 1287, repr. n. pl.: Dār Ḥamad wa-Miḥyū, 1973.

Ibn Saʿīd, *Rāyāt al-mubarrizīn* = Ibn Saʿīd al-Andalusī, *Rāyāt al-mubarrizīn wa-ghāyāt al-mumayyizīn*, ed. al-Nuʿmān ʿAbd al-Mutaʿāl al-Qāḍī. Cairo: Lajnat Iḥyāʾ al-Turāth al-Islāmī, 1393/1973.

Ibn al-Ṣāʾigh, *Mukhtār* = Abū l-Muʾayyad Muḥammad ibn al-Mujallī ibn al-Ṣāʾigh al-Jazarī al-ʿAntarī *al-Mukhtār al-sāʾigh min dīwān Ibn al-Ṣāʾigh*, MS (Taymūriyyah collection) accessible online at https://archive.org/download/M-tymoriyah-SHER/805.zip (the numbering refers to pages, not folios).

Ibn al-Ṣalāḥ, *Zur Kritik* = Ibn al-Ṣalāḥ, *Zur Kritik der Koordinaten-uberlieferung im Sternkatalog des Almagest*. ed. and tr. Paul Kunitzsch. (Abhandlung der Akademie der Wissenschaften in Göttingen, Philologisch-hist. Kl., Dritte Folge, 94) Göttingen: Vandenhoeck & Ruprecht, 1975.

Ibn Sallām al-Jumaḥī, *Ṭabaqāt* = Muḥammad Ibn Sallām al-Jumaḥī, *Ṭabaqāt fuḥūl al-shuʿarāʾ*, ed. Maḥmūd Muḥammad Shākir. Cairo: Dār al-Maʿārif, 1952.

Ibn Samajūn, *Jāmiʿ al-adwiya al-mufrada, Compendium of Simple Drugs*, ed. F. Sezgin. [Ser. C, Facsimile editions, 54] 4 vols. Frankfurt am Main: Institut für Geschichte der Arabisch-Islamischen Wissenschaften, 1990.

Ibn Sanāʾ al-Mulk, *Dār al-ṭirāz* = Abū l-Qāsim Hibat Allāh ibn Jaʿfar Ibn Sanāʾ al-Mulk, *Dār al-ṭirāz fī ʿamal al-muwashshaḥāt*, ed. Jawdat al-Rikābī. Damascus: no publ., 1368/1949.

Ibn Sanāʾ al-Mulk, *Dīwān*, ed. Muḥammad Ibrāhīm Naṣr and Ḥusayn Muḥammad Naṣṣār. 2 vols. Cairo: Dār al-Kātib al-ʿArabī, 1968–1969.

Ibn al-Shaʿʿār, *Qalāʾi* = Ibn al-Shaʿʿār, al-Mubārak Kāmil al-Dīn Abū al-Barakāt al-Mubārak, *Qalāʾid al-jumān fī farāʾid shuʿarāʾ hādhā l-zamān*, taḥqīq Kāmil Salmān al-Jubūrī. 9 vols. Beirut: Dar al-kutub al-ʿilmiyyah, 2005.

Ibn Shaddād, *Aʿlāq* = *al-Aʿlāq al-khaṭīrah fī dhikr umarāʾ al-Shām wa l-Jazīrah*, I, i: *Taʾrīkh Ḥalab*. ed. D. Sourdel. Beirut: Institut Français de Damas, 1953.

Ibn Shākir al-Kutubī, *Fawāt* = Ibn Shākir, *Fawāt* = Abū ʿAbd Allāh Muḥammad Ibn Shākir al-Kutubī, *Fawāt al-Wafayāt*, ed. Iḥsān ʿAbbās. 5 vols. Beirut: Dār Ṣādir, 1973–1974.

Ibn Shibl, *Dīwān* = Ibn Shibl al-Baghdādī, *Dīwan*, ed. ʿAbd al-Rāziq Ḥuwayzī. Amman: Arwiqah li-l-Dirāsāt wa-l-Nashr, 2016.

Ibn Sīnā, *De anima* = *Avicenna's De anima* (*Arabic Text*), ed. F. Rahman. London: Oxford University Press, 1959.

Ibn Sīnā, *Deliverance: Logic* = Asad Q. Ahmed (trans.), *Avicenna's Deliverance: Logic*. Oxford: Oxford University Press, 2011.

Ibn Sīnā, *Al-Maqūlāt* = Ibn Sīnā, *Al-Maqūlāt: commentary on Aristotle's Categories*, tr. and introduced by Allan Bäck. Munich: Philosophia, 2016.

Ibn Sīnā, *The Metaphysics of The Healing* = Avicenna, and Michael E. Marmura, *The*

Metaphysics of The Healing: A Parallel English-Arabic Text = al-Ilāhiyyāt min al-Shifāʾ. Provo, Utah: Brigham Young University Press, 2005.

Ibn Sīnā, *The Physics of The Healing* = Avicenna, and Jon McGinnis, *The Physics of The Healing / Al-Shifāʾ: Al-samāʿ Al-ṭabīʿī; a Parallel English-Arabic Text*. Provo, Utah: Brigham Young University Press, 2009.

Ibn Sīnā, *al-Najah* = Ibn Sīnā, *al-Najah min al-gharaq fī baḥr ḍalālāt*, ed. Muḥammad Taqī Dānishpazhūh. Tehran: Publications de l'Université de Tehran, 1985.

Ibn Sīnā, *Qānūn* (Būlāq) = Abū ʿAlī Ibn Sīnā, *al-Qānūn fī l-ṭibb*, 3 vols. Būlāq, AH 1294.

Ibn Sīnā, *Urjūzah fī asbāb al-ḥummayāt* = Ibn Sīnā, *Urjūzah fī asbāb al-ḥummayāt*, ed. Dāwūd Muzbān al-Thāmiri, *al-Mawrid*, 14/4 (1985), 243–266.

Ibn Sinān, *Rasāʾil* = Ibrāhīm Ibn Sinān, *Rasāʾil Ibn Sinān*. Ḥaydarābād al-Dakkan: Maṭbaʿat Jamʿiyyat Dāʾirat al-Maʿārif al-ʿUthmāniyyah, 1948.

Ibn Ṭabāṭabā, *ʿIyār* = Abū l-Ḥasan Muḥammad ibn Aḥmad Ibn Ṭabāṭabā, *ʿIyār al-shiʿr*, ed. ʿAbd al-ʿAzīz ibn Nāṣir al-Māniʿ. Riyadh: Dār al-ʿUlūm, 1405/1985.

Ibn Taghrībirdī, *al-Nujūm al-zāhirah* = Ibn Taghrībirdī, *al-Nujūm al-zāhirah fī mulūk Miṣr wa-l-Qāhirah*. 16 vols. Cairo: Maṭbaʿat Dār al-Kutub al-Miṣriyyah, 1929–1972.

Ibn Ṭāwūs, *Faraj al-mahmūm* = Abū l-Qāsim ʿAlī ibn Mūsā Ibn Ṭāwūs, *Faraj al-mahmūm fī tārīkh ʿulamāʾ al-nujūm*. Qum: Dār al-Dhakhāʾir, 1363 [1984].

Ibn al-Tilmīdh, *Dispensatory* = Oliver Kahl, *The Dispensatory of Ibn at-Tilmīḏ: Arabic text, English translation, study and glossaries*. Leiden: Brill, 2007.

Ibn al-Tilmīdh, *Maqālāh fī l-Faṣd* = Hibat Allāh ibn Ṣāʿid Ibn al-Tilmīdh, *Maqālah fī l-Faṣd*, ed. Ṣubḥī Maḥmūd Ḥammāmī. Aleppo: Jāmiʿat Ḥalab, Maʿhad al-Turāth al-ʿIlmī al-ʿArabī, 1997.

Ibn al-Tilmīdh, *al-Risālah al-Amīniyyah fī l-Faṣd* = Hibat Allāh ibn Ṣāʿid Ibn al-Tilmīdh, *al-Risālah al-Amīniyyah fī al-faṣd: irsāl al-dam wa-ikhrājihi bi-ṭarīqay al-faṣd wa-al-ḥijāmah*, ed. ʿAbd al-Qādir Aḥmad ʿAbd al-Qādir. Damascus: Dār al-Wathāʾiq, 2008.

Ibn al-Ṭiqṭaqā, *Fakhrī* = Muḥammad ibn ʿAlī ibn Ṭabāṭabā Ibn al-Tiqtaqā, *al-Fakhrī fī l-ādāb al-sulṭāniyyah wa-l-duwal al-Islāmiyyah*, ed. Hartwig Derenbourg. Paris: É. Bouillon, 1895.

Ibn al-Ṭiqṭaqā, *Fakhrī* (Beirut ed.) = Ibn al-Ṭiqṭaqā, Muḥammad ibn ʿAlī, *al-Fakhrī fī l-ādāb al-sulṭāniyyah wa-l-duwal al-islāmiyyah*. Beirut: Dār Ṣādir, n.d.

Ibn ʿUnayn, *Dīwān* = Sharaf al-Dīn Abū l-Maḥāsin Muḥammad ibn Naṣr Ibn ʿUnayn al-Anṣārī al-Dimashqī, *Dīwān*, ed. Khalīl Mardam. Beirut, n.d. (c. 1946).

Ibn Wāfid, *Kitāb al-adwiyah al-mufradah* = Ibn Wāfid, *Kitāb al-adwiyah al-mufradah* (*El libro de los medicamentos simples*), ed. and tr. Luisa Fernanda Aguirre de Cárcer. Madrid: CSIC, 1995.

Ibn Wāfid, *Kitāb al-wisād fī l-ṭibb* = Ibn Wāfid, *Kitāb al-wisād fī l-ṭibb* (*Libro de la almohada sobre medicina*), ed. and tr. Camilo Álvarez de Mortales y Ruiz Matas. Toledo: Diputación provincial de Toledo, 2006.

Ibn Waḥshiyyah, *Filāḥah*, ed. Fahd = *al-Filāḥah al-Nabaṭiyyah: al-tarjamah al-manḥū-*

lah ilá Ibn Waḥshiyyah, ed. Toufic Fahd. 3 vols. Damascus: al-Maʿhad al-ʿIlmī al-Faransī li-l-Dirāsāt al-ʿArabiyyah, 1993–1998.

Ibn al-Walīd, Madelung & Mayer, *Avicenna's Allegory* = Alī Ibn Muḥammad Ibn al-Walīd, Wilferd Madelung, and Toby Mayer, *Avicenna's Allegory on the Soul: An Ismaili Interpretation. An Arabic edition and English translation of ʿAlī b. Muḥammad b. al-Walīd's al-Risāla al-mufīda*, (Ismaili Texts and Translations Ser., Institute of Ismaili Studies, v. 22). London: I.B. Tauris in assoc. with The Institute of Ismaili Studies, 2016.

Ibn Ẓāfir al-Azdī, *Badāʾiʿ* = ʿAlī Ibn Ẓāfir al-Azdī, *Badāʾiʿ al-badāʾih*, ed. Muḥammad Abū l-Faḍl Ibrāhīm. Cairo: Maktabat al-Anglo al-Miṣriyyah, 1970.

Ibn Ẓāfir al-Azdī, *Gharāʾib al-tanbīhāt* = ʿAlī ibn Ẓāfir al-Azdī, *Gharāʾib al-tanbīhāt ʿalā ʿajāʾib al-tashbīhāt*, ed. Muḥammad Zaghlūl Sallām and Muṣṭafā al-Ṣāwī al-Juwaynī. Cairo: Dār al-Maʿārif, 1971.

Ibn Zuhr, Abū l-ʿAlāʾ, *Mujarrabāt* = Abū l-ʿAlāʾ Ibn Zuhr, *Kitāb al-Muŷarrabāt* (*Libro de las experiencias médicas*), ed. and tr. C. Álvarez Millán. Madrid: CSIC, 1994.

Ibn Zuhr, Abū l-ʿAlāʾ, *La Teḏkirà* (Colin) = Abū l-ʿAlāʾ ibn Zuhr, *La Teḏkirà d'Abû 'l-ʿAlâ'*, ed. Gabriel Colin. Paris: Publ. fac. des lettres d' Alger tome 45, 1911.

Ibn Zuhr, Abū Marwān, 'El Kitāb al-iqtiṣād' = Abū Marwān ibn Zuhr, 'El Kitāb al-iqtiṣād de Avenzoar según el Ms. n° 834 de la Biblioteca del Real Monasterio de San Lorenzo del Escorial', ed. R. Kuhne Brabant. Unpublished Ph.D. diss., Madrid, 1971.

Ibn Zuhr, Abū Marwān, *al-Aghdhiyah* (al-Khaṭṭābī) = Abū Marwān ibn Zuhr, *al-Aghdhiyah wa-l-adwiyah ʿinda muʾallifī l-Gharb al-Islāmī*, ed. Muḥammad al-ʿArabī al-Khaṭṭābī. Beirut: Dār al-Gharb al-Islāmī, 1990.

Ibn Zuhr, Abū Marwān, *Kitāb al-agdiya* (García Sánchez) = Abū Marwān ibn Zuhr, *Kitāb al-agdiya* (*Tratado de los alimentos*), ed. and tr. E. García Sánchez (Fuentes Arabico-Hispanas, 4). Madrid: CSIC, Instituto de Cooperación con el Mundo Árabe, 1992.

Ibn Zuhr, Abū Marwān, *Tadhkirah* = Abū Marwān ibn Zuhr, *Tadhkirah fī al-dawāʾ al-mus'hil*, in Muḥammad al-ʿArabī al-Khaṭṭābī, *al-Ṭibb wa-l-aṭibbāʾ fī l-Andalus al-Islāmiyyah*. Beirut: Dār al-Gharb al-Islāmī, 1988, i:286–303.

Ibn Zuhr, Abū Marwān, *Kitāb al-taysīr* (Khūrī)= Abū Marwān ibn Zuhr, *Kitāb al-taysīr fī l-mudāwāh wa-l-tadbīr*, ed. Mīshīl Khūrī. Damascus: Dār al-Fikr, 1983.

Ibn Zuhr, Abū Marwān, *Kitâb al-Taysîr* (tr. Bouamrane) = Abū Marwān ibn Zuhr, *Le traité medical Kitâb al-Taysîr*, tr. Fadila Bouamrane. Paris: Librairie philosophique J. Vrin, 2010.

Ibn Zuhr, Abū Marwān, *Kitāb al-taysīr* (Rūdānī) = Abū Marwān ibn Zuhr, *Kitāb al-taysīr fī l-mudāwāh wa-l-tadbīr*, ed. Muḥammad ʿAbd Allāh al-Rūdānī. Rabat: Akādīmiyyat al-Mamlakah al-Maghribiyyah, 1991.

al-Ibshīhī, *Mustaṭraf* = Shihāb al-Dīn Muḥammad ibn Aḥmad Abū l-Fatḥ al-Ibshīhī al-Maḥallī, *al-Mustaṭraf fī kull fann mustaẓraf*. 2 vols. Cairo: Muṣṭafā al-Bābī al-Ḥalabī, 1371/1952.

al-Ibshīhī, *Mustaṭraf* [B] = Shihāb al-Dīn Muḥammad ibn Aḥmad Abū l-Fatḥ al-Ibshīhī al-Muḥallā [sic], *al-Mustaṭraf fī kull fann mustaẓraf*. 2 vols. Beirut: Dār al-Aḍwā', 1415/1995.

al-Idrīsī, *Nuzhah*, = *Kitāb Nuzhat al-mushtāq fī dhikr al-amṣār wa-l-aqṭār wa-l-buldān wa-l-juzur wa-l-madā' in wa-l-āfāq*, ed. E. Cerulli, A. Bombaci et al. 9 fasc. Naples: Istituto Universitario Orientale, 1970–1984.

Ihsanoğlu, *Catalogue* = Ekmeleddin Ihsanoğlu et al., *Catalogue of Islamic medical manuscripts (in Arabic, Turkish & Persian) in the libraries of Turkey / Fihris makhṭūṭāt al-ṭibb al-Islāmī bi-l-lughāt al-ʿArabiyyah wa-l-Turkiyyah wa-l-Fārisiyyah fī maktabāt Turkiyā*. Istanbul: Markaz al-Abḥāth li-l-Tārīkh wa-l-Funūn wa-l-Thaqāfah al-Islāmiyyah, 1984.

Ikhwān al-Ṣafā', *On arithmetic and geometry* = Ikhwān al-Ṣafā', *On arithmetic and geometry: an Arabic critical edition and English translation of Epistles 1–2*, ed. and tr. by Nader El-Bizri. Oxford: Oxford University Press, 2012.

Ikhwān al-Ṣafā', *On Logic* = Ikhwān al-Ṣafā', *On logic: an Arabic critical edition and English translation of Epistles 10–14*, ed. and tr. by Carmela Baffioni. Oxford: Oxford University Press, 2010.

Ikhwān al-Ṣafā', *Rasā'il* = Ikhwān al-Ṣafā', *Rasā'il Ikhwān al-Ṣafā'*. 4 vols. Beirut: Dār Ṣādir, 1376/1957.

Ilberg, *Galens Praxis* = Johannes Ilberg, *Aus Galens Praxis. Ein Kulturbild aus der römischen Kaiserzeit*. Extract from *Neue Jahrbücher für das Klassische Altertum, Geschichte und Deutsche Literatur*, vol. 15, 276–312. Leipzig B.G. Teubner, 1905.

Ilberg, *Rufus* = Johannes Ilberg, *Rufus von Ephesos, ein griechischer Arzt in trajanischer Zeit* (Abhandlungen der sächsichen Akademie der Wissenschaften, 41. No. 1). Leipzig: Hirzel, 1930.

Inati, *Ibn Sīnā's Theodicy* = Shams Constantine Inati, *The Problem of Evil: Ibn Sina's Theodicy*. 2nd. ed. Piscataway, NJ: Gorgias Press, 2017.

Inati, *Physics and Metaphysics* = Shams Constantine Inati, *Ibn Sina's Remarks and Admonitions: Physics and Metaphysics: An Analysis and Annotated Translation*. New York, 2014.

al-Irbilī, *Tadhkirah* = al-Ṣāḥib Bahā' al-Dīn al-Munshi' al-Irbilī, *al-Tadhkirah al-Fakhriyyah*, ed. Ḥātim Ṣāliḥ al-Ḍāmin. Damascus: Dār al-Bashā'ir, 2004.

Irvine & Temkin, 'Akīlāos' = Judith T. Irvine and Oswei Temkin, 'Who was Akīlāos? A Problem in Medical Historiography', *Bulletin of the History of Medicine*, 77 (2003), 12–24.

Irwin, *Camel* = Robert Irwin, *Camel*. London: Reaktion Books, 2010.

ʿĪsā, Aḥmad *see* Aḥmad ʿĪsā

Isḥāq ibn ʿImrān, *Maqālah fī l-mālīkhūliyā* = Isḥāq ibn ʿImrān, *Maqāla fī-'l-mālīḫūliyā: vergleichende kritische arabisch-lateinische Parallelausgabe, deutsche Übersetzung des arabischen Textes, ausführliche Einleitung und arabischer wie lateinischer dro-*

genkundlicher Apparat. Constantini Africani libri duo di melancholia, ed. and tr. Karl Garbers. Hamburg: Buske, 1977.

al-Iṣfahānī, *al-Aghānī* = Abū l-Faraj al-Iṣfahānī, *al-Aghānī*. 24 vols. Cairo: Dār al-Kutub / al-Hayʾah al-Miṣriyyah al-ʿĀmmah, 1927–1974.

al-Iṣfahānī, *Kharīdah* (*al-ʿIrāq*) = ʿImād al-Dīn al-Iṣfahānī al-Kātib, *Kharīdat al-qaṣr wa-jarīdat al-ʿaṣr, al-Qism al-ʿIrāqī*, ed. Muḥammad Bahjah al-Atharī and Jamīl Saʿīd. 6 vols. Baghdād: al-Majmaʿ al-ʿIlmī al-ʿIrāqī, 1955–1981.

al-Iṣfahanī, *Kharīdah* (*al-Maghrib*) = ʿImād al-Dīn al-Kātib al-Iṣfahānī, *Kharīdat al-qaṣr wa-jarīdat al-ʿaṣr, qism shuʿarāʾ al-Maghrib*, ed. Muḥammad al-Marzūqī, Muḥammad al-ʿArūsī al-Maṭwī and al-Jīlānī ibn al-Ḥājj Yaḥyā. Tunis: al-Dār al-Tūnisiyyah, 1973.

al-Iṣfahānī, *Kharīdah* (*Miṣr*) = ʿImād al-Dīn al-Iṣfahānī al-Kātib, *Kharīdat al-qaṣr wa-jarīdat al-ʿaṣr, Qism shuʿarāʾ Miṣr*, ed. Aḥmad Amīn and Shawqī Ḍayf. 2 vols. Cairo: Dār al-Kutub, 1951.

al-Iṣfahānī, *Kharīdah* (*al-Shām*) = ʿImād al-Dīn al-Iṣfahānī al-Kātib, *Kharīdat al-qaṣr wa-jarīdat al-ʿaṣr, Qism shuʿarāʾ al-Shām*, ed. Shukrī Fayṣal. 4 vols. Damascus: Majmaʿ al-Lughah al-ʿArabiyyah bi-Dimashq, 1955–1968.

Iskandar, ʻAn autograph of Ibn al-Tilmīdhʼ = Albert Z. Iskandar, ʻAn autograph of Ibn al-Tilmīdh's marginal commentary on Ibn Sīnā's "Canon of Medicine"', *Le Muséon*, 90 (1977), 177–123.

Iskandar, 'Bibliographical Studies' = Albert Z. Iskandar, 'Bibliographical studies in medical and scientific Arabic works. Galen's fī ʾAmal al-tashrīh (On Anatomical procedures), the Alexandrian book entitled fīʾl-Tashrīh ilā ʾl-mutaʾallimīn (On Anatomy for students), and *Rhazes*' al-Kāfī fī ʾl-tibb (The Sufficient Book on Medicine)', *Oriens*, 25–26 (1976), 133–147.

Iskandar, *Descriptive List* = Albert Z. Iskandar, *A Descriptive List of Arabic Manuscripts on Medicine and Science at the University of California, Los Angeles*. Leiden: Brill, 1984.

Iskandar, *Examinations* = Albert Z. Iskandar, *Galeni De optimo medico cognoscendo libellus* (*Galen's Book on Examinations by Which the Best Physicians Are Recognized*) (Corpus Medicorum Graecorum, Suppl. Or. IV). Berlin: Akademie-Verlag, 1988.

Iskandar, 'Medical bibliography' = Albert. Z. Iskandar, 'The Medical bibliography of al-Rāzī', in George F. Hourani (ed.), *Essays on Islamic philosophy and science*. Albany, NY: State University of New York Press, 1975, 41–46.

Iskandar, 'Medical Curriculum' = Albert Z. Iskandar, 'An Attempted Reconstruction of the Late Alexandrian Medical Curriculum', *Medical History*, 20 (1976), 235–258.

Iskandar, 'Ar-Rāzī's medical writings' = Albert Z. Iskandar, 'A study of Ar-Rāzī's medical writings with selected texts and English translations'. 2 vols. D.Phil. diss., University of Oxford, 1959.

Iskandar, 'A Study' = Albert Z. Iskandar, 'A Study of al-Samarqandi's Medical Writings', *Le Muséon*, 85 (1972), 451–479.

Iskandar, *Taḥqīq* = Albert Z. Iskandar, 'Taḥqīq fī sinī al-Rāzī 'inda bad' ishtighālihi bi-l-ṭibb', *Al-Mashriq* (1960), 168–177.

Iskandar, *Wellcome* = Albert Z. Iskandar, *A Catalogue of Arabic Manuscripts on Medicine and Sciences in the Wellcome Historical Medical Library*. London: Wellcome Institute for the History of Medicine, 1967.

Istanbouli, *history* = M.N. Istanbouli, 'The history of Arabic medicine based on the work of Ibn Abi Usaybe'ah, 1203–270'. Thesis, Loughborough University of Technology, 1981 (unpublished).

Iṣṭakhrī, *al-Masālik wa-l-mamālik* = Abū Isḥāq al-Fārisī al-Iṣṭakhrī, *al-Masālik wa-l-mamālik / Viae Regnorum. Descriptio ditionis moslemicae*, ed. M.J. de Goeje. Leiden: Brill, 1873, repr. 1927.

Ivry, *al-Kindī's Metaphysics* = Alfred L. Ivry, *al-Kindī's Metaphysics; a translation of Ya'qub ibn Isḥāq al-Kindī's treatise "On first philosophy"* (*fī al-Falsafah al-ūlā*). Albany: State University of New York Press, 1974.

Jacquart, 'Note' = Danielle Jacquart, 'Note sur la traduction latine du *Kitāb al-Manṣūrī* de *Rhazès*', *Revue d'Histoire des Textes*, 24 (1994), 359–374.

Jacquart, 'Ibn Māsawayh' = Danielle Jacquart, 'Ibn Māsawayh, Yūḥannā', in *Encyclopaedia of the History of Science, Technology, and Medicine in Non-Western Cultures*, ed. Helaine Selin. Dordrecht: Kluwer, 1997, 424–425.

Jadon, 'Comparison of wealth … of the physicians of Ṣalāḥ Al-Dīn' = Samira Jadon, 'A comparison of the wealth, prestige, and medical works of the physicians of Ṣalāḥ Al-Dīn in Egypt and Syria', *Bulletin of the History of Medicine*, 43 (1970), 64–75.

Jadon, 'Ibn Butlān's Medical Manual for the Use of Monks' = Samira Yousef Jadon, 'The Arab Physician Ibn Butlān's (d. 1066) Medical Manual for the Use of Monks and Country People' [in Arabic]. 2 vols. unpublished Ph.D. diss., University of California, Los Angeles, 1968.

Jadon, 'Physicians of Syria during reign of Ṣalāḥ Al-Dīn' = Samira Jadon, 'The physicians of Syria during the reign of Ṣalāḥ Al-Dīn 570–589 A.H. / 1174–1193 A.D.', *Journal of the History of Medicine and Allied Sciences*, 25 (1970), 323–340,

Jahier & Noureddine, *Ibn abi Uçaibi'a* = Henri Jahier and Abdelkader Noureddine, *Ibn abi Uçaibi'a, 'Uyûn al-Anbâ' fî T'abaqât al-At'ibbâ'. Sources d'informations sur les classes des médicins.* XIII*e* *Chapitre: Médicins de l'Occident Musulman* (Publications de la faculté mixte de médecine et de pharmacie d'Alger, 4). Algiers: Farārīs, 1958.

al-Jāḥiẓ, *Bayān* = al-Jāḥiẓ, *al-Bayān wa-l-tabyīn*, ed. 'Abd al-Salām Muḥammad Hārūn. 4 vols. Cairo: Maktabat al-Khānjī, 1968.

al-Jāḥiẓ, *Bukhalā'* = al-Jāḥiẓ, *al-Bukhalā'*, ed. Ṭāhā al-Ḥājirī. Cairo: Dār al-Ma'ārif, n.d.

al-Jāḥiẓ, *Ḥayawān* = al-Jāḥiẓ, *al-Ḥayawān*, ed. 'Abd al-Salām Muḥammad Hārūn. 8 vols. Cairo: Muṣṭafā al-Bābī al-Ḥalabī, 1965–1969.

ps.-Jāḥiẓ, *Maḥāsin* = *al-Maḥāsin wa-l-aḍdād al-mansūb ilā Abī 'Uthmān 'Amr ibn Baḥr al-Jāḥiẓ / Le livre des beautés et des antithèses attribué à Abu Othman Amr ibn Bahr al-Djahiz de Basra*, ed. G. van Vloten. Leiden: Brill, 1898.

Janos, 'Abū Bishr Mattā ibn Yūnus's Cosmology' = Damien Janos, '"Active Nature" and Other Striking Features of Abū Bishr Mattā ibn Yūnus's Cosmology as Reconstructed from His Commentary on Aristotle's *Physics*', in Damien Janos (ed.), *Ideas in motion in Baghdad and beyond: philosophical and theological exchanges between Christians and Muslims in the third/ninth and fourth/tenth centuries*. Leiden: Brill, 2016, 135–177.

Janos, *Circles of Learning* = Damien Janos, *The Place to Go: Circles of Learning in Baghdad in the Ninth and Tenth Centuries C.E.* (Studies in Late Antiquity and Early Islam). Princeton: Darwin Press, 2014.

Janos, *Method* = Damien Janos, *Method, Structure, and Development in al-Fārābī's Cosmology* (Islamic Philosophy, Theology and Science. Texts and Studies, 85). Leiden: Brill, 2012.

Janssens, *Annotated Bibliography* = Jules Janssens, *An Anotated Bibliography on Ibn Sînâ (1970–1989), including Arabic and Persian Publications and Turkish and Russian References*. Leuven: Leuven University Press, 1991.

Janssens, *Annotated Bibliography: First Supplement* = Jules L. Janssens, *An Annotated Bibliography on Ibn Sīnā: First Supplement (1990–1994)*. Louvain-la-Neuve [Belgium]: Fédération Internationale des Instituts d'Études Médiévales, 1999.

Janssens, *Annotated Bibliography: Second Supplement* = Jules L. Janssens, *An Annotated Bibliography on Ibn Sīnā: Second Supplement (1995–2009)*. Louvain-la-Neuve [Belgium]: Fédération Internationale des Instituts d'Études Médiévales, 2017.

Janssens, 'Bahmanyār' = Jules Janssens, 'Bahmanyār b. Marzubān: A Faithful Disciple of Ibn Sīnā?', in David C. Reisman, with the assistance of Ahmed H. al-Rahim (eds.), *Before and After Avicenna: Proceedings of the First Conference of the Avicenna Study Group* (Islamic Philosophy, Theology, and Science, 52). Leiden: Brill, 2003, 177–197.

Janssens, 'Le Dānesh-Nāmeh d'Ibn Sīnā' = Jules Janssens, 'Le Dānesh-Nāmeh d'Ibn Sīnā: Un Texte À Revoir?' *Bulletin de Philosophie Médiévale*, 28 (1986), 163–177.

Janssens, 'Revision' = Jules Janssens, 'Bahmanyār, and his Revision of Ibn Sīnā's Metaphysical Project', *Medioevo*, 32 (2007), 99–117.

Jaouiche, *Le Livre du Qarasṭūn* = Khalil Jaouiche, *Le Livre du Qarasṭūn de Ṯābit ibn Qurra: étude sur l'origine de la notion de travail et du calcul du moment statique d'une barre homogène*. Leiden: Brill, 1976.

al-Jarīrī, *Jalīs* = Abū l-Faraj al-Muʿāfā ibn Zakariyyā al-Nahrawānī al-Jarīrī, *al-Jalīs al-ṣāliḥ al-kāfī wa-l-anīs al-nāṣiḥ al-shāfī*, ed. Muḥammad Mursī al-Khawlī and Iḥsān ʿAbbās. 4 vols. Beirut: ʿĀlam al-Kutub, 1413/1993.

Jastrow, *Dictionary of the Targumim* = Marcus Jastrow, *A Dictionary of the Targumim, the Talmud Babli and Yerushalmi, and the Midrashic Literature*. 2 vols. New York: Pardes, 1950.

al-Jawālīqī, *al-Muʿarrab* = Abū Manṣūr al-Jawālīqī Mawhūb ibn Aḥmad ibn Muḥammad ibn al-Khaḍir, *al-Muʿarrab min al-kalām al-aʿjamī ʿalā ḥurūf al-muʿjam*, ed. F. ʿAbd al-Raḥīm. Damascus: Dār al-Qalam, 1410/1990.

al-Jawharī, *al-Ṣiḥāḥ* = Abū Naṣr Ismāʿīl ibn Ḥammād al-Jawharī, *Tāj al-lughah wa-ṣiḥāḥ al-ʿArabiyyah*, ed. Aḥmad ʿAbd al-Ghafūr ʿAṭṭār. 6 vols. Beirut: Dār al-ʿIlm li-l-Malāyīn, 1990.

al-Jilyānī al-Andalusī, *al-Mubashshirāt* = ʿAbd al-Munʿim ibn ʿUmar ibn Ḥassān al-Jilyānī al-Andalusī, *K. al-Mubashshirāt wa-l-Qudsiyyāt*, ed. ʿAbd al-Jalīl Ḥasan ʿAbd al-Mahdī. Amman: Dār al-Bashīr, 1989.

al-Jilyānī al-Andalusī, *Tadbīj* = ʿAbd al-Munʿim ibn ʿUmar ibn Ḥassān al-Jilyānī al-Andalusī, *Dīwān al-tadbīj: Fitnat al-ibdāʿ wa-dhurwat al-imtāʿ*, ed. Kamāl Abū Dīb and Dalāl Bakhsh. [London] / Oxford: Dār al-Sāqī / Oryx, 2010.

Johnstone, 'Arabic Botany and Pharmacology' = Penelope Johnstone, 'Arabic Botany and Pharmacology with particular reference to the work of al-Zahrāwī and Ibn Juljul'. D. Phil. diss., University of Oxford 1972.

Jolin, 'Epilepsy in Medieval Islamic History'= Paula Jolin, 'Epilepsy in Medieval Islamic History. A thesis submitted to the Faculty of Graduate Studies and Research in partial fulfilment of the requirements of the degree of Master's of Arts', Montreal, McGill University, 1999.

Jonkers, *Textual Tradition* = Gijsbert Jonkers, *The Textual Tradition of Plato's* Timaeus *and* Critias. Leiden: Brill, 2017.

Joosse, ''Abd al-Laṭīf al-Baghdādī as a Philosopher and a Physician' = N. Peter Joosse, ''Abd al-Laṭīf al-Baghdādī as a Philosopher and a Physician: Myth or Reality, Topos or Truth?', in Peter Adamson (ed.): *In the Age of Averroes: Arabic Philosophy in the Sixth/Twelfth Century* (Warburg Institute Colloquia, 16). London / Turin: The Warburg Institute/Nino Aragno Editore, 2011, 27–43.

Joosse, 'Alchemy and Alchemists' = N. Peter Joosse, 'Unmasking the Craft'. 'Abd al-Latif al-Bagdadi's Views on Alchemy and Alchemists', in Anna Ayşe Akasoy and Wim Raven (eds.), *Islamic Thought in the Middle Ages. Studies in Text, Transmission and Translation, in Honour of Hans Daiber* (Islamic Philosophy, Theology, and Science, 75). Brill: Leiden, 2008, 301–317.

Joosse, 'Arabic Diatessaron' = N. Peter Joosse, 'An Introduction to the Arabic Diatessaron', *Oriens Christianus*, 83 (1999), 72–129.

Joosse, 'Aristotle's Politica' = N. Peter Joosse, 'Between Enigma and Paradigm: The Reception of Aristotle's *Politica* in the Near East: The Arabic and Syriac-Aramaic Versions', in Vasileios Syros (ed.), *Well Begun is Only Half Done. Tracing Aristotle's Political Ideas in Medieval Arabic, Syriac, Byzantine, and Jewish Sources*. Tempe, Arizona: ACMRS, 2011, 97–120.

Joosse, 'De geest is uit de fles' = N. Peter Joosse, ''De geest is uit de fles'. De middeleeuwse arts ʿAbd al-Laṭīf ibn Yūsuf al-Baghdādī: zijn medische werk en zijn bizarre affiliatie met het twintigste-eeuwse spiritisme', *Gewina* (*Tijdschrift voor de Geschiedenis der Geneeskunde, Natuurwetenschappen, Wiskunde en Techniek*), 30 (2007), 211–229.

Joosse, 'Pride and Prejudice' = N. Peter Joosse, '"Pride and Prejudice, Praise and Blame": 'Abd al-Laṭīf al-Baghdādī's Views on Good and Bad Medical Practitioners', in Arnoud Vrolijk and Jan P. Hogendijk (eds.): *O ye Gentlemen: Arabic Studies on Science and Literary Culture in Honour of Remke Kruk* (Islamic Philosophy, Theology and Science, Texts and Studies, 74). Leiden: Brill, 2007, 129–141.

Joosse, *Rebellious Intellectual* = N. Peter Joosse, *The Physician as a Rebellious Intellectual. The Book of the Two Pieces of Advice or* Kitāb al-Naṣīḥatayn *by 'Abd al-Laṭīf ibn Yūsuf al-Baghdādī (1162–1231). Introduction, Edition, and Translation of the Medical Section* (*Beihefte zur Mediaevistik*, 18). Frankfurt am Main / Bern: Peter Lang, 2014.

Joosse, 'Taqdimat' = N. Peter Joosse, 'A Newly-Discovered Commentary on the Hippocratic *Prognostic* by Barhebraeus: Its Contents and Its Place within the Arabic *Taqdimat al-ma'rifa* Tradition', *Oriens*, 41/2–4 (2013), 499–523.

Joosse & Pormann, 'Archery' = N. Peter Joosse and Peter E. Pormann, 'Archery, mathematics, and conceptualising inaccuracies in medicine in 13th century Iraq and Syria', *Journal of the Royal Society of Medicine*, 101 (2008), 425–427.

Joosse & Pormann, 'Decline and Decadence' = N. Peter Joosse and Peter E. Pormann, 'Decline and Decadence in Iraq and Syria after the age of Avicenna? 'Abd al-Laṭīf al-Baghdādī (1162–1231) between Myth and History', *Bulletin of the History of Medicine*, 84 (2010), 1–29.

Joosse & Pormann, 'Prognostic' = N. Peter Joosse and Peter E. Pormann, "Abd al-Laṭīf al-Baghdādī's *Commentary on Hippocrates' 'Prognostic'*: A Preliminary Exploration', in Peter E. Pormann (ed.): *Epidemics in Context. Greek Commentaries on Hippocrates in the Arabic Tradition* (Scientia Graeco-Arabica Band 8). Berlin: De Gruyter, 2012, 251–283.

Jouanna, *Hippocrates* = Jacques Jouanna, *Hippocrates*. Translated by M.B. DeBevoise. Baltimore / London: The Johns Hopkins University Press, 1999.

Jubūrī, *Ibn Shibl al-Baghdādī; ḥayātuhu wa-shi'ruh* = Suhā Yūnus Salmān Jubūrī, *Ibn Shibl al-Baghdādī; ḥayātuhu wa-shi'ruh*. Ammān: Dār Ghaydā', 2011.

Jum'ah, *K. al-muqaddimāt fī l-falsafah* = Ibn Rushd, *Kitāb al-muqaddimāt fī l-falsafah, aw al-Masā'il fī l-manṭiq wa-l-'ilm al-ṭabī'ī wa-l-ṭibb*, ed. As'ad Jum'ah. Tunis: Kulliyyat al-Ādāb wa-l-'Ulūm al-Insāniyyah bi-l-Qayrawān, 2008.

al-Jurāwī, *al-Ḥamāsah al-maghribiyyah* = Abū l-'Abbās Aḥmad ibn 'Abd al-Salām al-Jurāwī al-Tādilī, *al-Ḥamāsah al-maghribiyyah, mukhtaṣar kitāb Ṣafwat al-adab wa-nukhbat dīwān al-'Arab*, ed. Muḥammad Riḍwān al-Dāyah. 2 vols. Beirut: Dār al-Fikr al-Mu'āṣir / Damascus: Dār al-Fikr, 1411/1991.

al-Jurjānī, *Kināyāt* = al-Qāḍī Abū l-'Abbās Aḥmad ibn Muḥammad al-Jurjānī, *Kināyāt al-udabā' wa-ishārāt al-bulaghā'*, ed. Maḥmūd Shākir al-Qaṭṭān. Cairo: al-Hay'ah al-Miṣriyyah al-'Āmmah li-l-Kitāb, 2003.

Kaʿb ibn Zuhayr, *Qaṣīdat al-Burdah* = Kaʿb ibn Zuhayr, *Qaṣīdat al-Burdah*, in Kaʿb ibn Zuhayr, *Dīwān, ṣanʿat Abī Saʿīd al-Sukkarī*, ed. Mufīd Qumayḥah. Riyadh: Dār al-Shawwāf, 1989, 109–116.

Kaḥḥālah, *Muʿjam al-muʾallifīn* = ʿUmar Riḍā Kaḥḥālah, *Muʿjam al-muʾallifīn: tarājim muṣannifī al-kutub al-ʿarabiyyah*. Damascus: Maṭbaʿat al-Taraqqī, 1957.

Kahil, 'Une ophtalmologie arabe' = Najib Kahil, 'Une ophtalmologie arabe par un praticien du Caire du XIIIe siècle (VIIe de l-Hégire)', in *Comptes Rendus du Congrès International de Médecine Tropicale et d'Hygiène* (*Le Caire, 1928*) Cairo: Imprimerie Nationale, 1929i:241–260.

Kahl, *The Dispensatory* = Oliver Kahl, *The Dispensatory of Ibn al-Tilmīḏ: Arabic text, English translation, study and glossaries*. Leiden: Brill, 2007.

Kahl, 'The Pharmacological Tables of Rhazes' = Oliver Kahl, 'The Pharmacological Tables of Rhazes'. *Journal of Semitic Studies*, 56/2 (2011), 367–399.

Kahl, *Sanskrit, Syriac and Persian Sources* = Oliver Kahl, *The Sanskrit, Syriac and Persian Sources in the Comprehensive Book of Rhazes*. Leiden: Brill, 2015.

Kahl, *Treatise on the Errors of the Physicians* = Oliver Kahl (ed., tr.), *Yaʿqūb ibn Isḥāq al-Isrāʾīlī's 'Treatise on the Errors of the Physicians in Damascus': A critical edition of the Arabic text together with an annotated English translation* (*Journal of Semitic Studies*, Suppl. 10). Oxford: Oxford University Press, 2000.

Kahle, *Cairo Geniza* = Paul E. Kahle, *The Cairo Geniza*. 2nd ed. Oxford: Basil Blackwell, 1959.

Kahle, 'Ibn Samağūn und sein Drogenbuch' = Paul E. Kahle, 'Ibn Samağūn und sein Drogenbuch. Ein Kapitel aus den Anfänger der arabischen Medizin', in J.W. Fück (ed.) *Documenta Islamica Inedita*. Berlin: Akademie-Verlag, 1952, 25–44.

Kalbarczyk, *Predication and Ontology* = Alexander Kalbarczyk, *Predication and Ontology. Studies and Texts on Avicennian and Post-Avicennian Readings of Aristotle's "Categories"*. Berlin: De Gruyter, 2018.

Karbstein, *Namen der Heilmittel* = Andreas Karbstein, *Die Namen der Heilmittel nach Buchstaben. Edition eines arabisch-romanischen Glossars aus dem frühen 17. Jahrhundert* (Kölner Romanistische Arbeiten, Neue Folge, 81). Geneva: Librairie Droz, 2002.

Karimullah, 'Prolegomena' = Kamran I. Karimullah, 'Assessing Avicenna's (d. 428/1037) Medical Influence in Prolegomena to Post-Classical (1100–1900 CE) Medical Commentaries', *MIDÉO*, 32 (2017), 93–134.

Karimullah, 'Transformation of Galen's Textual Legacy' = Kamran I. Karimullah, 'Transformation of Galen's Textual Legacy from Classical to Post-Classical Islamic Medicine: Commentaries on the Hippocratic Aphorisms', *Intellectual History of the Islamicate World* 5(3) (2017), 311–358.

Karmi, 'The Arabic Medical *kunnāsh* in the 10th Century' = Ghada Karmi, 'The Arabic Medical *kunnāsh* in the 10th Century: Its Status, Significance and Tradition. A study based mainly on the book *Ghiná wa Muná* by Abū Manṣūr al-Qamarī (al-Qumrī)'. Unpublished Ph.D. thesis, University College, London, 1978.

Karmi, 'Compendium' = Ghada Karmi, 'A Mediaeval Compendium of Arabic Medicine: Abu Sahl al-Masihi's "Book of the Hundred"', *Journal for the History of Arabic Science*, 2/2 (1978), 270–290.

Karmi, 'State Control' = Ghada Karmi, 'State Control of the Physicians in the Middle Ages: An Islamic Model', in A.W. Russell (ed.), *The Town and State Physicians in Europe from the Middle Ages to the Enlightenment*. Wolfenbüttel: Herzog August Bibliothek, 1981, 63–84.

Karmi, 'Study based on *Ghiná wa Muná*' = Ghada Karmi, 'Arabic Medicine in the Tenth Century: A study based on the book *Ghiná wa Muná* by Abū Manṣūr al-Ḥasan b. Nūḥ al-Qumrī', in *Proceedings of the First International Symposium for the History of Arabic Science, April 5–12, 1975, Vol. II: Papers in European Languages*. Aleppo: Institute for the History of Arabic Science, 1978, 326–335.

Karpova & Rosenfeld, 'The treatise of Thābit ibn Qurra on sections of a cylinder, and on its surface' = L. Karpova and B.A. Rosenfeld, 'The treatise of Thābit *ibn* Qurra on sections of a cylinder, and on its surface', *Archives Internationales d'Histoire des Sciences*, 24 (1974), 66–72.

Käs, 'Galenbibliographie' = Fabian Käs, 'Eine neue Handschrift von Ḥunayn ibn Isḥāqs Galenbibliographie', *Zeitschrift für Geschichte der Arabisch-Islamischen Wissenschaften*, 19 (2010–2011), 135–193.

Kazimirski, *Dictionnaire* = Albert de Biberstein-Kazimirski, *Dictionnaire arabe-français contenant toutes les racines de la langue arabe: leurs dérivés, tant dans l'idiome vulgaire que dans l'idiome littéral, ainsi que les dialectes d'Alger et de Maroc*. 2 vols. Paris: Maisonneuve, 1860.

Kelly, *Creeds* = J.N.D. Kelly, *Early Christian Doctrines*, 5th ed. London: Continuum, 1977.

Kennedy, *The Prophet and the Age of the Caliphates* = Hugh Kennedy, *The Prophet and the Age of the Caliphates: The Islamic Near East from the Sixth to the Eleventh Century*. 3rd ed. Abingdon / New York: Routledge, 2016.

Kennedy, *Muslim Spain and Portugal* = Hugh Kennedy, *Muslim Spain and Portugal. A Political History of al-Andalus*. London: Longman, 1996.

Kennedy-Day, *Books of Definition in Islamic Philosophy* = Kiki Kennedy-Day, *Books of Definition in Islamic Philosophy: The Limits of Words*. London: Routledge Curzon, 2003.

Kennedy & Kennedy, *Coordinates of Localities from Islamic Sources* = E.S. Kennedy & M.H. Kennedy, *Geographical Coordinates of Localities from Islamic Sources* (Veröffentlichungen des Institutes für Geschichte der Arabisch-Islamischen Wissenschaften. Reihe A: Texte und Studien, 2). Frankfurt am Main: Institut für Geschichte der Arabisch-Islamischen Wissenschaften, 1987.

Kerner, 'The *Kitāb al-diryāq* in text and image' = Jaclynne J. Kerner, 'Art in the name of science: the *Kitāb al-diryāq* in text and image', in Anna Contadini (ed.), *Arab Painting: Text and Image in Illustrated Arabic Manuscripts*. Leiden: Brill, 2007, 25–39.

Kessel, 'Syriac medicine' = Gregory Kessel, 'Syriac Medicine', in Daniel King (ed.), *The Syriac World*. London / New York: Routledge, 2019, 439–459.

Key, 'Al-Rāghib al-Iṣfahānī' = Alexander Key, 'Al-Rāghib al-Iṣfahānī', in Terri DeYoung and Mary St. Germain (eds.), *Essays in Arabic Literary Biography, 925–1350*. Wiesbaden: Harrassowitz, 2011, pp. 298–306.

al-Khafājī, *Sirr al-faṣāḥah* = Abū Muḥammad ʿAbd Allāh ibn Muḥammad ibn Saʿīd ibn Sinān al-Khafājī, *Sirr al-faṣāḥah*, ed. ʿAlī Fūdah. Cairo: Maktabat al-Khānjī, 1350/1932.

Khalidi, *Arabic Historical Thought* = Tarif Khalidi, *Arabic Historical Thought in the Classical Period*. Cambridge: Cambridge University Press, 1994.

al-Khālidiyyān, *al-Tuḥaf wa-l-hadāyā* = Abū Bakr Muḥammad ibn Hāshim al-Khālidī and Abū ʿUthmān Saʿīd ibn Hāshim al-Khālidī, *Kitāb al-Tuḥaf wa-l-hadāyā*, ed. Sāmī al-Dahhān. Cairo: Dār al-Maʿārif, 1952.

Khalīfāt, *Ibn Hindū* = Saḥbān Khalīfāt, *Ibn Hindū: Sīratuhu, ārāʾuhu al-falsafiyyah, muʾallafātuhu. Dirāsah wa-nuṣūṣ*. Amman: al-Jāmiʿah al-Urdunniyyah, 1996.

Khalīfah, *Daʿwat al-qusūs* = Ibn Buṭlān, *'Kitāb Daʿwat al-qusūs'*, ed. Ighnāṭiyūs ʿAbduh Khalīfah, *Al-Mashriq*, 53 (1959), 3–36.

Khan, *Bibliography of al-Bīrūnī* = Ahmad Saeed Khan, *A bibliography of the works of Abū'l-Raiḥān al-Bīrūnī*. New Delhi: Indian National Science Academy, 1982.

Khan, 'Some Aspects' = Abdul Muid Khan, 'Some Aspects of the Arabic Writings of the Philosopher Ibn Sina', *Islamic Culture*, 25/1 (1951), 27–42.

al-Khaṭīb al-Baghdādī, *Tārīkh Baghdād* = Abū Bakr Aḥmad ibn ʿAlī ibn Thābit al-Khaṭīb al-Baghdādī, *Tārīkh Madīnat al-Salām wa-akhbār muḥaddithīhā* (…), ed. Bashshār ʿAwwād Maʿrūf. 17 vols. Beirut: Dār al-Gharb al-Islāmī, 1422/2001.

al-Khaṭīb al-Qazwīnī, *al-Īḍāḥ* = al-Khaṭīb al-Qazwīnī, *al-Īḍāḥ fī ʿulūm al-balāghah*, ed. Muḥammad ʿAbd al-Munʿim Khafājī. Beirut: Dār al-Kitāb al-Lubnānī, 1391/1971.

Kholeif, *Controversies* = Fatḥ Allāh Khulay (F. Kholeif), *Fakhr al-Dīn al-Rāzī fī bilād Mā warāʾ al-nahr* (*A Study on Fakhr al-Dīn Rāzī and his Controversies in Transoxania*). Beirut: Institut de Lettres Orientales de Beyrouth, 1966. [in Arabic].

al-Khwārazmī, *Mafātīḥ al-ʿulūm* = Abū ʿAbd Allāh Muḥammad ibn Aḥmad ibn Yūsuf al-Khwārazmī al-Kātib, *Mafātīḥ al-ʿulūm / Liber Mafâtîḥ al-olûm*, ed. G. van Vloten. Leiden: Brill, 1895.

al-Kindī, *Risālah fī jawāhir al-suyūf* = Robert G. Hoyland and Brian Gilmour, *Medieval Islamic swords and swordmaking: Kindi's treatise On swords and their kinds*; edition, translation, and commentary. [Cambridge]: Gibb Memorial Trust, c. 2006.

King & Samsó, 'Astronomical Handbooks' = David A. King and Julio Samsó (with a contribution by Bernard R. Goldstein), 'Astronomical Handbooks and Tables from the Islamic World (750–1900): an Interim Report', *Suhayl* 2 (2001), 9–105.

King, 'The New *materia medica* of the Islamicate Tradition' = Anya King, "The New *materia medica* of the Islamicate Tradition: The Pre-Islamic Context," *Journal of the American Oriental Society*, 135/3 (2015), 499–528.

Kingsley, 'Greek Origin' = Peter Kingsley, 'The Greek Origin of the Sixth-Century Dating of Zoroaster', *BSOAS*, 53/2 (1990), 245–265.

Klamroth, 'Auszüge' = Martin Klamroth, 'Ueber die Auszüge aus griechischen Schriftstellern bei al-Jaʿqūbī', *Zeitschrift der Deutschen Morgenländische Gesellschaft*, 40 (1886), 189–233.

Klein-Franke, 'Fatal Disease' = Felix Klein-Franke, 'What was the Fatal Disease of al-Malik al-Ṣāliḥ Najm al-Dīn Ayyūb?', in Moshe Sharon (ed.), *Studies in Islamic History and Civilization in Honour of Professor David Ayalon*. Jerusalem: Cana / Leiden: Brill, 1986, 153–157.

Klein-Franke, *Über die Heilung* = Felix Klein-Franke, *Über die Heilung der Krankheiten der Seele und des Körpers*. Beirut: Dar el-Machreq Editeurs, 1977.

Kohlberg & Kedar, *A Melkite Physician* = Etan Kohlberg and Benjamin Z. Kedar, 'A Melkite Physician in Frankish Jerusalem and Ayyubid Damascus: Muwaffaq al-Dīn Yaʿqūb b. Siqlāb', *Asian and African Studies*, 22 (1988), 113–126.

Köhler, *Wissenschaft unter den ägyptischen Fatimiden* = Bärbel Köhler, *Die Wissenschaft unter den ägyptischen Fatimiden*. Hildesheim: Georg Olms, 1994.

Kopf, *Ibn Abu Usaibi'ah* = Lothar Kopf, 'Ibn Abu [sic] Usaibi'ah: History of Physicians', Bethesda, MD, National Library of Medicine, MS C 294, box 1. [unpublished carbon-copy of a typed manuscript of a translation prepared before 1969, commissioned by NLM but never published. It covers all 15 chapters, but omits most of the poetry and contains many errors. It has recently been scanned and placed on the web without permission of author or library at http://www.tertullian.org/fathers/ibn_abi_usaibia_01.htm].

Kraemer, *Humanism* = Joel L. Kraemer, *Humanism in the Renaissance of Islam: The Cultural Revival During the Buyid Age*. 2nd rev. ed. Leiden: Brill, 1992.

Kraemer, 'Maimonides: An Intellectual Portrait' = Joel Kraemer, 'Moses Maimonides: An Intellectual Portrait', in Kenneth Seeskin (ed.), *The Cambridge Companion to Maimonides*. Cambridge, Cambridge University Press, 2005, 10–57.

Kraemer, *Philosophy* = Joel L. Kraemer, *Philosophy in the Renaissance of Islam: Abū Sulaymān al-Sijistānī and his Circle*. Leiden: Brill, 1986

Kraus, '*Akhlāq*' = Paul Kraus, 'The Book of Ethics by Galen', *Bull. Fac. Arts Univ. Egypt*, 5/1 (Arabic Section) (1937), 1–51.

Kraus & Walzer, *Galeni compendium Timaei* = Paul Kraus and Richard Walzer, *Galeni compendium Timaei Platonis aliorumque dialogorum synopsis quae extant fragmenta*. London: The Warburg Institute, 1951.

Kraye et al., *Pseudo-Aristotle in the Middle Ages* = J. Kraye, W.F. Ryan, C.B. Schmitt, C.B. (eds.), *Pseudo-Aristotle in the Middle Ages: The Theology and Other Texts*. London: The Warburg Institute, 1986.

Krishnamurthy, *Bhela-Saṃhitā* = Krishnamurthy, K.H. *Bhela-Saṃhitā: Text with English Translation, Commentary, and Critical Notes* (Haridas Ayurveda Series, 8). Varanasi, India: Chaukhambha Visvabharati, 2000.

Kruk, 'Chimaera' = Remke Kruk, "Abd al-Laṭīf al-Baghdādī's *Kitāb al-Ḥayawān*: A Chimaera?' in Anna A. Akasoy and Wim Raven (eds.), *Islamic Thought in the Middle Ages. Studies in Text, Transmission and Translation in Honour of Hans Daiber* (Islamic Philosophy, Theology, and Science, 75). Leiden: Brill, 2008, 345–362.

Kruk, 'Elusive Giraffes' = Remke Kruk, 'Elusive Giraffes: Ibn Abī l-Ḥawāfir's *Badāʾiʿ al-Akwān* and Other Animal Books', in Anna Contadini (ed.), *Arab Painting: Text and Image in Illustrated Arabic Manuscripts*. Leiden: Brill, 2007, 49–64.

Kruk, 'De goede arts' = Remke Kruk, 'De goede arts: ideaalbeeld en werkelijkheid in de middeleeuws-Arabische wereld', *Hermeneus*, 2 (1999), 140–148.

Kruk, 'Hedgehogs' = Remke Kruk, 'Hedgehogs and Their Chicks: A Case History of the Artistotelian Reception in Arabic Zoology', *Zeitschrift für Geschichte der Arabisch-Islamischen Wissenschaften*, 2 (1984), 205–234.

Kruk, 'Ibn abī l-Ashʿath's *Kitāb al-Ḥayawān*' = Remke Kruk, 'Ibn abī l-Ashʿath's *Kitāb al-Ḥayawān*: A Scientific Approach to Anthropology, Dietetics and Zoological Systematics', *Zeitschrift für Geschichte der Arabisch-Islamischen Wissenschaften*, 14 (2001), 119–168. Repr. in Peter Pormann (ed.), *Islamic medical and scientific tradition* (Critical Concepts in Islamic Studies). 4 vols. London: Routledge, 2011, no. 28.

Kruk, 'La zoologie aristotélicienne' = Remke Kruk, 'Aristote de Stagire – La zoologie aristotélicienne. Tradition arabe', in Goulet, *Dictionnaire*, Suppl., 329–334.

Kuhne, 'El kitāb al-Durŷ' = R. Kuhne, 'El kitāb al-Durŷ, prototipo árabe de la Capsula Eburnea', *al-Qanṭara*, 10 (1989), 299–327 (parts I–II), 11 (1990), 3–58 (part III).

Kuhne Brabant, 'Al-Rāzī on when and how to eat fruit' = Rosa Kuhne Brabant, 'Al-Rāzī on when and how to eat fruit', in Frederik de Jong (ed.), *Miscellanea Arabica et Islamica: dissertationes in Academia Ultrajectina prolatae anno MCMXC*. Leuven: Peeters, 1993, 164–174.

Kuhne Brabant, 'revision de la bibliografía de Abū l-ʿAlāʾ Zuhr' = R, Kuhne Brabant, 'Hacia una revision de la bibliografía de Abū l-ʿAlāʾ Zuhr', *al-Qanṭara*, 13/2 (1992), 581–585.

Kunitzsch & Ullmann, *Die Plejaden* = Paul Kunitzsch und Manfred Ullmann, *Die Plejaden in den Vergleichen der arabischen Dichtung* (Bayerische Akad. der Wiss., Phil.-hist. Kl., Sitzungsberichte, Jrg. 1992, Heft 4). Munich: Verlag der Bayerischen Akademie der Wissenschaften, 1992.

Kurd ʿAlī, *Kunūz al-ajdād* = Muḥammad Kurd ʿAlī, *Kunūz al-ajdād*. Damascus: al-Majmaʿ al-ʿIlmī al-ʿArabī, 1950.

Kushājim, *Dīwān* = Kushājim, *Dīwān*, ed. Khayriyyah Muḥammad Maḥfūẓ. Baghdad: Wizārat al-Iʿlām, 1390/1970.

Kuthayyir ʿAzzah, *Dīwān*, ed. Iḥsān ʿAbbās. Beirut: Dār al-Thaqāfah, 1971.

Labarta, 'Ibn Baklarish's *Kitāb al-Mustaʿīnī*' = Ana Labarta, 'Ibn Baklarish's *Kitāb al-Mustaʿīnī*: the Historical Context to the Discovery of a New Manuscript', in Burnett, *Ibn Baklarish's Book of Simples*, 15–26.

Labīd, *Dīwān* = Labīd ibn Rabīʿah al-ʿĀmirī, *Dīwān*. Beirut: Dār Ṣādir, 1966.

Lagardère, 'L'épître d'Ibn Bâjja' = Lagardère, V., 'L'épître d'Ibn Bâjja sur la conjonction de l'intellect avec l'esprit humain', *REI* 49 (1981) 175–195.

Laín Entralgo, 'De prisca medicina' = Pedro Laín Entralgo, 'El escrito De prisca medicina y su valor historiografico', *Emerita*, 12 (1944), 1–28.

Lameer, *Al-Fārābī and Aristotelian Syllogism* = Joep Lameer, *Al-Fārābī and Aristotelian Syllogism: Greek Theory and Islamic Practice* (Islamic Philosophy, Theology, and Science, 20). Leiden: Brill, 1994.

Lammer, *The Elements of Avicenna's Physics* = Andreas Lammer, *The Elements of Avicenna's Physics: Greek Sources and Arabic Innovations*. Berlin: De Gruyter, 2018.

Lamoreaux, *Early Muslim Tradition of Dream Interpretation* = John C. Lamoreaux, *The Early Muslim Tradition of Dream Interpretation*. Albany, NY: State University of New York Press, 2002.

Lamoreaux, John C. – *see also* Ḥunayn ibn Isḥāq, *Galen Translations*.

Lane, *Lexicon* = Edward William Lane, *An Arabic-English Lexicon*. 8 parts. London: Williams and Norgate, 1863–1893.

Lang, *John Philoponus* = Uwe M. Lang, *John Philoponus and the Controversies over Chalcedon in the Sixth Century*. Leuven: Peeters, 2001.

Lange & Mecit, *The Seljuqs* = Christian Lange and Songül Mecit, *The Seljuqs: Politics, Society and Culture*. Edinburgh: Edinburgh University Press, 2012.

Langermann, 'Babylonian and Indian Wisdoms' = Y. Tzvi Langermann, 'Babylonian and Indian Wisdoms in Islamicate Culture', *Oriens*, 46 (2018), 433–475.

Langermann, 'Maimonides and the Sciences' = Y. Tzvi Langermann, 'Maimonides and the Sciences', in Daniel H. Frank and Oliver Leaman (eds)., *The Cambridge Companion to Medieval Jewish Philosophy*. Cambridge: Cambridge University Press, 2003, 157–175.

Langermann, 'Masīḥ bin Ḥakam' = Y. Tzvi Langermann, 'Masīḥ bin Ḥakam, a Jewish-Christian (?) Physician of the Early Ninth Century', *Aleph*, 4 (2004), 283–297.

Langermann, 'L'oeuvre médicale de Maimonide, un aperçu general' = Y. Tzvi Langermann, 'L'oeuvre médicale de Maimonide, un aperçu general', in T. Lévy and R. Rashed (eds), *Maimonide philosophe et savante (1138–1204)*. Leuven: Peeters, 2004, 275–302.

Lapidus, *Muslim Cities* = Ira M. Lapidus, *Muslim Cities in the Later Middle Ages*. Cambridge, MA: Harvard University Press, 1967.

Latham, 'Isaac Israeli's *Kitāb al-Hummayat*' = J.D. Latham, 'Isaac Israeli's *Kitāb al-Hummayat* and the Latin and Castilian Texts', *Journal of Semitic Studies*, 14 (1969), 80–95.

Latham & Isaacs, *Isaac Judaeus* = J.D. Latham and H.D. Isaacs, *Isaac Judaeus: On Fevers (The Third Discourse: On Consumption), Together with an Appendix containing a Facsimile of the Latin Version of this Discourse (Venice, 1576)*. Cambridge: Pembroke Arabic Texts for the Cambridge Middle East Centre, 1981.

Laurenti, *Aristotele* = Renato Laurenti, *Aristotele. I frammenti dei dialoghi*. Naples: L. Loffredo, 1978.

Le Strange, *Baghdad during the Abbasid Caliphate* = Guy Le Strange, *Baghdad during the Abbasid Caliphate: from Contemporary Arabic and Persian Sources*. 2nd ed., Oxford: Clarendon Press, 1924.

Le Strange, *Collected Works* = Guy Le Strange, *Collected Works of Guy Le Strange: The Medieval Islamic World*. 4 vols. London: I.B. Tauris, 2014.

Leiser & al-Khaledy, *Questions and Answers* = Gary Leiser and Noury al-Khaledy (eds.), *Questions and Answers for Physicians. A Medieval Arabic Study Manual by ʿAbd al-ʿAzīz al-Sulamī*. Leiden: Brill, 2004.

Leiser, 'Endowment' = Gary Leiser, 'The Endowment of the al-Ẓāhiriyya in Damascus', *Journal of the Economic and Social History of the Orient*, 27 (1983), 33–55.

Leiser, 'Restoration of Sunnism' = Gary Leiser, 'The Restoration of Sunnism in Egypt: Madrasas and Mudarrisūn 495–647/1101–1249'. Ph.D. diss., University of Pennsylvania, 1976.

Leschhorn, *Ären* = W. Leschhorn, *Antike Ären. Zeitrechnung, Politik und Geschichte im Schwarzmeerraum und in Kleinasien nördlich des Tauros*. Stuttgart: Franz Steiner, 1993.

Lettinck, *Aristotle's Meteorology* = Paul Lettinck, *Aristotle's Meteorology and its Reception in the Arab world: with an Edition and Translation of Ibn Suwār's Treatise on Meteorological Phenomena and Ibn Bājja's Commentary on the Meteorology*. Leiden: Brill, 1999.

Lettinck, *Aristotle's Physics* = Paul Lettinck, *Aristotle's Physics and Its Perception in the Arabic World with an Edition of the Unpublished Parts of Ibn Bājja's Commentary on the Physics*. Leiden: Brill, 1994.

Lev, 'An early fragment of Ibn Jazlah's tabulated manual' = Efraim Lev, 'An early fragment of Ibn Jazlah's tabulated manual "*Taqwīm al-abdān*" from the Cairo Genizah (T-S Ar.41.137)', *Journal of the Royal Asiatic Society*, 24/2 (2014), 189–223.

Lev & Amar, *Materia Medica* = Efraim Lev and Zohar Amar, *Practical materia medica of the medieval eastern Mediterranean according to the Cairo Genizah* (Sir Henry Wellcome Asian Series, 7). Leiden: Brill, 2008.

Levey, *Early Arabic Pharmacology* = Martin Levey, *Early Arabic Pharmacology: An Introduction based on Ancient and Medieval Sources*. Leiden: E.J. Brill, 1973

Levey, 'Ibn Māsawaih and his Treatise on Simple Aromatic Substances' = Martin Levey, 'Ibn Māsawaih and his Treatise on Simple Aromatic Substances', *Journal of the History of Medicine*, 16 (1961), 394–410.

Levey, *Medical Formulary* = Martin Levey, *The Medical Formulary or ʿAqrābādhīn of al-Kindī, translated with a study of its materia medica*. Madison, WI: The University of Wisconsin Press, 1966.

Levey, *Substitute Drugs* = Martin Levey, *Substitute drugs in early Arabic medicine. With special reference to the texts of Māsarjawaih, al-Rāzī, and Pythagoras*. Stuttgart: Wissenschaftliche Verlagsgesellschaft, 1971.

Levey & al-Khaledy, *Medical Formulary of Al-Samarqandi* = Martin Levey & Noury al-Khaledy, *The Medical Formulary of Al-Samarqandi and the Relation of Early Arabic Simples to Those Found in the Indigenous Medicine of the Near East and India*. Philadelphia: University of Pennsylvania Press, 1967.

Lewicka, 'Diet as Culture' = Paulina B. Lewicka, 'Diet as Culture. On the Medical Context of Food Consumption in the Medieval Middle East', *History Compass*, 12 (2014), 607–617.

Lewicka, *Food of Medieval Cairenes* = Paulina B. Lewicka, *Food and Foodways of Medieval Cairenes: aspects of life in an Islamic metropolis of the eastern Mediterranean* (Islamic History and Civilization, 88). Leiden: Brill, 2011.

Lewicka, 'Medicine for Muslims' = Paulina B. Lewicka, 'Medicine for Muslims? Islamic Theologians, Non-Muslim Physicians and the Medical Culture of the Mamluk Near East', in Stephan Conermann (ed.), *History and Society during the Mamluk Period (1250–1517)* (Studies in the Annemarie Schimmel Research College, 1). Göttingen: V&R Unipress / Bonn: Bonn University Press, 2014, 83–106.

Lewicka, 'Restaurants and Inns' = Paulina B. Lewicka, 'Restaurants, Inns and Taverns That Never Were: Some Reflections on Public Consumption in Medieval Cairo', *Journal of the Economic and Social History of the Orient*, 48/1 (2005), 40–91.

LGPN = Peter M. Fraser, Elaine Matthews, et al., *Lexikon of Greek Personal Names*. Oxford: Oxford University Press, 1987–, available online at http://www.lgpn.ox.ac.uk/index.html.

Liddell & Scott, *Greek-English Lexicon* = Henry George Liddell and Robert Scott, *A Greek-English Lexicon. A New Edition*, Revised and Augmented by Henry Stuart Jones with Robert McKenzie. Oxford: Clarendon Press, 1940; repr. 1958.

Littré, *Hippocrate* = Émile Littré, *Oeuvres complètes d'Hippocrate*. 10 vols. Paris: Baillière, 1839–1861; repr. Amsterdam: Hakkert, 1962.

Liu & Monroe, *Ten Hispano-Arabic Strophic Songs* = Benjamin M. Liu and James T. Monroe, *Ten Hispano-Arabic Strophic Songs in the Modern Oral Tradition: Music and Texts*. Berkeley, California: University of California Press, 1989.

Livingston, 'Qusṭā ibn Lūqā's psycho-physiological treatise' = John W. Livingston, 'Qusṭā ibn Lūqā's psycho-physiological treatise on the difference between the soul and the spirit: (Fī l-farq bayn al-nafs wa l-rūḥ)', *Scripta Mediterranea*, 2 (1981), 53–77.

Llavero Ruiz, *Historia de la filosofía y de las ciencias* = Eloísa Llavero Ruiz, *Historia de la filosofía y de las ciencias, o, Libro de las categorias de las naciones / Kitāb ṭabaqāt al-umam*. Madrid: Editorial Trotta, 2000.

Llavero Ruiz, 'Generaciones de médicos y sabios' = Eloísa Llavero Ruiz, 'Generaciones de médicos y sabios', unpublished Ph.D. diss., Granada, 1979.

Loewenthal, *Sinnsprüche* = Abraham Loewenthal, *Honain ibn Ishâq, Sinnsprüche der*

Philosophen. Nach der hebräischen Uebersetzung Charisi's ins Deutsche übertragen und erläutert. Berlin: S. Calvary & Co., 1896.

Lomba Fuentes, 'Avempace *Tratado de la unión*' = Joaquin Lomba Fuentes, 'Avempace "Tratado de la unión del Intelecto con el hombre"', in *Anaquel de Estudios Árabes*, 11 (2000), 369–391.

Lomba Fuentes, 'Sobre el fin del hombre' = Joaquin Lomba Fuentes, 'Ibn Bāŷŷa (Avenpace) "Sobre el fin del hombre"', *Sharq al-Andalus*, 11–12 (1993), 467–482.

Lonie, *Hippocratic Treatises* = Iain M. Lonie, *The Hippocratic Treatises, "On Generation", "On the Nature of the Child", "Diseases IV"*. Berlin: De Gruyter, 1981.

Lorch, 'The astronomy of Jābir ibn Aflaḥ' = R.P. Lorch, 'The astronomy of Jābir ibn Aflaḥ', *Centaurus*, 19 (1975), 85–107.

Lorch, *On the sector-figure* = Richard Lorch, *On the sector-figure and related texts*. Augsburg: Rauner, 2008.

Lorch, 'Projection' = Richard Lorch, 'Ibn al-Ṣalāḥ's Treatise on Projection: A Preliminary Survey', in Menso Folkerts and Richard Lorch (eds.), *Sic itur ad astra: Studien zur Geschichte der Mathematik und Naturwissenschaften. Festschrift für den Arabisten Paul Kunitzsch zum 70. Geburtstag*. Wiesbaden: Harrassowitz, 2000, 401–408.

Lorch, *Risālat fī al-šakl al-qiṭāʿ* = Richard Lorch, *Risālat fī al-šakl al-qiṭāʿ wa nuṣūṣ mumāṯilat* [sic]. Frankfurt am Main: Institut für Geschichte der Arabisch-Islamischen Wissenschaften, 2001.

Lory, *La science des lettres* = Pierre Lory, *La science des lettres en islam*. Paris: Dervy, 2004.

Lüling, *Ein Anderer Avicenna* = Günter Lüling, 'Avicenna und seine buddhistische Herkunft' in *idem*, Zwei Aufsätze zur Religions- und Geistesgeschichte. Erlangen: Lüling, 1977.

Lyons, *The Arabian Nights* = *The Arabian Nights: Tales of 1001 Nights*, tr. by Malcolm C. Lyons, with Ursula Lyons, introd. and annot. by Robert Irwin. 3 vols. London: Penguin Classics, 2008.

Lyons, Malcolm C., *Galeni De partibus artis medicativae, De causis contentivis, De diaeta in morbis acutis sercundum Hippocratem libellorum versiones arabicas* (Corpus medicorum Graecorum, Suppl. Or. II). Berlin: Academia Scientiarum, 1969.

Lyons, Malcolm C., *Galeni in Hippocratis De officina medici commentariorum versionem arabicam …*, CMG (Corpus medicorum Graecorum Suppl. Or. I). Berlin: Academia Scientiarum, 1963.

Lyons, *Themistius' Commentary* = Malcolm C. Lyons, *An Arabic Translation of Themistius Commentary on Aristotles*, De anima. Oxford: Cassirer, 1973.

Lyons & Mattock, *Hippocrates, Ajinna* = Malcolm C. Lyons and John N. Mattock, *Kitāb al-Ajinna li-Buqrāṭ. Hippocrates: On Embryos* (*On the Sperm and On the Nature of the Child*). Cambridge: Published for the Cambridge Middle East Centre by Pembroke Arabic Texts, 1978.

Lyons & Mattock, *Hippocrates, Ṭabīʿat al-Insān* = Malcolm C. Lyons and John N. Mattock, *Kitāb Buqrāṭ fī Ṭabīʿat al-Insān. Hippocrates: On the Nature of the Man*. Cambridge: Published for the Cambridge Middle East Centre by Heffer, 1968.

Lloyd, *Revolutions of Wisdom* = G.E.R. Lloyd, *The Revolutions of Wisdom: Studies in the Claims and Practice of Ancient Greek Science*. Berkeley: University of California Press, 1987.

al-Maʿarrī, *The Epistle of Forgiveness* = al-Maʿarrī, *The Epistle of Forgiveness, or, A Pardon to Enter the Garden by Abū l-ʿAlāʾ al-Maʿarrī: Hypocrites, Heretics and Other Sinners*, ed. and tr. Geert Jan van Gelder and Gregor Schoeler (Library of Arabic Literature). New York: New York University Press, 2016.

al-Maʿarrī, *Luzūmiyyāt* = Abū l-ʿAlāʾ al-Maʿarrī, *al-Luzūmiyyāt* (*Luzūm mā lā yalzam*), ed. Amīn ʿAbd al-ʿAzīz al-Khānjī. 2 vols. Cairo: Maktabat al-Khānjī, 1924, repr. Cairo: Maktabat al-Khānjī / Beirut: Maktabat al-Hilāl, n.d.

al-Maʿarrī, *Saqṭ al-zand* = Abū l-ʿAlāʾ al-Maʿarrī, *Saqṭ al-zand*. Beirut: Dār Ṣādir, 1400/1980.

McCall Smith, *Finer Points* = Alexander McCall Smith, *The Finer Points of Sausage Dogs*. Edinburgh: Birlinn Limited, 2003; reprint in Alexander McCall Smith, *The 2½ Pillars of Wisdom: The Von Igelfeld Trilogy*. London: Little, Brown Book Group (Abacus), 2004.

McCarthy, *al-Taṣānīf al-mansūbah ilā Faylasūf al-ʿArab* = Richard Joseph McCarthy, *al-Taṣānīf al-mansūbah ilā Faylasūf al-ʿArab: baḥth bi-munāsabat iḥtifālāt Baghdād wa-l-Kindī*. Baghdad: Maṭbaʿat al-ʿĀnī, 1382/1962.

McGill, 'Ibn Abī Uṣaibiʿa's biographies' = Willis Alexander McGill, 'Ibn Abī Uṣaibiʿa's biographies of Ḥunain ibn Isḥāq, Isḥāq ibn Ḥunain, and Ḥubaish ibn al-Ḥasan: Ninth century physicians and translators of Baghdad'. Unpublished M.A. diss., Kennedy School of Missions, Hartford Seminary Foundation, 1944.

al-Madāʾinī, *al-Murdifāt min Quraysh* = Abū l-Ḥasan ʿAlī ibn Muḥammad al-Madāʾinī, *al-Murdifāt min Quraysh*, in *Nawādir al-makhṭūṭāt*, ed. ʿAbd al-Salām Hārūn. 2 vols. Cairo: Muṣṭafā al-Bābī al-Ḥalabī, 1972–1973, i:57–80.

Maddison & Savage-Smith, *Science, Tools and Magis* = F.R. Maddison and Emilie Savage-Smith, *Science, Tools and Magic. Part I: Body and Spirit, Mapping the Universe, Part II: Mundane Worlds* (The Nasser D. Khalili Collection of Islamic Art, 12). 2 vols. London / Oxford: Azimuth Editions / Oxford University Press.

Madelung, 'Abū Isḥāq al-Ṣābī on the Alids' = Wilferd Madelung, 'Abū Isḥāq al-Ṣābī on the Alids of Tabaristān and Gīlān', *Journal of Near Eastern Studies*, 26/1 (1967), 17–57.

Madelung, *Succession* = Wilferd Madelung, *The Succession to Muhammad. A Study of the Early Caliphate*. Cambridge: Cambridge University Press, 1997.

Mahdavī, *Muṣannafāt-i Ibn Sīnā* = Yaḥyā Mahdavī, *Fihrist-i Nuskhah'hā-yi Muṣannafāt-i Ibn Sīnā*. Tehran: s.n., 1954.

Maḥmūd al-Warrāq, *Dīwān* = Maḥmūd al-Warrāq, *Dīwān Maḥmūd al-Warrāq shāʿir al-ḥikmah wa-l-mawʿiẓah*, ed. Walīd al-Qaṣṣāb. Damascus: Muʾassasat al-Funūn, 1991.

Maíllo Salgado, *Libro de las categorias de las naciones* = Felipe Maíllo Salgao, *Libro de las categorias de las naciones. Vislumbres desde el Islam clásico sobre la filosofía y la ciencia.* Madrid: Akal Ediciones, 1999.

Maimonides, *Art of Cure* = Uriel S. Barzel, *The Art of Cure, Extracts from Galen.* Foreword by Fred Rosner and bibliography by Jacob I. Dienstag (Maimonides' Medical Writings, 5). Haifa: Maimonides Research Institute, 1992.

Maimonides, *On Asthma* = Maimonides, *On Asthma, A Parallel Arabic-English Edition,* ed. and tr. Gerrit Bos. Provo, Utah: Brigham Young University, 2002.

Maimonides, *Glossary of Drug Names* = 'Moses Maimonides' Glossary of Drug Names. Translated from Max Meyerhof's French Edition', *Memoirs of the American Philosophical Society,* 135. Philadelphia: American Philosophical Society, 1978. Repr. Haifa: Maimonides Research Institute, 1995 (Maimonides' Medical Writings, 7).

Maimonides, *Guide of the Perplexed* = Maimonides, *The Guide of the Perplexed,* ed. & tr. Shlomo Pines. 2 vols. Chicago: University of Chicago Press, 1963.

Maimonides, *Medical Aphorisms* = Maimonides, *Medical Aphorisms,* ed. & tr. Gerrit Bos. 3 vols. Provo, Utah: Brigham Young University Press, 2004–2010.

Maimonides, *On Hemorrhoids* = Maimonides, *On Hemorrhoids. A New Parallel Arabic-English Edition and Translation,* ed. & tr. Gerrit Bos, with Michael R. McVaugh. Provo, Utah: Brigham Young University Press, 2012.

Maimonides, *On Poisons* = Maimonides, *On Poisons and the Protection against Lethal Drugs,* ed. & tr. Gerrit Bos and Michael R. McVaugh. Provo, Utah: Brigham Young University Press, 2009.

Maimonides, *On the Regimen of Health* (Bos & McVaugh) = Maimonides, *On the Regimen of Health. A New Parallel Arabic-English Translation,* ed. and tr. by Gerrit Bos and Michael R. McVaugh (The Medical Works of Moses Maimonides, 12). Leiden: Brill, 2019.

Maimonides, *Regimen of Health* = Ariel Bar-Sela, Hebbel E. Hoff, and Elias Faris, 'Moses Maimonides' Two Treatises on the Regimen of Health: *Fī tadbīr al-ṣiḥḥa* and *Maqāla fī bayān baʿḍ al-aʿrāḍ wa-al-jawāb ʿanhā*', in *Transactions of the American Philosophical Society,* n.s., 54 pt. 4. Philadelphia: The American Philosophical Society, 1964.

Maimonides, *Three Treatises on Health* = Fred Rosner and Jacob Israel Dienstag, *Moses Maimonides' Three Treatises on Health.* Haifa: Maimonides Research Institute, 1990.

Majnūn Laylā, *Dīwān* = Majnūn Laylā, *Dīwān,* ed. ʿAdnān Zakī Darwīsh, Beirut: Dār Ṣādir, 1414/1994.

Mallett, *Reactions to the Franks in the Levant* = Alex Mallett, *Popular Muslim Reactions to the Franks in the Levant, 1097–1291.* Farnham, Surrey, UK: Ashgate, 2014.

Mallet, 'Le rappel de la voie … Abū Naṣr al-Fārābī' = D. Mallet, 'Le rappel de la voie à suivre pour parvenir au bonheur de Abū Naṣr al-Fārābī: Intr., trad., et notes', *Bulletin d'études orientales* 39–40 (1987–1988), 113–140.

al-Manṣūr Muḥammad, *Miḍmār* = al-Manṣūr Muḥammad ibn ʿUmar Ṣāḥib Ḥamāh, *Miḍmār al-ḥaqāʾiq wa-sirr al-khalāʾiq*, ed. Ḥasan Ḥabashī. Cairo: ʿĀlam al-Kutub, 1968.

Maqdisī, *K. al-Badʾ wa-l-taʾrīkh* = al-Muṭahhar ibn Ẓāhir al-Maqdisī, *Kitāb al-Badʾ wa-l-taʾrīkh*. Cairo: Maktabah al-Thaqāfah al-Dīniyyah, s.d.

al-Maqqarī, *Nafḥ al-ṭīb* = Aḥmad ibn Muḥammad al-Maqqarī al-Tilimsānī, *Nafḥ al-ṭīb min ghuṣn al-Andalus al-raṭīb*, ed. Iḥsān ʿAbbās. 8 vols. Beirut: Dār Ṣādir, 1388/1968.

al-Maqrīzī, *Ittiʿāẓ* = Aḥmad ibn ʿAlī ibn ʿAbd al-Qādir ibn Muḥammad al-Maqrīzī, *Ittiʿāẓ al-ḥunafāʾ bi-akhbār al-aʾimmah al-fāṭimiyyīn al-khulafāʾ*, ed. Jamāl al-Dīn Shayyāl, Muḥammad Ḥilmī and Muḥammad Aḥmad. Cairo: al-Lajnat Iḥyāʾ al-Turāth al-Islāmī, 1996.

al-Maqrīzī, *Khiṭaṭ* = Aḥmad ibn ʿAlī ibn ʿAbd al-Qādir ibn Muḥammad al-Maqrīzī, *Kitāb al-Mawāʿiẓ wa-l-iʿtibār bi-dhikr al-khiṭaṭ wa-l-āthār*. 2 vols. Cairo/Būlāq, 1270/1853.

al-Maqrīzī, *Musawwadat K. al-Mawāʿiẓ* = Aḥmad ibn ʿAlī ibn ʿAbd al-Qādir ibn Muḥammad al-Maqrīzī, *Musawwadat Kitāb al-Mawāʿiẓ wa-l-iʿtibār fī dhikr al-khiṭat wa-l-āthār / Le manuscrit autographe d'al-Mawāʿiẓ wa-l-iʿtibār fī dhikr al-khiṭaṭ wa-l-āthār*, ed. Fuʾād Sayyid. London: al-Furqān Islamic Heritage Foundation, 1995.

al-Maqrīzī, *Sulūk* = Aḥmad ibn ʿAlī ibn ʿAbd al-Qādir ibn Muḥammad al-Maqrīzī, *Kitāb al-sulūk li-maʿrifat duwal al-muluk*, ed. Muḥammad Muṣṭafā Ziyādah. 4 vols. Cairo: Maṭbaʿat Lajnat al-Taʾlīf wa-l-Tarjamah wa-l-Nashr, 1934–1972.

Marasco, 'Curriculum of Studies' = Gabriele Marasco, 'The Curriculum of Studies in the Roman Empire and the Cultural Role of Physicians', in Manfred Horstmanshoff (ed.), *Hippocrates and Medical Education*. Leiden / Boston: Brill, 2010, 205–222.

Marcinkowski, *Measures see* Hinz, *Islamische Masse*.

Margoliouth, 'The Legend of the Apostasy of Maimonides' = D. Margoliouth, 'The Legend of the Apostasy of Maimonides', *Jewish Quarterly Review* 13 (1901), 539–541.

Marín, 'El viaje a oriente de Abū Marwān al-Bāŷī' = M. Marín, 'El viaje a oriente de Abū Marwān al-Bāŷī (m. 635/1237)', in Manuela Marín (ed.), *Estudios onomástico-biográficos de al-Andalus*, 6. Madrid: CSIC, 1994, 273–303.

Marin & Waines, 'Ibn Sīnā on *Sakanjabīn*' = Manuela Marin and David Waines, 'Ibn Sīnā on *Sakanjabīn*', *Bulletin d'études orientales* 47 (1995), 81–97.

Maróth, *Correspondence* = Miklós Maróth, *The Correspondence between Aristotle and Alexander the Great: an Anonymous Greek Novel in Letters in Arabic Translation*. Piliscsaba: Avicenna Institute of Middle Eastern Studies, 2006.

al-Marrākushī, *Dhayl* = Muḥammad ibn Muḥammad ibn ʿAbd al-Malik al-Marrākushī, *al-Dhayl wa-l-takmilah li-kitabay al-mawṣūl wa-l-ṣilah*, ed. Iḥsān ʿAbbās, Muḥammad ibn Sharīfah, and Bashshār ʿAwwād Maʿrūf. 5 vols. Tunis: Dār al-Gharb al-Islāmī, 2012.

al-Marrākushī, *Muʿjib* = ʿAbd al-Wāḥid al-Marrākushī, *al-Muʿjib fī talkhīṣ akhbār al-Maghrib*, ed. Muḥammad Saʿīd al-ʿIryān and Muḥammad al-ʿArabī al-ʿIlmī. Cairo: Maṭbaʿah al-Istiqāmah, 1949.

Martin, 'La *Métaphysique*' = A. Martin, 'Aristote de Stagire – La *Métaphysique*. Tradition Syriaque et Arabe', in Goulet, *Dictionnaire*, i:528–534.

Martin, 'All Brutes are Subhuman' = John N. Martin, 'All Brutes Are Subhuman: Aristotle and Ockham on Private Negation', *Synthese*, 134/3 (2003), 429–461.

Martini Bonadeo, 'Abd al-Latif al-Baghdadi' = Cecilia Martini Bonadeo, 'Abd al-Latif al-Baghdadi', in *The Stanford Encyclopedia of Philosophy* (Fall 2015 Edition), available online at http://plato.stanford.edu/archives/fall2015/entries/al-baghdadi/.

Martini Bonadeo, 'La *Métaphysique*' = Cecilia Martini Bonadeo, 'Aristote de Stagire – La *Métaphysique*. Tradition syriaque et arabe', in Goulet, *Dictionnaire* Suppl., 259–264.

Martini Bonadeo, *Philosophical Journey* = Cecilia Martini Bonadeo, *'Abd al-Laṭīf al-Baġdādī's Philosophical Journey: From Aristotle's* Metaphysics *to the 'Metaphysical Science'* (Islamic Philosophy, Theology, and Science, 88). Leiden: Brill, 2013.

Di Martino, 'Aristote de Stagire – *Parva Naturalia*' = C. Di Martino, 'Aristote de Stagire – *Parva Naturalia*. Tradition arabe', in Goulet, *Dictionnaire*, Suppl., 375–378.

al-Marzūqī, *Dīwān al-ḥakīm Abī l-Ṣalt* = al-Marzūqī, Muḥammad, *Dīwān al-ḥakīm Abī l-Ṣalt Umayyah ibn 'Abd al-'Azīz al-Dānī 460–529*. Tunis: Dār Bū Salāmah, 1979.

al-Marzubānī, *Muʿjam al-shuʿarāʾ* = Abū 'Ubayd Allāh Muḥammad ibn 'Imrān al-Marzubānī, *Muʿjam al-shuʿarāʾ*, ed. 'Abd al-Sattār Aḥmad Farrāj. Cairo: Dār Iḥyāʾ al-Kutub al-'Arabiyyah, 1960.

al-Marzūqī, *Sharḥ* = Abū 'Alī Aḥmad ibn Muḥammad al-Marzūqī, *Sharḥ Dīwān al-Ḥamāsah*, ed. Aḥmad Amīn and 'Abd al-Salām Hārūn. 2 vols. repr. Beirut: Dār al-Jīl, 1991.

al-Masīḥī, *K. al-miʾah fī l-ṭibb* =Floreal Sanagustin (ed.), *K. al-miʾah fī l-ṭibb* [Le livre des cent questions en médecine d'Abū Sahl 'Īsa ibn Yaḥyā al-Masīḥī (m. ca 401h/1010 apr. J.-C.). 2 vols. Damascus: al-Maʿhad al-Faransī lil-Dirasāt al-'Arabiyyah, 2000.

Mason & Lupieri, *Golden Calf Traditions* = Eric F. Mason and Edmondo F. Lupieri (eds.), *Golden Calf Traditions in Early Judaism, Christianity, and Islam* (Themes in Biblical Narrative, 23). Brill: Leiden, 2018.

al-Masʿūdī, *L'avertissement* = al-Masʿūdī (*Maçoudi*), *Le livre de l'avertissement et de la révision*, tr. B. Carra de Vaux. Paris: L'imprimerie nationale, 1896.

al-Masʿūdī, *Murūj* = al-Masʿūdī, *Murūj al-dhahab wa-maʿādin al-jawhar*, ed. Barbier de Meynard & Pavet de Courteille, revue et corrigée par Charles Pellat. 7 vols. Beirut: al-Jāmiʿah al-Lubnāniyyah, 1966–1979.

al-Masʿūdī, *Prairies* = al-Masʿūdī, *Les prairies d'or*. Texte et traduction par C. Barbier de Meynard et Pavet de Courteille. 9 vols. Paris: Imprimerie impériale, 1861–1917.

al-Masʿūdī, *al-Tanbīh* = al-Masʿūdī, *K. al-tanbīh wa-l-ishrāf*, ed. M.J. de Goeje. Leiden: Brill, 1894.

al-Maʿṣūmī, 'Ibn al-Imām' = Muḥammad Ṣaghīr Ḥasan al-Maʿṣūmī, 'Ibn al-Imām, the Disciple of Ibn Bājjah', *The Islamic Quarterly* 5/3–4 (1959–60), 102–108.

al-Maʿṣūmī, 'On Human End' = Muḥammad Ṣaghīr Ḥasan al-Maʿṣūmī, 'Ibn Bājjah on Human End', *Islamic Studies*, 4 (1965), 121–136.

Mattern, *Galen* = Susan P. Mattern, *Galen & the Rhetoric of Healing*. Baltimore: The Johns Hopkins University Press, 2008.

Mattila, 'Philosophical Lives of Ibn al-Haytham and Ibn Riḍwān' = Janne Mattila, 'The Philosophical Lives of Ibn al-Haytham and Ibn Riḍwān: Autobiography as an expression of the philosophical way of life', *Studia Orientalia*, 114 (2013), 325–341.

Mattock, 'The Medical Muse' = John N. Mattock, 'The Medical Muse', in Concepción Vázquez de Benito & Miguel Ángel Manzano Rodríguez (eds.), *Actas XVI Congreso UEAI [Salamanca, Aug–Sept. 1992]*. Salamanca, n. publ., 1995, pp. 337–343.

Mattock 'Traits' = J.N. Mattock, 'A Translation of the Arabic Epitome of Galen's Book περὶ ἠθῶν', in S.M. Stern, A. Hourani, and V. Brown (eds.), *Islamic Philosophy and the Classical Tradition. Essays Presented … to Richard Walzer*. Oxford: Oxford University Press, 1972, 235–260.

al-Māwardī, *Aḥkām* = Abū l-Ḥasan ʿAlī ibn Muḥammad ibn Ḥabīb al-Māwardī, *al-Aḥkām al-sulṭāniyyah wa-l-wilāyāt al-dīniyyah*, ed. Aḥmad Mubārak al-Baghdādī. Kuwait: Maktabat Dār Ibn Qutaybah, 1409/1989.

al-Māwardī, *The Ordinances*, tr. Wahba = al-Māwardī, *The Ordinances of Government / Al-Aḥkām al-Sulṭāniyya w'al-Wilāyāt al-Dīniyya*, tr. Wafaa H. Wahba. Reading: Garnet, 2000.

al-Maydānī, *Majmaʿ* = Abū l-Faḍl Aḥmad ibn Muḥammad al-Maydānī, *Majmaʿ al-amthāl*, ed. Naʿīm Ḥusayn Zarzūr. 2 vols. Beirut: Dār al-Kutub al-ʿIlmiyyah, 1988.

Mayer, *Islamic Astrolabists* = Leo Mayer, *Islamic Astrolabists and Their Works*. Geneva: A. Kundig, 1956.

McCarthy, *al-Taṣānīf al-mansūbah ilā Faylasūf al-ʿArab* = Richard Joseph McCarthy, *al-Taṣānīf al-mansūbah ilá faylasūf al-ʿArab : baḥth bi-munāsabat iḥtifālāt Baghdād wa-al-Kindī*. Baghdād: Maṭbaʿat al-ʿĀnī, 1962.

Mecit, *The Rum Seljuqs* = Songül Mecit, *The Rum Seljuqs: Evolution of a Dynasty*. Abingdon / New York: Routledge, 2014.

Meinardus, *Christian Egypt* = Otto F.A. Meinardus, *Christian Egypt Ancient and Modern*. Cairo: Cahiers d'histoire égyptienne, 1965.

Melgunof, *Das südliche Ufer* = Gregorii Valerianovich Melgunof, *Das südliche Ufer des kaspischen Meeres oder Die nordprovinzen Persiens*. Leipzig: Leopold Voss, 1868.

Mercier, 'Geodesy' = R.P. Mercier, 'Geodesy', in J.B. Harley and David Woodward (eds.) *History of Cartography. Vol. II, Book 1: Cartography in the Traditional Islamic and South Asian Societies*. Chicago: Chicago University Press, 1992, 175–188.

Mertier, 'Aristote' = J. Mertier, 'Aristote de Stagire – Opuscules', in Goulet, *Dictionnaire*, Suppl., 489–491.

Meulenbeld, *History of Indian Medical Literature* = Gerrit Jan Meulenbeld. *A History of Indian Medical Literature*. 5 vols. Groningen: E. Forsten, 1999–2002.

Mewaldt, CMG = Johannes Mewaldt, *Galeni In Hippocratis De natura hominis commentaria tria* (Corpus Medicorum Graecorum, 9, 1). Leipzig / Berlin: Lips., 1914.

Meyerhof, 'An Arabic Compendium' = Max Meyerhof, 'An Arabic Compendium of Medico-Philosophical Definitions', *Isis*, 10 (1928), 340–349.

Meyerhof, 'Bayhaqī' = Max Meyerhof, "Alī al-Bayhaqī's Tatimmat Ṣiwān al-Ḥikma. A Biographical Work on Learned Men of the [*sic*] Islam', *Osiris*, 8 (1948), 122–117.

Meyerhof, 'Joannes Grammatikos (Philiponos)' = Max Meyerhof, 'Joannes Grammatikos (Philiponos) von Alexandrien und die arabische Medizin', *Mitteilungen des Deutschen Institut für ägyptische Altertumskunde in Kairo* 2 (1932), 1–21.

Meyerhof, 'Medieval Jewish Physicians' = Max Meyerhof, 'Medieval Jewish Physicians in the Near East from Arabic Sources', *Isis*, 28 (1938), 432–460.

Meyerhof, 'New Light' = Max Meyerhof, 'New Light on Ḥunain Ibn Isḥâq and his Period', *Isis*, 8 (1926), 685–724.

Meyerhof, 'Sultan Saladin's Physician' = Max Meyerhof, 'Sultan Saladin's Physician on the Transmission of Greek Medicine to the Arabs', *Bulletin of the History of Medicine*, 18 (1945), 169–178.

Meyerhof, *Ten Treatises* = Max Meyerhof, *The Book of the Ten Treatises on the Eye, Ascribed to Hunain ibn Is-Hâq (809–877 A. D.) The Earliest Existing Systematic Text-Book of Ophthalmology*. Cairo: Government Press, 1928. Repr. in Fuat Sezgin (ed.), *Augenheilkunde im Islam: Texte, Studien und Uebersetzungen*. 4 vols. Frankfurt am Main: Institut für Geschichte der Arabisch-Islamischen Wissenschaften, 1986, ii:1–518.

Meyerhof, 'Tooth-Worm' = Max Meyerhof, 'Arabic Tooth-Worm Stories', *Bulletin of the History of Medicine*, 17 (1945), 203–204.

Meyerhof, 'Über echte und unechte Schriften' = Max Meyerhof, 'Über echte und unechte Schriften Galens nach arabischen Quellen', *Sitzungsberichte der Preussischen Akademie der Wissenschaften, Philol.-hist. Kl.* Berlin, 1928, 531–548.

Meyerhof, 'Vorwort zur Drogenkinde des Bērūnī' = Max Meyerhof, 'Das Vorwort zur Drogenkinde des Bērūnī', *Quellen und Studien zur Geschichte der Naturwissenschaften und der Medizin*, 3 (1932), 1–52.

Meyerhof & Prüfer, 'Augenheilkunde des Jūḥannā ibn Māsawaih' = M. Meyerhof and C. Prüfer, 'Die Augenheilkund des Jūḥannā ibn Māsawaih', *Der Islam*, 6 (1915), 217–256.

Meyerhof & Sbath, *Le livre des questions* = Max Meyerhof and Paul Sbath, *Le livre des questions sur l'œil de Ḥonaïn ibn Isḥāq*, Cairo: Imprimerie de l'Institut français d'archéologie orientale, 1938.

Meyerhof & Schacht, *Theologus autodidactus* = Max Meyerhof and Joseph Schacht, *The Theologus autodidactus of Ibn al-Nafīs*. Oxford: Clarendon Press, 1968.

Meyerhof & Sobhī, *The Abridged Version* = Max Meyerhof and George Sobhy (Jūrjī Ṣubḥī), *The Abridged Version of "The Book of Simple Drugs" of Aḥmad ibn Muḥammad al-Ghâfiqî by Gregorius Abu'l-Farağ* (*Barhebraeus*). Cairo: The Egyptian University, Faculty of Medicine, 1932.

Micheau, 'Mécènes et médecins' = Micheau, Françoise, 'Mécènes et médecins à Baghdad au IIIe/IXe siécle. Les commanditaires des traductions de Galien par Ḥunayn ibn Isḥāq', in Danielle Jacquart (ed.), *Les voies de la science grecque. Études sur la transmission des textes de l'Antiquité au dix-neuvième siècle*. Geneva: Droz, 1997, 147–180.

Millás Vallicrosa, 'Los primeros tratados de astrolabio' = José María Millás Vallicrosa, 'Los primeros tratados de astrolabio en la España árabe', *Revista del Instituto Egipcio de Estudios Islámicos*, 3 (1955), 35–49 (Latin section), 47–76 (Arabic section).

Millás Vallicrosa, *Assaig d'història* = José María Millás Vallicrosa, *Assaig d'història de les idees físiques i matemàtiques a la Catalunya medieval*. Barcelona: Estudis Universitaris Catalans, 1931.

Miller, *The Birth of the Hospital* = Timothy S. Miller, *The Birth of the Hospital in the Byzantine Empire*. Baltimore: The Johns Hopkins University Press, 1985.

Miskawayh, *Tajārib* = Aḥmad Ibn Muḥammad Ibn Miskawayh & Leone Caetani, *The Tajârib Al-umam, Or, History of Ibn Miskawayh* (*Abu 'Ali Aḥmad B. Muḥammad*): *Reproduced in Facsimile from the Ms. at Constantinople in the Âyâ Ṣûfiyya Library*. Leiden: E.J. Brill / London: Luzac, 1909.

Miskawayh, *Tajārib al-umam* = *The Eclipse of the 'Abbasid Caliphate: Original Chronicles of the Fourth Islamic Century*, ed., tr., and elucidated by H.F. Amedroz and D.S. Margoliouth. 7 vols. Oxford: Blackwell, 1920.

Mitchell, 'Elective Surgery' = Piers D. Mitchell, 'Evidence for Elective Surgery in the Frankish States of the Near East in the Crusader Period (12th–13th Centuries)', in Florian Steger and Kay Peter Jankrift (eds.), *Gesundheit – Krankheit. Kulturtransfer medizinischen Wissens von der Spätantike bis in die Frühe Neuzeit*. Köln / Weimar / Wien: Böhlau, 2004, 121–138.

Mitchell, 'An evaluation of the leprosy of King Baldwin IV' = Piers D. Mitchell, 'An evaluation of the leprosy of King Baldwin IV of Jerusalem in the context of the mediaeval world', in B. Hamilton, (ed.) *The Leper King and his Heirs, Baldwin IV and the Crusader Kingdom of Jerusalem*. Cambridge University Press, 2000, 245–258.

Mitchell, *Medicine in the Crusades* = Piers D. Mitchell, *Medicine in the Crusades: Warfare, Wounds and the Medieval Surgeon*. Cambridge University Press, 2004.

Miquel, *La géographie humaine* = André Miquel, *La géographie humaine du monde musulman jusqu'au milieu du 11e siècle. Géographie arabe et représentation du monde: la terre et l'étranger*. 4 vols. Paris-La Haye: Mouton, 1967–1988.

Mobayyed, *Bimarestans of Aleppo* = Amer Rachid Mobayyed, *Bimarestans of Aleppo. The distinction of Islamic architecture*. Aleppo: al-Qalam al-ʿArabī, 2006.

Mohaghegh, *Fīlsūf-i Rayy* = Mehdi Mohaghegh, *Fīlsūf-i Rayy, Muḥammad ibn Zakariyā-yi Rāzī*. Tehran: Anjuman-i Āthār-i Millī, 1970.

Mohaghegh, *The Kitāb al-Shukūk ʿalā Jālīnūs of Muḥammad Ibn Zakariyyā Al-Rāzī* = Mehdi Mohaghegh, '*The Kitāb al-Shukūk ʿalā Jālīnūs of Muḥammad Ibn Zakariyyā Al-Rāzī*', in Wael B. Hallaq and Donald P. Little (eds.), *Islamic studies presented to Charles J. Adams*. Brill, Leiden, 1991, 107–116.

Mohaghegh, 'The title of a work of Rāzī with reference to al-tīn al-nīshābūrī' = Mehdi Mohaghegh, 'The title of a work of Rāzī with reference to al-tīn al-nīshābūrī', in *Proceedings of the XXIII International Congress of the History of Medicine, London … 1972*. 2 vols. London: Wellcome Institute of the History of Medicine, 1974, ii:1073–1076.

Monferrer Sala, 'Al-Yabrūdī' = Juan Pedro Monferrer Sala, 'Al-Yabrūdī', in *Christian-Muslim Relations. A Bibliographical History* (General Editor David Thomas). Available at Brill Online, 2013, at http://www.encquran.brill.nl/entries/christian-muslim-relations/al-yabrudi-COM_23417.

Monneret de Villard, *Qasr eš-Šamʿ* = U. Monneret de Villard, *Ricerche sulla topografia di Qasr eš-Šamʿ*, in *Bull. Soc. Géog. Egypte*, 12 (1923–1924), 205–232, 13 (1924–1925), 73–94.

Monroe, *Hispano-Arabic Poetry* = James T. Monroe, *Hispano-Arabic Poetry: A Student Anthology*. Berkeley, California: University of California Press, 1974.

Moosa, 'A new source on Aḥmad ibn al-Ṭayyib al-Sarakhsī' = Matti Moosa, 'A new source on Aḥmad ibn al-Ṭayyib al-Sarakhsī: Florentine MS Arabic 299', *Journal of the American Oriental Society*, 92/1 (1972), 19–24.

Moraux, *Listes* = Paul Moraux, *Les listes anciennes des ouvrages d'Aristote*. Louvain: Éditions Universitaires de Louvain, 1951.

Moreh, *Live Theatre* = Shmuel Moreh, *Live Theatre and Dramatic Literature in the Medieval Arabic World*. Edinburgh: Edinburgh University Press, 1992.

Morelon, *Oeuvres d'astronomie* = Régis Morelon, *Oeuvres d'astronomie: Thābit ibn Qurra; texte établi et traduit par Régis Morelon*. Paris: Belles Lettres, 1987.

Morelon, 'Le Corpus des manuscrits Arabes des œuvres d'astronomie de Thābit b. Qurra' = Régis Morelon, 'Le Corpus des manuscrits Arabes des œuvres d'astronomie de Thābit b. Qurra', in Yusuf Ibish (ed.), *Proceedings of the Fourth Conference of Al-Furqān Islamic Heritage Foundation, 29th–30th November 1997*. London: Al-Furqān Islamic Heritage Foundation, 1999, 115–130.

Morier, *The Adventures of Hajji Baba* = James Morier, *The Adventures of Hajji Baba of Ispahan*. Edited with Introduction and Notes by C.W. Stewart. [The World's Classics] London: Oxford Univerity Press, 1923.

Morray, *An Ayyubid Notable* = David Morray, *An Ayyubid Notable & his World. Ibn al-ʿAdīm and Aleppo as Portrayed in his Biographical Dictionary of People Associated with the City*. Leiden: Brill, 1994.

Mougin, 'Amawar (ou Mizwar)' = L. Mougin, 'Amawar (ou Mizwar)', in Gabriel Camps (ed.), *Encyclopédie Berbère*. Leuven: Peeters, 1987, iv:623–629.

Mourad, *La Physiognomonie arabe et le Kitāb al-firāsā de Fakhr al-Dīn al-Rāzī* = Youssef

Mourad, *La Physiognomonie arabe et le Kitāb al-firāsā de Fakhr al-Dīn al-Rāzī*. Paris: P. Geuthner, 1939.

Mousley, *Abdollatiphi Bagdadensis* = John Mousley, *Abdollatiphi Bagdadensis vita, auctore Ibn Abi Osaiba, e codicibus MSS. Bodleianis descripsit, et Latine vertit*. Oxford: e typographeo Clarendoniano, 1808.

Moussaid, 'Qawlihi fī sharḥ al-āthār' = Mohamed Moussaid, "Min qawlihi fī sharḥ al-āthār al-'ulwiyyah", in *Dafātīr majmū'āt al-baḥth fī l-falsafah al-islāmiyyah* [= *Les cahiers du groupe de recherche de philosophie islamique*, Fez, Centre des Études Ibn Rushd], 2 (1999), 7–81.

Moustafa & Sperl, 'The Cosmic Script' = Ahmed Moustafa, Stefan Sperl, Jeremy Henzell-Thomas, *The Cosmic Script: Sacred Geometry and the Science of Arabic Penmanship*. 2 vols. London: Thames & Hudson, 2014.

al-Mubarrad, *Kāmil* = Abū l-'Abbās Muḥammad ibn Yazīd al-Mubarrad, *al-Kāmil fī l-lughah wa-l-adab*, ed. 'Abd al-Ḥamīd Hindāwī. 4 vols. Beirut: Dār al-Kutub al-'Ilmiyyah, 1419/1999.

al-Mubashshir ibn al-Fātik, *Dicts and Sayings* = Mubashshir ibn al-Fātik, *Dicts and Sayings of the Philosophers, the Translations made by Stephen Scrope, William Worcester, and an anonymous translator*, ed. Curt F. Buehler. (Early English Text Society, no. 211). London: Oxford University Press, 1941.

al-Mubashshir ibn al-Fātik, *Mukhtār al-ḥikam* = *see* Badawī, *Al-Mubashshir ibn Fātik*.

Mufaḍḍaliyyāt = *al-Mufaḍḍaliyyāt*, ed. Aḥmad Muḥammad Shākir and 'Abd al-Salām Muḥammad Hārūn. 4th ed. Cairo: Dār al-Ma'ārif, n.d.

Muḥammad *et al.*, *Taqdimat al-ma'rifah* = Māhir 'Abd al-Qādir Muḥammad, 'Alī 'Abd al-Mu'ṭī Muḥammad and Muḥammad 'Abduh Maḥjūb (eds.), *Kitāb Sharḥ Taqdimat al-ma'rifah li-l-Dakhwār*. Alexandria: Dār al-Ma'rifah al-Jāmi'iyyah, 2000.

Muhanna, *The World in a Book* = Elias Muhanna, *The World in a Book: al-Nuwayrī and the Islamic Encyclopedic Tradition*. Princeton: Princeton University Press, 2018.

Mujeeb Khan, 'Islam' = M.A. Mujeeb Khan, Ch. vi: 'Islam' in Gary B. Ferngren and Ekaterina N. Lomperis (eds.), *Essential Readings in Medicine & Religion*. Baltimore: Johns Hopkins University Press, 2017, 162–187.

Müller, 'Das arabische Verzeichniss der Aristotelischen Schriften' = Müller, August, 'Das arabische Verzeichniss der Aristotelischen Schriften', in *Morgenländische Forschungen. Festschrift Herrn Professor Dr. H.L. Fleischer*. Leipzig: F.A. Brockhaus, 1875, 1–32.

Müller, 'Lesarten' = August Müller, 'Lesarten. Vergleichung der Recensionen. Verbesserungen', in Ibn Abī Uṣaybi'ah, *'Uyūn al-anbā' fī ṭabaqāt al-aṭibbā'*, ed. A. Müller. Vol. 2. Königsberg: Selbstverlag, 1884, pp. liv + 80. Repr. in Sezgin (ed.), *Studies on Ibn Abī Uṣaibi'a*, i:53–138.

Müller, *Nahrungsmittel* = Juliane Müller, *Nahrungsmittel in der arabischen Medizin: Das Kitāb al-Aġḏiya wa-l-Ashriba des Naǧīb ad-Dīn as-Samarqandī*. Leiden: Brill, 2017.

Müller, 'Text' = August Müller, *Über Text und Sprachgebrauch von Ibn Abî Uṣeibi'a's*

Geschichte der Ärzte (Sitzungsberichte der philosophisch-philologischen und historischen Classe der Königlich Bayerischen Akademie der Wissenschaften zu München 1884). Munich, 1885. Repr. in Sezgin (ed.), *Studies on Ibn Abī Uṣaibiʿa*, i:139–263.

Müller, 'Über Ibn Abī Oçeibiʿa' = August Müller, 'Über Ibn Abī Oçeibiʿa und seine Geschichte der Aerzte', in *Actes du 6ème Congrès International des Orientalistes tenu en 1883 à Leide, Deuxième partie, Section 1: Sémitique*, Leiden, 1885, 257–280. Repr. in Sezgin, *Studies* II, 191–214.

Munk, 'Notice sur Joseph Ben-Jehoudah' = Salomon Munk, 'Notice sur Joseph Ben-Jehoudah ou Aboul'hadjādj Yousouf ben Ya'hya al-Sabti al-Maghrebi, disciple de Maïmonide', *Journal Asiatique*, 3rd ser., no. 14 (1842), 5–70.

al-Muqaddasī, *Aḥsan al-taqāsim* = Muḥammad ibn Aḥmad al-Muqaddisī, *Aḥsan al-taqāsim*, ed. Shakir Luʿaybī. Abū Dhabi: Dār al-Suwaydī, 2003.

Mūsā ibn Maymūn *see* Maimonides.

Muslim, *Ṣaḥīḥ* = Muslim ibn Ḥajjāj, *al-Musnad al-ṣaḥīḥ*, ed. Muḥammad Fu'ād 'Abd al-Bāqī. 4 vols. Beirut: Dār Iḥyā' Turāth al-'Arabī, 1956–1972.

Muslim ibn al-Walīd, *Dīwān* = *Sharḥ Dīwān Ṣarīʿ al-Ghawānī Muslim ibn al-Walīd al-Anṣārī*, ed. Sāmī al-Dahhān. Cairo: Dār al-Maʿārif, 1985.

al-Mutanabbī, *Dīwān* = Abū l-Ṭayyib al-Mutanabbī, *Dīwān* (with commentary by al-Wāḥidī), ed. F. Dieterici. Berlin: Mittler, 1861.

al-Muṭarrizī, *al-Mughrib* = Nāṣir ibn 'Abd al-Sayyid al-Muṭarrizī, *al-Mughrib fī tartīb al-muʿrib*, ed. Maḥmūd Fākhūrī and 'Abd al-Ḥamīd Mukhtār. 2 vols. Aleppo: Maktabat Usāmah ibn Zayd, 1979–1982.

Nagy, *Abhandlungen* = Albino Nagy (ed.), *Die philosophischen Abhandlungen des Jaʿqūb ben Isḥāq al-Kindī*. Münster: Verlag der Aschendorffschen Buchhandlung, 1897.

Narshakhī, *Tārīkh-i Bukhārā* = Abū Bakr Muḥammad ibn Jaʿfar al-Narshakhī, *Tārīkh-i Bukhārā*. Tihrān: Kitābfurūshī-i Sanā'ī, [1317], 1938.

al-Nashshābī al-Irbilī, *Mudhākarah* = Abū l-Majd Asʿad ibn Ibrāhīm al-Nashshābī al-Irbilī al-Kātib, *al-Mudhākarah fī alqāb al-shuʿarā'*, ed. Shākir al-ʿĀshūr. Baghdad: Dār al-Shu'ūn al-Thaqāfiyyah al-ʿĀmmah, 1988.

Nāṣir-i Khusraw, *Zād al-musāfirīn* = Ḥakīm-i Nāṣir-i Khusraw, *Kitāb Zād al-musāfirīn*. Berlin: Kaviani, 1923.

Nasrallah, 'Abū-l-Faraj al-Yabrūdī' = Joseph Nasrallah, 'Abū-l-Faraj al-Yabrūdī, médecin chrétien de Damas xe–xie s.', *Arabica*, 23 (1976), 13–22.

Nasrallah, *Annals* = Nawal Nasrallah, *Annals of the Caliphs' Kitchens. Ibn Sayyār al-Warrāq's Tenth-Century Baghdadi Cookbook* (Islamic History and Civilization, 70). Leiden: Brill, 2007.

Nasrallah, *Histoire* = Joseph Nasrallah, *Histoire du mouvement littéraire dans l'église melchite du ve au xxe siècle*. 4 vols. Louvain: Peeters, 1979.

Nawādir al-makhṭūṭāt = *Nawādir al-makhṭūṭāt*, ed. 'Abd al-Salām Hārūn, 2 vols. Cairo: Maktabat Muṣṭafā al-Bābī al-Ḥalabī, 1972–1973; repr. Cairo, 1991.

al-Nawājī, *Ḥalbat al-kumayt* = Shams al-Dīn Muḥammad ibn al-Ḥasan al-Nawājī, *Ḥalbat al-kumayt fī l-adab wa-l-nawādir wa-l-fukāhāt al-mutaʿalliqah bi-l-khamriyyāt*. Cairo: al-Maktabah al-ʿAllāmiyyah, 1357/1938.

Nebbia, 'Ibn al-Haytham nel millesimo anniversario' = G. Nebbia, 'Ibn al-Haytham nel millesimo anniversario della nascita', *Physis*, 9 (1967), 165–214.

Negri, *Vita Gabrielis* = Salomon Negri, *Vita Gabrielis filii Bachtishuæ*, in John Freind, *The History of Physick: From the Time of Galen to the Beginning of the Sixteenth Century*, 2 vols. London: J. Walthoe, 1725–1726.

Nemoy, *Yale* = Leon Nemoy, 'Arabic Manuscripts in the Yale University Library', *Transactions of the Connecticut Academy of Arts and Sciences*, 40 (1956), 1–273.

Neubauer, 'Die Theorie vom *īqāʿ*' = Neubauer, 'Die Theorie vom *īqāʿ*: II. Übersetzung des *Kitāb Iḥṣāʾ al-īqāʿāt* von Abū Naṣr al-Fārābī', *Oriens*, 34 (1994), 103–173.

New DSB see DSB.

Newman, *The Sultan's Sex Potions* = Daniel L. Newman (ed.), *The Sultan's Sex Potions. Arab Aphrodisiacs in the Middle Ages* [= A Critical Edition of Naṣīr al-Dīn al-Ṭūsī's *Kitāb albāb al-bāhiyyah wa-l-tarākīb al-sulṭāniyyah*]. London: Saqi Books, 2014.

Nicholson, *A Literary History* = Reynold A. Nicholson, *A Literary History of the Arabs*. London: T.F. Unwin, 1907, repr. Cambridge: Cambridge University Press, 1956.

Nicholson, *Studies in Islamic Poetry* = R.A. Nicholson, *Studies in Islamic Poetry*. Cambridge: Cambridge University Press, 1921.

Nicolae, 'Ibn Jumayʿ' = Daniel S. Nicolae, 'A mediaeval court physician at work: Ibn Jumayʿ's commentary on the *Canon of Medicine*'. D.Phil. diss., University of Oxford, 2012. [available on-line].

Niebyl, 'Non-Naturals' = P.H. Niebyl, 'The Non-Naturals', *Bulletin of the History of Medicine*, 45/5 (1971), 486–492.

Niebyl, 'Old Age' = P.H. Niebyl, 'Old Age, Fever, and the Lamp Metaphor', *Journal of the History of Medicine and Allied Sciences*, 26 (1971), 351–368.

Niẓāmī ʿArūḍī Samarqandī, *Chahār Maqāla* = Niẓāmī ʿArūḍī Samarqandī, *Chahār Maqāla* (*The Four Discourses*). Translated into English by Edward G. Browne (E.J.W. Gibb Memorial series). London: Luzac 1921.

North, *Horoscopes and History* = J.D. North, *Horoscopes and History* (Warburg Institute Surveys and Texts XIII). London: The Warburg Institute 1986.

Northedge, *Historical Topography of Samarra* = Alastair Northedge, *The Historical Topography of Samarra* (Samarra Studies. 1). 2nd rev. ed. London: The British School of Archaeology in Iraq, 2007.

Northrup, 'Al-Bīmāristān al-Manṣūrī' = Linda S. Northrup, 'Al-Bīmāristān al-Manṣūrī – Explorations: The Interface Between Medicine, Politics and Culture in Early Mamluk Egypt', in Stephan Conermann (ed.), *History and Society during the Mamluk Period (1250–1517)*. (Studies in the Annemarie Schimmel Research College, 1). Göttingen: V&R Unipress / Bonn: Bonn University Press, 2014, 107–142.

al-Nuʿaymī, *al-Dāris* = ʿAbd al-Qādir ibn Muḥammad al-Nuʿaymī, *al-Dāris fī taʾrīkh al-madāris*, ed. Ibrāhīm Shams al-Dīn. 2 vols. Beirut: Dār al-Kutub al-ʿIlmiyyah, 1990.

Nusseibeh, *Avicenna's al-Shifāʾ* = Sari Nusseibeh, *Avicenna's Al-Shifāʾ: Oriental Philosophy*. London: Routledge, 2018.

Nutton, 'Avoiding Distress' = Vivian Nutton, 'Avoiding Distress', in P.N. Singer (ed.), *Galen: Psychological Writings*, Cambridge: Cambridge University Press, 2013, 43–106.

Nutton, *Galeni De praecognitione* = Vivian Nutton, *Galeni De praecognitione / Galen, A Treatise on Prognosis* (Corpus Medicorum Graecorum, 8, 1). Berlin: Akademie-Verlag, 1979.

Nutton, *Medicine* = Vivian Nutton, *Ancient Medicine*, 2nd ed. London: Routledge 2012.

Nutton, 'Murders and Miracles' = Vivian Nutton, 'Murders and Miracles: Lay Attitudes towards Medicine in Classical Antiquity', in Roy Porter (ed.), *Patients and Practitioners: Lay Perceptions of Medicine in Pre-Industrial Society*. Cambridge: Cambridge University Press, 1985, 23–54.

Nutton, 'Seeds' = Vivian Nutton, 'The Seeds of Disease: An Explanation of Contagion and Infection from the Greeks to the Renaissance', *Medical History*, 27 (1983), 1–34.

Nutton & Bos, *On Problematical Movements* = Vivian Nutton and Gerrit Bos, *Galen, On Problematical Movements*. Cambridge: Cambridge University Press, 2011.

al-Nuwayrī, *Nihāyat al-arab* = Shihāb al-Dīn Aḥmad ibn ʿAbd al-Wahhāb al-Nuwayrī, *Nihāyat al-arab fī funūn al-adab*. Cairo: Dār al-Kutub / al-Hayʾah al-Miṣriyyah al-ʿĀmmah li-l-Kitāb, 1342/1923–.

al-Nuwayrī, *Nihāyat al-arab* (Beirut) = Shihāb al-Dīn Aḥmad ibn ʿAbd al-Wahhāb al-Nuwayrī, Nihāyat al-arab fī funūn al-adab, ed. Mufīd Qamīhah et al., 33 vols. Beirut: Dār al-Kutub al-ʿIlmiyyah, 2004.

Olsson, 'Reputation' = Josh T. Olsson, 'The Reputation of Ḥunayn ibn Isḥāq in Contemporaneous and Later Sources', *Journal of Abbasid Studies*, 3 (2016), 29–55.

Omar, 'The Significance of the Colours of Banners' = Farouk Omar, 'The Significance of the Colours of Banners in the Early ʿAbbāsid Period', in F. Omar, *ʿAbbāsiyyat: Studies in the History of the Early ʿAbbāsids*. Baghdad: University of Baghdad, 1976, 148–154.

Orfali, 'Works' = B. Orfali, 'The Works of Abū Manṣūr al-Thaʿālibī (350–429/961–1039)', *Journal of Arabic Literature*, 40 (2009), 273–318.

Ottaviani, *De radiis* = Didier Ottaviani, *De radiis / Théorie des arts magiques*. Paris: Editions Allia, 2003.

Ouyang, 'Literature and Thought' = Wen-Chin Ouyang, 'Literature and thought: Re-reading al-Tawḥīdī's transcription of the debate between logic and grammar', in Maurice A. Pomerantz and Aram A. Shahin (eds.), *The heritage of Arabo-Islamic learning: studies presented to Wadad Kadi* (Islamic History and Civilization, 122). Leiden: Brill, 2016, 444–460.

Overwien, Oliver, *Hippocratis De Humoribus / Hippokrates. Über die Säfte* (Corpus Medicorum Graecorum, 1/3, 1). Berlin: De Gruyter, 2014.

Paavilainen, *Medieval Pharmacotherapy* = H.M. Paavilainen, *Medieval Pharmacotherapy, Continuity and Change: Case Studies from Ibn Sīnā and Some of His Late Medieval Commentators*. Leiden: Brill, 2009.

Pahlitzsch, *Aerzte ohne Grenzen* = Johannes Pahlitzsch, 'Aerzte ohne Grenzen: Melkitische, jüdische und samaritanische Aerzte in Aegypten und Syrien zur Zeit der Kreuzzüge', in Florian Steger und Kay Peter Jankrift (eds.): *Gesundheit-Krankheit. Kulturtransfer medizinischen Wissens von der Spätantike bis in die Frühe Neuzeit*. Köln-Weimar-Wien: 2004, 101–119.

Pahlitzsch, *Melkites in Fatimid Egypt* = Johannes Pahlitzsch, 'The Melkites in Fatimid Egypt and Syria (1021–1171)', *Medieval Encounters*, 21 (2015), 485–515.

Pavlov, *Abu'l Barakat al-Baghdadi's metaphysical philosophy* = Moshe M. Pavlov, *Abu'l Barakat al-Baghdadi's metaphysical philosophy: The Kitab Al-Mu'tabar*. London: Routledge, 2017.

Pavlov, *Abu'l Barakat al-Baghdadi's scientific philosophy* = Moshe M. Pavlov, *Abu'l Barakat al-Baghdadi's scientific philosophy: The Kitab Al-Mu'tabar*. London: Routledge, 2016.

Peacock & Yildiz, *The Seljuks of Anatolia* = A.C.S. Peacock and Sara Nur Yildiz (eds.), *The Seljuks of Anatolia: The Court and Society in the Medieval Middle East*. London / New York: I.B. Tauris, 2013.

Pellegrin, 'Corpus biologique' = Pierre Pellegrin, 'Aristote de Stagire – Corpus biologique', in Goulet, *Dictionnaire*, i:472–481.

Penelas, *Kitāb Hurūsiyūs* = Penelas, Mayte, *Kitāb Hurūsiyūs (traducción árabe de las* Historiae adversus paganos *de Orosio)*. Madrid: Universidad Autónoma, 1998.

Penn, *Envisioning Islam* = Michael Philip Penn, *Envisioning Islam. Syriac Christians and the Early Muslim World*. Philadelphia, Pennsylvania: University of Pennsylvania Press, 2015.

Perho, *The Prophet's Medicine* = Irmeli Perho, *The Prophet's Medicine: A Creation of the Muslim Traditionalist Scholars* (Studia Orientalia, 74). Helsinki: The Finnish Oriental Society, 1995.

Perry, *Medieval Arab Cookery* = Charles Perry, A.J. Arberry, Maxime Rodinson, *Medieval Arab Cookery: Essays and Translations*. Totness, Devon: Prospect Books, 2001.

Peters, *Aristoteles Arabus* = Francis E. Peters, *Aristoteles Arabus: the oriental translations and commentaries of the Aristotelian Corpus*. Leiden: Brill, 1967.

Petit, *Galien, Le médecin. Introduction* = Caroline Petit, *Galien, tome III. Le médecin. Introduction*. Paris: Les Belles Lettres, 2009.

Petraitis, *The Arabic version of Aristotle's Meteorology* = Casimir Petraitis, *The Arabic version of Aristotle's Meteorology: a critical edition with an introduction and Greek-Arabic glossaries*. Beirut: Dār al-Mashriq, 1967.

Pfaff, *Kommentar* = Franz Pfaff, 'Kommentar des Proklos zu Platons Timaios c. 43 (89e–90c)', in J.M. Schmutte (ed.), *Galeni de consuetudinibus* (Corpus Medicorum Graecorum, Suppl. III). Leipzig: Teubner, 1941.

Phillips, *Life and Legend of the Sultan Saladin* = Jonathan Phillips, *The Life and Legend of the Sultan Saladin*. London: The Bodley Head, 2019.

Pinault, *Hippocratic Lives* = Jody Rubin Pinault, *Hippocratic Lives and Legends*. Leiden: Brill, 1992.

Pines, 'Aristotle's *Politics*' = Shlomo Pines, 'Aristotle's *Politics* in Arabic Philosophy', in *Israel Oriental Studies* 5 (1975) 150–160 = S. Pines, *Studies in Arabic Versions of Greek Texts and in Mediaeval Science*. Jerusalem Magnes Press / Leiden: Brill, 1986, 146.

Pines, 'La loi naturelle' = Shlomo Pines, 'La loi naturelle et la société. La doctrine politico-théologique d'Ibn Zur'a, philosophe chrétien de Bagdad', in U. Heyd (ed.), *Studies in Islamic History and Civilisation* (Scripta Hierosolymitana, 9). Jerusalem: The Hebrew University of Jerusalem, 1961, 154–190.

Pines, 'On the Pseudo-Platonic Quotation' = Shlomo Pines, 'On the Pseudo-Platonic Quotation in the *Kuzari* and the *K. al-Nawāmīs*', in Shlomo Pines, *Studies in the History of Jewish Thought*. Jerusalem: Magnes Press, 1997, 236–239.

Pines, 'Rāzī critique' = Shlomo Pines, 'Rāzī critique de Galien', in *Actes du VIIIe Congrès international d'Histoire des Sciences* (Paris, 1953) 480–487; repr. in *Studies in Arabic Versions of Greek Texts and Medieval Science* (*The Collected Works of Shlomo Pines*, 2). Jerusalem: Magnes Press / Leiden: Brill, 1986, 256–263.

Pingree, *Astral Omens to Astrology* = David Pingree, *From Astral Omens to Astrology: From Babylon to Bīkāner*. Rome: Istituto Italiano per l'Africa e l'Oriente, 1997.

Pingree, *Census of the Exact Sciences in Sanskrit* = David Pingree, *Census of the Exact Sciences in Sanskrit*. 5 vols. Philadelphia, PA: American Philosophical Society, 1970–1994.

Pingree, 'Plato's Hermetic Book of the Cow' = David Pingree, 'Plato's Hermetic Book of the Cow', *Transactions of the American Philosophical Society*, 104 (2014), 463–475.

Pingree, 'The Ṣābians of Ḥarrān' = David Pingree, 'The Ṣābians of Ḥarrān and the Classical Tradition', *International Journal of the Classical Tradition*, 9/1 (2002), 8–35.

Pingree, *The Sayings of Abū Maʿshar* = David Pingree, 'The *Sayings of Abū Maʿshar* in Arabic, Greek, and Latin', in G. Marchetti, O. Rignani, V. Sorge (eds.), *Ratio et superstitio: Essays in Honor of Graziella Federici Vescovini*. Turnhout: Brepols, 2003, 41–57.

Pingree, *Thousands* = David Pingree, *The Thousands of Abū Maʿshar*. London: The Warburg Institute, 1968.

Pomerantz, *Licit Magic: The Life and Letters of al-Ṣāḥib b. ʿAbbād* = Maurice Pomerantz, *Licit Magic: The Life and Letters of al-Ṣāḥib b. ʿAbbād (d. 385/995)* (Islamic History and Civilization, 146). Leiden: Brill, 2017.

Pormann, 'Arabo-Islamic Tradition' = Peter E. Pormann, 'Arabo-Islamic Tradition', in Peter E. Pormann (ed.), *The Cambridge Companion to Hippocrates*. Cambridge: Cambridge University Press, 2018, 340–361.

Pormann, 'Al-Fārābī' = Peter E. Pormann, 'Al-Fārābī, the Melancholic Thinker and Philosopher Poet', *Journal of the American Oriental Society*, 135/2 (2015), 209–224.

Pormann, 'Isḥāq ibn 'Imrān' = Peter E. Pormann, 'Isḥāq ibn 'Imrān on "Scholarly Melancholy"', in Peter E. Pormann (ed.), *Rufus of Ephesus, On Melancholy*. Tübingen: Mohr Siebeck, 2008, 289–294.

Pormann, 'Islamic Hospitals' = Peter E. Pormann, 'Islamic Hospitals in the time of al-Muqtadir', in J. Nawas (ed.), *Abbasid Studies II: Occasional Papers of the School of 'Abbasid Studies, Leuven, 28 June – 1 July 2004*. Leuven: Dudley Peters, 2009, 337–382.

Pormann, 'Al-Kaskarī' = Peter E. Pormann, 'Al-Kaskarī (10th cent.) and the quotations of classical authors: a philological study', in I. Garofalo, A. Lami and A. Roselli (eds.), *Sulla tradizione indiretta dei testi medici greci. Atti del II Seminario Internazionale di Sienna, Certosa di Pontignano, 19–20 Settembre 2008*. Pisa / Rome: Fabrizio Serra Editore, 2009, 105–138.

Pormann, *Mirror of Health* = Peter E. Pormann, *The mirror of health: Discovering Medicine in the Golden Age of Islam*. London: Royal College of Physicians, 2013.

Pormann, 'Philosophical Topics' = Peter E. Pormann, 'Philosophical Topics in Medieval Arabic Medical Discourse. Problems and Prospects', in Adamson & Pormann, *Philosophy and Medicine*, 10–33.

Pormann, 'Al-Rāzī on the benefits of sex' = Peter E. Pormann, 'Al-Rāzī on the benefits of sex: a clinician caught between philosophy and medicine', in Arnoud Vrolijk and Jan P. Hogendijk (eds.), *O ye gentlemen: Arabic studies on science and literary culture, in honour of Remke Kruk*. Leiden: Brill, 2007, 115–127. Repr. in Peter E. Pormann (ed.) *Islamic medical and scientific tradition*. 4 vols. London & New York: Routledge, 2011, ii:134–145.

Pormann, *Rufus of Ephesus* = Peter E. Pormann (ed.). *Rufus of Ephesus: On Melancholy*. Tübingen: Mohr Siebeck, 2008.

Pormann, 'Theory and Practice' = Peter E. Pormann, 'Theory and Practice in the early Hospitals in Baghdad – Al-Kaškarī on rabies and melancholy', *Zeitschrift für Geschichte der Arabisch-Islamischen Wissenschaften*, 15 (2002–2003), 197–248.

Pormann, 'Yuḥannā ibn Sarābiyūn' = Peter E. Pormann, 'Yuḥannā ibn Sarābiyūn: Further Studies into the Transmission of his Works', *Arabic Sciences and Philosophy*, 14 (2004), 233–262.

Pormann et al., *1001 Cures* = Peter E. Pormann et al (eds). *1001 Cures, Contributions in Medicine & Healthcare from Muslim Civilisation*. Manchester: Foundation for Science, Technology and Civilisation, 2018.

Pormann et al. 'Palladius' = Peter E. Pormann et al., 'The Enigma of the Arabic and Hebrew Palladius', *Intellectual History of the Islamicate World*, 5 (2017), 252–310.

Pormann & Joosse, 'Aphorisms' = Peter E. Pormann and N. Peter Joosse, 'Commentaries on the Hippocratic Aphorisms in the Arabic Tradition: The Example of Melancholy', in Peter E. Pormann (ed.): *Epidemics in Context. Greek Commentaries on Hippocrates in the Arabic Tradition* (Scientia Graeco-Arabica, 8). Berlin / Boston: De Gruyter, 2012, 211–249.

Pormann & Savage-Smith, *Medieval Islamic Medicine* = Peter E. Pormann and Emilie Savage-Smith, *Medieval Islamic Medicine*. Edinburgh: Edinburgh University Press, 2007.

Pormann & Selove, *Two New Texts* = Peter E. Pormann and Emily Selove, 'Two new texts on medicine and natural philosophy by Abū Bakr al-Rāzī'. *Journal of the Americam Oriental Society*, 137/2 (2017), 279–299.

Porphyry, *Introduction* = Porphyry, *Introduction*; translated with a commentary by Jonathan Barnes. Oxford: Clarendon Press, 2003.

Porter, Saif & Savage-Smith, 'Medieval Islamic Amulets' = Venetia Porter, Liana Saif, and Emilie Savage-Smith, 'Medieval Islamic Amulets, Talismans, and Magic', in *A Companion to Islamic Art and Architecture. Volume I: From the Prophet to the Mongols*, ed. Finbarr Barry Flood and Gülru Necipoğlu. Hoboken, NJ: Wiley Blackwell, 2017, 521–557.

Pradines & Rahmat Khan, 'Fāṭimid gardens' = Stephanie Pradines & Sher Rahmat Khan, 'Fāṭimid gardens: archaeological and historical perspectives', *Bulletin of the School of Oriental and African Studies*, 79 (2016), 473–502.

Price, *Rituals and Power* = Simon R.F. Price, *Rituals and Power. The Roman Imperial Cult in Asia Minor*. Cambridge: Cambridge University Press, 1984.

Prioreschi, *History of Medicine* = Plinio Prioreschi, *History of Medicine. Vol. IV: Byzantine and Islamic Medicine*. Omaha: Horatius Press, 2001.

Prüfer & Meyerhof, 'Die angebliche Augenheilkunde des Thabit b. Qurra' = C. Prüfer and M. Meyerhof, 'Die angebliche Augenheilkunde des Thabit b. Qurra', *Centralblatt für praktische Augenheilkunde*, 35 (1911), 4–8, 38–41.

Ptolemy, *Tetrabiblos* = Ptolemy, *Tetrabiblos*, ed. and tr. by F.E. Robbins (Loeb Classical Library). Cambridge, Mass.: Harvard University Press, 1940.

PUA = Prosopografía de los Ulemas de al-Andalus (Online Database).

Puertas, *Linajes de poder* = Jiménez Puertas, *Linajes de poder en la Loja islámica. De los Banu Jalid a los Alatares (siglos VIII–XV)*. Granada: Fundación Ibn al-Jaṭīb, 2009.

Puig, *Libro de la Generación y la Corrupción* = Josep Puig, *Avempace. Libro de la Generación y la Corrupción*. Madrid: CSIC, 1995.

Puig, 'Materials on Averroes' = Josep Puig, 'Materials on Averroes's Circle', *Journal of Near Eastern Studies*, 51/4 (1992), 241–260.

Pummer, *Early Christian Authors* = Reinhard Pummer, *Early Christian Authors on Samaritans and Samaritanism: Texts, Translations and Commentary*. Tübingen: Mohr Siebeck, 2002.

Qabīṣī, *Introduction to Astrology* = Abū al-Ṣaqr ʿAbd al-ʿAzīz ibn ʿUthmān al-Qabīṣī, *Al-Qabīṣī (Alcabitius): The Introduction to Astrology. Editions of the Arabic and Latin texts and an English translation*, ed. Charles Burnett, Keiji Yamamoto, and Michio Yano (Warburg Institute Studies and Texts, 2). London: The Warburg Institute, 2004.

al-Qāḍī, 'Biographical Dictionaries' = Wadād al-Qāḍī, 'Biographical Dictionaries: Inner Structure and Cultural Significance', in George N. Atiyeh (ed.), *The Book in the Islamic World: The Written Word and Communication in the Middle East*. Albany: State University of New York Press, 1995, 93–122.

al-Qāḍī, '*Kitāb Ṣiwān al-ḥikma*' = Wadād al-Qāḍī, '*Kitāb Ṣiwān al-ḥikma*: Structure, Composition, Authorship and Sources', *Der Islam*, 58 (1981), 87–124.

al-Qāḍī l-Nuʿmān, *Iftitāḥ* = al-Qāḍī l-Nuʿmān, *K. Iftitāḥ al-daʿwah*, ed. Faraḥāt al-Dashrāwī. Tunis: al-Sharika al-Tūnisiyyah li-l-Tawzīʿ, 1986.

Qalʿah'jī & al-Wafāʾī, *Amrād al-ʿayn* = Muḥammad Rawwās Qalʿah'jī & Muḥammad Ẓāfir al-Wafāʾī, *Amrāḍ al-ʿayn wa-muʿālajātuhā min kitābay al-ḥikmah taʾlīf Abī l-Ḥasan ibn Muḥammad al-Ṭabarī wa-Firdaws al-ḥikmah taʾlīf ʿAlī ibn Sahl Rabban al-Ṭabarī*. London: al-Furqan Islamic Heritage Foundation, 1998.

al-Qālī, *Amālī* = Abū ʿAlī Ismāʿīl ibn al-Qāsim al-Qālī, *al-Amālī*. 2 vols. Cairo: Dār al-Kutub al-Miṣriyyah, 1926, repr. Beirut: Dār al-Āfāq al-Jadīdah, n.d.

al-Qalqashandī, *Ṣubḥ* = Abū l-ʿAbbās Aḥmad al-Qalqashandī, *Ṣubḥ al-aʿshā fī ṣināʿat al-inshā*. 14 vols. Cairo: Dār al-Kutub, 1913–1922.

al-Qamarī, *K. al-Tanwīr* (ed. Karmi) = Abū Manṣūr al-Ḥasan ibn Nūḥ al-Qamarī, *Kitāb al-Tanwīr fī al-iṣṭilāḥāt al-ṭibbiyyah*, ed. Ghādah Ḥasan al-Karmī. Riyadh: Maktab al-Tarbiyah al-ʿArabī li-Duwal al-Khalīj, 1991.

al-Qazwīnī, *ʿAjāʾib* (Wüstenfeld) = Zakariyyā ibn Muḥammad ibn Maḥmūd al-Qazwīnī, *ʿAjāʾib al-makhluqāt wa-gharāʾib al-mawjūdāt / Kosmographie*, ed. Ferdinand Wüstenfeld. Göttingen: Verlag der Dieterichschen Buchhandlung, 1849.

al-Qazwīnī, *Āthār al-bilād* = Zakariyyā ibn Muḥammad ibn Maḥmūd al-Qazwīnī, *Āthār al-bilād wa-akhbār al-ʿibād*. Beirut: Dār Ṣādir, n.d.

al-Qifṭī, *Muḥammadūn* = Abū l-Ḥasan ʿAlī ibn Yūsuf al-Qifṭī, *al-Muḥammadūn min al-shuʿarāʾ*, ed. Muḥammad ʿAbd al-Sattār Khān, 2 vols. Hyderabad: Dāʾirat al-Maʿārif al-ʿUthmāniyyah, 1966–1967.

al-Qumrī *see* al-Qamarī.

Qusṭā ibn Lūqā, 'Abhandlung über die Ansteckung' = Hartmut Fähndrich, 'Abhandlung über die Ansteckung von Qusṭā ibn Lūqā', *Abhandlungen für die Kunde des Morgenlandes*, 48/2 (1987), 1–44.

Qusṭā ibn Lūqā, *Les Arithmétiques* = Diophante, *Les Arithmétiques: texte établi et traduit par Roshdi Rashed*. Paris: Belles Lettres, 1984.

Qusṭā ibn Lūqā, *Books IV to VII of Diophantus' Arithmetica* = Jacques Sesiano, *Books IV to VII of Diophantus' Arithmetica in the Arabic translation attributed to Qusṭā ibn Lūqā*. New York: Springer-Verlag, 1982.

Qusṭā ibn Lūqā, *Das Buch über die Geschlechtlichkeit* = N.A. Barhoum, *Das Buch über die Geschlechtlichkeit (Kitāb fī l-bāh) von Qusṭā Ibn Lūqā*. Erlangen: Seminar für Geschichte der Medizin, 1974.

Qusṭā ibn Lūqā, *Une correspondance islamo-chrétienne* = Qusṭā ibn Lūqā, *Une corres-*

pondance islamo-chrétienne entre Ibn al-Munağğim, Ḥunayn ibn Isḥāq et Qusṭā ibn Lūqā, ed. and tr. Khalil Samir and Paul Nwyia. Turnhout: Brepols, 1981.

Qusṭā ibn Lūqā, *Introduction to Geometry* = J.P. Hogendijk, 'The Introduction to Geometry by Qusṭā ibn Lūqā: Translation and Commentary', *Suhayl*, 8 (2008), 163–221.

Qusṭā ibn Lūqā, *'Livre des caractères'* = Paul Sbath, 'Le Livre des caractères de Qostâ ibn Loûqâ, grand savant et célèbre médecin chrétien au IXe siècle', *Bulletin de l'Institut d'Égypte*, 23 (1941), 103–169.

Qusṭā ibn Lūqā, *Medical Regime* = Gerrit Bos, *Qusṭā ibn Lūqā's Medical Regime for the Pilgrims to Mecca, The* Risāla fī tadbīr safar al-ḥajj, *edited with translation and commentary*. Leiden: E.J. Brill, 1992.

Qusṭā ibn Lūqā, *On Numbness* = Lena Ambjörn, *Qusṭā ibn Lūqā On Numbness. A Book on Numbness, its Kinds, Causes and Treatment according to the opinion of Galen and Hippocrates*. Stockholm: Almqvist & Wiksell International, 2000.

Qusṭā ibn Lūqā, 'On Sleeplessness' = Oliver Kahl, 'Qusṭā ibn Lūqā On Sleeplessness', *Journal of Semitic Studies*, 43 (1998), 311–326.

Qusṭā ibn Lūqā, 'On the Use of Celestial Globe' = W.H. Worrell, 'Qusta ibn Luqa on the Use of the Celestial Globe', *Isis*, 35 (1944), 285–293

Quṭb al-Dīn al-Shīrāzī, *Sharḥ Ḥikmat al-Ishrāq* = Maḥmūd ibn Masʿūd Kāzarūnī maʿrūf bi-Quṭb al-Dīn Shīrāzī, *Sharḥ-i Ḥikmat al-Ishrāq-i Suhravardī*. ed. ʿAbd Allāh Nūrānī and Mahdī Muḥaqqiq. Tehran: Mu'assasah-'i Muṭālaʿāt-i Islāmī, Dānishgāh-i Tihrān, Dānishgāh-i Mak'gīl, 2001.

Raby, 'Terra Lemnia' = Julian Raby, 'Terra Lemnia and the Potteries of the Golden Horn: An Antique Revival under Ottoman Auspices', in Stephanos Efthymiadis, Claudia Rapp, and Dimitris Tsougarakis (eds.), *Bosphorus: Essays in Honour of Cyril Mango Presented in Oxford, 6 July, 1995* (Byzantinische Forschungen). Amsterdam: Hakert, 1995, 305–342.

Ragab, *Medieval Islamic Hospital* = Ahmed Ragab, *The Medieval Islamic Hospital: Medicine, Religion, and Charity*. Cambridge University Press, 2015.

Ragab, *Piety and Patienthood* = Ahmed Ragab, *Piety and Patienthood in Medieval Islam*. London / New York: Routledge, 2018.

Ragep & Wallis, *The Herbal of al-Ghāfiqī* = F. Jamil Ragep & Faith Wallis (eds.), *The Herbal of al-Ghāfiqī: A facsimile edition of MS 7508 in the Osler Library of the History of Medicine, McGill University, with Critical Essays*. Montreal: McGill-Queen's University Press, 2014.

Raggetti, *ʿĪsā ibn ʿAlī's Book* = Lucia Raggetti, *ʿĪsā ibn ʿAlī's Book on the Useful Properties of Animal Parts: Edition, translation and study of a fluid tradition* (Science, Technology, and Medicine in Ancient Cultures). Berlin: De Gruyter, 2018.

al-Rāghib al-Iṣbahānī, *Muḥāḍarāt* = Abū l-Qāsim Ḥusayn ibn Muḥammad al-Rāghib al-Iṣbahānī, *Muḥāḍarāt al-udabāʾ wa-muḥāwarāt al-shuʿarāʾ*. 2 vols. Būlāq: Maṭbaʿat Ibrāhīm al-Muwayliḥī, AH 1287.

al-Rahim, *The Creation of Philosophical Tradition* = Ahmed H. al-Rahim, *The Creation of Philosophical Tradition: Biography and the Reception of Avicenna's Philosophy from the Eleventh to the Fourteenth Centuries*. Wiesbaden: Harrassowitz, 2018.

Rajak, *Translation* = T. Rajak, *Translation and Survival. The Greek Bible of the Jewish Diaspora*. Oxford: Oxford University Press, 2009.

Rante & Afround, *Rayy: from its Origins to the Mongol Invasion* = Rocco Rante and Ghadir Afround, *Rayy: from its Origins to the Mongol Invasion. An Archaeological and Historiographical Study* (Arts and archaeology of the Islamic world, 4). Leiden: Brill, 2015.

Rapoport & Savage-Smith, *Egyptian Guide* = Yossef Rapoport and Emilie Savage-Smith, *An Eleventh-Century Egyptian Guide to the Universe: The 'Book of Curiosities,' edited and translated*. Leiden: Brill, 2014.

Rapoport & Savage-Smith, *Lost Maps of the Caliphs* = Yossef Rapoport and Emilie Savage-Smith, *Lost Maps of the Calilphs: Drawing the world in eleventh-century Cairo*. Oxford: Bodleian Library / Chicago: University of Chicago Press, 2018.

Rapoport, *Rural Society* = Yossef Rapoport, *Rural Society in Medieval Islam*, online resource, at http://www2.history.qmul.ac.uk/ruralsocietyislam/index.html.

Rapoport, *Rural Economy and Tribal Society* = Yossef Rapoport, *Rural Economy and Tribal Society in Islamic Egypt: A Study of al-Nābulusī's* 'Villages of the Fayyum' (The Medieval Countryside, 19). Turnhout. Brepols Publishers, 2018.

Rashed, 'Configuration of the universe' = Rashed, Roshdi, 'The configuration of the universe: A book by al-Ḥasan ibn al-Haytham?', *Revue d'histoire des sciences*, 60 (2007), 47–63.

Rashed, '*De Generatione et Corruptione*' = Marwan Rashed, 'Aristote de Stagire – De Generatione et Corruptione. Tradition arabe', in Goulet, *Dictionnaire*, Suppl., 304–314.

Rashed, *Mathématiques infinitésimales* I = Roshdi Rashed, *Les mathématiques infinitésimales du IXe au XIe. Vol I: Fondateurs et commentateurs: Banū Mūsā, Ibn Qurra, Ibn Sinān, al-Khāzin, al-Qūhī, Ibn al-Samḥ, Ibn Hūd*. London: Al-Furqān Islamic Heritage Foundation, 1993.

Rashed, *Mathématiques infinitésimales* II = Roshdi Rashed, *Les mathématiques infinitésimales du IXe au XIe. Vol II: On Ibn al-Haytham*. London: Al-Furqān Islamic Heritage Foundation, 1993.

Rashed, *Mathématiques infinitésimales* IV =Roshdi Rashed, *Les mathématiques infinitésimales du IXe au XIe siècle. Vol. IV*. London: Al-Furqān Islamic Heritage Foundation, 2002.

Rashed, *Mathématiques infinitésimales* V = Roshdi Rashed, *Les mathématiques infinitésimales du IXe au XIe. Vol V: Ibn al-Haytham, astronomie, géométrie sphérique et trigonométrie*. London: Al-Furqān Islamic Heritage Foundation, 2006.

Rashed, *New evidence* = Marwan Rashed, 'New evidence on the critique of the Qur'ānic miracle at the end of the third/ninth century: *Qusṭā ibn Lūqā* vs. the Banū al-

Munajjim', in Peter Adamson (ed.), *In the age of al-Fārābī: Arabic philosophy in the fourth/tenth century*. London: Warburg Institute / Turin: Nino Aragno, 2008, 277–293.

Rashed, 'Priorité' = Marwan Rashed, 'Priorité de l'ΕΙΔΟΣ ou du ΓΕΝΟΣ entre Andronicos et Alexandre: vestiges arabes et grecs inédits', *Arabic Sciences and Philosophy*, 14 (2004), 9–63.

Rashed, *Thābit ibn Qurra* = Roshdi Rashed (ed.), *Thābit ibn Qurra: Science and Philosophy in Ninth-century Baghdad*. New York / Berlin: W. de Gruyter, 2009.

Rashed, *Theodosius's On Habitations* = Roshdi Rashed, 'Theodosius's On Habitations', *Journal for the History of Astronomy*, 44/2 (2013), 226–227.

Rashed & Jolivet, *Œuvres Philosophiques et Scientifiques* = Roshdi Rashed et Jean Jolivet, *Œuvres philosophiques et scientifiques d'Al-Kindī*. 2 vols. Leiden: Brill, 1997–1998.

Rāshid, *Ibrāhīm Ibn Sinān* = Rushdī Rāshid & Hélène Bellosta, *Ibrāhīm Ibn Sinān, Logique Et Géométrie Au xe Siècle*. Leiden: Brill, 2000.

Raven, '*De Mundo*' = Wim Raven, 'Aristote de Stagire – *De Mundo*. Tradition syriaque et arabe', in Goulet, *Dictionnaire*, Suppl. 481–483.

Raymond, *Cairo* = André Raymond, *Cairo*. Cambridge, MA: Harvard University Press, 2000.

Raynaud, *Ibn al-Haytham's On the Shape of the Eclipse* = Dominique Raynaud, *A Critical Edition of Ibn al-Haytham's* On the Shape of the Eclipse. *The First Experimental study of the Camera Obscura*. Cham, Switzerland: Springer, 2016.

Raynaud, 'Ibn al-Haytham's Solar Eclipse Record' = Dominique Raynaud, 'A Tentative Astronomical Dating of Ibn al-Haytham's Solar Eclipse Record', *Nuncius*, 29 (2014), 324–358.

al-Rāzī, *al-Ḥāwī* (Beirut) = Abū Bakr Muḥammad ibn Zakariyyā al-Rāzī, *al-Ḥāwī fī l-ṭibb*. ed. Haytham Khalīfah Tuʿaymī. 23 vols in 7. Beirut: Dār Iḥyā' al-Turāth al-ʿArabī, 2002.

al-Rāzī, *al-Ḥāwī* (ed. Ḥarbī) = Abū Bakr Muḥammad ibn Zakariyyā al-Rāzī, *al-Ḥāwī fī l-ṭibb*, ed. Khālid Ḥarbī. 10 vols. Alexandria: Dār al-Wafā', 2012.

al-Rāzī, *al-Ḥāwī* (Hyderabad) = Abū Bakr Muḥammad ibn Zakariyyā al-Rāzī, *al-Ḥāwī fī l-ṭibb*. 23 vols. Hyderabad: Dā'irat al-Maʿārif al-ʿUthmāniyyah, 1955–1971.

al-Rāzī, *K. fī l-Judarī wa-l-ḥaṣbah* (Greenhill) = Abū Bakr al-Rāzī, *A Treatise on the Small-Pox and Measles*, ed. and tr. W.A. Greenhill. London: Sydenham Society, 1848; repr. Birmingham, AL: Classics of Medicine Library, 1987.

al-Rāzī, *al-Mudkhal* = María de la Concepción Vázquez de Benito, *Libro De La Introducción Al Arte De La Medicina O "isagoge"*. Salamanca: Ediciones Universidad de Salamanca, Instituto Hispano-Árabe de Cultura, 1979.

al-Rāzī, *al-Mudkhal al-taʿlīmī* = Abū Bakr Muḥammad ibn Zakariyyā al-Rāzī, *al-Mudkhal al-ta'līmī, yā, Rāhnamā-yi āmūzish-i ṣanʿat-i kīmiyā*, ed. Ḥasan ʿAlī Shaybānī. Tihrān: Intishārāt-i Danishgāh-i Tihrān, 1346 [1967/68].

al-Rāzī, *M. fī l-Niqris* = Abū Bakr Muḥammad ibn Zakariyyā al-Rāzī, *Maqālah fī l-Niqris*, ed. Yūsuf Zaydān. Alexandria: Bibliotheca Alexandrina, 2003.

al-Rāzī, *al-Qūlanj* = al-Rāzī, *Kitāb al-Qūlanj* (*Le livre de la colique*), *édition et traduction*, ed. and tr. Ṣubhī Maḥmūd Ḥammāmī. Aleppo: Maʿhad al-Turāth al-ʿIlmī al-ʿArabī, 1983.

al-Rāzī, *al-Shukūk* = Abū Bakr Muḥammad Ibn Zakariyyā al-Rāzī, *Kitāb al-Shukūk ʿalā Jālīnūs*, ed. Mehdi Mohaghegh. Tehran: Maʿhad al-Dirāsāt al-Islāmiyyah, Jāmiʿat Ṭihrān; Kuala Lumpur: al-Maʿhad al-ʿĀlī al-ʿĀlamī li-l-Fikr wa-l-Ḥaḍārah al-Islāmiyyah, 1993.

al-Rāzī, *Traité sur le calcul* (de Koning) = al-Rāzī, *Traité sur le calcul dans les reins et dans la vessie*, traduction accompagnée du texte par P. de Koning. Leiden: Brill, 1896.

RE = *Realencyclopädie der classischen Altertumswissenschaft*. Munich: A. Druckenmüller, 1894–1978.

Reinink, 'Ṣharbokht bar Msargis' = G.J. Reinink, 'Ṣharbokht bar Msargis', in Sebastian P. Brock (ed.), *Gorgias Encyclopedic Dictionary of the Syriac Heritage*. Piscataway, NJ: Gorgias Press, 2011, 371.

Reiske, *Excerpta* = J.J. Reiske, *Excerpta ex Ibn Abí Uszeibi'ae historia medicorum illustrium*. Copenhagen, Det Kongelige Bibliotek, MS N. 31 (CLXXI). Unpublished Latin translations made in 1746 of a few extracts.

Reisman, 'Plato's *Republic* in Arabic' = David C. Reisman, 'Plato's *Republic* in Arabic: A Newly Discovered Passage', *Arabic Sciences and Philosophy*, 14/2 (2004), 263–300.

Reisman, *The Making of the Avicennian Tradition* = David Column Reisman, *The Making of the Avicennian Tradition. The Transmission, Contents, and Structure of Ibn Sīnā's* Al-Mubāḥaṭāt (The Discussions). [IPTS vol. 49] Leiden: Brill, 2002.

Rescher, *Bibliography* = Nicholas Rescher, *Al-Farabi: An Annotated Bibliography*. Pittsburgh: University of Pittsburgh Press, 1962.

Rescher, *Galen and the Syllogism* = Nicholas Rescher, *Galen and the Syllogism. An Examination of the Thesis that Galen Originated the Fourth Figure of the Syllogism in the Light of New Data from Arabic Sources*. Pittsburgh: University of Pittsburgh Press, 1966.

Rescher, *Al-Kindī* = Nicholas Rescher, *Al-Kindī. An Annotated Bibliography*. Pittsburgh: University of Pittsburgh Press, 1964.

Rescher, *Short Commentary* = Nicholas Rescher, *Al-Farabi's Short Commentary on Aristotle's Prior Analytics*. Pittsburgh: University of Pittsburgh Press, 1963.

Rescher & Marmura, *Refutation of Galen* = Nicholas Rescher and Michael Marmura, *The Refutation by Alexander of Aphrodisias of Galen's Treatise on the Theory of Motion*. Islamabad: Islamic Research Institute, 1965.

Reynolds, *Interpreting the Self* = Dwight F. Reynolds et al., *Interpreting the Self: Autobiography in the Arabic Literary Tradition*. Berkeley: University of California Press, 2001.

Richards, *History* = D.S. Richards: *The rare and excellent history of Saladin or al-Nawādir al-Sultāniyya wa'l-Maḥāsin al-Yūsufiyya by Bahā' al-Dīn Ibn Shaddād.* Aldershot: Ashgate, 2001.

Richards (tr.), *The Chronicle of Ibn al-Athīr* = D.S. Richards (tr.), *The Chronicle of Ibn al-Athīr for the Crusading Period*, II. Aldershot: Ashgate, 2007.

Richardson, *Difference and Disability* = Kristina L. Richardson, *Difference and Disability in the Medieval Islamic World: Blighted Bodies.* Edinburgh: Edinburgh University Press, 2012.

Richter, 'Beiträge zur Geschichte der Pocken' = Paul Richter, 'Beiträge zur Geschichte der Pocken bei den Arabern', *Sudhoffs Archiv*, 5 (1912), 311–331.

Richter-Bernburg, 'Iran's Contribution to Medicine' = Lutz Richter-Bernburg, 'Iran's Contribution to Medicine and Veterinary Science in Islam AD100–900/AD 700–1500', in John A.C. Greppin, Emilie Savage-Smith, and John L. Gueriguian (eds.), *The Diffusion of Greco-Roman Medicine in the Middle East and the Caucasus*, New York: Caravan Press, 1999, 139–168.

Richter-Bernburg, 'Magie' = Lutz Richter-Bernburg, 'Magie zwischen galenischer und prophetischer Medizin', in Sebastian Günther and Dorothee Pielow (eds.), *Die Geheimnisse der oberen und die unteren Welt: Magie im Islam zwischen Glaube und Wissenschaft* (Islamic History and Civilization, 158). Leiden: Brill, 2018, 178–192.

Richter-Bernberg, 'Pseudo-Ṯābit, Pseudo-Rāzī, Yūḥannā b. Sarābiyūn' = Lutz Richter-Bernberg, 'Pseudo-Ṯābit, Pseudo-Rāzī, Yūḥannā b. Sarābiyūn', *Der Islam* 60 (2009), 48–77.

Richter-Bernburg, 'Thābit ibn Qurra and Ibrāhīm ibn Zahrūn' = Lutz Richter-Bernburg, 'Thābit ibn Qurra and Ibrāhīm ibn Zahrūn and their sons: Two Ṣābian families of physicians', in *Proceedings, I. International Congress on the History of Turkish-Islamic Science and Technology.* Istanbul: İstanbul Teknik Üniversitesi, Bilim ve Teknoloji Tarihi Enstitüsü, 1981, 105–113.

Ritter & Plessner, *Picatrix* = Pseudo-Majrīṭī, *"Picatrix": Das Ziel des Weisen von Pseudo-Mağrīṭī.* Translated into German from the Arabic by Hellmut Ritter and Martin Plessner. London: Warburg Institute, 1962.

Robinson, 'The conquest of Khūzistān' = Chase F. Robinson, 'The Conquest of Khūzistān: A Historiographical Reassessment', *Bulletin of the School of Oriental and African Studies*, 67 (2004), 14–39.

Rocca, 'Trepanation' = Julius Rocca, 'Galen and the Uses of Trepanation', in Robert Arnott, Stanley Finger and C.U.M. Smith (eds.), *Trepanation: History, Discovery, Theory.* Lisse: Swets & Zeitlinger, 2003, 253–271.

Roggema, 'Abdisho bar Bahrīz' = Barbara H. Roggema, 'Abdisho bar Bahrīz', in Sebastian P. Brock et al. (eds.), *Gorgias Encyclopedic Dictionary of the Syriac Heritage.* Piscataway, NJ: Gorgias Press, 2011, 2–3.

Roggema, 'Iyob of Edessa' = Barbara H. Roggema, 'Iyob of Edessa', in: Sebastian P. Brock et al. (Eds.), *Gorgias Encyclopedic Dictionary of the Syriac Heritage*. Piscataway, NJ: Gorgias Press, 2011, 225–226.

Roscher, *Lexicon* = Wilhem H. Roscher, *Ausführliches Lexicon der griechischen und römischen Mythologie*. Leipzig: B.G. Teubner, 1884–1937.

Rosenthal, *Aḥmad b. aṭ-Tayyib as-Saraḫsî* = Rosenthal, *Aḥmad b. aṭ-Tayyib as-Saraḫsî: A scholar and littérateur of the ninth century*. New Haven, Conn.: American Oriental Society, 1943.

Rosenthal, 'An Ancient Commentary' = Franz Rosenthal, 'An Ancient Commentary on the Hippocratic Oath', *Bulletin of the History of Medicine*, 30 (1956), 52–87; repr. in Rosenthal, *Science and Medicine*, item III.

Rosenthal, 'Arabische Autobiographie' = Franz Rosenthal, 'Die arabische Autobiographie', in Franz Rosenthal, G. von Grünebaum and W.J. Fischel (eds.), *Studia Arabica I* (Analecta Orientalia, 14). Rome: Pontificium Institutum Biblicum, 1937, 3–40.

Rosenthal, *Classical Heritage* = Franz Rosenthal, *The Classical Heritage in Islam*. London: Routledge and Kegan Paul, 1975 (tr. by Emile and Jenny Marmorstein of *Das Fortleben der Antike im Islam*, Zürich, 1965).

Rosenthal, 'Defense of Medicine' = Franz Rosenthal, 'The Defense of Medicine in the Medieval Muslim World', *Bulletin of the History of Medicine* 43 (1969), 519–532. Reprinted in Franz Rosenthal, *Man Versus Society in Medieval Islam*, ed. Dimitri Gutas. Leiden / Boston: Brill 2015, 1011–1025.

Rosenthal, 'Eleventh-Century List of Works of Hippocrates' = Franz Rosenthal, 'An Eleventh-Century List of the Works of Hippocrates', *Journal of the History of Medicine and Allied Sciences*, 28 (1973), 156–165; repr. in Rosenthal, *Science and Medicine*, item VII.

Rosenthal, 'Isḥāq b. Ḥunayn' = Franz Rosenthal, 'Isḥāq b. Ḥunayn's *Ta'rīḫ al-aṭibbā'*', *Oriens*, 7 (1954), 55–80; repr. in Rosenthal, *Science and Medicine*, item II.

Rosenthal, "Istanbul Materials for Al-Kindî and As-Saraḫsî." = Franz Rosenthal, "From Arabic Books and Manuscripts VI Istanbul Materials for Al-Kindî and As-Saraḫsî." *Journal of the American Oriental Society*, 76/1 (1956), 27–31.

Rosenthal, '"Life is short, the Art is long"' = Franz Rosenthal, '"Life is short, the Art is long": Arabic Commentaries on the First Hippocratic Aphroism', *Bulletin of the History of Medicine*, 40 (1966), 226–245; repr. in Rosenthal, *Science and Medicine*, item V.

Rosenthal, 'Poetry and Architecture' = Franz Rosenthal, 'Poetry and Architecture: The *Bādhanj*', *Journal of Arabic Literature*, 8 (1977), 1–19.

Rosenthal, 'Prolegomena to an Abortive Edition' = Franz Rosenthal, 'Al-Mubashshir ibn Fātik: Prolegomena to an Abortive Edition', *Oriens*, 13–14 (1960–1961), 132–158.

Rosenthal, 'Pythagorean Documents', I = Franz Rosenthal, 'Some Pythagorean Documents Transmitted in Arabic. Continued', in *Orientalia*, 10 (1941), 383–395.

Rosenthal, 'Pythagorean Documents', II = Franz Rosenthal, 'Some Pythagorean Documents Transmitted in Arabic', in *Orientalia*, 10 (1941), 104–115.

Rosenthal, 'As-Sarakhsī (?) on the Appropriate Behavior for Kings' = Franz Rosenthal. 'From Arabic Books and Manuscripts, XVI: As-Sarakhsī (?) on the Appropriate Behavior for Kings' *Journal of the American Oriental Society*, 115/1 (1995), 105–109.

Rosenthal, 'As-Saraḫsî on Love' = Franz Rosenthal, 'From Arabic Books and Manuscripts, VIII: As-Saraḫsî on Love', *Journal of the American Oriental Society* 81/3 (1961) 222–224.

Rosenthal, 'A Short Treatise' = Franz Rosenthal, 'A Short Treatise on the Meaning of the Names of Some Greek Scholars Attributed to al-Fârâbî', *Journal of the American Oriental Society*, 62/1 (1942), 73–74.

Rosenthal, *Science and Medicine* = Franz Rosenthal, *Science and Medicine in Islam: A Collection of Essays* (Collected Studies, 330). Aldershot: Variorum, 1975.

Rosenthal, *Technique and Approach* = Franz Rosenthal, *The Technique and Approach of Muslim Scholarship*. Rome: Pontificium Institutum Biblicum, 1947.

Rosenthal & Walzer, *Alfarabius* = Franz Rosenthal and Richard Walzer, *Alfarabius de Platonis philosophia*. London: Warburg Institute, 1943.

Rudolph, 'Abū Naṣr al-Fārābī' = Ulrich Rudolph, 'Abū Naṣr al-Fārābī', in Ulrich Rudolph and Renate Würsch (eds.), *Philosophie in der islamischen Welt*. Basel: Schwabe Verlag, 2012, 363–457.

Rudolph, *Doxographie* = Ulrich Rudolph, *Die Doxographie des Pseudo-Ammonios: ein Beitrag zur Neuplatonischen Überlieferung im Islam*. Wiesbaden: F. Steiner, 1989.

Rudolph, Hansberger & Adamson, *Philosophy in the Islamic World* = Ulrich Rudolph, Rotraud Hansberger and Peter Adamson (eds.), *Philosophy in the Islamic World. Volume 1: 8th–10th Centuries*. [Handbook of Oriental Studies: Near and Middle East, vol. 115] Leiden: Brill, 2017.

Ruhāwī, *Adab al-ṭabīb* (Levey) = Martin Levey, *Medical ethics of medieval Islam: with special reference to al-Ruhāwī's Practical ethics of the physician* (Transactions of the American Philosophical Society, New Ser., 57, pt. 3). Philadephia: American Philosophical Society, 1967.

Ruhāwī, *Adab al-ṭabīb* (facs.) = Isḥāq ibn 'Alī l-Ruhāwī, *Kitāb Adab al-ṭabīb, Conduct of the Physician* (Ser. C, Facsimile editions, 18). Frankfurt am Main: Institut für Geschichte der Arabisch-Islamischen Wissenschaften, 1985 [Facsimile reproduction of MS 1658 Selimiye Library, Edirne].

Ruhāwī, *Adab al-ṭabīb* (Riyadh edn.) = Isḥāq ibn 'Alī l-Ruhāwī, *Kitāb Adab al-ṭabīb*, ed. Kamāl al-Sāmarrā'ī and Dāwūd Salmān 'Alī. Repr. Tehran: Mu'assasa-i Muṭāla'āt-i Tārīkh-i Pizishkī Ṭibb-i Islāmī, 2008.

Ruland, *Die arabischen Fassungen* = Hans-Jochen Ruland, *Die arabischen Fassungen von zwei Schriften des Alexander von Aphrodisias. Über die Vorsehung und Über das liberum arbitrium*. Saarbrücken: Universität des Saarlandes, 1976.

Ruland, *Zwei arabische Fassungen* = Hans-Jochen Ruland, *Zwei arabische Fassungen der Abhandlung des Alexander von Aphrodisias über die Universalia ⟨Quaestio I 11 a⟩*. Göttingen: Vandenhoeck and Ruprecht, 1979.

Ruska, *Das Steinbuch* = Julius Ruska, *Das Steinbuch des Aristoteles mit literargeschichtlichen Untersuchungen*. Heidelberg: Carl Winter, 1912.

Sabra, 'One Ibn al-Haytham or two?' = A.I. Sabra, 'One Ibn al-Haytham or two? An Exercise in Reading the Bio-Bibliographical Sources', *Zeitschrift für Geschichte der Arabisch-Islamischen Wissenschaften*, 12 (1998), 1–50.

Sabra, 'One Ibn al-Haytham or two? Conclusion' = A.I. Sabra, 'One Ibn al-Haytham or two? Conclusion', *Zeitschrift für Geschichte der Arabisch-Islamischen Wissenschaften*, 15 (2002–2003), 95–108.

Sabra, *Optics* = A.I. Sabra, *The Optics of Ibn al-Haytham. Books I–III On Direct Vision. Translated with Introduction and Commentary*. 2 vols. London: The Warburg Institute, 1989.

Sabra, 'Simplicius Proof' = A.I. Sabra, 'Simplicius's Proof of Euclid's Parallels Postulate', *Journal of the Warburg and Courtauld Institutes*, 32 (1969), 1–24.

Sabra, 'Solution of Difficulties Concerning *Iltifāf*' = A.I. Sabra, 'Ibn al-Haytham's Treatise: Solution of Difficulties Concerning the Movement of Solution of Difficulties Concerning *Iltifāf*', *Journal for the History of Arabic Science*, 3 (1979), 388–420.

Sābūr ibn Sahl, *ʿAḍudī recension* = *Sābūr ibn Sahl's Dispensatory in the Recension of the ʿAḍudī Hospital*, ed. and tr. Oliver Kahl (Islamic Philosophy, Theology, and Science, 78). Leiden: Brill, 2009.

Sābūr ibn Sahl, *Dispensatorium* = Sābūr ibn Sahl, *Dispensatorium parvum / al-Aqrābādhīn al-ṣaghīr*, analysed, ed. and annot. by Oliver Kahl (Islamic Philosophy, Theology, and Science, 16). Leiden: Brill, 1994.

Sābūr ibn Sahl, *Small Dispensatory* = Oliver Kahl, *Sābūr ibn Sahl. The Small Dispensatory. Translated from the Arabic, together with a study and glossaries* (Islamic philosophy, theology, and science, 53). Leiden: Brill 2003.

al-Ṣafadī, *Nakt al-himyān* = Ṣalāḥ al-Dīn Khalīl ibn Aybak al-Ṣafadī, *Nakt al-himyān fī nukat al-ʿumyān*, ed. Aḥmad Zakī. Cairo: al-Maṭbaʿah al-Jamāliyyah, 1329/1911.

al-Ṣafadī, *Nuṣrat al-thāʾir* = Ṣalāḥ al-Dīn Khalīl ibn Aybak al-Ṣafadī, *Nuṣrat al-thāʾir ʿalā l-Mathal al-sāʾir*, ed. Muḥammad ʿAlī Sulṭān. Damascus: Majmaʿ al-Lughah al-ʿArabiyyah, [1971, date of preface].

al-Ṣafadī, *Tawshīʿ* = Ṣalāḥ al-Dīn Khalīl ibn Aybak al-Ṣafadī, *Tawshīʿ al-tawshīḥ*, ed. Albīr Ḥabīb Muṭlaq. Beirut: Dār al-Thaqāfah, 1966.

al-Ṣafadī, *Wāfī* = al-Ṣafadī, Ṣalāḥ al-Dīn Khalīl ibn Aybak, *al-Wāfī bi-l-Wafayāt / Das biographische Lexikon des Ṣalāhaddīn Ḫalīl ibn Aibak aṣ-Ṣafadī* (Bibliotheca Islamica, 6). 30 vols. Beirut / Wiesbaden / Berlin: Franz Steiner – Klaus Schwarz, 1931–2005.

Sahlān, *Mukhtaṣar fī l-ṭīb* = Paul Sbath, '*Mukhtaṣar fī l-ṭīb*, abrégé sur les arômes par Sahlān ibn Kaisān, médecin chrétien melchite égyptien du calife al-Aziz mort en 990', *Bulletin de l'Institut d'Egypte*, 26 (1943–1944), 183–213.

Sahner, 'From Augustine to Islam' = Christian C. Sahner, 'From Agustine to Islam: Translation and History in the Arabic Orosius', *Speculum*, 88/4 (2013), 905–931.

Sahner, *Christian Martyrs under Islam* = Christian C. Sahner, *Christian Martyrs under Islam: Religious Violence and the Making of the Muslim World*. Princeton: Princeton University Press, 2018.

Ṣāʿid al-Andalusī, *Science in the Medieval World* = Ṣāʿid al-Andalusī, *Science in the Medieval World* "Book of the Categories of Nations", tr. and ed. by Semaʿan I. Salem and Alok Kuma. [History of Science Series, No. 5] Austin: University of Texas Press, 1991.

Ṣāʿid al-Andalusī, *Ṭabaqāt* (Bū ʿAlwān) = Ṣāʿid al-Andalusī, *Ṭabaqāt al-umam*, ed. Ḥayāt Bū ʿAlwān. Beirut: Dār al-Ṭalīʿah, 1985.

Ṣāʿid al-Andalusī, *Ṭabaqāt* (Cheikho) = Abū l-Qāsim Ṣāʿid ibn Aḥmad ibn Ṣāʿid al-Andalusī, *Kitāb Ṭabaqāt al-umam*, ed. Louis Cheikho. Beirut: al-Maṭbaʿah al-Kāthūlīkiyyah li-l-Ābāʾ al-Yasūʿiyyīn, 1912.

Saʿīd ibn al-Biṭrīq, *Annalenwerk* (Breydy) = Michel Breydy, *Das Annalenwerk des Eutychios von Alexandrien: ausgewählte Geschichten und Legenden kompiliert von Saʿīd bin Baṭrīq um 935 A.D.* (Corpus Scriptorum Christianorum Orientalium, Scriptores Arabici, 44, 45). Louvain: E. Peeters, 1985.

Saʿīd ibn al-Biṭrīq, *Etudes* (Breydy) = Michel Breydy, *Études sur Saʿīd ibn Baṭrīq et ses sources* (Corpus scriptorum Christianorum Orientalium, 450; Corpus Scriptorum Christianorum Orientalium, Subsidia, 69). Louvain: E. Peeters, 1983.

Saʿīd ibn al-Biṭrīq, *Naẓm al-*jawhar (Selden/Pococke) = Saʿīd ibn al-Biṭrīq, *Naẓm al-jawhar* / Contextio gemmarum, sive, Eutychii patriarchæ Alexandrini annales, ed. John Selden, Latin tr. by Edward Pococke. 2 vols. Oxford, 1654–1691.

Saʿīd ibn Hibat Allāh, *al-Ḥudūd wa-l-furūq* = Abū l-Ḥasan Saʿīd ibn Hibat Allāh ibn al-Ḥusayn, *al-Ḥudūd wa-l-furūq*, ed. Ghulām ʿAlī Yaʿqūbī. Beirut: Majmaʿ al-Buḥūth al-Islāmiyyah li-l-Dirāsāt wa-l-Nashr, 1995.

Saʿīd ibn Hibat Allāh, *Khalq al-insān* = Abū l-Ḥasan Saʿīd ibn Hibat Allāh ibn al-Ḥusayn, *Khalq al-insān*, ed. Yaḥyā Murād. Beirut: Manshūrāt Maḥmūd ʿAlī Bayḍūn, Dār al-Kutub al-ʿIlmiyyah, 2003.

Saʿīd ibn Hibat Allāh, *al-Mughnī* = Abū l-Ḥasan Saʿīd ibn Hibat Allāh ibn al-Ḥusayn, *al-Mughnī fī tadbīr al-amrāḍ wa-maʿrifat al-ʿilal wa-l-aʿrāḍ*, ed. Muḥammad Y. Zakkūr. Jedda: Dār al-Minhāj li-l-Nashr wa-l-Tawzīʿ, 2011.

Saidan, *The Arithmetic of al-Uqlīdisī* = A.S. Saidan, *The Arithmetic of Al-Uqlīdisī: The Story of Hindu-Arabic Arithmetic as told in Kitāb al-Fuṣūl fī al-Ḥisāb al-Hindī, by Abū al-Ḥasan, Aḥmad Ibn Ibrāhīm al-Uqlīdisī, written in Damascus in the year 341 (A.D. 952/3)*. Dordrecht: Reidel, 1978.

St. Augustine, *De Quantitate Animae* = St. Augustine, *The greatness of the soul; The teacher*, tr. and annotated by J.M. Colleran. Westminster, Md: Newman Press / London: Longmans, Green & Co., 1950.

Saint-Laurent, *Missionary Stories and the Formation of the Syriac Churches* = Jeanne-Nicole Mellon Saint-Laurent, *Missionary Stories and the Formation of the Syriac Churches*. Oakland, California: University of California Press, 2015.

al-Sakhāwī, *al-Maqāṣid al-ḥasanah* = Shams al-Dīn Muḥammad ibn ʿAbd al-Raḥmān al-Sakhāwī, *al-Maqāṣid al-ḥasanah fī bayān kathīr min al-aḥādīth al-mushtahirah ʿalā l-alsinah*, ed. ʿAbd Allāh Muḥammad al-Ṣadīq and ʿAbd al-Wahhāb ʿAbd al-Laṭīf. Beirut: Dār al-Kutub al-ʿIlmiyyah, 1979.

Salazar, *War Wounds* = Christine F. Salazar, *The Treatment of War Wounds in Graeco-Roman Antiquity*. Leiden: Brill, 2000.

Saliba, 'Al-Bīrūnī and the Science of his Time' = George Saliba, 'Al-Bīrūnī and the Science of his Time', in M.J.L. Young, J.D. Latham, and R.B. Serjeant (eds.), *Religion, Learning and Science in the 'Abbasid Period* (Cambridge History of Arabic Literature). Cambridge: Cambridge University Press, 1990, 405–423.

Saliba, *Mu'ayyad al-Dīn al-ʿUrḍī* = George Saliba, *The Astronomical Work of Mu'ayyad al-Dīn al-ʿUrḍī (Kitāb al-Hay'a): A Thirteenth Century Reform of Ptolemaic Astronomy* (*in Arabic with English introduction*). Beirut: Markaz Dirāsāt al-Waḥdah al-ʿArabiyyah, 1990.

Sālim, *K. Jālīnūs ilā Ghulūqun* = Salīm Sālim, *K. Jālīnūs ilā Ghulūqun fī l-ta'attī li-shifā' al-amrāḍ*. Cairo: al-Hay'ah al-Miṣriyyah al-ʿĀmmah li-l-Kitāb, 1982.

Salmon, *Topographie du Caire … la Birkat al-Fīl* = Georges Salmon, *Études sur la topographie du Caire: la Kalʿat al-Kabsh et la Birkat al-Fīl* (Institut français d'archéologie orientale du Caire, Mémoires, 7). Cairo: Imprimerie de l'Institut français d'archéologie orientale, 1902.

Salvesen, 'Yaʿqub of Edessa' = Alison G. Salvesen, 'Yaʿqub of Edessa', in Sebastian P. Brock (ed.), *Gorgias Encyclopedic Dictionary of the Syriac Heritage*. Piscataway, NJ: Gorgias Press, 2011, 432–433.

Samama, *Les médecins* = Everlyne Samama, *Les médecins dans le monde grec: sources épigraphiques sur la naissance d'un corps médical*. Geneva: Librairie Droz, 2003.

al-Samʿānī, *Ansāb* = Abū Saʿd ʿAbd al-Karīm ibn Muḥammad al-Samʿānī, *al-Ansāb*, ed. ʿAbdallāh ʿUmar al-Bārūdī. 5 vols. Beirut: Dār al-Jinān, 1988.

Samīr, 'Ḥunayn ibn Isḥāq *M. fī l-ājāl*' = Ḥunayn b. Isḥāq, *Maqāla fī l-ājāl*, ed. Samīr Khalīl Samīr, in *al-Mashriq*, 65 (1991), 403–425.

Samīr, *Ḥunayn ibn Isḥāq* = Samīr Khalīl Samīr, *Ḥunayn ibn Isḥāq fī l-aʿmār wa-l-ājāl*. Beirut: Dār al-Mashriq, 2001.

Samīr, '*Maqālat*' = Samīr Khalīl Samīr, '*Maqālat Ḥunayn ibn Isḥāq fī kayfiyyat idrāk ḥaqīqat al-diyānah*', *al-Mashriq*, 71 (1997), 345–363.

Samīr, 'La place d'Ibn at-Ṭayyib dans la pensée arabe' = Samīr Khalīl Samīr, 'La place d'Ibn at-Ṭayyib dans la pensée arabe', *Journal of Eastern Christian Studies*, 58 (2006), 177–193.

Samīr, 'Un traité perdu de Ḥunayn ibn Isḥāq' = Samīr Khalīl Samīr, 'Un traité perdu de Ḥunayn ibn Isḥāq retrouvé dans la 'Somme' d'Ibn al-ʿAssāl', in *Aram*, 3 (1991), 171–192.

Samuel, *Chronology* = A.E. Samuel, *Greek and Roman Chronology. Calendars and Years in Classical Antiquity*. Munich: Beck, 1972.

Samsó & Berrani, 'World Astrology' = Julio Samsó and Hamid Berrani, 'World Astrology in eleventh-century al-Andalus: The Epistle on *Tasyīr* and the Projection of Rays by al-Istijjī', *Journal of Islamic Studies*, 10 (1999), 293–231.

al-Ṣanawbarī, *Dīwān* = Aḥmad [ibn] Muḥammad ibn al-Ḥasan al-Ḍabbī al-Ṣanawbarī, *Dīwān*, ed. Iḥsān ʿAbbās. Beirut: Dār Ṣādir, 1998.

Sanguinetti, 'Extraits' = B.R. Sanguinetti, 'Extraits de l'ouvrage arabe d'Ibn Aby Ossaïbiʿah sur l'histoire des médicins, traduction française, accompagnée de notes', *Journal Asiatique*, 5th series, 3 (1854), 230–291; 4 (1854), 177–213; 5 (1855), 401–469; 6 (1855), 129–190; 8 (1856), 175–196, 316–353. Repr. in Sezgin, *Studies* III, 16–309.

al-Sarī al-Raffāʾ, *Muḥibb* = al-Sarī ibn Aḥmad al-Raffāʾ, *al-Muḥibb wa-l-maḥbūb wa-l-mashmūm wa-l-mashrūb*, ed. Miṣbāḥ al-Ghalāwunjī and Mājid Ḥasan al-Dhahabī. 4 vols. Damascus: Majmaʿ al-Lughah al-ʿArabiyyah bi-Dimashq, 1986.

al-Sarrāj, *Maṣāriʿ* = Abū Muḥammad Jaʿfar ibn Aḥmad al-Sarrāj al-Qāriʾ, *Maṣāriʿ al-ʿushshāq*. 2 vols. Beirut: Dār Ṣādir, n.d.

Sato, *State* = Tsugitaka Sato, *State and Rural Society in Medieval Islam*. Leiden: Brill, 1997.

Saunders, *Life of Muwaffiq ad-Din* = R.H. Saunders, *Healing through Spirit Agency by the Great Persian Physician Abduhl Latif ("The Man of Baghdad") and information concerning The Life Hereafter of the deepest interest to all enquirers and students of Psychic phenomena*. London: Hutchinson & Co, 1927.

Savage-Smith, 'Attitudes toward dissection' = 'Attitudes toward dissection in medieval Islam,' *Journal of the History of Medicine and Allied Sciences*, 50 (1995), 68–111. Repr, in P.E. Pormann (ed.), *Islamic Medical and Scientific Tradition*. London: Taylor & Francis, 4 vols., 2010, no. 11.

Savage-Smith, *Celestial Globes* = Emilie Savage-Smith, *Islamicate Celestial Globes: Their History, Construction, and Use* (Smithsonian Studies in History and Technology, 46). Washington, DC: Smithsonian Institution, 1985.

Savage-Smith, 'Frankish study of Arabic medical texts' = Emilie Savage-Smith, 'New evidence for the Frankish study of Arabic medical texts in the Crusader period', *Crusades*, 5 (2006), 99–112.

Savage-Smith, 'Galen's Account of the Cranial Nerves' = Emilie Savage-Smith, 'Galen's Account of the Cranial Nerves and the Autonomic Nervous System', *Clio Medica*, 6 (1971), 77–98, 173–194.

Savage-Smith, 'Galen's Lost Ophthalmology' = Emilie Savage-Smith, 'Galen's Lost Ophthalmology and the *Summaria Alexandrinorum*', in Vivian Nutton (ed.), *The Unknown Galen*, (Bulletin of the Institute of Classical Studies, Dictionnaire Suppl. 77). London: Institute of Classical Studies, University of London, 2002, 121–138.

Savage-Smith, 'Ibn al-Nafīs's *Perfected Book on Ophthalmology*' = E. Savage-Smith, 'Ibn al-Nafīs's *Perfected Book on Ophthalmology* and his treatment of trachoma and its sequelae', *Journal for the History of Arabic Science*, 4 (1980), 147–206. Repr. in Peter

Pormann (ed.), *Islamic medical and scientific tradition* (Critical Concepts in Islamic Studies). 4 vols. London: Routledge, 2011, no. 14.

Savage-Smith, 'Medicine in Medieval Islam' in *The Cambridge History of Science. Vol. 2: Medieval Science*, ed. David C. Lindberg & Michael H. Shank. Cambridge: Cambridge University Press, 2013, 139–167.

Savage-Smith, *NCAM-1* = Emilie Savage-Smith, *A New Catalogue of Arabic Manuscripts in the Bodleian Library, Oxford, Vol. 1: Medicine*. Oxford: Oxford University Press, 2011.

Savage-Smith, *NLM* = Emilie Savage-Smith, *Islamic Medical Manuscripts at the National Library of Medicine* [on-line catalogue available at www.nlm.nih.gov/hmd/arabic].

Savage-Smith, 'The Practice of Surgery in Islamic Lands' = Emilie Savage-Smith, 'The Practice of Surgery in Islamic Lands: Myth and Reality', *Social History of Medicine*, 13 (2000), 307–321.

Savage-Smith, 'Sources for editing a medieval Arabic surgical tract' = Emilie Savage-Smith, 'Some sources and procedures for editing a medieval Arabic surgical tract', *History of Science*, 14 (1976), 245–264.

Savage-Smith, 'The Universality and Neutrality of Science' = Emilie Savage-Smith, 'The Universality and Neutrality of Science', in Michael G. Morony (ed.), *Universality in Islamic Thought. Rationalism, Science and Religious Belief*. London: I.B. Tauris, 2014, 157–192.

Savage-Smith, 'The Working Files of Rhazes' = Emilie Savage-Smith, 'The Working Files of Rhazes: Are the *Jāmiʿ* and the *Ḥāwī* Identical?', in R. Hansberger, M. Afifi al-Akiti, and C. Burnett (eds.), *Medieval Arabic Thought. Essays in Honour of Fritz Zimmermann* (Warburg Studies and Texts, 4), London: Warburg Institute / Turin: Nino Aragno Editore, 2012, 163–180.

Savage-Smith, 'Were the Four Humours Fundamental' = Emilie Savage-Smith, 'Were the Four Humours Fundamental to Medieval Islamic Medical Practice?', in Elisabeth Hsu and Peregrine Horden (eds.), *The Body in Balance: Humoral Theory in Practice* (Epistemologies of Healing, 6). Oxford: Berghahn Books, 2013, 89–106.

Savage-Smith & Smith, 'Islamic Geomancy' = E. Savage-Smith & Marion B. Smith, 'Islamic Geomancy and a Thirteenth-Century Divinatory Device: Another Look', in E. Savage-Smith (ed.), *Magic and Divination in Early Islam* (The Formation of the Classical Islamic World, 42). Aldershot: Ashgate, 2004, 211–276.

Sawa, *Rhythmic theories* = George Dimitri Sawa, *Rhythmic theories and practices in Arabic writings to 339 AH/950 CE: annotated translations and commentaries*. Ottawa, Ont.: Institute of Mediaeval Music, c. 2009.

Sbath, 'Le formulaire des hôpitaux d'Ibn abil Bayan' = Paul Sbath, 'Le formulaire des hôpitaux d'Ibn abil Bayan, médicin du bimaristan annacery au Caire au xiiie siècle', *Bulletin de l'Institut d'Égypte*, 15 (1932), 9–78

Sbath, 'Le livre des temps d'Ibn Massawaih' = Paul Sbath, 'Le livre des temps d'Ibn Massawaih', *Bulletin de l'Institut d'Egypte*, 25 (1933), 235–257.

Sbath, *Vingt traités* = *Vingt traités philosophiques et apologétiques d'auteurs arabes Chrétiens du IXe au XIVe siècle*, ed. Paul Sbath. Cairo: Friedrich, 1929.

Sbath & Avierinos, 'Sahlān Ibn Kaysān et Raŝīd al-Dīn Abū Ḥulayqa' = Paul Sbath and Christos D. Avierinos, *Sahlān Ibn Kaysān et Raŝīd al-Dīn Abū Ḥulayqa: deux traités medicaux*' (Publications de l'Institut français d'Archéologie orientale du Caire, Textes arabes et études islamiques, 10). Cairo: Imprimerie de l'Institut français d'archéologie oriental, 1953.

Scalenghe, *Disability* = Sara Scalenghe, *Disability in the Ottoman Arab World, 1500–1800*. Cambridge: Cambridge University Press, 2014.

Schacht & Meyerhof, *Medico-Philosophical Controversy* = Joseph Schacht and Max Meyerhof, *The Medico-Philosophical Controversy between Ibn Butlan of Baghdad and Ibn Ridwan of Cairo. A Contribution to the History of Greek Learning Among the Arabs* (Faculty of Arts publ. 13). Cairo: Cairo University, 1937.

Schefer, *Sefer Nameh* = Charles Henri Auguste Schefer, *Sefer Nameh: relation du voyage de Nassiri Khosrau en Syrie, en Palestine, en Egypte, en Arabie et en Perse, pendant les années de l'hégire 437–444* (*AD 1045–1052*). Paris, 1881, repr. Amsterdam: Philo Press, 1970.

Schibli, *Pherekydes of Syros* = Hermann P. Schibli, *Pherekydes of Syros*. Oxford: Clarendon Press, 1990.

Schmidl, 'al-ʿUrḍī' = Petra G. Schmidl, "ʿUrḍī, Muʾayyad (al-Milla wa-) al-Dīn (Muʾayyad ibn Barīk [Burayk]) al-ʿUrḍī (al-ʿĀmirī al-Dimashqī)', in Thomas Hockey (ed.), *The Biographical Encyclopedia of Astronomers*. 4 vols. New York: Springer, 2007, ii:1161–1162.

Schmitz, *Bildung und Macht* = Thomas Schmitz, *Bildung und Macht: Zur sozialen und politischen Funktion der zweiten Sophistik in der griechischen Welt der Kaiserzeit*. Munich: C.H. Beck, 1997.

Schmucker, *Materia Medica* = Werner Schmucker, *Die pflanzliche und mineralische Materia medica im Firdaus al-Ḥikma des Ṭabarī*. Bonn: im Selbstverlag des Orientalischen Seminars der Universität Bonn, 1969.

Schnell, 'Die Kugel' = Hans Schnell, 'Die Kugel mit dem Schemel (Kurra ḏât al Kursî) nach Alfons X. von Kastilien und seinen arabischen Quellen'. Inaugural dissertation, Erlangen, 1924.

Schoonheim, '*Météorologiques*' = Pieter L. Schoonheim, 'Aristote de Stagire – *Météorologiques*. Tradition syriaque, arabe et latine', in Goulet, *Dictionnaire*, Suppl., 324–328.

Schramm, *Ibn al-Haytham* = Matthias Schramm, *Ibn al-Haythams Weg zur Physik*. Wiesbaden: Franz Steiner, 1963.

Schramm, 'Frederick II of Hohenstaufen and Arabic Science' = Matthias Schramm, 'Frederick II of Hohenstaufen and Arabic Science', *Science in Context*, 14 (2001), 289–312.

Schwarb, 'Early Kalām' = Gregor Schwarb, 'Early *Kalām* and the Medical Tradition', in Adamson & Pormann, *Philosophy and Medicine*, 104–169.

Seeskin, *Cambridge Companion to Maimonides* = Kenneth Seeskin (ed.), *The Cambridge Companion to Maimonides*. Cambridge, Cambridge University Press, 2005.

Sesiano, *Diophantus* = Jacques Sesiano, *Books IV to VII of Diophantus' Arithmetica in the Arabic translation attributed to Qusṭā Ibn Lūqā* (Sources in the History of Mathematics and Physical Sciences, 3). New York: Springer Verlag, 1983.

Sellheim, 'Das Lächeln' = Rudolph Sellheim, 'Das Lächeln des Propheten', in *Festschrift für Ad. E. Jensen*, ed. Eike Haberland et al. Munich: Renner, 1964, 621–630.

Selove, 'Medicine' = Emily Selove, 'Medicine, *Mujūn*, and Microcosm in *Ḥikāyat Abī l-Qāsim al-Baghdādī*', *Journal of Abbasid Studies*, 2 (2015), 107–118.

Sengers, *Women and Demons* = Gerda Sengers, *Women and Demons. Cult Healing in Islamic Egypt*. Leiden: Brill, 2003.

Sepmeijer, 'Ibn al-Tayyib's Commentary on Matthew' = Floris Sepmeijer, 'Ibn al-Tayyib's Commentary on Matthew 1–9:32–34', *Parole de l'Orient*, 25 (2000), 557–564.

Serikoff, *Wellcome Catalogue* = Nicolai Serikoff, *Arabic Medical Manuscripts of the Wellcome Library: A Descriptive Catalogue of the Ḥaddād Collection*. Leiden: Brill, 2005.

Serrano, 'Ibn Rushr al-Ḥafīd (Averroes) and his exile' = Delfina Serrano, 'Ibn Rushd al-Ḥafīd (Averroes) and his exile to Lucena: Jewish ancestry, genealogy and forced conversion', *al-Qanṭara*, 38/2 (2017), 131–152.

Şeşen et al., *Turkish Manuscripts* = R. Şeşen, C. Akpınar, and C. İzgi, *Fihris makhṭūṭāt al-ṭibb al-Islāmī bi-l-lughāt al-ʿArabiyyah wa-l-Turkiyyah wa-l-Fārisiyyah fī maktabāt Turkiyā*. Istanbul, 1984.

Sesiano, 'Un Complément de Ṯābit Ibn Qurra Au Peri Diaireseōn d'Euclide' = J. Sesiano, 'Un Complément de Ṯābit Ibn Qurra Au Peri Diaireseōn d'Euclide'. *Zeitschrift für Geschichte der Arabisch-Islamischen Wissenschaften*, 4 (1987–1988), 149–159.

Seymore, *Life of Ibn Riḍwān* = Jennifer Ann Seymore, 'The Life of Ibn Riḍwān and His Commentary on Ptolemy's *Tetrabiblos*'. Unpublished Ph.D. diss., Columbia University, 2001.

Sezgin, GAS III = Fuat Sezgin, *Geschichte des arabischen Schrifttums. III: Medizin – Pharmazie – Zoologie – Tierheilkunde bis ca 430 H.* Leiden: Brill, 1978.

Sezgin, GAS III N = Fuat Sezgin, 'Nachträge [zu Band III]' in Sezgin, *Geschichte des arabischen Schrifttums*, III, 411–414.

Sezgin, GAS III N5 = Fuat Sezgin, 'Nachträge zu Band III' in Sezgin, *Geschichte des arabischen Schrifttums*, V, 404–416.

Sezgin, GAS III N7 = Fuat Sezgin, 'Nachträge zu Band III' in Sezgin, *Geschichte des arabischen Schrifttums*, VII, 375–391.

Sezgin, GAS IV = Fuat Sezgin, *Geschichte des arabischen Schrifttums. IV: Alchemie–Chemie–Botanik–Agrikultur bis ca 430 H.* Leiden: Brill, 1971.

Sezgin, GAS V = Fuat Sezgin, *Geschichte des arabischen Schrifttums. V: Mathematik bis ca 430 H.* Leiden: Brill, 1974.

Sezgin, GAS VI = Fuat Sezgin, *Geschichte des arabischen Schrifttums. VI: Astronomie bis ca 430 H.* Leiden: E.J. Brill, 1978.

Sezgin, GAS VII = Fuat Sezgin, *Geschichte des arabischen Schrifttums. VII: Astrologie – Meteorologie und Verwandtes bis ca 430 H.* Leiden: Brill, 1979.

Sezgin, *Presentation* = Abu 'l-Qāsim al-Zahrāwī, *A Presentation to Would-Be Authors on Medicine*. Facsimile ed. Fuat Sezgin. Frankfurt am Main: Institut für Geschichte der Arabisch-Islamischen Wissenschaften, 1986.

Sezgin, *Studies* I = Fuat Sezgin (ed.), *Studies on Ibn Abī Uṣaibiʿa (d. 1270) and his* 'Uyūn al-anbā' fī ṭabaqāt al-aṭibbā', *Collected and reprinted, I* (Islamic Medicine, 3). Frankfurt am Main: Institut für Geschichte der Arabisch-Islamischen Wissenschaften, 1995.

Sezgin, *Studies* II = Fuat Sezgin (ed.), *Studies on Ibn Abī Uṣaibiʿa (d. 1270) and his* 'Uyūn al-anbā' fī ṭabaqāt al-aṭibbā', *Collected and reprinted, II* (Islamic Medicine, 4). Frankfurt am Main: Institut für Geschichte der Arabisch-Islamischen Wissenschaften, 1995.

Sezgin, *Studies* III = Fuat Sezgin (ed.), *Studies on Ibn Abī Uṣaibiʿa (d. 1270) and his* 'Uyūn al-anbā' fī ṭabaqāt al-aṭibbā', *Collected and reprinted, III* (Islamic Medicine, 5). Frankfurt am Main: Institut für Geschichte der Arabisch-Islamischen Wissenschaften, 1995.

Sezgin, *al-Zahrawī: Texts and Studies* = Fuat Sezgin (ed.), *Abu 'l-Qāsim al-Zahrāwī. Texte und Studien*, in collaboration with Māzin ʿAmāwī, Carl Ehrig-Eggert, and Eckhard Neubauer. Frankfurt am Main: Institut für Geschichte der Arabisch-Islamischen Wissenschaften, 1996.

Sezgin, *al-Kindī: Texts and Studies* = Fuat Sezgin (ed.), *Abū Yūsuf Yaʿqūb ibn Isḥāq al-Kindī (d. after 256/870): texts and studies*, collected and reprinted by Fuat Sezgin; in collaboration with Mazen Amawi, Carl Ehrig-Eggert, Eckhard Neubauer. Frankfurt am Main: Institut für Geschichte der Arabisch-Islamischen Wissenschaften, 1999.

Sezgin, *ar-Rāzī: Texts and Studies* = Fuat Sezgin (ed.), *Muḥammad ibn Zakarīyā' ar-Rāzī (d. 313/925): texts and studies*, collected and reprinted by Fuat Sezgin; in collaboration with Mazen Amawi, Carl Ehrig-Eggert, Eckhard Neubauer. Frankfurt am Main: Institute for the History of Arabic-Islamic Science at the Johann Wolfgang Goethe University, 1995.

Sezgin, *The Banu Musa and Thabit ibn Qurra: Texts and Studies* = Fuat Sezgin (ed.), *The Banu Musa and Thabit ibn Qurra: their works in western translations and adaptations: texts and studies*, collected and reprinted by Fuat Sezgin; in collaboration with Carl Ehrig-Eggert, and Eckhard Neubauer. Frankfurt am Main: Institut für Geschichte der Arabisch-Islamischen Wissenschaften, 2006.

Sezgin, *The School of Baghdad (4th–5th/10–11th cent.) and its achievements* = Fuat Sezgin (ed.), *The School of Baghdad (4th–5th/10–11th cent.) and its achievements: Matta ibn*

Yunus, Yahya ibn ʿAdi, Ibn Zurʿa, Ibn Suwar, Ibn as-Samh: texts and studies, collected and reprinted by Fuat Sezgin; in collaboration with Carl Ehrig-Eggert, and Eckhard Neubauer. Frankfurt am Main: Institut für Geschichte der Arabisch-Islamischen Wissenschaften, 2000.

Sezgin, *Empedocles* = Fuat Sezgin (ed.), (*Pseudo-*)*Empedocles in the Arabic Tradition: Texts and Studies*. Frankfurt am Main: Institut für Geschichte der Arabisch-Islamischen Wissenschaften, 2000.

al-Shābushtī, *Diyārāt* = Abū l-Ḥasan ʿAlī ibn Muḥammad al-Shābushtī, *al-Diyārāt*, ed. Kūrkīs (Gūrgīs) ʿAwwād. Beirut: Dār al-Rāʾid al-ʿArabī, 1986.

Shafīʿ, *Tatimmat Ṣiwān* = M. [ʿAlī ibn Zayd al-Bayhaqī,] Shafīʿ, *Kitāb Tatimmat Ṣiwān al-ḥikmah*. 2 vols. Lahore: no publ., 1935 [a continuation of the *Ṣiwān al-ḥikmah*, falsely attributed to Abū Sulaymān al-Sijistānī].

al-Shahrastānī, *Milal* = Abū l-Fatḥ Muḥammad ibn ʿAbd al-Karīm al-Shahrastānī, *K. al-Milal wa-l-niḥal*, ed. Aḥmad Fahmī Muḥammad. 3 vols. Beirut: Dār al-Kutub al-ʿIlmiyyah, 1992.

al-Shahrastānī, *Nihāyat al-iqdām* = al-Shahrastānī, *Nihāyat al-iqdām fī ʿilm al-kalām*, ed. Alfred Guillaume, repr. Cairo: Maktabat al-Thaqāfah al-Dīniyyah, 2009.

Shahrazūrī, *Nuzhat al-arwāḥ*, ed. Khūrshīd = Shams al-Dīn al-Shahrazūrī, *Nuzhat al-arwāḥ wa-rawḍat al-afrāḥ*, ed. Khūrshīd Aḥmad. Hyderabad: Dāʾirat al-Maʿārif al-ʿUthmāniyyah, 1976.

Shahrazūrī, *Nuzhat al-arwāḥ*, ed. Suwayrib = al-Shahrazūrī, Shams al-Dīn, *Nuzhat al-arwāḥ wa-rawḍat al-afrāḥ*, ed. ʿAbd al-Karīm ʿUmar Abū Suwayrib. Paris: Dar Byblion, 2007 [1st ed. Tripoli 1988].

Shaykh Mufīd, *Irshād* = Muḥammad ibn Muḥammad al-Mufīd, *al-Irshād fī maʿrifat ḥujaj Allāh ʿalā l-ʿibād*. 2 vols. Qum, 1372/1993.

Shaykh Mufīd, *The Book of Guidance* = Shaykh al-Mufīd, *Kitāb al-irshād: The Book of Guidance into the Lives of the Twelve Imams*. tr. I.K.A. Howard. London: Balagha books; The Muhammadi Trust, 1981.

Shefer-Mossensohn & Hershkovitzh, 'Early Muslim Medicine' = M. Shefer-Mossensohn and K. Abou Hershkovitzh, 'Early Muslim Medicine and the Indian Context: A Reinterpretation', *Medieval Encounters*, 19 (2013), 277–299.

Shihadeh, *Doubts on Avicenna* = Ayman Shihadeh, *Doubts on Avicenna: A Study and Edition of Sharaf al-Dīn al-Masʿūdī's Commentary on the Ishārāt* (Islamic Philosophy, Theology, and Science, 95). Brill, Leiden, 2016.

Shubbar, *Ṭibb al-aʾimmah* = al-Sayyid ʿAbd Allāh Shubbar, *Ṭibb al-aʾimmah* (*ʿalayhim al-salām*). [Qum?]: Dār al-Iʿtiṣām, n.d.

Sibṭ ibn al-Jawzī, *Mirʾāt al-zamān* = Sibṭ ibn al-Jawzī, *Mirʾāt al-zamān fī taʾrīkh al-aʿyān*, vol. VIII [parts 1 & 2]. Hyderabad: Dāʾirat al-Maʿārif al-ʿUthmāniyyah, 1951–1952.

Ṣibt ibn al-Jawzī, *Mirʾāt al-zamān* (ed. Barakāt) = Ṣibt ibn al-Jawzī, *Mirʾāt al-zamān fī tawārikh al-zamān*, ed. Muḥammad Barakāt et al. Damascus. 23 vols. Dār al-Risālah al-ʿĀlamiyyah, 2013.

Sibṭ Ibn al-Taʿāwīdhī, *Dīwān* = Abū l-Fatḥ Muḥammad ibn ʿUbayd Allāh Sibṭ Ibn al-Taʿāwīdhī, *Dīwān*, ed. D.S. Margoliouth. Cairo: Maktabat al-Muqtaṭaf, 1903, repr. Beirut: Dār Ṣādir, 1967.

Siddiqi, *Medical Literature* = Muhammad Zubayr Siddiqi, *Studies in Arabic and Persian Medical Literature*. Calcutta: Calcutta University, 1959.

Sideras, *De renum et vesicae morbis* = Alexander Sideras, *Rufi Ephesii De renum et vesicae morbis* (Corpus Medicorum Graecorum, 3.1). Berlin: Akademie-Verlag, 1977.

Sideras, 'Rufus von Ephesos' = Alexander Sideras, 'Rufus von Ephesos und sein Werk im Rahmen der antiken Medizin', in Hildegard Temporini, and Wolfgang Haase (eds.), *Aufstieg und Niedergang der Römischen Welt*, vol. 2, Principat 37.2. Berlin: De Gruyter, 1994, 1077–1253.

Sideras, '*Testament*' = Alexander Sideras, 'Nochmals das sogenannte *Testamentum des Hippokrates* (unter Berücksichtigung zweier neuer Handschriften)', *Sudhoffs Archiv*, 90 (2006), 161–202.

Siggel, *Die Propädeutischen Kapitel* = Alfred Siggel, *Die Propädeutischen Kapitel aus dem Paradies der Weisheit über die Medizin dess ʿAli b. Sahl Rabban aṭ-Ṭabarī*. Wiesbaden: In Kommission bei F. Steiner, 1953.

Signes Codoñer, 'La diplomacia del libro en Bizancio' = Juan Signes Codoñer, 'La diplomacia del libro en Bizancio: algunas reflexiones en Torno a la posible entrega de libros griegos a los árabes en los siglos VIII–X', *Scrittura e civiltà*, 20 (1996), 153–187.

al-Sijistānī, *Ṣiwān al-ḥikmah* = Abū Sulaymān al-Manṭiqī al-Sijistānī, *Ṣiwān al-ḥikmah wa-thalāth rasāʾil*, ed. ʿAbd al-Raḥmān Badawī. Tehran: Bunyād-i Farhang-i Īrān, 1974 / Paris: Dar Biblyon, 2004 [The work is falsely attributed to al-Sijistānī].

Silverstein, *Postal Systems* = Adam Silverstein, *Postal Systems in the Pre-Modern Islamic World*. Cambridge: Cambridge University Press, 2007.

Simon, *Sieben Bücher* = Max Simon, *Sieben Bücher Anatomie des Galen*. 2. vols. Leipzig, J.C. Hinrichs'sche Buchhandlung, 1906.

Singer, *Galen* = Peter N. Singer (ed.), *Galen: Pyschological Writings*. Cambridge: Cambridge University Press, 2013.

al-Sīrāfī, *Mā yaḥtamilu l-shiʿr min al-ḍarūrah* = Abū Saʿīd al-Ḥasan ibn ʿAbd Allāh al-Sīrāfī, *Mā yaḥtamilu l-shiʿr min al-ḍarūrah*, ed. ʿAwaḍ ibn Ḥamad al-Qūzī. [Riyadh:] Dār al-Maʿārif, 1991.

Sizgorich, 'Monks and their Daughters' = Thomas Sizgorich, 'Monks and their Daughters' in Margaret Cormack (ed.), *Muslims and Others in Sacred Space*. Oxford: Oxford University Press, 2013, 193–216.

Skreslet, 'Greeks in Medieval Islamic Egypt' = Stanley H. Skreslet II, 'The Greeks in Medieval Islamic Egypt: A Melkite Dhimmī Community under the Patriarch of Alexandria (640–1095)', Ph.D. diss., 1987, Yale University.

Smith, *Hippocrates* = Wesley D. Smith (ed., tr.), Hippocrates, *Pseudepigraphic Writings: Letters – Embassy – Speech from the Altar – Decree*. (Studies in Ancient Medicine, 2). Leiden: Brill, 1990.

Solomon, *Ptolemy Harmonics* = Jon Solomon, *Ptolemy Harmonics: translation and commentary*, by Jon Solomon (Mnemosyne, biblioteca classica Batava, Suppl. 203). Leiden: Brill, 2000

Sophronius, *Narratio miraculorum sanctorum Cyri et Joannis*, ed. N. Fernández Marcos, (*Los Thaumata de Sofronio: contribución al estudio de la incubatio cristiana*). Madrid: Instituto Antonio de Nebrija, 1975.

Sournia & Troupeau, 'Médicine arabe' = Jean-Charles Sournia and Gérard Troupeau, 'Médecine arabe: biographies critiques de Jean Mésué (VIIIe siècle) et du prétendu 'Mésué le Jeune' (xe siècle)', *Clio Medica*, 3 (1968), 109–117.

Speca, *Hypothetical Syllogistic and Stoic logic* = Anthony Speca, *Hypothetical Syllogistic and Stoic logic* (Philosophia Antiqua, 87). Leiden: Brill, 2001.

Sperl & Moustafa, *The Cosmic Script* = Stefan Sperl and Ahmed Moustafa, *The Cosmic Script: sacred geometry and the science of Arabic penmanship*. London: Thames & Hudson, 2014.

Speziale, *Soufisme* = Fabrizio Speziale, *Soufisme: religion et médicine en Islam indien*. Paris: Karthala, 2010.

Spies, *Anatomie und Chirurgie … nach Ibn al-Quff* = Otto Spies, *Anatomie und Chirurgie des Schādels, inbesondere der Hals-, Nasen- und Ohrenkrankheiten nach Ibn al-Quff*. Berlin: Walter de Gruyter, 1971.

Spies, 'Al-Kindī's treatise on the cause of the blue colour of the sky' = Otto Spies, 'Al-Kindī's treatise on the cause of the blue colour of the sky', *Journal of the Bombay Branch of the Royal Asiatic Society*, 13 n.s. (1937), 7–19.

Spitzer, 'The Hebrew Translations of the *Sod ha-Sodot*' = Spitzer, 'The Hebrew Translations of the *Sod ha-Sodot*', in Kraye et al., *Pseudo-Aristotle in the Middle Ages*, 34–54.

Stanford Encyclopedia of Philosophy = *Stanford Encyclopedia of Philosophy*. Stanford: Center for the Study of Language and Literature (on-line resource at: https://plato.stanford.edu/search/searcher.py).

Stearns, *Contagion* = Justin K. Stearns, *Infectious Ideas: Contagion in Premodern Islamic and Christian Thought in the Western Mediterranean*. Baltimore: Johns Hopkins University Press, 2011.

Steingass, *Persian-English dictionary* = F.J. Steingass, *A comprehensive Persian-English dictionary: including the Arabic words and phrases to be met with in Persian literature, being Johnson and Richardson's Persian, Arabic, and English dictionary revised, enlarged, and entirely reconstructed*. London: W.H. Allen, 1892. Repr. London: Routledge and Kegan Paul, 1984.

Steinschneider, *Al-Farabi* = Moritz Steinschneider, *Al-Farabi* (*Alpharabius*) *des arabis-*

chen Philosophen Leben und Schriften (Mémoires de l'Académie Impériale des Sciences de Saint-Pétersbourg, series 7, vol. 13, no. 4). St. Petersburg, 1869.

Steinschneider, 'Arabische Aerzte' = Steinschneider, Moritz, 'Arabische Aerzte und deren Schriften', *Zeitschrift der Deutschen Morgenländischen Gesellschaft*, 31 (1877), 758–765.

Steinschneider, *Polemische und Apologetischer Literatur* = Moritz Steinschneider, *Polemische und apologetische Literatur in arabischer Sprache zwischen Muslimen, Christen und Juden, nebst Anhängen verwandten Inhalts*. Leipzig: F.A. Brockhaus, 1877.

Stern, 'A Collection' = Samuel Miklos Stern, 'A collection of treatises by ʿAbd al-Laṭīf al-Baghdādī', *Islamic Studies*, 1 (1962), 53–70. Repr. in Fritz W. Zimmermann (ed.), S.M. Stern, *Medieval Arabic and Hebrew thought*. London: Variorum, 1983, No. XVIII.

Stern, 'Early Ismāʿīlī Missionaries' = Samuel Miklos Stern, 'The Early Ismā'Īlī Missionaries in North-West Persia and in Khurāsān and Transoxania', *Bulletin of the School of Oriental and African Studies*, 23/1 (1960), 56–90.

Stern, 'Ibn Masarra' = Samuel Miklos Stern, 'Ibn Masarra, follower of Pseudo-Empedocles – an illusion', in *Actas IV congresso de estudios árabes et islâmicos*. Leiden: Brill, 1971, 325–337.

Stern, 'Ibn al-Samḥ' = Samuel Miklos Stern, 'Ibn al-Samḥ', *Journal of the Royal Asiatic Society*, 1956, 31–44.

Stern, 'Notes on al-Kindi's Treatise on Definitions' = Samuel Miklos Stern, 'Notes on al-Kindi's Treatise on Definitions.' *Journal of the Royal Asiatic Society*, 1–2 (1959), 32–43.

Stern, *Studies in early Ismāʿīlism* = Samuel Miklos Stern, *Studies in Early Ismāʿīlism*. Jerusalem: Magnes, Hebrew University / Leiden: Brill, 1983.

Stetkevych, 'Toward an Arabic Elegiac Lexicon' = Jaroslav Stetkevych, 'Toward an Arabic Elegiac Lexicon: The Seven Words of the *Nasīb*', in Suzanne Pinckney Stetkevych (ed.), *Reorientations / Arabic and Persian Poetry*. Bloomington: Indiana University Press, 1994, 58–129.

Stetkevych, *The Mute Immortals Speak* = Suzanne Pinckney Stetkevych, *The Mute Immortals Speak: Pre-Islamic Poetry and the Poetics of Ritual*. Ithaca / London: Cornell University Press, 1993.

Stilt, *Islamic Law in Action* = Kristen Stilt, *Islamic Law in Action. Authority, Discretion, and Everyday Experiences in Mamluk Egypt*. Oxford: Oxford University Press, 2011.

Stowasser, *The Day Begins at Sunset* = Barbara Freyer Stowasser, *The Day Begins at Sunset: Perceptions of Time in the Islamic World*. London: I.B. Tauris, 2014.

Stowasser, *Medieval Egypt* = Karl Stowasser (tr. and ann.), *Medieval Egypt: al-Khiṭāṭ of Aḥmad ibn ʿAlī al-Maqrīzī, Part 1*. N. pl.: CreateSpace, 2014.

Strauss, 'Das Giftbuch des Šānāq' = Bettina Strauss, 'Das Giftbuch des Šānāq. Eine literaturgeschichtliche Untersuchung', *Quellen und Studien zur Geschichte der Naturwissenschaften*, 4/2 (1934), 1–65 + 66 pp. Arabic text.

Strohmaier, 'Asklepios und das EI' = Gotthard Strohmaier, 'Asklepios und das EI. Zur Ikonographie in einem arabisch erhaltenen Kommentar zum Hippokratischen Eid', in Ruth Stiehl and Hans Erich Stier (eds.), *Beiträge zur Alten Geschichte und deren Nachleben. Festschrift für Franz Altheim.* 2 vols. Berlin: De Gruyter, 1970, ii:143–153.

Strohmaier, *Avicenna* = Gotthard Strohmaier, *Avicenna*, 2nd ed., Munich: C.H. Beck, 2006.

Strohmaier, CMG = Gotthard Strohmaier, *Galeni De partium homoeomerium differentia libelli versionem arabicam* … (Corpus Medicorum Graecarum, Dictionnaire Suppl. Or. 3). Berlin: Akademie-Verlag, 1970.

Strohmaier, 'Die griechischen Götter' = Gotthard Strohmaier, 'Die griechischen Götter in einer christlich-arabischen Übersetzung. Zum Traumbuch des Artemidor in der Version des Ḥunain ibn Isḥāḳ', in Franz Altheim und Ruth Stiehl (eds.), *Die Araber in der Alten Welt*, V/1. Berlin: De Gruyter, 1968, 127–162.

Strohmaier, 'Ḥunain' = Gotthard Strohmaier, 'Ḥunain ibn Isḥāq und die Bilder', *Klio*, 43–45 (1965), 525–533.

Stroumsa, *Freethinkers of Medieval Islam* = Sarah Stroumsa, *Freethinkers of Medieval Islam. Ibn al-Rāwandī, Abū Bakr al-Rāzī, and Their Impact on Islamic Thought* (Islamic Philosophy, Theology, and Science, 35). Leiden: Brill, 1999.

Stroumsa, *Maimonides in his World* = Sarah Stroumsa, *Maimonides in his World: Portrait of a Medieval Thinker*. Princeton: Princeton University Press, 2009.

al-Subkī, *Ṭabaqāt al-Shāfiʿiyyah* = Tāj al-Dīn al-Subkī, *Ṭabaqāt al-Shāfiʿiyyah al-Kubrā*, ed. Muḥammad Muḥammad al-Ṭanāḥī and ʿAbd al-Fattāḥ Muḥammad al-Ḥulw. 10 vols. Cairo: Dār Iḥyāʾ al-Kutub al-ʿArabiyyah, 1964–1976.

al-Sukkarī, *Sharḥ ashʿār al-Hudhaliyyīn* = Abū Saʿid al-Ḥasan ibn al-Ḥusayn al-Sukkarī, *Sharḥ ashʿār al-Hudhaliyyīn*, ed. ʿAbd al-Sattār Aḥmad Farrāj. 3 vols. Cairo: Dār al-ʿUrūbah, 1965.

al-Ṣūlī, *Awrāq* = Abū Bakr Muḥammad ibn Yaḥyā ibn al-ʿAbbās ibn Muḥammad ibn Ṣūl, *al-Awrāq*, ed. J. Heyworth Dunne. 3 vols. London: Luzac, 1936. Repr. Beirut: Dār al-Masīrah, 1982. (I: *Akhbār al-shuʿarāʾ al-muḥdathīn*, II: *Akhbār al-Rāḍī li-llāh wa-l-Muttaqī li-llāh*, III: *Ashʿār awlād al-khulafāʾ*)

al-Ṣūlī, *Life and Times of Abū Tammām* = Abū Bakr Muḥammad ibn Yaḥyā l-Ṣūlī, *The Life and Times of Abū Tammām*, ed. and tr. Beatrice Gruendler (Library of Arabic Literature). New York: New York University Press, 2015.

Suter, *Abhandlungen* = Heinrich Suter, *Abhandlungen zur Geschichte der mathematischen Wissenschaften mit Einschluss ihrer Anwendungen*. Leipzig: Teubner, 1900.

al-Suyūṭī, *Bughyah* = Jalāl al-Dīn ʿAbd al-Raḥmān al-Suyūṭī, *Bughyat al-wuʿāh fī ṭabaqāt al-lughawiyyīn wa-l-nuḥāh*, ed. Muḥammad Abū l-Faḍl Ibrāhīm. 2 vols. Beirut: Dār al-Fikr, 1979.

al-Suyūṭī, *Ḥusn al-muḥāḍarah* = Jalāl al-Dīn ʿAbd al-Raḥmān al-Suyūṭī, *Ḥusn al-muḥā-*

ḍarah fī tārīkh Miṣr wa-l-Qāhirah, ed. Muḥammad Abū l-Faḍl Ibrāhīm. 2 vols. Cairo: ʿĪsā al-Bābī al-Ḥalabī, 1967–1968.

al-Suyūṭī, *Tārīkh al-khulafāʾ* = Jalāl al-Dīn al-Suyūṭī, *Tārīkh al-khulafāʾ*, ed. Muḥammad Muḥyī l-Dīn ʿAbd al-Ḥamīd. Beirut: Dār al-Jīl, 1988.

Swain, 'Beyond the Limits' = Simon Swain, 'Beyond the Limits of Greek Biography: Galen from Alexandria to the Arabs', in Brian McGing and Judith Mossman (eds.), *The Limits of Ancient Biography*. Swansea: The Classical Press of Wales, 2006, 395–433.

Swain, *Economy, Family and Society* = Simon Swain, *Economy, Family, and Society from Rome to Islam: A Critical Edition, English Translation, and Study of Bryson's 'Management of the Estate'*. Cambridge: Cambridge University Press, 2013.

Swain, *Hellenism and Empire* = Simon Swain, *Hellenism and Empire. Language, Classicism, and Power in the Greek world, AD 50–250*. Oxford: Oxford University Press, 1996.

Swain, *Seeing the Face* = Simon Swain (ed.), *Seeing the Face, Seeing the Soul. Polemon's* Physiognomy *from Classical Antiquity to Medieval Islam*. Oxford: Oxford University Press, 2007.

Swain, *Themistius, Julian and Greek Political Theory* = Simon Swain, *Themistius, Julian, and Greek Political Theory under Rome: Texts, Translations and Studies of Four Key Words*. Cambridge: Cambridge University Press, 2013.

Swanson, 'A curious correspondence' = Mark N. Swanson, 'A curious and delicate correspondence: the Burhān of Ibn al-Munajjim and the Jawāb of Ḥunayn ibn Isḥāq', *Islam and Christian-Muslim Relations*, 22 (2011), 173–183.

Swift Riginos, *Platonica* = Alice Swift Riginos, *Platonica: The Anecdotes Concerning the Life and Writings of Plato*. Leiden, Brill, 1976.

Tabbaa, *Constructions* = Yasser Tabbaa, *Constructions of Power and Piety in Medieval Aleppo*. University Park, Pennsylvania: Pennsylvania State University Press, 1997.

al-Ṭabarī, *Tārīkh* (ed. de Goeje) = Abū Jaʿfar Muḥammad ibn Jarīr al-Ṭabarī, *Annales quos scripsit Abu Djafar Mohammed ibn Djarir at-Tabari/Tārīkh al-rusul wa-al-mulūk*. cum aliis edidit M.J. de Goeje. 14 vols. Leiden: Brill, 1879–1901.

al-Ṭabari, *Taʾrīkh* (Bosworth) = C. Edmund Bosworth (tr.), *The ʿAbbasid Caliphate in Equilibrium: The Caliphates of Musa al-Hadi and Harun al-Rashid, A.D. 785–809/A.H. 169–193* (The History of al-Tabari, 30). Albany: State University of New York, 1989.

al-Ṭabari, *Taʾrīkh Reunification* (Bosworth) = C. Edmund Bosworth (tr.), *The Reunification of the ʿAbbasid Caliphate: the Caliphate of al-Maʾmūn* (The History of al-Tabari, 32). Albany: State University of New York, 1987.

al-Ṭabari, *Taʾrīkh Storm and Stress* (Bosworth) = C. Edmund Bosworth (tr.), *Storm and Stress along the Northern Frontiers of the ʿAbbāsid Caliphate: The Caliphate of al-Muʿtasim A.D. 833–842/A.H. 218–227* (The History of al-Tabari, 33). Albany: State University of New York, 1987.

al-Ṭabari, *Taʾrīkh* (Kraemer) = J. Kraemer (tr.), *The History of al-Tabari, vol. XXXIV: Incip-*

ient Decline: The Caliphates of al-Wāthiq, al-Mutawakkil, and al-Muntaṣir, A.D. *841–863/*A.H. *227–224* (The History of al-Tabari, 34). Albany: State University of New York, 1989.

Tabatabaei et al., 'Razi's description and treatment of facial paralysis' = S.M. Tabatabaei, A. Kalantar-Hormozi, and M. Asadi, 'Razi's description and treatment of facial paralysis', *Archives of Iranian Medicine*, 14/1 (2011), 73–75.

al-Ṭabrisī, *Majmaʿ al-bayān* = Amīn al-Islām Abū ʿAlī al-Faḍl ibn al-Ḥasan al-Ṭabrisī, *Majmaʿ al-bayān fī tafsīr al-Qurʾān*. 10 vols., Beirut: Dār al-ʿUlūm, 2005.

Taha, 'Compilation of Pharmacological Ideas' = Mohamad Bin Taha, 'The Compilation of Pharmacological Ideas during the ʿAbbāsid Caliphate with Special Reference to a Section of *al-Ḥāwī* of al-Rāzī'. M.Phil. thesis, University of St Andrews, 1988.

Tahānawī, *Kashshāf* (Wajih, al-Haqq, Kadir) = Muḥammad Aʿlā ibn ʿAlī al-Tahānawī, *Kashshāf iṣṭilāḥāt al-funūn, Dictionary of Terms used in the Sciences of the Musalmans*, ed. M. Wajih, Abd al-Haqq, and Gholam Kadir. 2 vols. Calcutta: W.N. Lees' Press, 1862.

Tahānawī, *Kashshāf* (Jawdat) = Muḥammad Aʿlā ibn ʿAlī al-Tahānawī, *Kashshāf iṣṭilāḥāt al-funūn*, ed. Aḥmad Jawdat [Ahmet Cevdet Paşa]. Al-Astānah (= Istanbul): Maṭbaʿat Iqdām bi-Dār al-Khilāfah, 1317/1899–1900.

Talmon-Heller, *Islamic Piety* = Daniella Talmon-Heller, *Islamic Piety in Medieval Syria. Mosques, Cemeteries and Sermons under the Zangids and Ayyūbids (1146–1260)* (Jerusalem Studies in Religion and Culture, 7). Leiden: Brill, 2007.

al-Tamīmī, *K. māddat al-baqāʾ* (Shaʿar) = al-Tamīmī al-Maqdisī, *Māddat al-baqāʾ fī iṣlāḥ fasād al-hawāʾ wa-l-taḥarruz min ḍarar al-awbāʾ*, ed. Yahyā Shaʿar. Cairo: The Institute of Arabic Manuscripts, 1999.

al-Tanūkhī, *Ende gut, alles gut* (Hottinger) = Arnold Hottinger, *Ende gut, alles gut: Das Buch der Erleichterung nach der Bedrängnis*. Zürich: Manesse, 1979.

al-Tanūkhī, *Faraj* = Abū ʿAlī al-Muḥassin ibn ʿAlī al-Tanūkhī, *al-Faraj baʿd al-shiddah*, ed. ʿAbbūd al-Shāljī. 5 vols. Beirut: Dār Ṣādir, 1978.

al-Tanūkhī, *Nishwār al-muḥāḍarah* = Abū ʿAlī al-Muḥassin ibn ʿAlī al-Tanūkhī, *Nishwār al-muḥāḍarah wa-akhbār al-mudhākarah*, ed. ʿAbbūd al-Shāljī. 8 vols. Beirut: no publ., 1971–1973.

al-Tanūkhī, *Il Solievo* (Ghersetti) = Antonella Ghersetti, *Il sollievo dopo la distretta* = (*al-Farağ baʿd al-šidda*). Milano: Ariele, 1995.

al-Tawḥīdī, *Baṣāʾir* = Abū Ḥayyān ʿAlī ibn Muḥammad ibn al-ʿAbbās al-Tawḥīdī, *al-Baṣāʾir wa-l-dhakhāʾir*, ed. Wadād al-Qāḍī. 9 vols. Beirut: Dār Ṣādir, 1988.

al-Tawḥīdī, *Imtāʿ* = Abū Ḥayyān al-Tawḥīdī, *al-Imtāʿ wa-l-muʾānasa*, ed. Aḥmad Amīn and Aḥmad al-Zayn. 3 vols. Cairo: Lajnat al-Taʾlīf wa-l-Tarjamah wa-l-Nashr, 1939–1953.

al-Tawḥīdī, *Muqābasāt* = Abū Ḥayyān al-Tawḥīdī, *al-Muqābasāt*, ed. Muḥammad Tawfīq Ḥusayn. Beirut: Dār al-Ādāb, 1989.

Taylor, *Righteous* = Christopher S. Taylor, *In the Vicinity of the Righteous: Ziyāra and the Veneration of Muslim Saints in Late Medieval Egypt* (Islamic History and Civilization, 22). Leiden: Brill, 1999.

Taylor & López-Farjeat, *The Routledge Companion to Islamic philosophy* = Richard C. Taylor and Luis Xavier López-Farjeat (eds.), *The Routledge Companion to Islamic Philosophy*. New York: Routledge, 2016.

Temkin, *Double Face of Janus* = Owsei Temkin, *The Double Face of Janus and Other Essays in the History of Medicine*. Baltimore: The Johns Hopkins University Press, 1977.

Teule, 'Ibn al-Khammār' = Herman G.B. Teule, 'Ibn al-Khammār', in David Thomas and Alexander Mallet (eds.), *Christian-Muslim Relations: A Bibliographic History 900–1050*. Leiden: Brill, 557–560.

al-Thaʿālibī, *Fiqh al-lughah* = Abū Manṣūr ʿAbd al-Malik ibn Muḥammad ibn Ismāʿīl al-Thaʿālibī, *Fiqh al-lughah wa-asrār al-ʿarabiyyah*. Cairo, AH 1318, repr. Beirut: Maktabat al-Ḥayāt, n.d.

al-Thaʿālibī, *Iʿjāz* = Abū Manṣūr ʿAbd al-Malik ibn Muḥammad ibn Ismāʿīl al-Thaʿālibī, *al-Iʿjāz wa-l-ījāz*, ed. Iskandar Āṣāf. Cairo: al-Maṭbaʿah al-ʿUmūmiyyah, 1897.

al-Thaʿālibī, *Khāṣṣ al-khāṣṣ* = Abū Manṣūr ʿAbd al-Malik ibn Muḥammad ibn Ismāʿīl al-Thaʿālibī, *Khāṣṣ al-khāṣṣ*. Beirut: Dār Maktabat al-Ḥayāt, n.d.

al-Thaʿālibī, *Kitāb man ghāba ʿanhu l-muṭrib* = Abū Manṣūr ʿAbd al-Malik ibn Muḥammad ibn Ismāʿīl al-Thaʿālibī al-Nīsābūrī, [*Kitāb*] *Man ghāba ʿanhu l-muṭrib*, ed. al-Nabawī ʿAbd al-Wāḥid Shaʿlān. Cairo: Maktabat al-Khānjī, 1984.

al-Thaʿālibī, *Laṭāʾif al-luṭf* = Abū Manṣūr ʿAbd al-Malik ibn Muḥammad al-Nīsābūrī al-Thaʿālibī, *Laṭāʾif al-luṭf*, ed. ʿUmar al-Asʿad. Beirut: Dār al-Masīrah, 1980.

al-Thaʿālibī, *Muntaḥal* = Abū Manṣūr al-Thaʿālibī, *al-Muntaḥal* ed. Aḥmad Abū ʿAlī. Alexandria: al-Maktabah al-Tijāriyyah, 1901.

al-Thaʿālibī, *Nathr al-nazm* = Abū Manṣūr ʿAbd al-Malik ibn Muḥammad ibn Ismāʿīl al-Thaʿālibī, *Nathr al-naẓm wa-ḥall al-ʿaqd*. Damascus: Maṭbaʿat Maʿārif al-Wilāyah, AH 1300.

al-Thaʿālibī, *Taḥsīn al-qabīḥ* = Abū Manṣūr ʿAbd al-Malik ibn Muḥammad ibn Ismāʿīl al-Thaʿālibī, *Taḥsīn al-qabīḥ wa-taqbīḥ al-ḥasan*, ed. Shākir al-ʿĀshūr. Baghdad: Wizārat al-Awqāf, 1981.

al-Thaʿālibī, *al-Tamthīl wa-l-muḥāḍarah* = al-Thaʿālibī, *al-Tamthīl wa-l-muḥāḍarah*, ed. Muḥammad. ʿAbd al-Fattāḥ al-Ḥulw. Cairo: Dār Iḥyāʾ al-Kutub al-ʿArabiyyah, 1961.

al-Thaʿālibī, *Tatimmat al-Yatīmah* = Abū Manṣūr ʿAbd al-Malik al-Thaʿālibī, *Tatimmat Yatīmat al-dahr fī maḥāsin ahl al-ʿaṣr*, ed. Mufīd Muḥammad Qumayḥah. Beirut: Dār al-Kutub al-ʿIlmiyyah, 1983.

al-Thaʿālibī, *Tawfīq* = Abū Manṣūr ʿAbd al-Malik ibn Muḥammad ibn Ismāʿīl al-Nīsābūrī al-Thaʿālibī, *al-Tawfīq li-l-talfīq*, ed. Hilāl Nājī and Zuhayr Zāhid. Beirut: ʿĀlam al-Kutub, 1996.

al-Thaʿālibī, *Thimār al-qulūb* = Abū Manṣūr ʿAbd al-Malik ibn Muḥammad ibn Ismāʿīl al-Nīsābūrī al-Thaʿālibī, *Thimār al-qulūb fī l-muḍāf wa-l-mansūb*, ed. Muḥammad Abū l-Faḍl Ibrāhīm. Cairo: Dār al-Maʿārif, 1985.

al-Thaʿālibī, *Yatīmah* = Abū Manṣūr ʿAbd al-Malik ibn Muḥammad ibn Ismāʿīl al-Thaʿālibī, *Yatīmat al-dahr fī maḥāsin ahl al-ʿaṣr*, ed. Muḥammad Muḥyī l-Dīn ʿAbd al-Ḥamīd. 4 vols. Cairo: Maktabat al-Ḥusayn al-Tijāriyyah, 1366/1947.

al-Thaʿālibī, *Yatīmah* (ed. Qumayḥa) = Abū Manṣūr ʿAbd al-Malik ibn Muḥammad ibn Ismāʿīl al-Thaʿālibī, *Yatīmat al-dahr fī maḥāsin ahl al-ʿaṣr*, ed. Mufīd Muḥammad Qumayḥah. 5 vols. Beirut: Dār al-Kutub al-ʿIlmiyyah, 1983.

Thābit ibn Qurrah, *al-Baṣar wa-l-baṣīrah* = Thābit ibn Qurrah al-Ḥarrānī, *al-Baṣar wa-al-baṣīrah fī ʿilm al-ʿayn wa-ʿilalihā wa-mudāwātihā*. ed. Muḥammad Rawwās Qalʿah'jī, and Muḥammad Ẓāfir Wafāʾī. al-Riyāḍ: Sharikat ʿUbaykān, 1991.

Thābit ibn Qurrah, *al-Dhakhīrah* = Thābit ibn Qurrah *al-Dhakhīrah fī ʿilm al-ṭibb: muʿālajat al-amrāḍ bi-al-aʿshāb*. ed. Aḥmad Farīd al-Mazīdī. Beirut: Manshūrāt Muḥammad ʿAlī Bayḍūn; Dār al-Kutub al-ʿIlmiyyah, 1998.

Thābit ibn Qurrah, *Three treatises* = Sih risālah az s̱ābit ibn Qurrah: Sāʿat'hā-yi āftābī, Ḥarakkat-i khūrshīd vā māh, Chahārdah vajhī-i muḥāṭ dar kurah. Tihrān: Markaz-i Pizhūhishī-i Mīrās̱-i Maktūb bā hamkārī-i Pizhūhishkadah-ʾi Tārīkh-i ʿIlm-i Dānishgāh-i Tihrān, 2014.

Theodosius, *De habitationibus* = Theodosius, *De habitationibus: Arabic and medieval Latin translations*. ed. Paul Kunitzsch and Richard Lorch. Munich: Verlag der Bayerischen Akademie der Wissenschaften in Kommission beim Verlag C.H. Beck, 2011.

Thesleff, *Pythagorean Texts* = Holger Thesleff, *The Pythagorean Texts of the Hellenistic Period*. Åbo: Åbo Akademi, 1965.

Thies, *Diabetes* = Hans-Jürgen Thies, *Der Diabetestraktat ʿAbd al-Laṭīf al-Baġdādī's. Untersuchungen zur Geschichte des Krankheitsbildes in der arabischen Medizin*. Bonn: Selbstverlag des Orientalischen Seminars der Universität Bonn, 1971.

Thies, *Die Lehren der arabischen Mediziner Tabari und Ibn Hubal* = Dorothee Thies, *Die Lehren der arabischen Mediziner Tabari und Ibn Hubal über Herz, Lunge, Gallenblase und Milz* (Beiträge zur Sprachen und Kulturgeschichte des Orients, 20). Bonn: Verlag für Orientkunde Dr. H. Vorndran, 1968.

Thillet, *Traité de la providence* = Pierre Thillet, *Traité de la providence = Peri pronoias*. Alexandre d'Aphrodise; version arabe de Abū Bišr Mattā ibn Yūnus; introduction, édition et traduction de Pierre Thillet. Lagrasse: Verdier, 2003.

Thom, *Golden Verses* = Johan C. Thom, *The Pythagorean Golden Verses* (Religions in the Graeco-Roman World, 123). Leiden: Brill, 1995.

Thom, *Cosmic Order* = Johan C. Thom, *Cosmic Order and Divine Power. Pseudo-Aristotle, On the Cosmos*. Tübingen: Mohr Siebeck, 2014.

Thomann, 'La Tradition Arabe de la *Physiognomonie* d'Aristote' = Johannes Thomann, 'Aristote de Stagire – La Tradition Arabe de la *Physiognomonie* d'Aristote', in Goulet, *Dictionnaire*, Suppl., 499–505.

Thomann, 'Ein al-Fārābī zugeschriebener Kommentar'= Johannes Thomann, 'Ein al-Fārābī zugeschriebener Kommentar zum Almagest (Hs. Tehran Mağlis 6531)', *Zeitschrift für Geschichte der arabisch-islamischen Wissenschaften*, 19 (2010–2011), 35–76.

Thomann, 'Al-Fārābī's Kommentar' = Johannes Thomann, 'Al-Fārābī's Kommentar zum Almagest in sekundärer Überlieferung bei Ibn al-Ṣalāḥ: Ein vorläufiger Bericht', *Asiatische Studien*, 69 (2015), 99–113.

Thomas & Roggema, *Christian-Muslim Relations* = David Thomas and Barbara Roggema (eds.), *Christian Muslim Relations: A Bibliographical History. Vol. 1 (600–900)*. Leiden: Brill, 2009.

Tibbetts, 'Beginnings of a Cartographic Tradition' = Gerald B. Tibbetts, 'The Beginnings of a Cartographic Tradition' in J.B. Harley and David Woodward (eds.), *History of Cartography. Vol. II, Book 1: Cartography in the Traditional Islamic and South Asian Societies*. Chicago: Chicago University Press, 1992, 90–107.

Tibi, *Medicinal Use of Opium* = Selma Tibi, *The Medicinal Use of Opium in Ninth-Century Baghdad* (Sir Henry Wellcome Asian Series, 5). Leiden: Brill, 2006.

al-Tibrīzī, *al-Wāfī fī l-ʿarūḍ wa-l-qawāfī* = Yaḥyā ibn ʿAlī al-Tibrīzī, *al-Wāfī fī l-ʿarūḍ wa-l-qawāfī*. Aleppo: al-Maktaba al-ʿArabiyya, 1970.

Tillier, 'Prisons' = Mathieu Tillier, 'Prisons et autorités urbaines sous les Abbassides', *Arabica*, 55 (2008), 387–408.

Toomer, *Apollonius' Conics* = G.J. Toomer, *Apollonius' Conics, Books V to VII, the Arabic translation of the lost Greek original in the version of the Banū Mūsā*. New York: Springer, 1990.

Toorawa, 'Autobiography' = Shawkat M. Toorawa, 'The Autobiography of ʿAbd al-Laṭīf al-Baghdādī', in Reynolds, *Interpreting the Self*, 156–164.

Toorawa, *Ibn Abī Ṭāhir Ṭayfūr* = Shawkat M. Toorawa, *Ibn Abī Ṭāhir Ṭayfūr and Arabic Writerly Culture. A Ninth-Century Bookman in Baghdad*. London: Routledge, 2005.

Toorawa, 'Language and Male Homosocial Desire' = Shawkat M. Toorawa, 'Language and Male Homosocial Desire in the Autobiography of ʿAbd al-Laṭīf al-Baghdādī', *Edebiyāt: Special Issue – Arabic Autobiography*, n.s. 7/2, (1997), 251–265.

Toorawa, 'A Portrait' = Shawkat M. Toorawa, 'A Portrait of ʿAbd al-Laṭīf al-Baghdādī's Education and Instruction', in Joseph E. Lowry, Devin J. Stewart and Shawkat M. Toorawa (eds.), *Law and Education in Medieval Islam. Studies in Memory of George Makdisi*. Cambridge: The E.J.W. Gibb Memorial Trust, 2004, 91–109.

Toorawa, 'Travel in the medieval Islamic world' = Shawkat M. Toorawa, 'Travel in the medieval Islamic world: the importance of patronage, as illustrated by ʿAbd al-Latif al-Baghdadi (d. 629/1231) (and other littérateurs)', in Rosamund Allen (ed.), *Eastward Bound: Travel and travellers 1050–1550*. Manchester / New York: Manchester University Press, 2004, 53–70.

Tor, *Violent Order* = Deborah G. Tor, *Violent Order: Religious Warfare, Chivalry, and the ʿAyyār Phenomenon in the Medieval Islamic World*. Würzburg: Ergon, 2007.

Tornero, 'Dos epístolas de Avempace' = E. Tornero, 'Dos epístolas de Avempace sobre el móvil y sobre la facultad impulsiva', *al-Qanṭara*, 4 (1983), 6–21.

Totelin, 'Mithridates' = Lawrence M.V. Totelin, 'Mithridates' Antidote – A Pharmacological Ghost', *Early Science and Medicine*, 9/1 (2004), 1–19.

Toulouse, 'Ptolémée'= Stéphane Toulouse, 'Ptolémée *al-Ġarīb*', in Goulet, *Dictionnaire*, vol 5a, 1744–1747.

Touwaide, 'Translation and Transliteration' = Alain Touwaide, 'Translation and Transliteration of Plant Names in Ḥunayn ibn Isḥāq's and Iṣṭifān b. Bāsil's [*sic*] Arabic Version of Dioscorides, *De Materia Medica*', *al-Qanṭara*, 30 (2009), 557–580.

Tramontana, '*Khubz* as Iqṭā'' = Felicita Tramontana, '*Khubz* as Iqṭāʿ in Four Authors from the Ayyubid and Early *Mamluk* Periods', *Mamlūk Studies Review* 16, (2012), 103–122.

Travaglia, *Magic, Causality and Intentionality* = Pinella Travaglia, *Magic, causality and intentionality: the doctrine of rays in al-Kindi*. [Florence]: SISMEL, Edizioni del Galluzzo, 1999.

Treiger, 'Unpublished Texts (1)' = Alexander Treiger, 'Unpublished Texts from the Arab Orthodox Tradition (1): On the Origins of the Term 'Melkite' and On the Destruction of the Maryamiyya Cathedral in Damascus', *Chronos* (*Revue d'Histoire de l'Université de Balamand*), 29 (2014), 7–37.

Treiger, 'Unpublished Texts (2)' = Alexander Treiger, 'Unpublished Texts from the Arab Orthodox Tradition (2): Miracles of St. Eustratius of Mar Saba (written ca. 860)', *Chronos* (*Revue d'Histoire de l'Université de Balamand*), 33 (2016), 7–20.

Tritton, *Catalogue RCP* = Arthur Stanley Tritton, 'Catalogue of Oriental Manuscripts in the Library of the Royal College of Physicians', *Journal of the Royal Asiatic Society* (1951), 182–194.

Trombley, *Hellenic Religion* = Frank Trombley, *Hellenic Religion and Christianization c. 370–529*. 2 vols. Leiden: Brill, 1993.

Troupeau, 'Le premier traité arabe de diététique' = Gérard Troupeau, 'Le premier traité arabe de diététique: le *Kitāb ḫawāṣṣ al-aġḏiyah* de Yūḥannā Ibn Māsawayh', *Medicina nei Secoli Arte e Scienza*, 7 (1995), 121–139.

Troupeau, 'Le livre des temps de Jean Ibn Māsawayh' = Gérard Troupeau, 'Le livre des temps de Jean Ibn Māsawayh: Traduit et annoté', *Arabica*, 15 (1968), 113–142.

Tubbs et al., *Historical vignette* = R.S. Tubbs, M. Loukas, M.M. Shoja, M. Ardalan, and W.J. Oakes. 'Ibn Jazlah and His 11th Century Accounts (Taqwim Al-Abdan Fi Tadbir Al-Insan) of Disease of the Brain and Spinal Cord. Historical Vignette', *Journal of Neurosurgery Spine*, 9/3 (2008), 314–317.

al-Ṭughrāʾī, *Dīwān* = al-Ṭughrāʾī, *Dīwān*. Constantinople: Maṭbaʿat al-Jawāʾib, AH 1300.

Turner, *Inquisition* = John P. Turner, *Inquisition in Early Islam. The Competition for Political and Religious Authority in the Abbasid Empire*. London / New York: I.B. Taurus, 2013.

al-Ṭurṭūshī, *Sirāj* = al-Ṭurṭūshī, *Sirāj al-mulūk*, ed. Jaʿfar al-Bayatī. London: Riad El-Rayyes, 1990.

Tyan, *Organisation judiciaire* = Emile Tyan, *Histoire de l'organisation judiciaire en pays d'Islam*. 2nd ed. Leiden: Brill, 1960.

ʿUbayd Allāh ibn Bukhtīshūʿ, *On Apparent Death* = ʿUbayd Allāh ibn Bukhtīshūʿ, *ʿUbaidallāh Ibn Buḫtīšūʿ on Apparent Death. The* Kitāb Taḥrīm dafn al-aḥyāʾ, *Arabic Edition and English Translation*. Ed. and tr. Oliver Kahl; Hebrew Supplement by Gerrit Bos. Leiden / Boston: Brill, 2018.

ʿUbayd Allāh ibn Jibrīl, *al-Rawḍah al-ṭibbiyyah* = Paul Sbath, *Ar-Raoudat at-tibbiyya* (*Le jardin medical*) *par Ubaîd-Allah Ben Gibraîl Ben Bakhtichoû, chrétien décédé en 1058*. Cairo: H. Friedrich & Co., 1905.

Ülken, *İbn Sina risâlerei* = *Ibn Sina risâleleri. Les opuscules d'Ibn Sina. Publiés sous la direction de Hilmi Ziya Ülken*. 3 vols. Ankara: Türk Tarih Kurumu Basımevi, 1953.

Ullmann, 'Arabische Überlieferung' = Manfred Ullmann, 'Die arabische Überlieferung der Schriften des Rufus von Ephesos', in Hildegard Temporini and Wolfgang Haase, *Aufstieg und Niedergang der Römischen Welt*, vol. 2, Principat 37.2. Berlin: De Gruyter, 1994, 1293–1349.

Ullmann, *Die Gelbsucht* = Manfred Ullmann, *Die Schrift des Rufus von Ephesos über die Gelbsucht in arabischer und lateinischer Übersetzung*. Göttingen: Vandenhoeck & Ruprecht, 1983.

Ullmann, *Das Gespräch* = Manfred Ullmann, *Das Gespräch mit dem Wolf* (Beiträge zur Lexikographie des Klassischen Arabisch, Nr. 2 – Bayerische Akad. der Wiss., Phil.-hist. Kl., Sitzungsberichte, Jrg. 1981, Heft 2). Munich: C.H. Beck / Verlag der Bayerischen Akademie der Wissenschaften, 1981.

Ullmann, *Islamic Medicine* = Manred Ullmann, *Islamic Medicine*. Edinburgh: Edinburgh University Press, 1978.

Ullmann, *Medizin* = Manfred Ullmann, *Die Medizin im Islam* (Handbuch der Orientalistik, Abteilung I, Ergänzungsband vi, Abschnitt 1). Leiden: Brill, 1970.

Ullmann, *Natur- und Geheimwissenschaften* = Manfred Ullmann, *Die Natur- und Geheimwissenschaften im Islam* (Handbuch der Orientalistik, Abteilung I, Ergänzungsband vi, Abschnitt 2). Leiden: Brill, 1972.

Ullmann, 'Neues zu den diätetischen Schriften' = Manfred Ullmann, 'Neues zu den diätetischen Schriften des Rufus von Ephesos', *Medizinisches Journal*, 9/1 (1974), 23–40.

Ullmann, *Nikomachische Ethik* = Manfred Ullmann, *Die Nikomachische Ethik des Aristoteles in arabischer Übersetzung*. 2 vols. Wiesbaden: Harrassowitz, 2011–2012.

Ullmann, 'Tadhkira' = Manfred Ullmann, 'Die Tadhkira des Ibn as-Suwaidī, eine wichtige Quelle zur Geschichte der griechisch-arabischen Medizin und Magie', *Der Islam*, 54 (1977), 33–65.

Ullmann, *Untersuchungen* = Manfred Ullmann, *Untersuchungen zur arabischen Überlieferung der Materia medica des Dioskurides*. Wiesbaden: Harrassowitz, 2009.

Ullmann, *Der verstohlene Blick* = Manfred Ullmann, *Der verstohlene Blick: zur Metaphorik des Diebstahls in der arabischen Sprache und Literatur*. Wiesbaden: Harrassowitz 2017.

Ullmann, *WGAÜ* = Manfred Ullmann, *Wörterbuch zu den griechisch-arabischen Übersetzungen des 9. Jahrhunderts*. Wiesbaden: Otto Harrassowitz, 2002.

Ullmann, *WGAÜ Supplement* = Manfred Ullmann, *Wörterbuch zu den griechisch-arabischen Übersetzungen des 9. Jahrhunderts. Supplement*. 2 vols. Wiesbaden: Otto Harrassowitz, 2002.

Ullmann, *WKAS*: *see WKAS*.

Ullmann, 'Yūḥannā ibn Sarābiyūn' = Ullmann, Manfred, 'Yūḥannā ibn Sarābiyūn: Untersuchungen zur Überlieferungsgeschichte seiner Werke', *Medizinhistorisches Journal*, 6/4 (1971), 278–296.

Umayyah Abū l-Ṣalt, *al-Risālah al-Miṣriyyah see* Abū l-Ṣalt Umayyah.

Usāmah ibn Munqidh, *al-Badīʿ* = Usāmah ibn Munqidh, *al-Badīʿ fī naqd al-shiʿr*, ed. Aḥmad Aḥmad Badawī and Ḥāmid ʿAbd al-Majīd. Cairo: Maktabat Muṣṭafā l-Bābī al-Ḥalabī, 1960.

Vagelpohl, *Aristotle's Rhetoric* = Uwe Vagelpohl, *Aristotle's Rhetoric in the East. The Syriac and Arabic Translation and Commentary Tradition* (Islamic Philosophy, Theology, and Science, 76). Leiden: Brill, 2008.

Vagelpohl, 'Translator's Workshop' = Uwe Vagelpohl, 'In the Translator's Workshop', *Arabic Sciences and Philosophy*, 21 (2011), 249–288.

Vagelpohl, *Galeni In Hipp. Epid.* = Uwe Vagelpohl, *Galeni In Hippocratis Epidemiarum librum I commentariorum I–III versionem arabicam …* (Corpus medicorum Graecorum, Suppl. Or., 1), Berlin: De Gruyter, 2014.

Vagelpohl & Swain, *Galeni In Hipp. Epid.* = Uwe Vagelpohl with Simon Swain, *Galeni In Hippocratis Epidemiarum librum II commentariorum I–VI versionem arabicam …* (Corpus medicorum Graecorum, Suppl. Or., 2). 2 vols. Berlin: De Gruyter, 2016.

van Berkel, 'Accountants and Men of Letters' = Maaike L. van Berkel, 'Accountants and Men of Letters: Status and Position of Civil Servants in Early Tenth-Century Baghdad'. Ph.D. diss., University of Amsterdam, 2003.

van Berkel *et alii*, *Crisis and Continuity* = Maaike L.M. van Berkel, Nadia Maria El Cheikh, Hugh Kennedy and Letizia Osti, *Crisis and Continuity at the Abbasid Court: Formal and Informal Politics in the Caliphate of al-Muqtadir (295–320/908–932)* (Islamic History and Civilization, 102). Leiden/Boston: Brill, 2013.

van Bladel, *The Arabic Hermes* = Kevin van Bladel, *The Arabic Hermes: From Pagan Sage to Prophet of Science*, Oxford: Oxford University Press, 2009.

van Bladel, 'Bactrian Background of the Barmakids' = Van Bladel, Kevin, 'The Bactrian Background of the Barmakids', in Anna Akasoy, Charles Burnett, and Roni Yoeli-Tlalim (eds.), *Islam and Tibet – Interactions along the Musk Routes*. Farnham: Ashgate, 2011, 43–88.

van Bladel, 'Eutychius of Alexandria (Sa'īd ibn Batrīq)' = Kevin van Bladel, 'Eutychius of Alexandria (Sa'īd ibn Batrīq), *ANNALS*: On Justinian's Constructions at Mount Sinai', in Daniel Caner et al., *History and Hagiography from the Late Antique Sinai* (Translated Texts for Historians, 23). Liverpool: Liverpool University Press, 2010, 277–282.

van der Eijk, *Diocles* = Philip van der Eijk, *Diocles of Carystus. A Collection of the Fragments with Translation and Commentary*. 2 vols. Leiden: Brill, 2000–2001.

van Ess, 'Al-Fārābī and Ibn al-Rēwandī' = van Ess, 'Al-Fārābī and Ibn al-Rēwandī', *Hamdard Islamicus*, 3/4 (1980), 3–15.

van Ess, *Der Eine und das Andere* = Josef van Ess, *Der Eine und das Andere. Beobachtungen an Islamischen Häresiographischen Texten*. Berlin / Boston: De Gruyter, 2011.

van Ess, *Theologie und Gesellschaft* = Josef van Ess, *Theologie und Gesellschaft im 2. und 3. Jahrhundert Hidschra. Eine Geschichte des religiösen Denkens im frühen Islam*, 6 vols. Berlin: De Gruyter, 1991–1997.

van Gelder, *Anthology* = *Classical Arabic Literature: A Library of Arabic Literature Anthology*, selected amd tr. by Geert Jan van Gelder (Library of Atabic Literature). New York: New York University Press, 2013.

van Gelder, 'Antidotes and Anecdotes' = Geert Jan van Gelder, 'Antidotes and Anecdotes: A Literary History of Medicine from 13th-Century Syria.' in Petra M. Silverstein, James E. Montgomery, and Geert Jan van Gelder, *Wit and Wisdom in Classical Arabic Literature*. Leiden: Leiden University Press, 2015, 57–75.

van Gelder, 'Arabic Banqueters' = Geert Jan van Gelder, 'Arabic Banqueters: Literature, Lexicography and Reality', in Rika Gyselen (ed.), *Banquets d'Orient* (= *Res Orientales*, 4). Bures-sur Yvette: Groupe pour l'étude de la civilisation du Moyen-Orient, 1992, 85–93.

van Gelder, *The Bad and the Ugly* = Geert Jan van Gelder, *The Bad and the Ugly: Attitudes towards Invective Poetry* (Hijā') *in Classical Arabic Literature*. Leiden: Brill, 1988.

van Gelder, 'Beautifying the Ugly' = Geert Jan van Gelder, 'Beautifying the Ugly and Uglifying the Beautiful: The Paradox in Classical Arabic Literature', *Journal of Semitic Studies*, 48/2 (2003), 321–351.

van Gelder, *Close Relationships* = Geert Jan van Gelder, *Close Relationships. Incest and Inbreeding in Classical Arabic Literature*. London: I.B. Tauris, 2005.

van Gelder, 'Conceit' = Geert Jan van Gelder, 'The Conceit of Pen and Sword: On an Arabic Literary Debate' in *Journal of Semitic Studies*, 32 (1987), 329–360.

van Gelder, *The Doubts of Ibn al-Shibl al-Baghdādī* = Geert Jan van Gelder, *The Doubts of Ibn al-Shibl al-Baghdādī (d. 474/1081–1082), Poet, Philosopher, and Physician* (The Margaret Weyerhaeuser Jewett Chair of Arabic, Occasional Papers). Beirut: American University of Beirut, 2014

van Gelder, 'Encumbering Trifles' = Geert Jan van Gelder, 'al-Mutanabbī's Encumbering Trifles', *Arabic and Middle Eastern Literatures*, 2/1 (1999), 5–19.

van Gelder, *God's Banquet* = Geert Jan van Gelder, *God's Banquet. Food in Classical*

Arabic Literature. New York: Columbia University Press, 2000 (publ. in the UK as *Of Dishes and Discourse: Classical Arabic Literary Representations of Food*. Richmond, Surrey, 2000).

van Gelder, 'Joking Doctor' = Geert Jan van Gelder, 'The Joking Doctor: Abū l-Ḥakam 'Ubayd Allāh Ibn al-Muẓaffar (d. 549/1155)', in Concepción Vázquez de Benito & Miguel Ángel Manzano Rodríguez (eds.), *Actas XVI Congreso UEAI [Salamanca, Aug–Sept. 1992]*. Salamanca, n.publ., 1995, 217–228.

van Gelder, 'Mixtures of Jest and Earnest' = Geert Jan van Gelder, 'Mixtures of Jest and Earnest in Classical Arabic Literature', *Journal of Arabic Literature* 23:3 (1992), 83–108, 169–190.

van Gelder, *Sound and Sense* = Geert Jan van Gelder, *Sound and Sense in Classical Arabic Poetry*. Wiesbaden: Harrassowitz, 2012.

van Lit, 'Measurement' = L.W.C. van Lit, '"The Measurement of the Circle" of Archimedes in Naṣīr al-Dīn al-Ṭūsī's revision of the "middle books (taḥrīr al-mutawassiṭāt)"'. Bachelor Thesis, Department of Mathematics, Utrecht University. 2008

van Renterghem, *Les élites bagdadiennes* = Vanessa van Renterghem, *Les élites bagdadiennes au temps des Seldjoukides: Étude d'histoire sociale*. Beirut: IFPO, 2015.

Väth, *Geschichte artuqidischen Fürstentümer* = Gerhard Väth, *Die Geschichte der artuqidischen Fürstentümer in Syrien und der Ǧazīra'l Furātīya (496–812/1002–1409)*. (Islamkundliche Untersuchungen, 121). Berlin: Klaus Schwarz, 1987.

Vázquez de Benito, *Commentaria Averrois in Galilenum* (1984) = María Concepción Vázquez de Benito, *Commentaria Averrois in Galilenum*. Madrid: Consejo Superior de Investigaciones Científicas, Instituto Miguel Asín, 1984.

Vázquez de Benito, *La medicina de Averroes* = María Concepción Vázquez de Benito, *La medicina de Averroes: comentarios a Galeno*. Zamora: Colegio Universitario de Zamora, 1987.

Velusamy, 'Guest Stars: historical supernovae and remnants' = T. Velusamy, 'Guest Stars: historical supernovae and remnants' in *History of Oriental Astronomy (IAU Colloquium 91)*, Cambridge: Cambridge University Press, 1987, 265–270.

Vernet, 'Los médicos andaluces' = Juan Vernet, 'Los médicos andaluces en el "Libro de las generaciones de médicos" de Ibn Ŷulŷul', *Anuario de Estudios Medievales*, 5 (1968), 445–462.

Vernia, *Tratado de los medicamentos simples* = Pedro Vernia, *Tratado de los medicamentos simples. Abū-s-Ṣalt Umayya 1068–1134*. Alicante: Colegio Oficial de Farmacéuticos de la Provincia de Alicante, 1999.

Viguera Molins, 'Trujillo en las crónicas árabes' = María Jesús Viguera Molins, 'Trujillo en las crónicas árabes', in *Actas del Congreso Trujillo medieval*. Trujillo: Real Academia de Extremadura de las Letras y las Artes, 2002, 185–223.

Viladrich, 'Dos capítulos' = Mercè Viladrich, 'Dos capítulos de un libro perdido de Ibn al-Samḥ', *Al-Qanṭara*, 7 (1986), 5–11.

Viladrich, *El "Kitāb al-ʿamal bi-l-asṭurlāb"* = Mercè Viladrich, *El "Kitāb al-ʿamal bi-l-asṭurlāb" (Llibre de l'ús de l'astrolabi) d'Ibn al-Samḥ*. Barcelona: Institut d'Estudis Catalans, 1986.

Vizcaíno, 'Los al-Bāŷī al-Lajmī' = Juan M. Vizcaíno, 'Los al-Bāŷī al-Lajmī de Sevilla', in Manuela Marín and Jesús Zanón (eds.), *Estudios-onomástico-biográficos de al-Andalus*, 5. Madrid: CSIC, 1992, 433–466.

von Staden, *Herophilus* = H. von Staden, *Herophilus. The Art of Medicine in Early Alexandria*. Cambridge: Cambridge University Press, 1989.

Wagner, *Abū Nuwās* = Ewald Wagner, *Abū Nuwās. Eine Studie zur arabischen Literatur der frühen ʿAbbāsidenzeit*. Wiesbaden: Franz Steiner, 1965.

Waines, '*Murrī*' = David Waines, '*Murrī*: the tale of a condiment', *al-Qanṭara*, 12 (1991), 371–388.

Waines & Marín, 'Muzawwar' = David Waines and Manuela Marín, '*Muzawwar*: Counterfeit fare for fasts and fevers', *Der Islam*, 69/2 (1992), 289–301.

Wakelnig, 'Other Arabic Version' = Elvira Wakelnig, 'The Other Arabic Version of Proclus' *De aeternitate mundi*. The Surviving First Eight Arguments', *Oriens*, 40 (2012), 51–95.

Wakelnig, *A Philosophy Reader from the Circle of Miskawayh* = Elvira Wakelnig, *A Philosophy Reader from the Circle of Miskawayh: Text, Translation and Commentary*. Cambridge: Cambridge University Press, 2014.

Walbridge, *Epitomes* = John Walbridge, *The Alexandrian Epitomes of Galen*. Vol. 1, Provo, Utah: Brigham Young University, 2014.

Walker, *The Emperor and the World* = Alicia Walker, *The Emperor and the World: Exotic Elements and the Imaging of Middle Byzantine Imperial Power, Ninth to Thirteenth Centuries C.E.* Cambridge: Cambridge University Press, 2012.

Walker, 'The Fatimid Caliph al-ʿAziz and His Daughter Sitt al-Mulk' = Paul E. Walker, 'The Fatimid Caliph al-ʿAziz and His Daughter Sitt al-Mulk: A Case of Delayed but Eventual Succession to Rule by a Woman', *Journal of Persianate Studies* 4 (2011) 30–44.

Waly, 'Drei Kapitel' = Hamed Waly, 'Drei Kapitel aus der Ärztegeschichte des Ibn Abī Oṣaibiʿa'. Inaug.-diss., Friedrich-Wilhelms-Universität zu Berlin. Berlin: G. Schade, 1910. Repr. in Sezgin, *Studies* II, 215–259.

Walzer, *Experience* = Richard Walzer, *Galen on Medical Experience*. London: Oxford University Press, 1944.

Walzer, *Galeni De septimestri partu* = Richard Walzer, 'Galens Schrift über die Siebenmonatskinder', *Rivista di Studi Orientali*, 15 (1934), 323–357.

Walzer, *Greek into Arabic* = Richard Walzer, *Greek into Arabic: Essays on Islamic Philosophy*. Oxford: B. Cassirer, 1962.

Walzer, *Jews and Christians* = Richard Walzer, *Galen on Jews and Christians*. London: Oxford University Press, 1949.

al-Washshāʾ, *Muwashshā* = Abū l-Ṭayyib Muḥammad ibn Isḥāq al-Washshāʾ, *al-Muwashshā* / Abû 'ṭ-Ṭayyib Muḥammad ibn Isḥâq al-Waśśâ, *Kitâb al-Muwaśśâ*, ed. Rudolph E. Brünnow. Leiden: Brill, 1886.

Watson, 'In Defence of the Lemon' = Alasdair Watson (ed. and tr.), 'In Defence of the Lemon: Ibn Jumayʿ and his predecessors on a peculiarly Egyptian drug', in Myriam Wissa (ed.), *Non-muslims in medieval Islamic culture*. Edinburgh: Edinburgh University Press (forthcoming, 2020).

Watson, *Theriac & Mithridatum* = Gilbert Watson, *Theriac and Mithridatum: A Study in Therapeutics*. London: Wellcome Historical Medical Library, 1966.

Watts, E., 'The Enduring Legacy' = E. Watts, 'The Enduring Legacy of the Iatrosophist Gessius', *Greek, Roman and Byzantine Studies*, 49 (2009), 113–133.

Weaver, 'What wasn't an encyclopaedia' = James Weaver, 'What Wasn't an Encyclopaedia in the Fourth Islamic Century?', *Asiatische Studien* 71 / 3 (2017), 959–999.

Webb, *Imagining the Arabs* = Peter Webb, *Imagining the Arabs. Arabic Identity and the Rise of Islam*. Edinburgh: Edinburgh University Press, 2017.

Webb, 'An inaugural dissertation on the colic' = An inaugural dissertation on the colic, submitted to the examination of the Rev. John Ewing, S.S.T.P. [sic] provost, the trustees and medical professors, of the University of Pennsylvania. On the twenty-second day of May, 1798, for the degree of Doctor of Medicine. By William Webb, of Virginia, member of the Philadelphia Medical Society.

Weiss, *Legal Education* = Bernard Weiss, 'Medieval Islamic Legal Education as Reflected in the Works of Sayf al-Dīn al-Āmidī', in Joseph E. Lowry, Devin J. Stewart and Shawkat M. Toorawa (eds.), *Law and Education in Medieval Islam. Studies in Memory of George Makdisi*. Cambridge: E.J.W. Gibb Memorial Trust, 2004, 110–127.

Weisser, 'Die hippokratische Lehre von den Siebenmonatskindern bei Galen und Tābit ibn Qurra' = Ursula Weisser, 'Die hippokratische Lehre von den Siebenmonatskindern bei Galen und Tābit ibn Qurra', *Archiv für Geschichte der Medizin* (*Südhoffs*), 63 (1979), 209–238.

Weisser, 'Thābit Ibn Qurra's Epitome of Galen's Book on Seven-Month Children' = Ursula Weisser, 'Thābit Ibn Qurra's Epitome of Galen's Book on Seven-Month Children', *Studies in History of Medicine*, 6 (1982), 67–75; also in *Journal for the History of Arabic Science*, 7 (1983), 141–150.

Wellmann, Dioscorides = Max Wellmann, *Pedanii Dioscuridis Anazarbei De materia medica libri quinque*. 3 vols. Berlin: Weidmann, 1907.

Wensinck, *Concordance* = Arent Jan Wensinck, Jan Just Witkam, and Wim Raven (eds.), *Al-Muʿjam al-mufahras li-alfāẓ al-Ḥadīth al-Nabawī: ʿan al-kutub al-sittah wa-ʿan Musnad al-Dārimī wa-Muwaṭṭaʾ Mālik wa-Musnad Aḥmad ibn Ḥanbal / Concordance et indices de la tradition musulmane*. 8 vols. Leiden: Brill, 1936–1988.

Werthmuller, *Coptic Identity* = Kurt J. Werthmuller, *Coptic Identity and Ayyubid Politics in Egypt 1218–1350*. Cairo: The American University in Cairo Press, 2010.

Whitmarsh, *Second Sophistic* = Tim Whitmarsh, *The Second Sophistic*. Oxford: Oxford University Press, 2005.

Wiedemann, 'Fragen aus dem Gebiete der Naturwissenschaften' = E. Wiedemann, 'Fragen aus dem Gebiete der Naturwissenschaften, gestellt von Friedrich II, dem Hohenstaufen', *Archiv für Kulturgeschichte*, 11 (1914), 483–485.

Wiedemann, 'Ibn al-Haiṯam, ein arabischer Gelehrter' = E. Wiedemann, 'Ibn al-Haiṯam, ein arabischer Gelehrter', in *Beiträge zur wissenschaftlichen und praktischen Medizin* (Sonderausg. d. II. Teils d. Festschrift für J. Rosenthal). Leipzig: Thieme, 1906, 149–178.

Wiedemann & Frank, *Sonnenuhren* = E. Wiedemann and J. Frank, *Über die Konstruktion der Schattenlinien auf horizontalen Sonnenuhren von Ṯâbit ben Qurra*. Frankfurt am Main: Institut für Geschichte der Arabisch-Islamischen Wissenschaften, 1997. [first publ. Copenhagen: Høst, 1922].

Wiedemann & Hauser, 'Uhren' = E. Wiedemann and F. Hauser, 'Über die Uhren im Bereich der islamischen Kultur', in *Nova Acta der Kaiserl. Leop. Deutschen Akad. der Naturforscher*, 103 (1918), 167–272.

Wilcox, 'Qusṭā ibn Lūqā' = Judith Wilcox, 'Qusṭā ibn Lūqā and the Eastward Diaspora of Hellenic Medicine', in John A.C. Greppin, Emilie Savage-Smith, and John L. Gueriguian (eds.), *The Diffusion of Greco-Roman Medicine into the Middle East and the Caucasus*. Delmar, NY: Caravan Books, 1999, 73–128.

Wilson, *Dictionary of Astrology* = James Wilson, *A Complete Dictionary of Astrology*. London: William Hughes, 1819.

Wisnovsky, 'Yāhyā b. 'Adī's Discussion' = Robert Wisnovsky, 'Yaḥyā b. 'Adī's Discussion of the Prolegomena to the Study of a Philosophical Text', in Michael Cook (ed.), *Law and Tradition in Classical Islamic Thought: Studies in Honor of Professor Hossein Modarressi*. Basingstoke: Palgrave Macmillan, 2013, 187–201.

Witakowsky, 'Chronicle' = Witold Witakowsky, 'The Chronicle of Eusebius: Its Type and Continuation in Syriac Historiography', *Aram*, 12 (2000) 419–437.

Witkam, 'Reflections on al-Maqrīzī's Biographical Dictionary' = Jan Just Witkam, 'Reflections on al-Maqrīzī's Biographical Dictionary', in Obada Kohela (Ed.), *On History and Islamic Civilisation. Essays in Honour of Ayman Fu'ād Sayyid*. Cairo: al-Dār al-Miṣriyyah al-Lubnāniyyah, 2014, 93–114.

WKAS = *Wörterbuch der klassischen arabischen Sprache*, ed. Manfred Ullmann et al., 2 vols. Wiesbaden: Otto Harrassowitz, 1970–.

Wolfson, 'Infinite and Privative Judgements in Aristotle' = H.A. Wolfson, 'Infinite and Privative Judgements in Aristotle, Averroes, and Kant', *Philosophy and Phenomenological Research*, 8/2 (1947), 173–187.

Wolfson, *The Philosophy of the Kalam* = H.A. Wolfson, *The Philosophy of the Kalam*. Cambridge, Mass.: Harvard University Press, 1976.

Wollock, *Noblest Animate Motion* = Jeffrey L. Wollock, *The Noblest Animate Motion*.

Speech, Physiology and Medicine in Pre-Cartesian Linguistic Thought. Amsterdam: J. Benjamins, 1997.

Wolska-Conus, 'Stéphanos' = Wanda Wolska-Conus, 'Stéphanos d'Athènes et Stéphanos d'Alexandrie. Essai d'identification et de biographie', *Revue des études byzantines*, 47 (1989), 5–89.

Wright, *Grammar* = William Wright, *A Grammar of the Arabic Language*, tr. from the German of Caspari and ed. with numerous additions and corrections, 3rd ed. rev. by William Robertson Smith and Michael Jan de Goeje, 2 vols. Cambridge: Cambridge University Press, 1896–1898.

Wujastyk, *Roots of Ayurveda* = Dominik Wujastyk, *The Roots of Ayurveda: Selections from the Ayurvedic Classics*. New Delhi / London: Penguin Books, 1998.

Wüstenfeld, *Geschichte* = F. Wüstenfeld, *Geschichte der arabischen Aerzte und Naturforscher*. Göttingen: Dieterich, 1840.

Yaḥyā ibn ʿAdī, *Tahdhīb* = *The Refinement of Character or* Tahdhīb al-akhlāq. Parallel Arabic-English tr. by Sidney H. Griffith. Provo, Utah: Brigham Young University Press, 2002.

Yaḥyā ibn Saʿīd al-Anṭākī, *Histoire* (Kratchkovsky & Vasiliev) = Ignace Kratchkovsky and Aleksandr Aleksandrovich Vasiliev (eds., tr.), *Histoire de Yahyā Ibn Saʿīd d'Antioche, continuateur de Saʿid-Ibn-Bitriq* (Patrologia Orientalis, 23, pt. 3, fasc. 2). Paris: Firmin-Didot, 1932, 347–520.

Yaḥyā ibn Saʿīd al-Anṭākī, *Histoire* (Micheau & Tropeau) = Ignace Kratchkovsky (ed), Françoise Micheau, Gérard Tropeau (trns.), *Histoire de Yahyā Ibn Saīd d'Antioche, édition critique du texte arabe* et traduction française (Patrologia Orientalis, 47, fasc.4). Turnhout: Brepols, 1997.

Yaḥyā ibn Saʿīd al-Anṭākī, *Ta'rīkh al-Anṭākī* = Yaḥyā ibn Saʿīd ibn Yaḥyā, *Ta'rīkh al-Anṭākī al-maʿrūf bi-ṣilat Ta'rīkh Ūtīkhā*, ed. ʿUmar ʿAbd al-Salām Tadmurī. Tripoli: Jarūs Baras. 1990.

Yaman, *Prophetic Niche* = Hikmet Yaman, *Prophetic Niche in the Virtuous City* (Islamic Philosophy, Theology, and Science, 81). Leiden: Brill, 2011.

Yano & Viladrich, '*Tasyīr* Computation' = M. Yano and M. Viladrich, '*Tasyīr* Computation of Kūšyār ibn Labbān', *Historia Scientarum*, 41 (1991), 1–16.

al-Yaʿqūbī, *Kitāb al-Buldān* = al-Yaʿqūbī, Aḥmad ibn Abī Yaʿqūb, *Kitāb al-Buldān*, ed. M.J. de Goeje (Bibliotheca Geographorum Arabicorum, 7). Leiden: Brill, 1892. Repr. in *Ibn Rusta's Kitāb al-Aʿlāq al-nafisa and Kitāb al-Buldān by al-Yaʿqūbī. M.J. de Goeje's Classic Editions (1892)*. Leiden: Brill, 2014.

al-Yaʿqūbī, *Tārīkh* = Aḥmad ibn Abī Yaʿqūb ibn Wāḍiḥ al-Yaʿqūbī, *Tārīkh*, 2 vols. Beirut: Dār Bayrūt, n.d.

al-Yaʿqūbī, *The Works* = Aḥmad ibn Abī Yaʿqūb ibn Wāḍiḥ al-Yaʿqūbī, *The Works of Ibn Wāḍiḥ al-Yaʿqūbī : An English translation*, ed. Matthew S. Gordon, Chase F. Robinson, Everett K. Rowson and Michael Fishbein. (Islamic History and Civilization, 152) 3 vols. Leiden: Brill, 2017.

Yāqūt, *Muʿjam al-buldān* (Wüstenfeld) = Yāqūt ibn ʿAbd Allāh al-Rūmī, *Muʿjam al-buldān*, ed. Ferdinand Wüstenfeld. 6 vols. Leipzig: F.A. Brockhaus, 1866–1870.

Yāqūt, *Muʿjām al-buldān* (Beirut 1995) = Yāqūt, *Muʿjām al-buldān*. 7 vols. Beirut: Dār Ṣādir, 1995.

Yāqūt, *Muʿjam al-udabāʾ* = Yāqūt al-Ḥamawī, *Muʿjam al-udabāʾ*, ed. Aḥmad Farīd Rifāʿī. 20 vols. Cairo, 1936–1938, repr. Beirut: Iḥyāʾ al-Turāth al-ʿArabī, n.d.

Yāqūt, *Muʿjam al-udabāʾ* (1993) = Yāqūt al-Ḥamawī, *Muʿjam al-udabāʾ: Irshād al-arīb ilā maʿrifat al-adīb*, ed. Iḥsān ʿAbbās. 7 vols. Beirut: Dār al-Gharb al-Islāmī, 1993.

Young, 'Arabic Biographical Writing' = M.J.L. Young, 'Arabic Biographical Writing', Ch. 11 in M.J.L. Young, J.D. Latham and R.B. Serjeant (eds.), *Religion, Learning and Science in the ʿAbbasid Period* (The Cambridge History of Arabic Literature), Cambridge: Cambridge University Press, 1990 168–187.

Yūḥannā ibn Māsawayh, *Le Livre des axiomes médicaux* = Yūḥannā ibn Māsawayh, *Le Livre des axiomes médicaux*; édition du texte arabe et des versions latines avec traduction française et lexique par Danielle Jacquart et Gérard Troupeau. Geneva: Droz / Paris: Champion, 1980.

al-Yūnīnī, *Dhayl Mirʾāt al-zamān* = Quṭb al-Dīn Mūsā ibn Muḥammad al-Yūnīnī, *Dhayl Mirʾāt al-zamān*. 3 vols. Hyderabad: Dāʾirat al-Maʿārif al-ʿUthmāniyyah, 1955–1960.

al-Zabīdī, *Tāj al-ʿarūs* = Murtaḍā al-Ḥusaynī al-Zabīdī, *Tāj al-ʿarūs min jawāhir al-Qāmūs*, ed. ʿAbd al-Sattār Farrāj et al. 40 vols. Kuwait: Maṭbaʿat Ḥukūmat al-Kuwayt, 1965–2001.

Zaborowski, 'Abū Shākir as a modern Christian expert' = Jason R. Zaborowski, 'Arab Christian physicians as interreligious mediators. Abū Shākir as a modern Christian expert', *Islam and Christian-Muslim Relations*, 22 (2011), 185–196.

al-Zajjājī, *Amālī* = Abū l-Qāsim ʿAbd al-Raḥmān ibn Isḥāq al-Zajjājī, *Amālī al-Zajjājī*, ed. ʿAbd al-Salām Hārūn. Beirut: Dār al-Jīl, 1987.

Zakeri, *Persian wisdom in Arabic garb* = Mohsen Zakeri, *Persian wisdom in Arabic garb. ʿAlī b. ʿUbayda al-Rayḥānī (D. 219/834) and his Jawāhir al-kilam wa-farāʾid al-ḥikam* (Islamic Philosophy, Theology and Science 66). Leiden: Brill 2007, i:59–73.

Zaki, *Early Arabic Bookmaking Techniques* = Mahmoud Zaki, 'Early Arabic bookmaking techniques as described by *al-Rāzī* in his recently rediscovered *Zīnat al-Katabah*', *Journal of Islamic Manuscripts*, 2/2 (2011), 223–234, 238.

al-Zamakhsharī, *Al-Kashshāf* = *Al-Kashshāf ʿan ḥaqāʾiq al-tanzīl*. 2 vols. Calcutta: W.N. Lees' Press, 1856.

al-Zamakhsharī, *Rabīʿ al-abrār* = Maḥmūd ibn ʿUmar al-Zamakhsharī, *Rabīʿ al-abrār wa-nuṣūṣ al-akhbār*, ed. Salīm al-Nuʿaymī. 4 vols. Baghdad: Maṭbaʿat al-ʿĀnī, 1976–1982.

al-Zamakhsharī, *Rabīʿ al-abrār* (1992) = Maḥmūd ibn ʿUmar al-Zamakhsharī, *Rabīʿ al-abrār wa-nuṣūṣ al-akhbār*, ed. ʿAbd al-Amīr Muhannā. 5 vols. Beirut: Muʾassasat al-Aʿlamī li-l-Maṭbūʿat, 1992.

al-Zawzanī, *Ḥamāsat al-ẓurafāʾ* = Abū Muḥammad ʿAbd Allāh ibn Muḥammad al-Zawzanī al-Abdalakānī, *Ḥamāsat al-ẓurafāʾ min ashʿār al-muḥdathīn wa-l-qudamāʾ*, ed. Khalīl ʿImrān al-Manṣūr. Beirut: Dār al-Kutub al-ʿIlmiyyah, 2002.

ZGAIW = *Zeitschrift für Geschichte der Arabischen-Islamischen Wissenschaften*.

Zimmermann, 'Chronology' = F.W. Zimmermann, 'The Chronology of Isḥāq ibn Ḥunayn's *Taʾrīḫ al-aṭibbāʾ*', *Arabica*, 21 (1974), 324–330.

Zimmermann, 'Origins' = Fritz. W. Zimmermann, 'The Origins of the so-called *Theology of Aristotle*', in Kraye et al. (eds.), *Pseudo-Aristotle in the Middle Ages*, 110–240.

Ziriklī, *al-Aʿlām* = Khayr al-Dīn al-Ziriklī, *al-Aʿlām: Qāmūs tarājim li-ashhar al-rijāl wa-l-nisāʾ min al-ʿArab wa-l-mustaʿribīn wa-l-mustashriqīn*. 5th ed. 8 vols. Beirut: Dār al-ʿIlm li-l-Malāyīn, 1970.

al-Ziriklī, *al-Aʿlām* (2002) = Khayr al-Dīn al-Ziriklī, *al-Aʿlām: Qāmūs tarājim li-ashhar al-rijāl wa-l-nisāʾ min al-ʿArab wa-l-mustaʿribīn wa-l-mustashriqīn*. 15th ed., 8 vols. Beirut: Dār al-ʿIlm li-l-Malāyīn, 2002.

Zonta, 'De Lapidibus' = Mauro Zonta, 'Aristote de Stagire – Pseudo-Aristotle, De Lapidibus', in Goulet, *Dictionnaire*, Suppl. 652–654.

Zonta, 'Les Éthiques' = Mauro Zonta, 'Aristote de Stagire – Les Éthiques: Tradition syriaque et arabe', in Goulet, *Dictionnaire*, Suppl. 191–198.

Zurayk, *The Refinement of Character* = Constantine K. Zurayk, *The Refinement of Character: A Translation from the Arabic of Aḥmad ibn-Muḥammad Miskawayh's Tahdhīb al-akhlāq*. Beirut: American University of Beirut, 1968.

General Index to Vol. 1

All references are to pages. The Arabic article (a)l-, in all positions, is ignored for the alphabetical order. Names in lists and appendices (pp. 68–75, 201–213) are not included. Provenances of manuscripts are not included. Readers should consult the main indexes in the present volume for references to people, places, and subjects in the *'Uyūn al-anbā'* itself.

Glossary of Weights and Measures

Note: All weights and measures varied with time and place. Only approximate values can be given today. In some cases the measure is both obvious and vague, as in *kaff* meaning a 'handful' of an ingredient. The classic study of weights and measures in Islamic lands is the 1955 study of Hinz, *Islamische Masse*; it implies, however, a greater precision than existed in earlier times. See also, Mercier, 'Geodesy'; Chipman, *Pharmacists in Mamluk Cairo*, 88–93; and in *EI*[2] art. 'Makāyīl (A.)' for measures of capacity, 'Mawāzīn' for weights, and 'Misāḥa' for length and surface measurement, as well as art. 'Dīnār' (G.C. Miles) and 'Dirham' (G.C. Miles).

bashīzajāt Meaning uncertain; a measurement of girth. The ending -*ajāt* shows that it is from Persian. According to Steingass (*Persian-English dictionary*), a *pishēza* is something small: 'the scale of a fish; a small, thin piece of money; the skirt of a tent; a thin layer of metal between the handle and the blade of a knife'. Perhaps in Ch. 8.8.5, the distance between the perforations of a belt.

cubit (*dhirāʿ*) An ancient unit of length that may have originated in Egypt close to 5,000 years ago, approximately equivalent to the length of the human arm from elbow to fingertip. The Egyptian cubit (ca. 52 cm) was used to calibrate the nilometer at Roda, an island in the Nile near Cairo. Other definitions of a 'cubit' were also used, some defining it as 24 digits or 32 digits.

dānaq or ***dāniq*** A small coin, one-sixth of a dirham in value, as well as the corresponding weight (roughly half a gram). It is derived from Middle Persian *dānag*, 'seed, grain'.

dinar (*dīnār*) A gold coin weighing about 4.25 grams. Its name derived from the Roman *denarius*. First shaped by the Umayyad coin reform in 77/696, 'dinar' is still the name of various national currencies.

dirham (*dirham*) A basic unit of weight (slightly over 3 grams) as well as a silver coin. As a monetary unit, its value varied greatly. It dates back to the Greek drachma, which, borrowed as *darāhim*, sounded like an Arabic plural and thus gave rise to the singular form *dirham*. It formed the basis of Islamic weighing in the pre-modern era. Dirhams are still used today in a number of Arab countries. The *shujāʿī* or *tājī* dirham was a coin minted in Baghdād by ʿAḍud al-Dawlah, bearing the inscription *ʿAḍud al-Dawlah wa-Tāj al-Millah Abū Shujāʿ*; see Abbott, 'Two Būyid Coins'.

faddān A commonly agreed quantity of land, loosely reckoned as the quantity of land which a yoke of oxen can plough in one day.

fitr The distance between the ends of the outstretched thumb and index finger. It is to be distinguished from the *shibr*, the span of the hand.

kurr A unit of quantity, defined as six ass-loads of a commodity. See Lane, *Lexicon*, s.v.

mann A measure of capacity as well as a unit of weight, both of which varied considerably depending on time and place. It was usually about two *raṭl*s – that is, approximately 1.6 kg.

mīl The *mīl* is defined as 4,000 *dhirāʿ* (usually translated 'cubit'). The length of the cubit varied at different times and places, but if we take 50 cm as an approximation of a *dhirāʿ*, then a *mīl* would have measured about 2 km.

mithqāl A basic unit of weight usually equivalent to 1 + 3/7 dirhams, or nearly 4.4 grams, though its value varied greatly. It was also a gold coin of that weight.

parasang (*farsakh*) An ancient Iranian unit of distance, roughly equal to 3.5 miles or 5.5 km. The Arabic *farsakh* (from the Parthian *frasakh*, Old Persian *parāthanga*, cf. Greek *parasangēs*, modern Persian *farsang*) was most often defined as equal to 3 Arabic miles (*mīl*) or 12,000 cubits.

qinṭār A unit of weight. In Egypt the value ranged between 45 kg (the *qinṭār fulfulī* of Alexandria) and 96.7 kg (the *qintār jarwī*).

raṭl or ***riṭl*** A measure of capacity as well as a unit of weight, both of which varied considerably depending on time and place. The Baghdad *raṭl* weighed approximately 400 grams and as a measure of liquids or capacity it was equivalent to approximately 400 ml. Depending on locality and time, a *raṭl* could vary greatly from anywhere between 120 grams to 2.5 kg. The word is derived from Greek *litron* (cf. 'litre').

shibr A unit of length equivalent to the span of a person's hand, or the maximum distance between the tip of the thumb and the little finger.

tapia (*ṭābiyah*) A unit of measure, used exceptionally (see Ch. 13.64.2.2). It was roughly equivalent to 10 spans. The *tapia* originates from a Maghrib term for a panel of wall made with mud pressed and dried in a wooden-frame; in Spanish it is *tapia*, Portugese *taipa*.

Index of Verses to Vol. 2

This index covers the edition of the Arabic text of *ʿUyūn al-anbāʾ* and provides an alphabetic list of initial rhyme words of all poems, the metre and the number of verses for each poem, the poet, and the section numbers. Names of poets not mentioned by IAU but given in the English annotation are between square brackets. If no name is given in the index, the poet is unknown.

فهرس القوافي

أسماء الشعراء الغير المذكورين في النصّ مكتوبة بين قوسين []

م

فهرس أنصاف الأبيات

Index of Qur'anic Quotations and Allusions to Vols. 2 and 3

Non-literal quotations and allusions are marked with an asterisk; (p) refers to allusions and quotations in poetry. For some longer poems a line number is given in addition.

The index covers the edition, translation, and annotation of the *'Uyūn* but not the essays and material in the first volume of *A Literary History of Medicine*. The numbers provided in the entries refer first to the chapter number, then the biography number, and then, when relevant, to numbered subsections and sub-subsections within that biography. The numerical references apply to both the edited Arabic text and the annotated English translation.

Index of Personal Names to Vols. 2 and 3

The Arabic article (a)l-, in all positions, is ignored for the alphabetical order. The numbers provided in the entries refer first to the chapter number, then the biography number, and then, when relevant, to numbered subsections and sub-subsections within that biography. The numerical references apply to both the edited Arabic text and the annotated English translation. A bold number refers to an entry entirely devoted to the person named. After a number, n refers to a footnote, 'no.' refers to listed titles or sayings, and (p) refers to names mentioned in poems. Arabic names are often complex and Ibn Abī Uṣaybiʿah regularly uses different parts of names to refer to the same person. Many cross-references in this index will help to some extent, but they are by no means exhaustive.

The index covers the edition, translation, and annotation of the *ʿUyūn* but not the essays and material in the first volume of *A Literary History of Medicine*.

Index of Place Names to Vols. 2 and 3

The Arabic article (a)l-, in all positions, is ignored for the alphabetical order. The numbers provided in the entries refer first to the chapter number, then the biography number, and then, when relevant, to numbered subsections and sub-subsections within that biography. The numerical references apply to both the edited Arabic text and the annotated English translation. After a number, n refers to a footnote, 'no.' refers to listed titles or sayings, and (p) refers to places mentioned in poems. Cross references are supplied for places with names common in English (e.g., Dimashq *see* Damascus).

The index covers the edition, translation, and annotation of the *ʿUyūn* but not the essays and material in the first volume of *A Literary History of Medicine*.

Subject Index to Vols. 2 and 3

The Arabic article (a)l-, in all positions, is ignored for alphabetical order. The numbers provided in the entries refer first to the chapter number, then the biography number, and then, when relevant, to numbered subsections and sub-subsections within that biography. The numerical references apply to both the edited Arabic text and the annotated English translation. After a number, n refers to a footnote, 'no.' refers to listed titles or sayings, and (p) refers to names mentioned in poems.

Booklists have been indexed only where titles are of particular interest or relevance. The index covers the edition, translation, and annotation of the *ʿUyūn* but not the essays and material in the first volume of *A Literary History of Medicine*.

www.ingramcontent.com/pod-product-compliance
Lightning Source LLC
LaVergne TN
LVHW011101300926
849160LV00008B/250
* 9 7 8 9 0 0 4 6 9 6 5 5 6 *